Seismic Exploration Methods

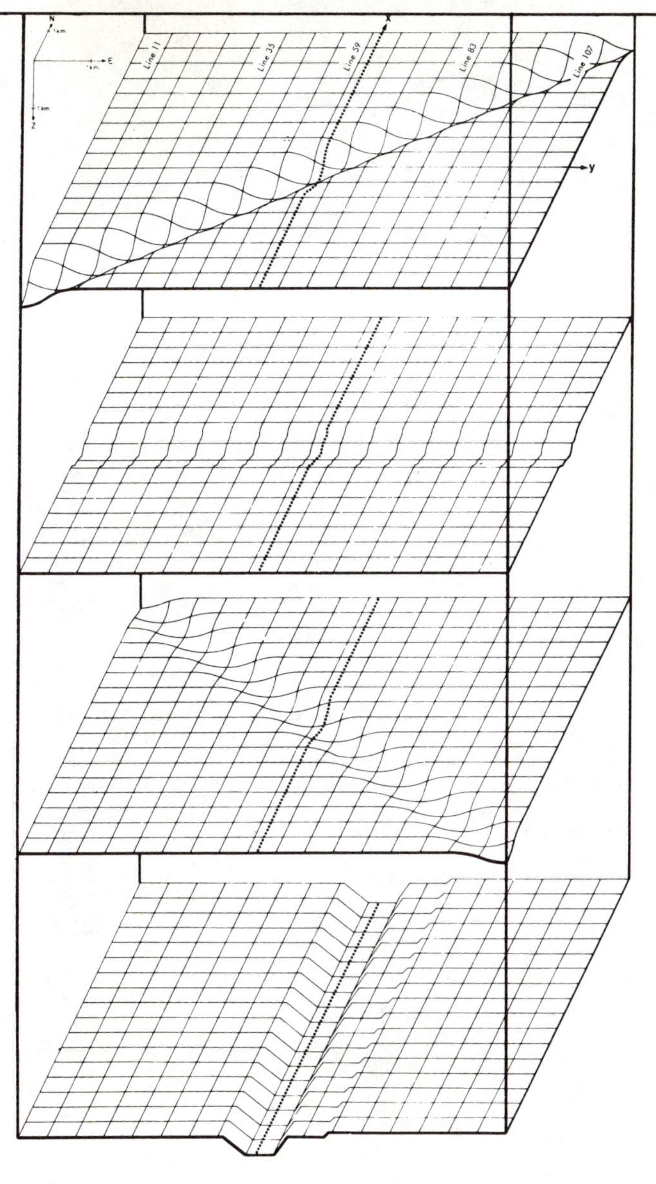

Seismic Exploration Methods

R.L. Sengbush

President
Pexcon International Inc.

IHRDC

BOSTON

Library of Congress Cataloging in Publication Data

Sengbush, R. L. (Ray L.), 1921–
Seismic exploration methods.

Bibliography: p.
Includes index.
1. Seismic prospecting. I. Title.

TN269.S42 1983 622′.159 82-81559

ISBN 0-934634-21-1

Printed in the United States of America

Interior design by Outside Designs
Cover design by Diane Sawyer

Preface

This book describes the seismic methods used in geophysical exploration for oil and gas in a comprehensive, nonrigorous, mathematical manner. I have used it and its predecessors as a manual for short courses in seismic methods, and it has been extensively revised time and again to include the latest advances in our truly remarkable science. I once called it, "Advanced Seismic Interpretation," but the geophysicists who attended the courses always wondered when I was going to start discussing interpretation. They discovered at the end that I never did discuss interpretation as they knew it. No mention was made of reflection picking, posting times, mapping, contouring, and things they already knew perfectly well. Instead, I discussed Fourier transforms, sampling theory, impulse responses, distortion operators, Wiener filters, noise in *f-k* space, velocity spectra, wave-equation migration, and direct detection of hydrocarbons as each of these topics appeared on the seismic scene. I wanted the geophysicists to think beyond the routine of interpretation, to develop a better understanding of why seismic sections look as they do, to have a better feel for what digital processing is doing, for good or evil, to the seismic data. I attempted to stretch their minds. Whitehead said it best: "A mind once stretched by a new idea can never shrink to its former dimension." May this book be a successful mind-stretcher.

R. L. Sengbush

Dedication

I dedicate this book to the many geophysicists with whom I have worked in the 35 years that I have spent learning and applying the art and science of geophysics in the search for oil and gas, and to my family, who has traveled with me along the sometimes tortuous and always rewarding path toward knowledge in my chosen field.

Perhaps the reader will be interested in the story of how I became a geophysicist. The seed was planted in my mind by Bill Dabney, a friend and former classmate at the Naval Radar School at M.I.T., who wrote me a letter in November 1946 telling about his job with the Magnolia Petroleum Company as observer on a seismic crew in Athens, Texas. I had no idea what a seismic crew did for an oil company, so I went to the public library in Racine, Wisconsin. (I was in Racine working as a chemist, developing auto waxes after having been unsuccessful in getting a job in radar.) In the library, I took down the volume of the Encyclopedia Brittanica containing the S's and looked up *seismic*. The first thing I saw was the picture of the wavefronts being reflected by the rock layers. "Just like radar," I thought. "I'm going to try to get a job like Dabney's," I vowed. In his next letter, he said he liked his job but was going to quit and sell ball bearings. So I wrote to Magnolia, was interviewed by Dayton Clewell in Chicago, and was soon on my way to Dallas as Dabney's replacement on the seismic crew in Athens, Texas. That was the beginning of 25 years as a research geophysicist for Magnolia and its parent, Mobil Oil.

During that time, I worked with many persons who helped shape my career. I wish to acknowledge some of them now:

My first field crew: Randy Simon, party chief; Stan Heaps, computer; Carl Richards, observer; Charley Ballard, shooter; and Van Funderburk, surveyor.

My successors as observer: Warren Hicks, J. W. Miller, and Joyce Wiler.

My supervisors and leaders in research projects: Ed White, Milt Dobrin, Dan Feray, Hal Frost, Frank McDonal, and Manus Foster.

My fellow research workers: Bob Watson, Bill Ruehle, Phil Lawrence, Frank Angona, C. D. McClure, Norm Guinzy, Clyde Kerns, and Al Musgrave.

In 1972, I left Mobil and formed the consulting company PEXCON with Norris Harris, a fellow geophysicist who, first among those at Mobil, recognized and used bright spots. In my ten years of consulting, the geophysicists with whom I was associated who stand out in my memory include Harry Barbee, George Ball, Jim Ducas, Bert Pronk, Ray Doan, Jim Lyon, Enders Robinson, Bill Voskamp, and Charles Saxon.

It has been my pleasure and a distinct honor and privilege to have worked with this talented cross section of geophysicists. Each has contributed to my knowledge, and each, directly or indirectly, has made a contribution to this book that cites me as the author.

I also dedicate this book to my family: to my wife, Earlene, who packed up and left pastoral Wisconsin with me and our eight-month-old daughter, Lynn, for the wilds of Texas and the adventurous life of a doodlebugger; and to our children, Lynn, Bill, and Larry, who never stopped believing in me. Finally, I wish to remember my father, now gone, and mother, and brothers Glen and Bob, with whom I shared the good life growing up on a small Wisconsin farm. To all of them, I say thank you from the bottom of my heart.

I wish to thank Bonnie McFerren, who typed and retyped the final stages of the book, and Debbie Dureell, who drafted and redrafted many of the final figures.

R. L. Sengbush
Houston, Texas
September 18, 1982

Contents

Seismic Exploration Methods

Chapter 1

Introduction to Seismic Methods

INTRODUCTION

The seismic method is based on the propagation of elastic waves in the earth and their reflection and refraction due to changes in the earth's velocity-density distribution. Active sources of energy are required; dynamite, air guns, and chirp signal generators are the most widely used sources. Detection of faint pressure or particle motion at or near the surface is achieved by use of sensitive pressure gauges or geophones. The received signal is amplified, recorded (usually by digital methods), processed (usually in a digital computer), and displayed in a form that is interpretable in terms of geologic structure, stratigraphy, and hydrocarbon content.

Two seismic methods coexist—one based on reflections and the other based on refractions. In the reflection method, which is the most explicit, displays of reflection data look like slices through the earth. It is important, however, to understand why such a simplistic viewpoint often fails. The refraction method peels off the layers, divulging the gross features of the velocity distribution with depth.

REFLECTION METHOD

A pictorial view of the reflection method shows wavefronts of seismic energy progressing outward from the source, with reflections at discontinuities in the geologic section. The received energy is recorded at or near the surface on a spread of detectors located at various distances from the source. The arrival times of reflections are observable, and the distances traveled can be cal-

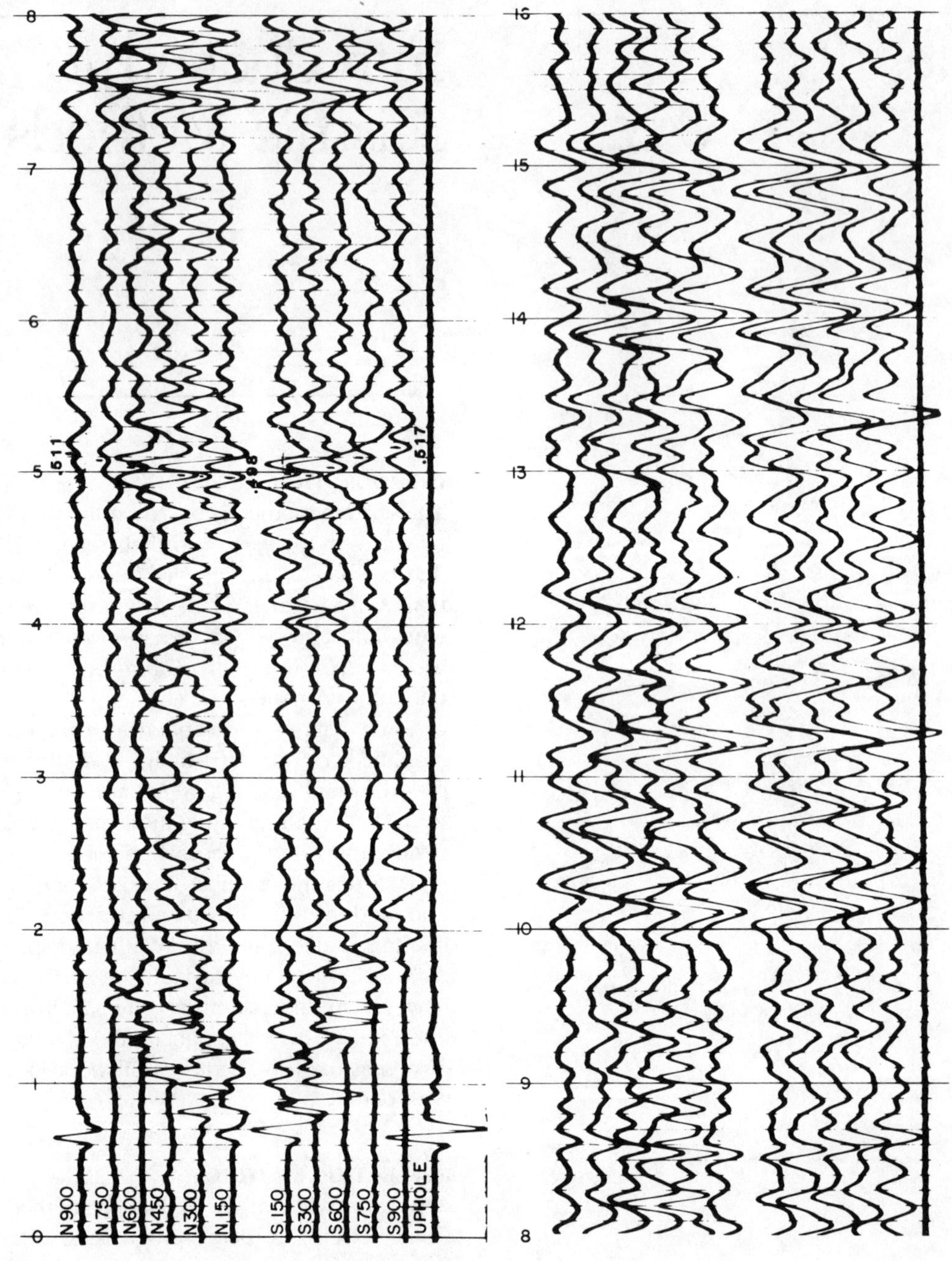

FIGURE 1.1. *Uncorrected seismic record.*

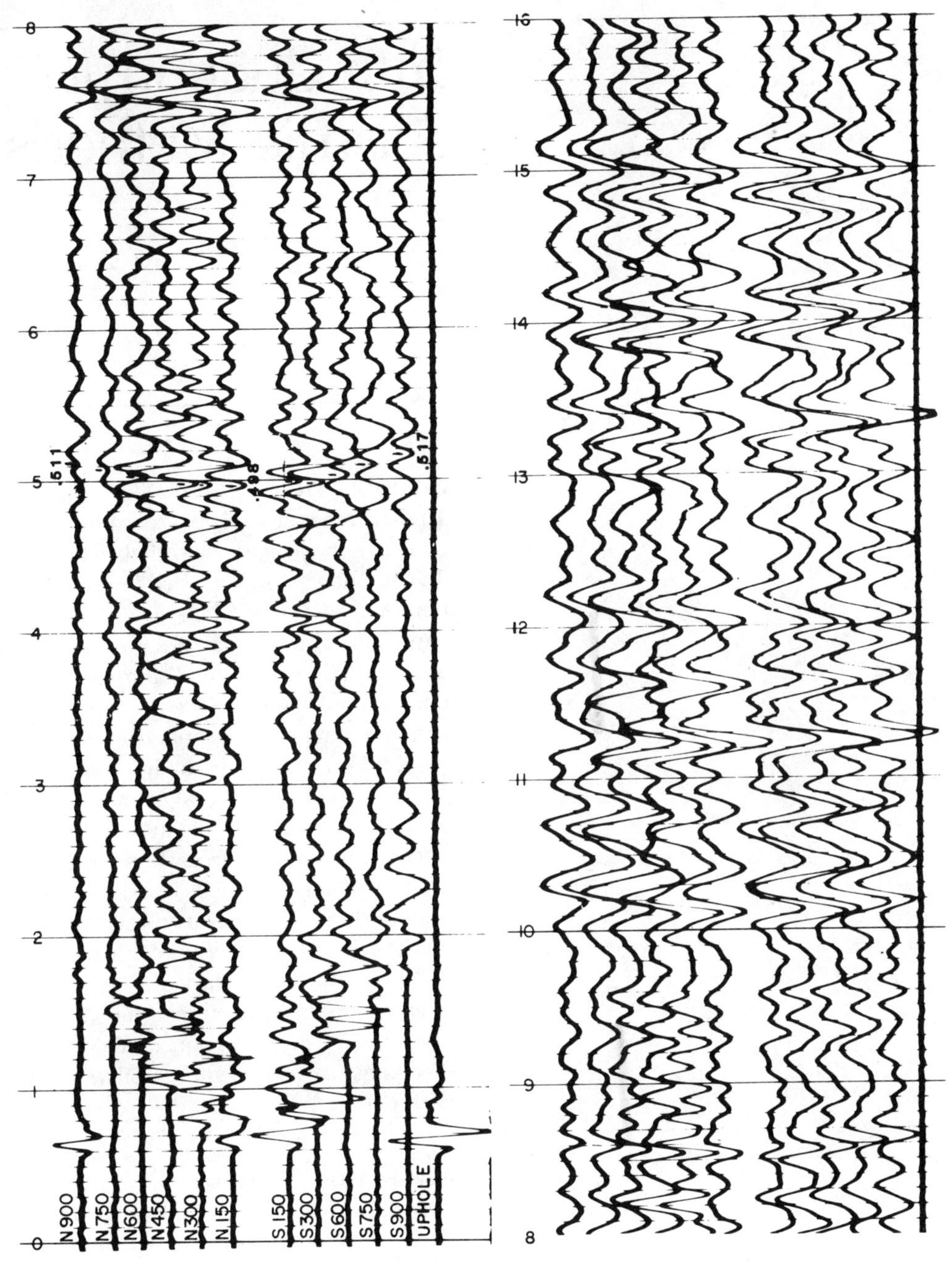

FIGURE 1.1. *Uncorrected seismic record.*

Chapter 1

Introduction to Seismic Methods

INTRODUCTION

The seismic method is based on the propagation of elastic waves in the earth and their reflection and refraction due to changes in the earth's velocity-density distribution. Active sources of energy are required; dynamite, air guns, and chirp signal generators are the most widely used sources. Detection of faint pressure or particle motion at or near the surface is achieved by use of sensitive pressure gauges or geophones. The received signal is amplified, recorded (usually by digital methods), processed (usually in a digital computer), and displayed in a form that is interpretable in terms of geologic structure, stratigraphy, and hydrocarbon content.

Two seismic methods coexist—one based on reflections and the other based on refractions. In the reflection method, which is the most explicit, displays of reflection data look like slices through the earth. It is important, however, to understand why such a simplistic viewpoint often fails. The refraction method peels off the layers, divulging the gross features of the velocity distribution with depth.

REFLECTION METHOD

A pictorial view of the reflection method shows wavefronts of seismic energy progressing outward from the source, with reflections at discontinuities in the geologic section. The received energy is recorded at or near the surface on a spread of detectors located at various distances from the source. The arrival times of reflections are observable, and the distances traveled can be calculated if the velocity distribution of the beds traveled through is known.

A typical uncorrected land record from the Williston Basin is shown in Figure 1.1. The source is a three-pound charge of dynamite at a depth of 200 ft. By locating the shot point in the middle of a spread of detectors, a split-spread record is produced. In marine work, using a towed cable that contains a spread of detectors, and in some land work, the source is offset from the end of the spread, producing end-on records.

In the usual case on land, where low-velocity weathered material overlies a higher-velocity subweathering, the first arrivals on each trace travel a refraction path from the source to the detector. In sand/shale sequences, an abrupt velocity change from about 2000 ft/sec to about 6000 ft/sec occurs at the natural water table. The presence of limestones and other high-velocity rocks near the surface will, of course, influence the ray paths of the first breaks. On marine data, the first breaks arrive directly through the water, depending on the velocity of the bottom sediments and the water depth, until the critical distance for refractions from the water bottom is exceeded.

Reflections can be identified in Figure 1.1 by the in-phase lineups across the record, at about 0.498, 0.745, 1.011, 1.125, 1.403, and 1.498 sec. A given reflection arrives progressively later as the offset (the source-to-detector separation) increases, because the length of the travel path to a given reflector increases progressively with offset. This effect is known as *normal moveout* (NMO). The reflection lineups are somewhat erratic about the NMO curves because of near-surface variations.

STATIC AND DYNAMIC CORRECTIONS

With the reflection method, it is desirable to know the depths (or arrival times) vertically below a datum. This requires application of two corrections to the seismic field data, a static correction and a dynamic correction.

The static correction overcomes the effects of changes in elevation at source and receiver locations, changes in source and receiver depths, variations in velocity and thickness of the weathered zone on land, and variations of water depth and subbottom velocity on marine data. For a given source-receiver location, the static correction is a constant time shift applied to the corresponding seismic trace.

The static correction moves the source and receiver to a datum. Given a source at depth h below a surface point whose elevation is e_s, for example, and receiver and datum elevations e_r and e_d, suppose that the source is beneath the weathering and in a subweathered zone with constant velocity V_1. The time correction T_{sd} to move the source to datum is

$$T_{sd} = [e_d - (e_s - h)]/V_1. \tag{1.1}$$

By recording the uphole time T_u (the travel time from the source to the surface), the receiver at zero offset can be moved to the source depth by subtracting the uphole time from the record time. The receiver can then be moved to datum by the time correction T_{sd}. The time correction T_{rd} to move the zero-offset receiver to datum is then

$$T_{rd} = -T_u + T_{sd}. \tag{1.2}$$

The correction to move both source and zero-offset receiver to datum is the static correction T_s:

$$T_s = 2T_{sd} - T_u. \tag{1.3}$$

The static correction applies to receivers close to the source location and is measurable at each source point. Receiver points intermediate to the source points usually require an additional correction because of changes in elevation or weathering thickness and velocity. First-break arrivals are often used in determining corrections at points intermediate to the source points. Under the assumptions that the base of the weathering is flat and that the average weathering velocity V_0 is constant, elevation changes will result in an additional correction of $(e_s - e_r)/V_0$.

The near-surface velocity distribution is usually determined by an uphole survey, measuring the direct arrival time at the surface as a function of shot depth. The depth of the base of the weathering and the average weathering and subweathering velocities are obtained from this survey.

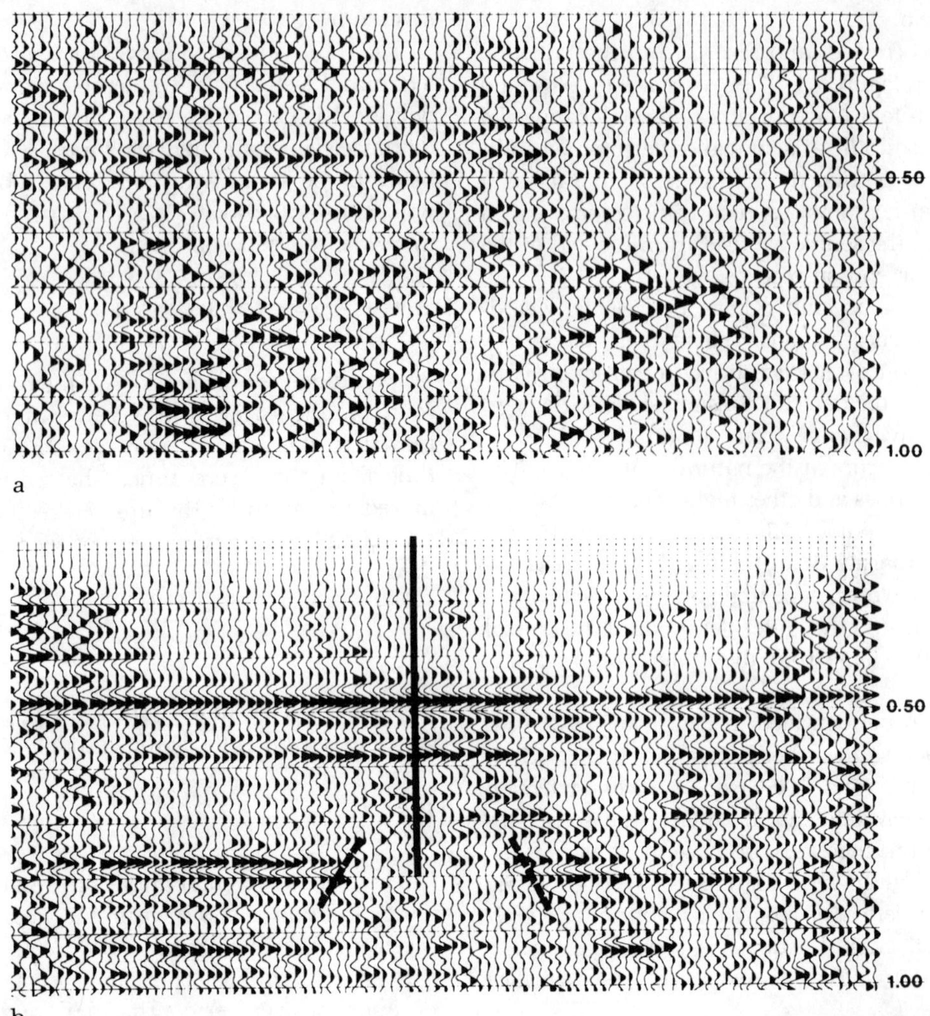

FIGURE 1.2. *Optimal data processing for Michigan reefs. (a) Preliminary stack. (b) Final stack. (From McClintock 1975, courtesy of AAPG.)*

Static corrections are always critical on land data and sometimes critical on marine data, and they constitute one of the most significant problems in reducing the data to a meaningful form for interpretation. The importance of applying proper static corrections is illustrated in a Michigan reef study by McClintock (1975) (see Figure 1.2). Exploration for reefs in Michigan is complicated by the rolling topography and variations in the velocity and thickness of the glacial drift covering the area. The improvement between the preliminary and final sections came largely from statics that were carefully picked, both automatically and by hand. By interpretation of the final section, the reef was detected by the absence of reflections. The discovery well produced 650 bbl of oil and 560,000 ft^3 of gas per day.

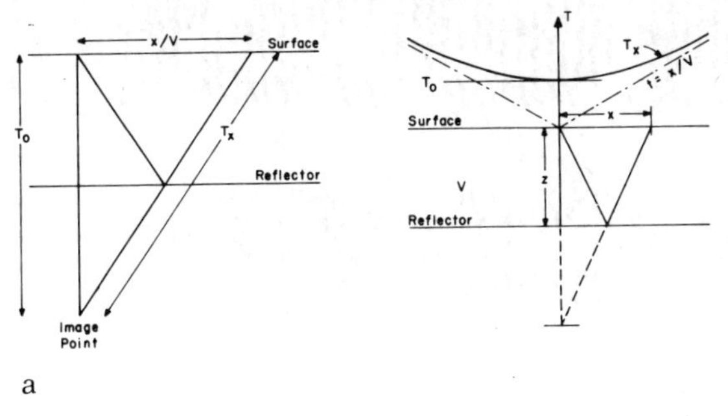

a

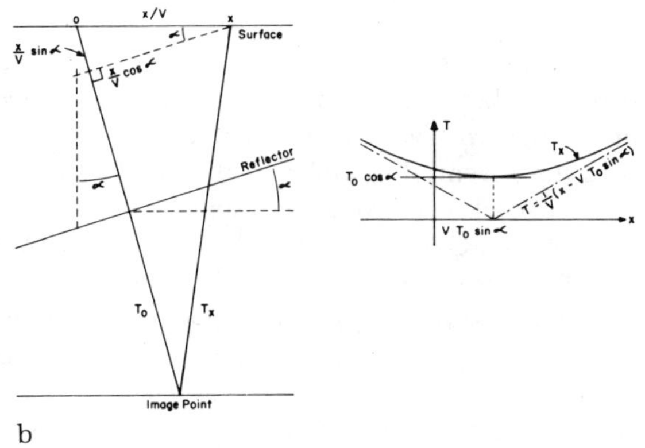

b

FIGURE 1.3. *Normal moveout correction on seismic data. (a) No dip. (b) Dipping beds.*

The dynamic correction overcomes the effect of offset distance between source and receiver. Increasing the offset increases the arrival time of a reflection from a given layer as a result of the increased distance traveled to the reflector and back to the surface. This geometric effect, known as normal moveout, is shown in Figure 1.3. In the no-dip case, with constant velocity V to the reflector, the reflection time at offset distance x is T_x and the reflection time at zero offset is T_0. The direct arrival time at offset distance x is x/V. These three arrival times constitute the sides of a right triangle obtained by using the source point, the image point of the source, and the receiver point as the three vertices of the triangle. By the Pythagorean theorem, then,

$$T_x^2 = T_0^2 + (x/V)^2. \tag{1.4}$$

The time-distance curve of T_x versus x is a hyperbola with vertex at $T = T_0$ and $x = 0$. The asymptotes to the hyperbola are $T = \pm x/V$. The difference between T_x and T_0 is the normal movement Δ, given by

$$\Delta = \sqrt{T_0^2 + (x/V)^2} - T_0. \tag{1.5}$$

For a given reflector (a given T_0), the NMO increases as x increases. At a given offset distance, the NMO decreases as the reflector depth increases; that is, the correction on a given trace changes with record time. Thus, the correction is dynamic rather than static.

In the dipping case, with dip angle α, the hyperbola is shifted, with its vertex located at $T = T_0 \cos \alpha$ and $x = VT_0 \sin \alpha$. The NMO is not symmetrical about $x = 0$, as it is in the no-dip case.

Parabolic approximations to the hyperbolic moveout expressions are obtained by substituting $T_0 + \Delta$ for T_x, expanding the resulting $(T_0 + \Delta)^2$ term, and neglecting Δ^2. In the no-dip case,

$$\Delta = \frac{1}{2T_0}\left(\frac{x}{V}\right)^2; \tag{1.6}$$

and, in the dipping case,

$$\Delta = \frac{1}{2T_0}\left(\frac{x}{V}\right)^2 - \frac{x}{V}\sin \alpha. \tag{1.7}$$

The velocity can be calculated from the field data using the normal moveout equation (eq. 1.4), as the other terms in that equation are known or measurable on the field records. Consider the reflection near 0.5 sec on the record in Figure 1.1. The trough on the near traces has $T_0 = 0.498$ sec. The average arrival time of this trough on the two end traces at 900 ft distance is 0.515 sec. Thus for $x = 900$ ft, $T_x = 0.515$ sec. Substituting these values into equation (1.4), which has been written to solve for velocity, gives

$$V = \frac{x}{\sqrt{T_x^2 - T_0^2}} = 6840 \text{ ft/sec}. \tag{1.8}$$

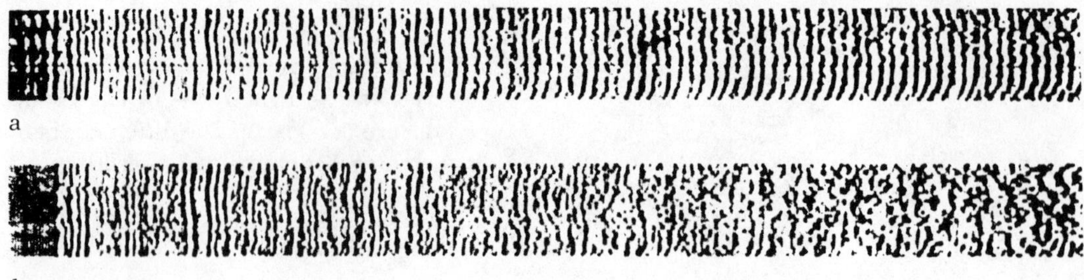

FIGURE 1.4. *Deconvolution of reverberated marine data. (a) Before deconvolution. (b) After deconvolution. (From Peacock and Treitel 1969, courtesy Geophysics.)*

Using this velocity, an estimate of the depth of this reflector is obtained as follows:

$$z = VT_0/2 = 1708 \text{ ft.} \tag{1.9}$$

The velocity obtained from the NMO equation is now called the stacking velocity and is recognized as an estimate of Dix's root-mean-square (rms) velocity; in the early days before Dix, it was considered an estimate of the average velocity. This topic is covered more completely in Chapter 8.

WIENER DECONVOLUTION

The raw seismic data collected in the field are often severely distorted by multiple reflections in the near surface. These distortions are called ghosts (Van Melle and Weatherburn, 1953) and reverberations (Backus, 1959). Once the phenomena of ghosts and reverberations were understood and formulated in terms of filter theory, numerous techniques were developed to suppress these distortions. Elongated directive sources that enhance primary reflections and suppress ghosts were the first effective technique developed for land exploration, but this technique was cumbersome and expensive to use in the field (Musgrave, Ehlert, and Nash, 1958). A much more serious problem than ghosts was the severe distortions occurring on marine data, due to reverberations within the water layer. In many cases, the reverberations were so strong that it was impossible to make a valid interpretation of the subsurface. No operational field technique could be found to suppress these reverberations, which meant that they must be accepted on the data and suppressed later by processing. Backus (1959) developed the first analog processing scheme that was effective in suppressing reverberations on marine data, but the technique did not work well in some areas.

With the advent of digital recording and processing, the first major breakthrough in suppressing distortions on seismic data came through the application of Wiener's (1949) optimal filter theory. This technique has become known as Wiener deconvolution. The technique could never be applied without recourse to digital data processing.

Deconvolution can be described quite simply by stating that the unknown distorted source wavelet that produces the field data is replaced in processing with a desired wavelet that does not contain any distortions; this can be done without knowing anything about the source wavelet or the distortions. There is no doubt that the rapid switch from analog to digital methods was a direct result of this one technique. Wiener deconvolution was a tremendous victory for digital processing.

An example of reverberated marine data before and after deconvolution is shown in Figure 1.4. Before deconvolution, the seismic data are sinusoidal and do not give a clue about the geological layering in the subsurface. This example shows the virtually impossible task faced by seismic interpreters in mapping structures on marine data before Wiener deconvolution.

On land data where ghosting is severe, Wiener deconvolution effectively removes the ghost reflections and leaves the primaries, and it does so without requiring any precise information about the mechanism that created the ghosts. Deconvolution also compensates for variations in the source waveform that occur from record to record, thereby producing records that have more consistent character.

In the processing sequence, deconvolution is applied before NMO, because NMO differentially stretches the data, which causes spectral changes in the data.

HORIZONTAL STACKING

Consider seismic data collected in the following way (see Figure 1.5). An in-line spread of n traces, with equal spacing Δd is used with the source offset from the nearest trace by Δd. (Any arbitrary but fixed offset distance may be used.) After the first shot is taken, the entire system is moved by $\Delta s = \Delta d/2$ for the second shot, $2\Delta s$ for the third shot, and so on. In the no-dip case, the following raypaths to a given interface have a common reflection point: Trace 1 on Record 1, Trace 2 on Record 2, . . . , Trace n on Record n. The offset distances for these traces are Δd, $2\Delta d$, . . . , $n\Delta d$. The reflection from this interface will have increasing NMO on the successive traces. After correcting the set of traces for NMO, the traces are added together; that is, they are said to be horizontally stacked (Mayne, 1962). Because n traces have been stacked, the stacking is said to be n-fold.

The fold depends on the number of traces and the ratio $\Delta d/\Delta s$. In the foregoing case, the ratio equals 2 and the fold equals n. If the ratio is unity, then the fold equals $n/2$. The general relation is that the fold equals $(n/2)(\Delta d/\Delta s)$, which holds for any $\Delta d/\Delta s$ ratio for which the fold has an integer value less than or equal to n.

Horizontal stacking serves two purposes in regard to signal-to-noise improvement. First, in the normal case where the velocity increases with depth, a surface multiple that arrives at the same time as a primary reflection will have greater NMO because it sees a lower average velocity than the primary. Hence, after correcting the primary for NMO, there will be a residual NMO on the multiple. Stacking the traces will enhance the primary, because it is lined up by applying its NMO, and will suppress the multiple, because of its out-of-phase alignment resulting from the residual NMO. The second purpose of horizontal stacking is that random noise present on the data will be suppressed, as compared to signal. The ratio of signal amplitude to rms noise power increases as the square root of the fold, provided that the signal is identical on each trace and that the random noises on the traces are mutually uncorrelated. Correlation of signal and noise causes the signal-to-noise ratio to be less than the square root of the fold.

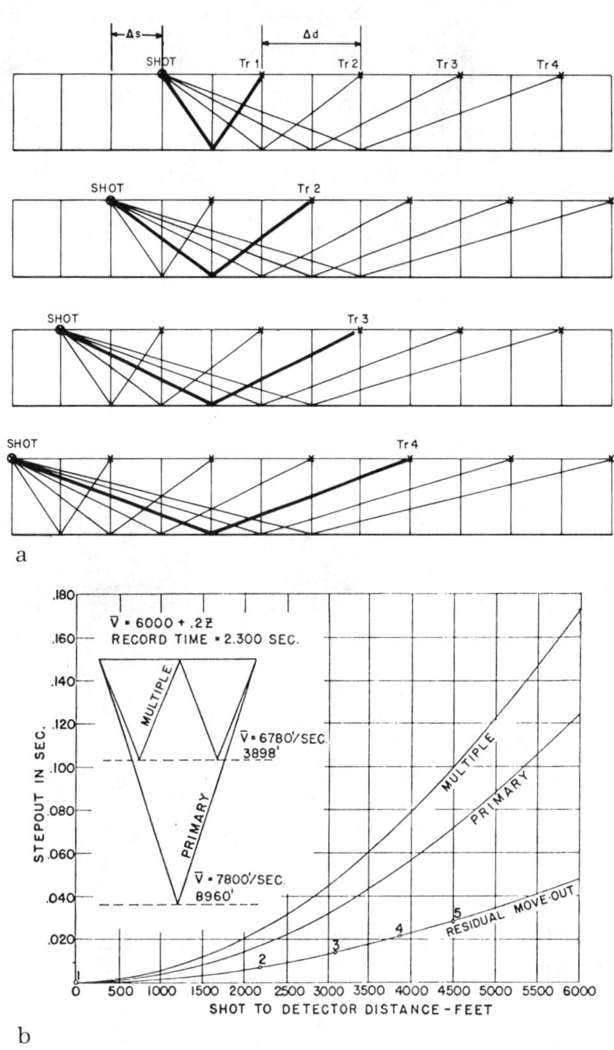

FIGURE 1.5. *Common depth point (CDP) procedures. (a) Field procedure for data collection,* $\Delta s = \Delta d/2$. *(b) Normal moveout of primary and multiple that have same arrival time: (from Mayne 1962, courtesy of Geophysics).*

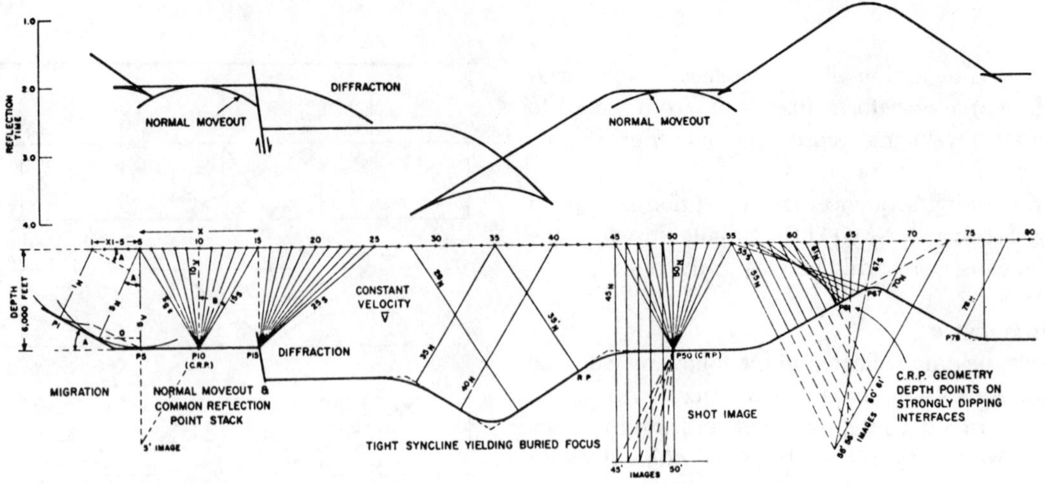
REFLECTION TIME
DIFFRACTION
NORMAL MOVEOUT
NORMAL MOVEOUT
DEPTH 6,000 FEET
CONSTANT VELOCITY
DIFFRACTION
MIGRATION
NORMAL MOVEOUT & COMMON REFLECTION POINT STACK
5' IMAGE
TIGHT SYNCLINE YIELDING BURIED FOCUS
SHOT IMAGE
C.R.P. GEOMETRY DEPTH POINTS ON STRONGLY DIPPING INTERFACES
IMAGES

a

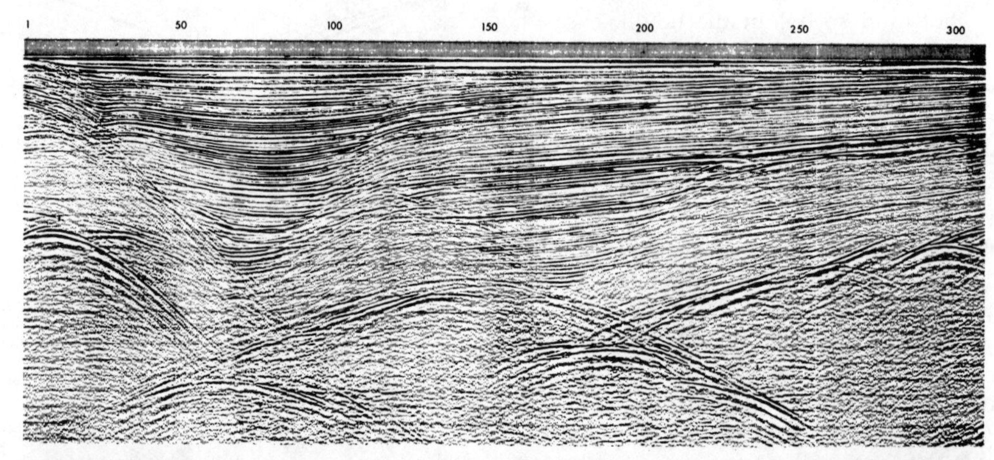

b

MIGRATION

After deconvolution and static and dynamic corrections are applied, the data are horizontally stacked and displayed as a record section in time as a function of surface location along a seismic line. The data stacked at each surface point are plotted vertically below that point. This display will be a true time-slice through the earth only if all layers have no dip and if no discontinuities or structures exist. Fortunately, this is not the usual case, and the time-record section is a more complicated transformation of the subsurface layering. On the time section, faults produce diffractions, synclines produce multibranch reflections if their foci are beneath the surface, and anticlines appear broader than they actually are. An example from Rockwell (1971) in Figure 1.6(a) shows these effects diagrammatically.

In order to display a true slice through earth, it is necessary to migrate the data to place reflections at the originating surfaces, to compress diffractions back to their originating points, and to move multibranches into their true structural position. The example in Figure 1.6(b) from the Gulf of Alaska shows the seismic data before and after migration (Reilly and Greene, 1976). The section after migration is a good approximation of a time-slice through the earth. Converting the time scale to depth, by applying the velocity distribution to the time section, will give a good approximation to a depth-slice through the earth.

The diffraction moveout is always greater than the NMO (Figure 1.7), as shown by the following development. For a source directly above the fault discontinuity, the geometry in Figure 1.8 holds. The time from source to reflector is $T_0/2$, and the time directly from source to detector at distance x is x/V. Consequently, the time from the diffraction point to the detector at x is

$$\sqrt{\left(\frac{x}{V}\right)^2 + \left(\frac{T_0}{2}\right)^2}.$$

FIGURE 1.6. *Seismic time-record sections are complicated transformations of a time-slice through the earth. (a) Effects of faults, synclines, and anticlines. (From Rockwell 1971, courtesy of Oil and Gas Journal.) (b) Seismic data before and after migration. (From Reilly and Green 1976, courtesy of Seiscom-Delta Report.)*

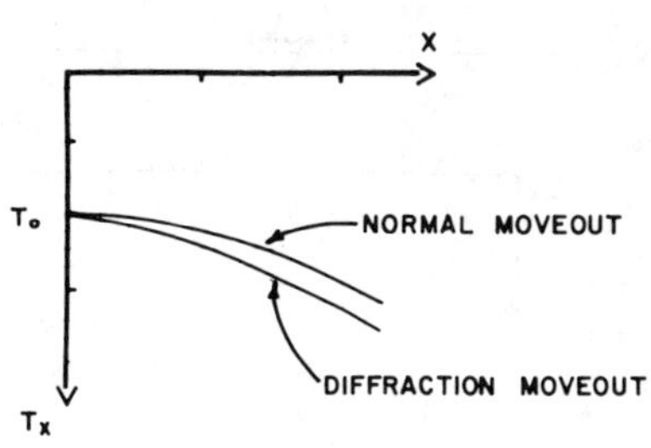

FIGURE 1.7. *Comparison of NMO with diffraction moveout.*

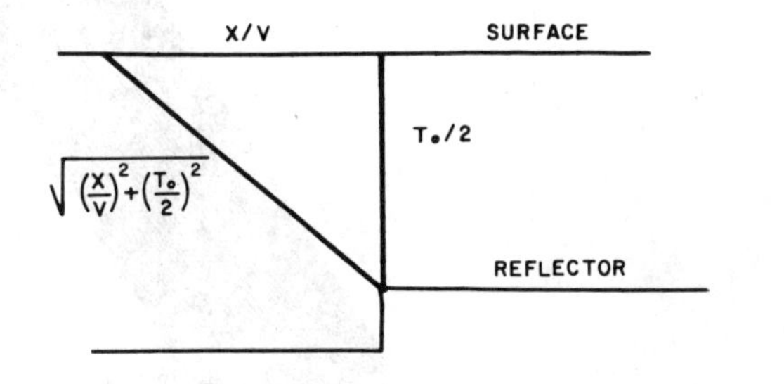

FIGURE 1.8. *Diffraction moveout.*

Thus, the diffraction arrival time at x is

$$T_d = \frac{T_0}{2} + \sqrt{\left(\frac{x}{V}\right)^2 + \left(\frac{T_0}{2}\right)^2}. \tag{1.10}$$

Squaring equations (1.10) and (1.4) and subtracting gives

$$T_d^2 - T_x^2 = T_0\left[\sqrt{\left(\frac{x}{V}\right)^2 + \left(\frac{T_0}{2}\right)^2} - \frac{T_0}{2}\right]. \tag{1.11}$$

It is apparent that the term in brackets is zero for $x = 0$ and positive for all $x > 0$. Therefore, T_d is always greater than T_x for all $x > 0$.

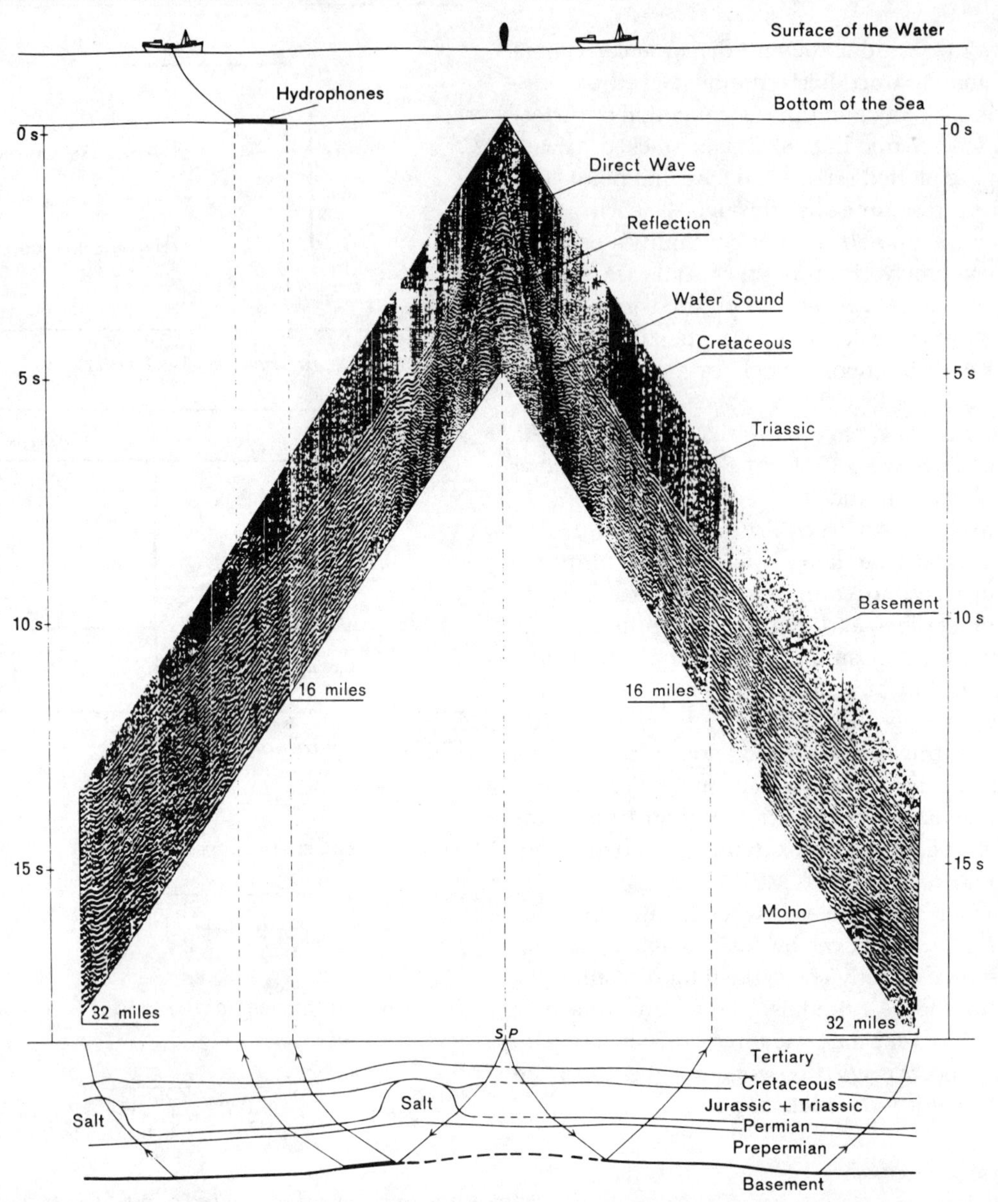

FIGURE 1.9. *Refraction profile. (From Prakla-Seismos Brochure, courtesy of Geophysics.)*

REFRACTION METHOD

The refraction method depends on seismic raypaths being bent at velocity discontinuities in accordance with Snell's law. A compressional plane wave that impinges on a boundary separating two media with different acoustic impedances is partially reflected and partially transmitted. The angle of reflection equals the angle of incidence. For nonnormal incidence, the transmitted wave is refracted into the lower medium, in accordance with Snell's law,

$$\frac{\sin a_1}{V_1} = \frac{\sin a_2}{V_2}. \tag{1.12}$$

The incident angle is a_1 and the refracted angle is a_2. With $V_2 > V_1$, the refracted wave is bent away from the normal. It is bent toward the normal if $V_2 < V_1$. When $a_2 = 90°$, the raypath in the lower medium is along the interface rather than into the second medium. The incident angle that causes $a_2 = 90°$ is called the first critical angle (a_p), or the critical angle for compressional waves, given by

$$\sin a_p = V_1/V_2. \tag{1.13}$$

In the normal case, where the velocity increases with depth, the first arrival will travel a refraction path that gets deeper into the earth as the offset distance increases. From the data, the refraction velocity can be determined in each layer and the depth to each refracting horizon can be calculated. To compensate for dip, it is necessary to shoot the refraction profiles from both ends. Details of the computations are given by Dix (1952), Slotnik (1959), and Dobrin (1976). *Seismic Refraction Prospecting*, edited by A. W. Musgrave (1967), a publication of the Society of Exploration Geophysicists (SEG), deals with refraction methods, field techniques, and case histories.

The refraction method is a reconnaissance method, in that it gives gross features of layering rather than the detailed features obtainable by the reflection method. Refractions have been used very effectively in outlining salt and shale masses. In areas where the reflection data are not usable because of noise, surface conditions, or unknown causes, refractions may give useful information, because first arrivals can always be detected given sufficient charge size.

An example from Prakla-Seismos (1972) in Figure 1.9 shows a 32-mile split-spread refraction in the North Sea. The first arrivals dig progressively deeper into the earth and are identified as Cretaceous, Triassic, and, finally, Basement. The Moho is identified as a later-arriving event.

SUMMARY

The highlights of the reflection seismic method discussed here include two of the most critical aspects of data processing—static corrections and normal moveout. These two corrections have always been important. They were the geophysicists' bread and butter in the days of paper records, when geophysicists had to fight their way through these corrections before a sensible interpretation could be made. It was impossible to make a casual interpretation from a collection of paper records. The digital revolution and seismic record sections changed all that. Now almost anyone can look at a record section and make a casual interpretation. Behind these sections, however, someone has made decisions about static corrections and normal moveout. The importance of these corrections must not be underestimated while concentrating on deconvolution or some other exotic processing technique. Inattention to these corrections or mistakes in applying them are the most prevalent causes of data becoming useless, and, in some cases, no one is even aware of it.

Chapter 2

Seismic Wave Propagation

INTRODUCTION

Seismic wave propagation is founded on the principles and laws of classical physics: Hooke's law relating stress and strain, the theory of elasticity, the principles of conservation of energy and momentum, Newton's laws of motion, and the wave equation. Some of the greatest physicists and mathematicians have worked on the problems of wave propagation in elastic media, including Lamb, Love, Rayleigh, Stonley, Poisson, Kirchhoff, and Cagniard.

The mathematical difficulty in solving the wave equation in heterogeneous material such as the earth forced geophysicists to rely on simpler concepts. One such concept is based on Fermat's principle, by which the study of wave propagation reduces to the study of raypaths along which traveltimes are minimal. Snell's law, which defines the minimal time paths, is a consequence of Fermat's principle.

Another simpler model is the communication theory model, in which the seismic signal process is the convolution of a deterministic source pulse with the random reflectivity of the earth. This model has given good insight into the actual reflection process in the real earth and has led to powerful processing techniques, such as deconvolution.

Significant recent developments by Claerbout (1976) have brought the wave equation back into prominence. The most significant result derived from his work has been the recent development of wave equation migration of seismic data, which is a decided improvement over previous methods.

This chapter is a simplified discussion of the highlights of wave propagation. The mathematics of wave propagation in the simple case for plane waves in a homogeneous, isotropic medium is given in some detail in Appendix E.

ELASTICITY

The seismic method depends on the propagation of waves in an elastic medium. This requires, first, a mathematical description of the properties of an elastic medium. Consider a medium that is homogeneous and isotropic; *homogeneous* means that the medium has the same properties at all points, and *isotropic* means that it has the same properties in all directions.

The theory of elasticity is centered on the concepts of stress and strain. *Strain* is the deformation per unit length or per unit volume. If the body is elastic, the strain will be associated with internal restoring forces within the material. *Stress* is the internal force per unit area.

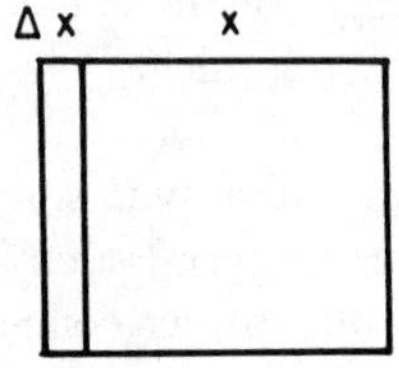

FIGURE 2.1. *Longitudinal strain.*

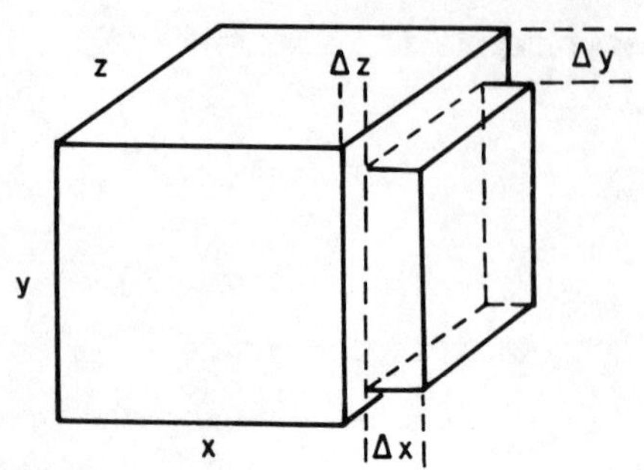

FIGURE 2.2. *Volume strain.*

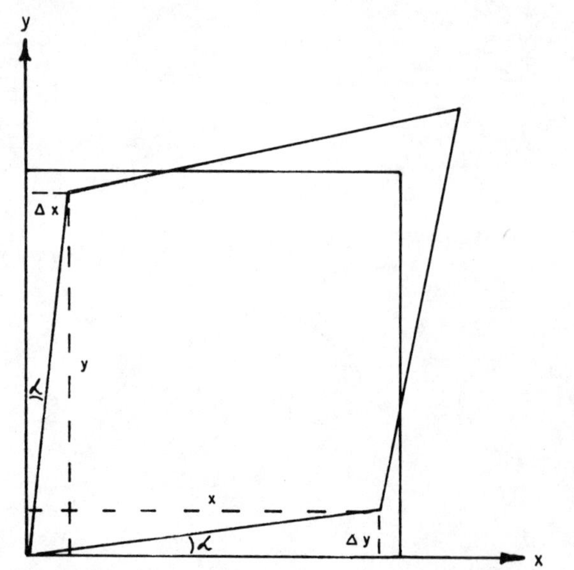

FIGURE 2.3. *Shear strain.*

Hooke's law states that stress is proportional to strain. The constant of proportionality is called the *elastic constant*, or *elastic modulus*, of the material to the particular type of stress-strain. *Young's modulus* (E) applies to longitudinal stress-strain, such as the stretching of a wire. The strain is given by the change in length per unit length, and Hooke's law becomes stress $= E\Delta x/x$ (see Figure 2.1). *Bulk modulus* (k), or incompressibility, applies to volume stress-strain, such as the compression of a body by hydrostatic pressure. The strain is given by the change in volume per unit volume, and Hooke's law becomes stress (pressure) $= k\Delta\text{vol/vol}$ (see Figure 2.2). *Shear modulus* (μ), or rigidity, applies to changes in shape without change in volume. The shear strain in the y-direction is the change in x per unit length in y-direction, and Hooke's law becomes shear stress $= \mu\Delta x/y$ (see Figure 2.3). *Poisson's ratio* (σ) is a measure of the change in geometrical shape of a body subjected to stress. When an elastic body is stretched, a lateral contraction accompanies the longitudinal extension. Poisson's ratio is defined as the ratio of the contraction to the extension,

$\sigma = (\Delta y/y)/(\Delta x/x)$ (see Figure 2.4).

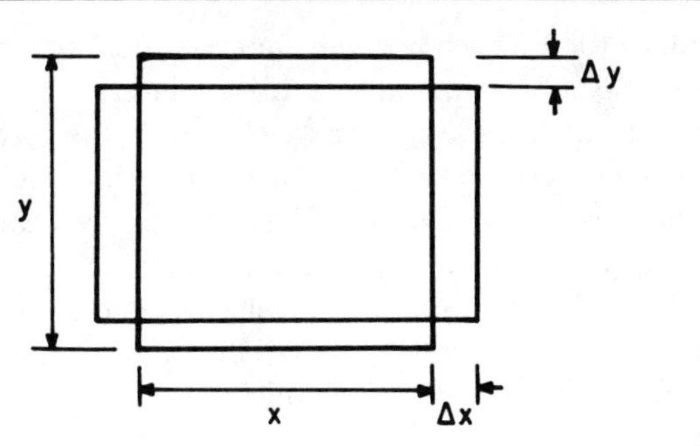

FIGURE 2.4. *Poisson's ratio.*

In a homogeneous and isotropic medium, only two elastic constants are required to define the stress-strain relations. Bulk modulus (k) and shear modulus (μ) may be used, or Young's modulus (E) and Poisson's ratio (σ). A third pair of constants, the Lamé constants (λ and μ), is frequently used for mathematical simplicity. As the Lamé constant, μ, is identical to the shear modulus, there are five elastic constants. Only two of the five constants are independent. When two of the constants are chosen, the others can be written in terms of the chosen two. The following gives the relations between the elastic moduli:

$$\lambda = k - \frac{2}{3}\mu, \qquad \lambda = \frac{\sigma E}{(1 + \sigma)(1 - 2\sigma)},$$
$$\mu = \frac{E}{2(1 + \sigma)}, \qquad \sigma = \frac{\lambda}{2(\lambda + \mu)}. \tag{2.1}$$

ELASTIC WAVES

HOMOGENEOUS AND ISOTROPIC MEDIA

Only two types of waves can propagate in the interior of a homogeneous and isotropic material: *compressional* and *shear* waves. Compressional waves have particle motion in line with the direction of propagation, while shear waves have particle motion transverse to the direction of propagation. The compressional velocity (V_p) and shear velocity (V_s) are given in terms of the elastic constants k, λ, and μ, and the density ρ:

$$V_p = \sqrt{\frac{k + 4/3 \cdot \mu}{\rho}} = \sqrt{\frac{\lambda + 2\mu}{\rho}} \tag{2.2}$$

$$V_s = \sqrt{\frac{\mu}{\rho}}. \tag{2.3}$$

Combining equations (2.2) and (2.3) gives k in terms of the velocities and density:

$$k = \rho\left[V_p^2 - \frac{4}{3}V_s^2\right]. \tag{2.4}$$

An ideal solid is defined by the property $\lambda = \mu$. In this case, $V_p = \sqrt{3}V_s$, and Poisson's ratio equals 1/4.

In a fluid, shear waves do not propagate because the shear modulus is zero. Poisson's ratio for a fluid is 1/2.

Poisson's ratio can be determined from measurements of the shear and compressional velocities:

$$\sigma = \frac{\left(\frac{V_p}{V_s}\right)^2 - 2}{2\left(\frac{V_p}{V_s}\right)^2 - 2} \tag{2.5}$$

V_p/V_s	σ
1.41	0.0
1.50	0.1
1.63	0.2
1.87	0.3
2.45	0.4
∞	0.5

Experimental measurements show that most consolidated rocks have Poisson's ratios in the range 0.25 to 0.35, while unconsolidated materials have values between 0.40 and 0.45. Measurements by White and Sengbush (1953) in near-surface formations in Dallas County, Texas, showed velocities in ft/ms of $V_p = 8.5$ and $V_s = 3.5$ for the Austin Chalk and $V_p = 6.0$ and $V_s = 1.25$ for the Eagleford shale, giving Poisson's ratios of 0.398 and 0.477, respectively.

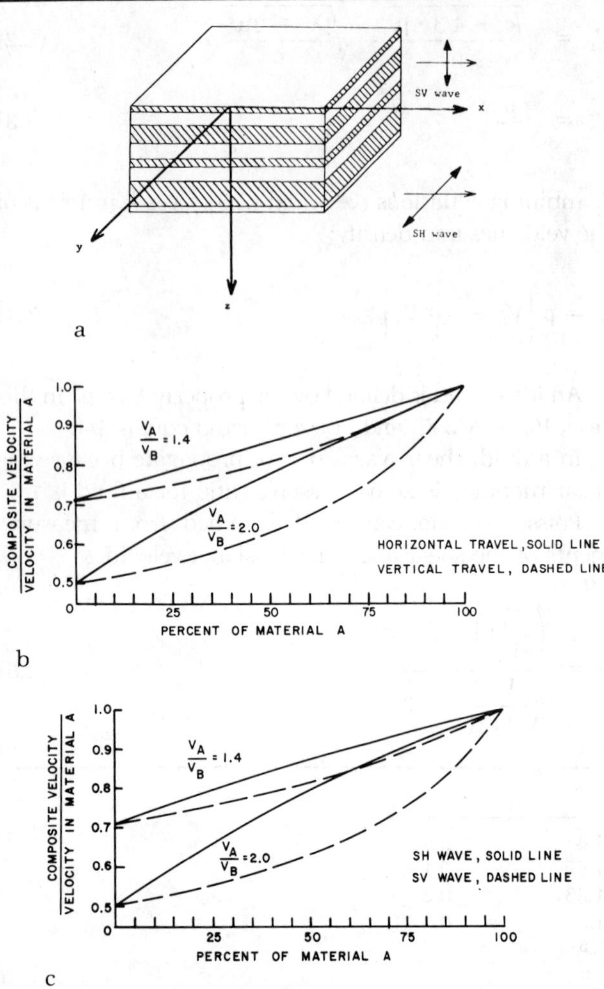

FIGURE 2.5. *Waves in transversely isotropic medium. (a) Transversely isotropic medium. (b) Comparison of compressional velocity in horizontal and vertical directions. (c) Comparison of velocity of horizontally traveling* SH- *and* SV-*waves. (From White and Angona, Journal of Acoustical Society of America, vol. 27, n. 2, 1955, courtesy of Journal of Acoustical Society of America.)*

LAYERED MEDIA

Stonley (1949) described wave propagation in a medium made up of horizontal isotropic layers. In such a medium, the elastic properties change in the vertical direction but are constant in each transverse layer. This medium is an idealization of the real layered earth.

In a transversely isotropic medium, the velocities of the compressional and shear waves depend on the direction of particle motion. The velocity is maximum when the particle motion is parallel to the layering and minimum when it is transverse. The property by which the velocity depends on direction of propagation is called *anisotropy*.

White and Angona (1955) calculated the velocities in a transversely isotropic medium that consists of alternate layers of two different materials (Figure 2.5). Comparison of compressional velocities in the vertical and horizontal direction shows the horizontal velocity to be larger than the vertical velocity, with the maximum difference between vertical and horizontal velocity occurring when the medium has equal parts of the two materials. The maximum difference is about 5 percent when the velocity ratio in the two materials is 1.4, and 23 percent when the ratio is 2.0.

Two types of shear waves exist in transversely isotropic media, with characteristic velocities for horizontal travel that depend on the direction of particle motion with respect to the direction of the laminations. The *SV*-waves have particle motion transverse to the laminations, in the *SH*-waves, it is parallel to the laminations. The comparison of horizontally traveling *SV*- and *SH*-waves shows the *SH*-velocity to be 6 percent higher than the *SV* velocity when the velocity ratio in the two materials is 1.4, and 25 percent higher when the ratio is 2.0. Stonley (1949) has shown that vertically traveling shear waves travel at *SV* velocity, even though its particle motion is parallel to the laminations.

REFLECTION AND TRANSMISSION COEFFICIENTS

When a wavefront strikes an interface at less than critical angle, the wave is partially reflected and partially transmitted into the second medium. Upon striking the interface at normal incidence, an incident plane wave with unit amplitude produces a reflection whose amplitude R depends on the density-velocity (ρV) product of each medium. The ρV product is called the *acoustic impedance* (Z) of a medium. R is called the *reflection coefficient* and is defined by

$$R = \frac{Z_2 - Z_1}{Z_2 + Z_1}, \tag{2.6}$$

where the subscript 1 refers to the incident medium and the subscript 2 refers to the medium containing the transmitted wave. The amplitude T of the transmitted wave is called the *transmission coefficient*, defined by

$$T = 1 - R = \frac{2Z_1}{Z_2 + Z_1}. \tag{2.7}$$

R is bounded in value by $(-1, +1)$ and T by $(0, 2)$.

The reflection coefficient at an interface is defined by incident waves traveling downward. An interface with reflection coefficient R for downward traveling waves has reflection coefficient $-R$ for upward traveling waves. The transmission coefficient for a downward traveling wave is $T_d = 1 - R$; for an upward traveling wave, it is $T_u = 1 + R$. The transmission loss caused by two-way travel across an interface, T_{2w}, is the product of T_d and T_u:

$$T_{2w} = (1 - R)(1 + R) = 1 - R^2. \tag{2.8}$$

In a layered medium with unit incident amplitude in the first layer, the reflection amplitude from the nth interface, as measured in the first layer, is affected by the two-way transmission losses through all interfaces above the nth interface:

$$\begin{aligned} A_n &= R_n\,(1 - R_1^2)(1 - R_2^2)\ldots(1 - R_{n-1}^2) \\ &= R_n \prod_{k=1}^{n-1} (1 - R_k^2), \end{aligned} \tag{2.9}$$

where A_n is the reflection amplitude from the nth interface as measured in the first layer, R_n is the reflection coefficient at the nth interface, and $1 - R_k^2$ is the two-way loss at the kth interface, for $k = 1, 2, \ldots, n - 1$.

Reflection and transmission coefficients apply to discrete layers, which can be described by a set of acoustic impedances, $Z_1, Z_2, \ldots, Z_n$. When the material has a continuously varying acoustic impedance $Z(t)$, where t is two-way traveltime, then the foregoing reflection coefficients do not apply. In the continuous case, the set of reflection coefficients is replaced by the reflectivity function $r(t)$, which is half the derivative of the logarithm of the acoustic impedance (Peterson, Fillipone, and Coker, 1955):

$$r(t) = \frac{1}{2}\frac{d \ln Z(t)}{dt}. \tag{2.10}$$

Foster (1975) shows that, in a continuous medium, the two-way transmission loss observed at the surface is zero. Foster concludes: "We have shown that the transmission effects in a continuous one-dimensional model are unsubstantial, and that *at the surface are not present at all.* To the extent that they significantly change the result of calculations based upon discrete approximations, *the calculations are at fault.*" He is saying that sampling does not create layering. Sampling a continuous $r(t)$ to get the time series $\{r_i\}$ differs from a discrete-layer case that has $\{R_i\}$ as a set of reflection coefficients because of the 2-way transmission losses present with discrete layers, which results in the time series given by equation (2.9) in the discrete-layer case.

RAYPATH THEORY

HUYGENS' PRINCIPLE

Wave propagation in three dimensions is generally quite a complicated process when considered in detail, especially when matter is present to absorb, reflect, refract, or scatter radiation. Waves propagate with different velocities. At discontinuous interfaces between media, there are boundary conditions to be imposed on the disturbance and its derivatives. These effects may be handled exactly in simple cases, but in three dimensions, only the most elementary geometries can be treated exactly. Thus, approximate descriptions of wave propagation that contain some physical truth without the full mathematical complexity of the exact solutions of the wave equation are desired. One such approximation is Huygens' principle. This principle tells how to calculate the position in space of a disturbance at a later time if it

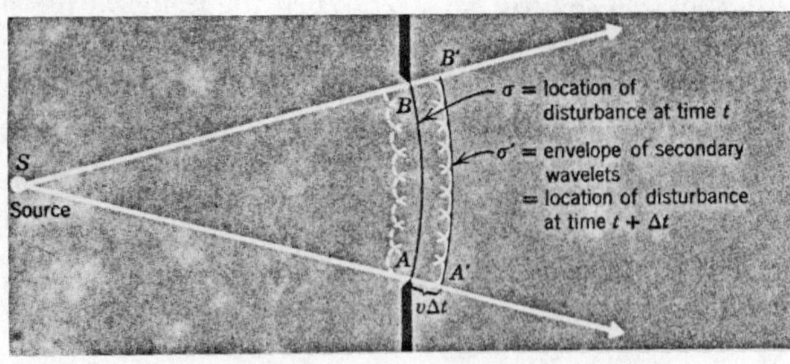

FIGURE 2.6. *Law of rectilinear propagation using Huygens' principle.*

is known at an earlier time. It does not give quantitative information about the resulting amplitude.

Huygens' principle contains the rules about the way waves propagate. Consider the wavefront σ at time t resulting from a disturbance at source S, as shown in Figure 2.6. To determine the wavefront at time $t + \Delta t$, Huygens' principle says "construct secondary spherical wavelets of radius $V\Delta t$ originating at each point on σ and form the superposition of these secondary wavelets to produce the disturbance at time $t + \Delta t$." The wavelets are most dense in the region included between the lines *SA* and *SB*. This region is called the region of geometrical brightness because the rays travel outward from the source S in straight lines, and this region will be illuminated, whereas the regions outside, the regions of geometrical shadow, will not be illuminated. The new wavefront is the envelope of all the outgoing secondary wavelets. Each point on the envelope is tangent to one and only one Huygens' wavelet. The envelope ends abruptly at the edge of the geometrical shadow. The backward-moving wavelets also have an envelope that would propagate toward the source. These are neglected in Huygens' prescription without any justification beyond an appeal to the experimental fact that the back wave envelope does not occur in nature. More advanced theory is needed to explain diffraction effects in the shadow zone. This requires the wave equation.

The notion of rays follows quite simply from the foregoing idealized picture. Rays are directed lines that are always perpendicular to the surface occupied by the disturbance at a given time and point along the direction of its motion. Three laws of geometrical wave propagation are usually stated in terms of rays:

(1) *The law of rectilinear propagation:* Rays in homogeneous media propagate in straight lines.
(2) *The law of reflection:* At an interface between two different homogeneous, isotropic media, an incident disturbance is partially reflected, and the reflected ray is in the plane of incidence (the plane determined by the incident ray and the normal to the surface). The angle it makes with the normal (the angle of reflection) equals the angle made by the incident ray with the normal (the angle of incidence).
(3) *The law of refraction:* At an interface between media with different properties, there is a refracted ray into the second medium, lying in the plane of incidence, making an angle with the normal that obeys Snell's law.

The laws of geometric wave propagation can be reformulated using Fermat's principle that a disturbance follows the least time path. Snell's law is a consequence of Fermat's principle.

SNELL'S LAW

In the case of oblique incidence of plane waves on a plane boundary, four boundary conditions must be satisfied. Assuming welded contact, the normal and tangential displacements and stresses must be equal on both sides of the boundary. To satisfy the boundary conditions for nonnormal incidence, it is necessary that an incident compressional wave produce both reflected and transmitted shear waves, as well as reflected and transmitted compressional waves.

Snell's law of refraction gives the angles of reflection and transmission as a function of the incident angle and the shear and compressional velocities:

$$\frac{\sin a_1}{V_{p1}} = \frac{\sin a_2}{V_{p2}} = \frac{\sin b_1}{V_{s1}} = \frac{\sin b_2}{V_{s2}}. \tag{2.11}$$

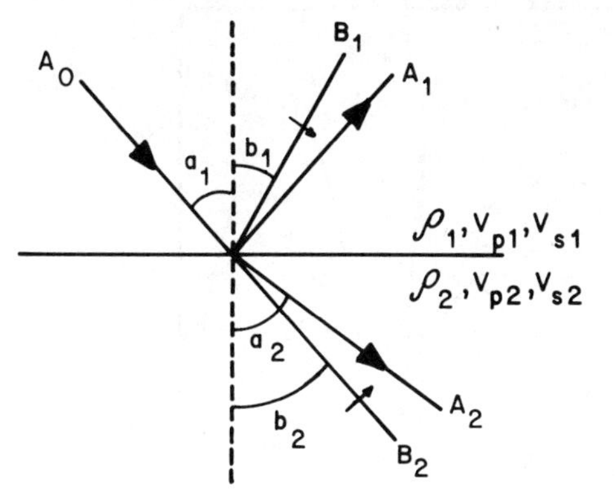

FIGURE 2.7. *Snell's law.*

The incident and reflected compressional angles are a_1, the transmitted compressional angle is a_2, and the reflected and transmitted shear angles are b_1 and b_2. The particle displacements for compressional waves are in the direction of the raypaths and, for shear waves, transverse to the direction of the path and directed toward the boundary. The amplitudes of the waves are designated A_0, incident compressional; A_1, reflected compressional; B_1, reflected shear; A_2, transmitted (refracted) compressional; and B_2, transmitted (refracted) shear (see Figure 2.7).

With a compressional source, if the compressional velocity in medium 1 is less than the compressional velocity in medium 2, then there exists an angle a_1 such that, beyond that angle, no compressional energy enters the second medium. To find this angle, let $a_2 = 90°$ in equation (2.11). This angle, called the first critical angle, a_p, or the critical angle for compressional waves, is given by

$$\sin a_p = \frac{V_{p1}}{V_{p2}}. \tag{2.12}$$

Also, with a compressional source, if the compressional velocity in medium 1 is less than the shear velocity in medium 2, then there exists an angle a_1 such that, beyond that angle, no shear energy enters the second medium. To find this angle, let $b_2 = 90°$ in equation (2.11). This angle, called the second critical angle, a_s, or the critical angle for *P-S* conversion, is given by

$$\sin a_s = \frac{V_{p1}}{V_{s2}}. \tag{2.13}$$

By applying Snell's law to the boundary conditions, one obtains four equations—one for each boundary condition—that must be satisfied simultaneously. These are known as the Zoeppritz equations. A complete discussion of their derivation can be obtained in such texts as Macelwane and Sohon (1936). These equations are listed in the order corresponding to the boundary conditions—normal and tangential displacement, normal and tangential stress:

$$(A_0 - A_1)\cos a_1 + B_1 \sin b_1 = A_2 \cos a_2 + B_2 \sin b_2$$

$$(A_0 + A_1)\sin a_1 + B_1 \cos b_1 = A_2 \sin a_2 - B_2 \cos b_2$$

$$(A_0 + A_1)\cos 2b_1 - B_1\left(\frac{V_{s1}}{V_{p1}}\right)\sin 2b_2$$

$$= A_2\left(\frac{\rho_2 V_{p2}}{\rho_1 V_{p1}}\right)\cos 2b_2 + B_2\left(\frac{\rho_2 V_{s2}}{\rho_1 V_{p1}}\right)\sin 2b_2$$

$$\rho_1 V_{s1}^2\left[(A_0 - A_1)\sin 2a_1 - B_1\left(\frac{V_{p1}}{V_{s1}}\right)\cos 2b_1\right]$$

$$= \rho_2 V_{s2}^2\left[A_2 \frac{V_{p1}}{V_{p2}}\sin 2a_2 - B_2 \frac{V_{p1}}{V_{s2}}\cos 2b_2\right]. \tag{2.14}$$

To use these equations, one must first use Snell's law to calculate the angles a_2, b_1, and b_2 for each angle of incidence. These angles are then substituted into the Zoeppritz equations, along with the densities, velocities, and incident amplitude. This gives a system of four equations in the four unknown amplitudes, which is then solved for the amplitudes.

The Zoeppritz equations contain normal incidence of plane compressional waves on a plane boundary as a special case. In this case, the tangential displacement and stress are zero; hence, B_1 and B_2 vanish. No shear waves are generated at the interface. Also, all angles equal zero; hence, all sine terms vanish and all cosine terms equal unity. Making these substitutions gives the

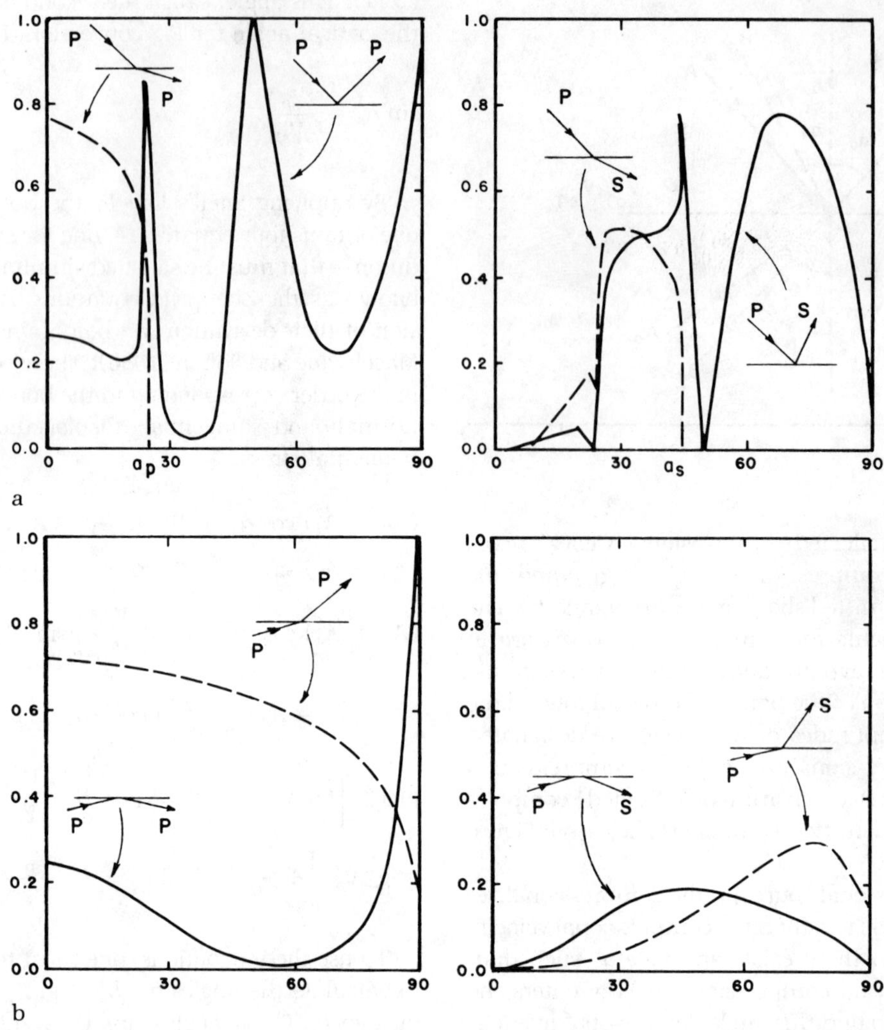

FIGURE 2.8. *Reflected and transmitted energy for $Z_2/Z_1 = 3.04$. (a) Incident* P*-wave in low-velocity medium. (b) Incident* P*-wave in high-velocity medium. (From Nafe 1957, vol. 47:3, pp. 205–220, courtesy of the Bulletin of the Seismological Society of America.)*

reflection and transmission coefficients for normal incidence of plane compressional waves, equations (2.6) and (2.7).

EXAMPLE 1 (FROM NAFE, 1957)

Figure 2.8(a) shows the energy in the reflected and transmitted components for a case where an incident P-wave of unit energy strikes a boundary whose parameters are given by $V_{p2}/V_{p1} = 2.5$, $\rho_2/\rho_1 = 1.22$, and $\sigma = 0.25$ (which implies $V_p = \sqrt{3}V_s$ in each medium); and Figure 2.8(b) shows the components for the reciprocal case (approach from the opposite direction).

When the incident P-wave is in the low-velocity medium—Figure 2.8(a)—there are several interesting points:

(1) At normal incidence, there is never any P-S conversion to shear, so the energy is partitioned into the reflected and the transmitted P-waves. In this example, the energy in the reflected P-wave is of magnitude 0.25, and in the transmitted P-wave, 0.75. The total energy must be unity to satisfy the law of conservation of energy. The P-wave amplitudes are given by

$$R = \frac{3.04 - 1}{3.04 + 1} = 0.51,$$

$$T = 1 - R = 0.49. \quad (2.15)$$

(2) No compressional energy enters the lower medium at angles beyond the first critical angle a_p:

$$a_p = \sin^{-1}\left(\frac{V_{p1}}{V_{p2}}\right) = \sin^{-1}\frac{1}{2.5} = 24°. \quad (2.16)$$

(3) No shear energy enters the lower medium at angles beyond the second critical angle a_s:

$$a_s = \sin^{-1}\left(\frac{V_{p1}}{V_{s2}}\right) = \sin^{-1}\left(\frac{\sqrt{3}}{2.5}\right) = 44°. \quad (2.17)$$

(4) The reflected energy is predominantly compressional at 24° (the first critical angle), 48°, and 90°, and the reflected energy is predominantly shear at 44° (the second critical angle) and 70°. Near 30°, the reflected P energy is very small because most of the energy is either reflected or transmitted shear energy.

When the P-wave is incident in the high-velocity medium—Figure 2.8(b)—there are no critical angles, and the curves are much smoother. The following points are noteworthy:

(1) At normal incidence, the energy in the reflected and transmitted P-waves is independent of direction of travel and so is 0.25 for the reflected and 0.75 for the transmitted P-waves. The reflection and transmission coefficients for normal incidence are given by

$$R = \frac{1 - 3.04}{1 + 3.04} = -0.51,$$

$$T = 1 - R = 1.51. \quad (2.18)$$

(2) The reflected P-wave has minimum energy at about 60° because of the large amount of energy in the other components.

(3) The reflected P-S conversion has maximum energy at about 50°, and the transmitted P-S conversion has maximum energy at about 75°.

EXAMPLE 2 (FROM GUTENBERG, 1944)

A second series of graphs is shown in Figure 2.9 for the case in which $V_{p2}/V_{p1} = 1.286$, $\rho_2/\rho_1 = 1.103$, and $\sigma = 0.25$ in both media. The acoustic impedance ratio is 1:4.

In Figure 2.9(a), the incident P-wave is in the low-velocity medium; in Figure 2.9(b), it is in the high-velocity medium, just as in the Nafe example, but here the graphs are given by $\sqrt{E/E_i}$, whereas Nafe plotted E/E_i. For normal incidence, $\sqrt{E/E_i}$ for the reflected P-wave is identical to the absolute value of the reflection coefficient R, which is equal to 0.17 in this example.

When the incident P-wave is in the low-velocity medium—Figure 2.9(a)—the curves are less complicated than in the Nafe example, because the shear velocity in the second medium is less than the compressional velocity in the first medium, and the second critical angle does not exist.

$$V_{s2} = (1.286/\sqrt{3})\ V_{p1}. \quad (2.19)$$

The first critical angle is

$$a_p = \sin^{-1}(1/1.103) = 65°. \quad (2.20)$$

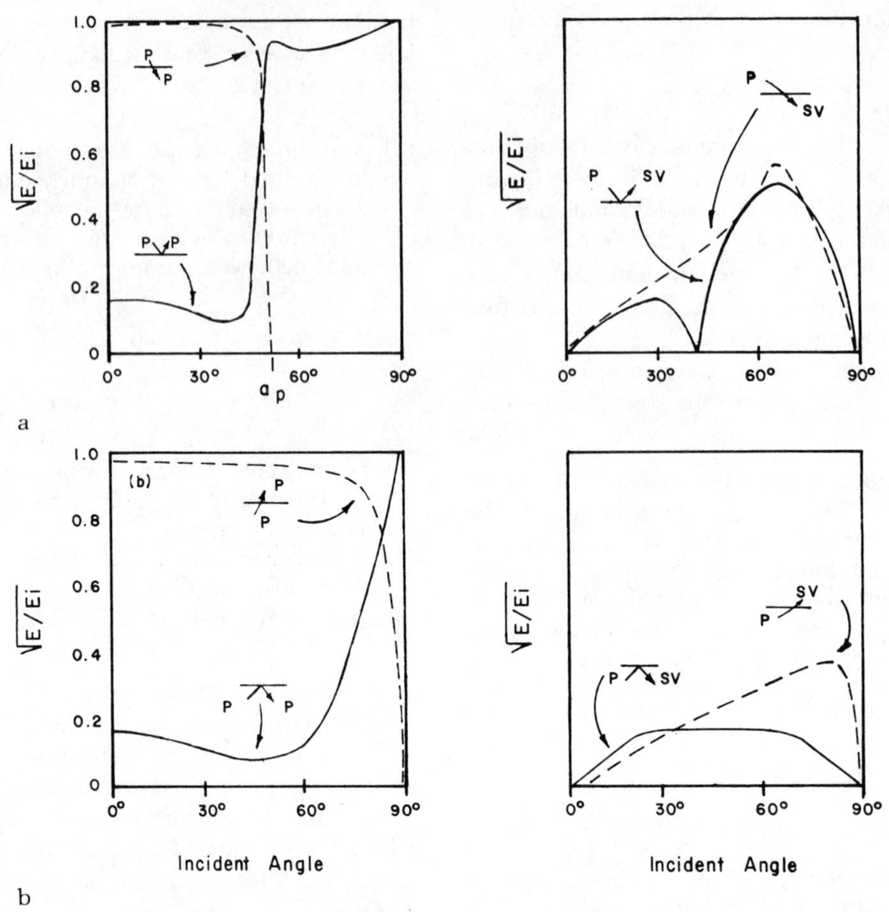

FIGURE 2.9. *Reflected and transmitted waves when* $Z_2/Z_1 = 1.4$. *(a) Incident* P-*wave in low-velocity medium. (b) Incident* P-*wave in high-velocity medium. (From Gutenberg 1944, vol. 34:2, pp. 85–102, courtesy of the Bulletin of the Seismological Society of America.)*

Probably the most noteworthy observation is the large increase in *P*-wave reflected energy past the first critical angle.

CHANGE IN REFLECTION CHARACTER WITH INCIDENT ANGLE

The change in character of the reflected *P*-wave due to change in incident angle was studied by Malinovskaya (1957). Theoretical waveforms obtained in four different three-layer cases are shown in Figure 2.10. The density and velocity ratios are indicated for each case. Figure 2.10(a) has contrasts, similar to those in the Gutenberg example, that show the strong increase in amplitude for angles greater than the first (and only) critical angle, in agreement with Gutenberg's curve in Figure 2.9(a). The density and velocity contrasts in Figure 2.10(c) are the same as in the Nafe example. The amplitude decreases between the two critical angles and increases past the second critical angle in agreement with Nafe's curve in Figure 2.8(a). In each example, the amplitude varies without phase shift for angles less than the first critical angle; however, beyond this angle, the reflection exhibits both amplitude and phase variations.

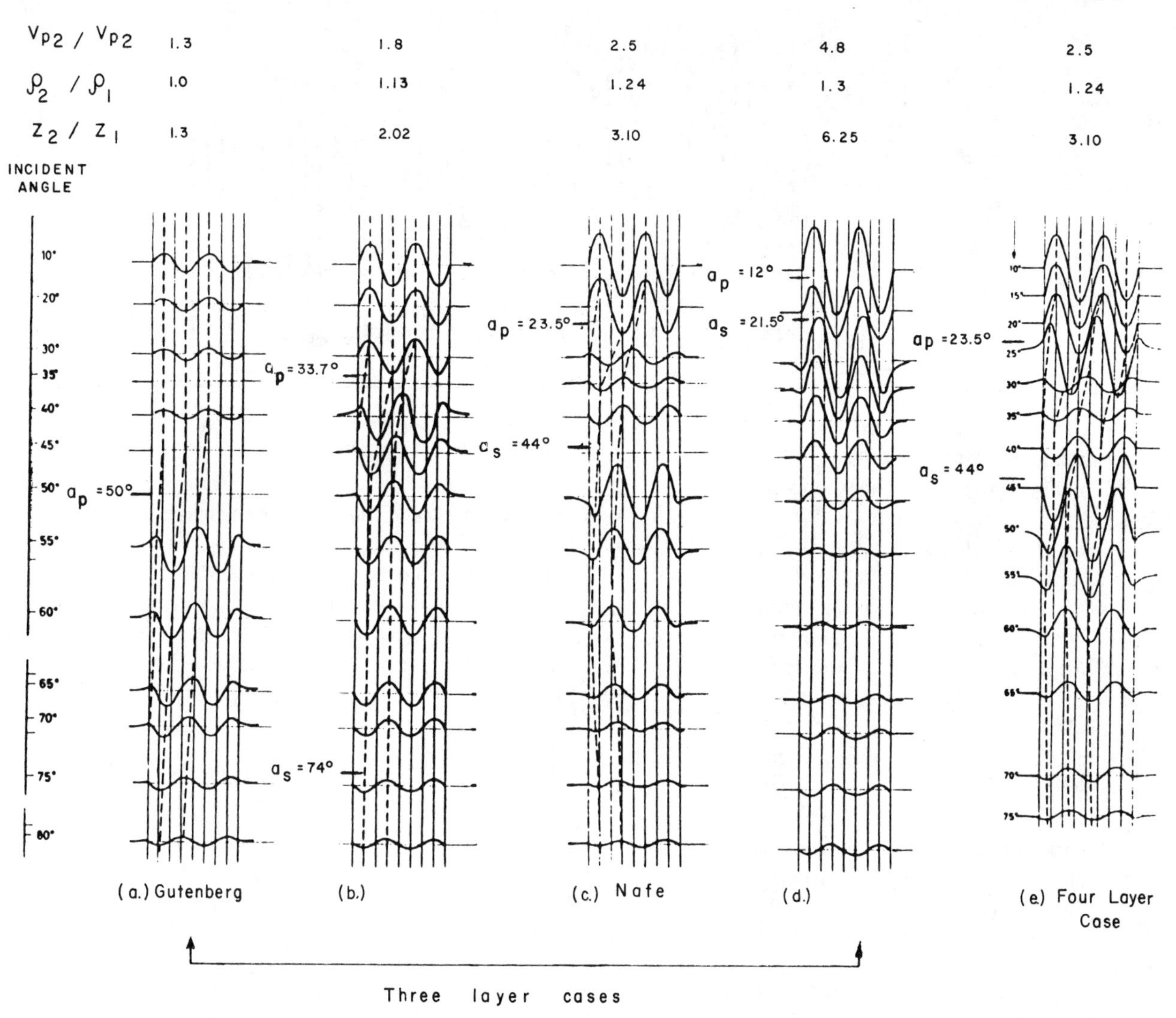

FIGURE 2.10. *Theoretical reflections as a function of incident angle. (From Malinovskaya 1957, courtesy of Geophysics.)*

A four-layer calculation by Malinovskaya, in which the layering in Figure 2.10(c) is modified by adding another layer at the surface, shows similar theoretical reflection waveform in the four-layer case, Figure 2.10(e), to that in the three-layer case, Figure 2.10(c). This indicates that the character and amplitude of a reflection are controlled by the elastic properties of the medium on each side of the reflecting boundary and are little affected by transit through the layers above.

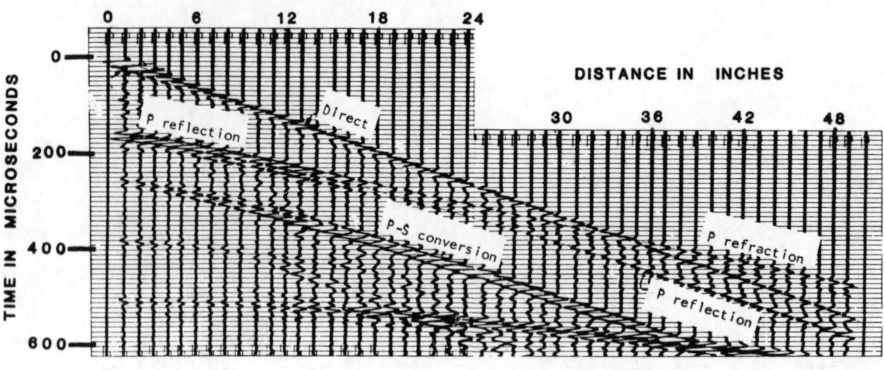

FIGURE 2.11. P-S *waves on refraction model. (From Angona 1960, courtesy of Geophysics.)*

P-S CONVERTED REFLECTIONS

One conclusion drawn from the Zoeppritz curves is that strong *P-S* reflections, known as *converted reflections*, may be expected at longer spread distance if there are large velocity contrasts in the subsurface and if the attenuation of shear waves above the reflectors is not too great. Ricker and Lynn (1950) mapped shallow reflections in Oklahoma, Louisiana, and Mississippi using *P-S* conversions. Vasil'ev (1957) observed this type of wave on field records.

P-S conversions are well demonstrated in models that have low attenuation of shear waves. Figure 2.11 shows a record from a plexiglass over copper model (Angona, 1960). At small angles, the *P-S* conversion is very small, whereas, at the larger angles, the *P-S* conversion becomes larger than the *P* reflection.

In the foregoing model, it is possible to determine the compressional velocity of plexiglass and copper from refraction measurements and also to estimate the plexiglass compressional velocity from the normal moveout of the reflection from the top of the copper. At small offset distances, the first arrival is the direct wave through the plexiglass. As the offset distance increases, the compressional reflection from the plexiglass/copper interface eventually bifurcates into a refraction, and a reflection, with the reflection increasingly delayed behind the refraction. Eventually, this refraction becomes the first arrival, because the velocity of copper is greater than the velocity of plexiglass. The compressional velocities of plexiglass and copper are derived experimentally from measurements of the respective refraction moveouts. The moveouts over the 49 in. from source to most-distant detector is 538 μs for plexiglass and 331 μs for copper (as measured along the tangent to the refraction in the latter case), giving velocities of 7.59 and 12.34 ft/ms, respectively. Substituting these velocities into equation (2.12) gives the critical angle for compressional waves,

$$a_p = \sin^{-1}\left(\frac{7.59}{12.34}\right) = 36^\circ. \qquad (2.21)$$

The velocity of plexiglass can also be found experimentally from the NMO. Substituting $T_0 = 149$ μs, $x = 49$ in., and $T_x = 559$ μs into equation (1.8) gives $V_{nmo} = 7.58$ ft/ms.

DIFFRACTIONS

Diffraction refers to the propagation of waves in a manner not predicted by the laws of geometrical wave propagation. Waves do not travel in straight lines when they pass an obstacle; rather, they curl around obstacles. They do not form sharp shadows. The waves are said to be diffracted by the obstacle. The phenomenon of diffraction owes its appearance to a limitation of the incident waveform caused by such things as a slit in optics, a brickwall in acoustics, and a fault in seismology.

The diffraction of light waves at the edge of an opaque half-space is shown at the top of Figure 2.12. The light

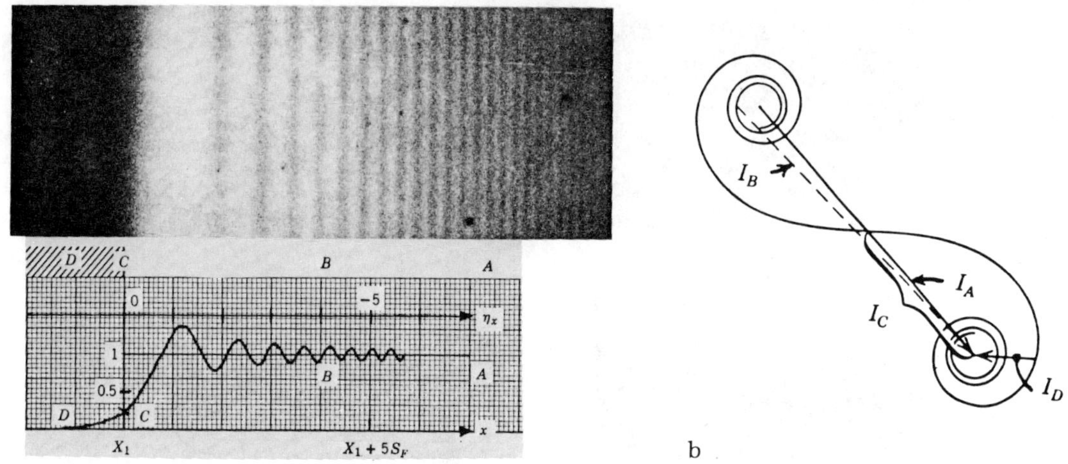

FIGURE 2.12. *Diffraction of light by an opaque straight edge. (a) Photograph and plot of light intensity. (b) Graphic solution of Fresnel integral using Cornu spiral. (From Klein 1970, copyright John Wiley & Sons; courtesy of John Wiley & Sons.)*

intensity as a function of distance from the edge is plotted in the middle of that figure. The intensity at the edge is ¼ the incident intensity. To the left of the edge, the intensity decreases monotonically. To the right, the intensity increases and becomes oscillatory. The Cornu spiral shown at the bottom of the figure shows a graphic solution to the Fresnel integral that gives the intensity.

The foregoing discussion introduces the subject of diffraction. Application of the wave equation and satisfaction of boundary conditions for relatively simple geometries for electromagnetic wave propagation has been successful. For elastic wave propagation, however, where both compressional and shear waves are involved, the diffraction problem becomes formidable. Knopoff (1956) gives a formal solution of the wave equation that yields the diffraction of elastic waves through a large aperture bounded by opaque walls in a solid material.

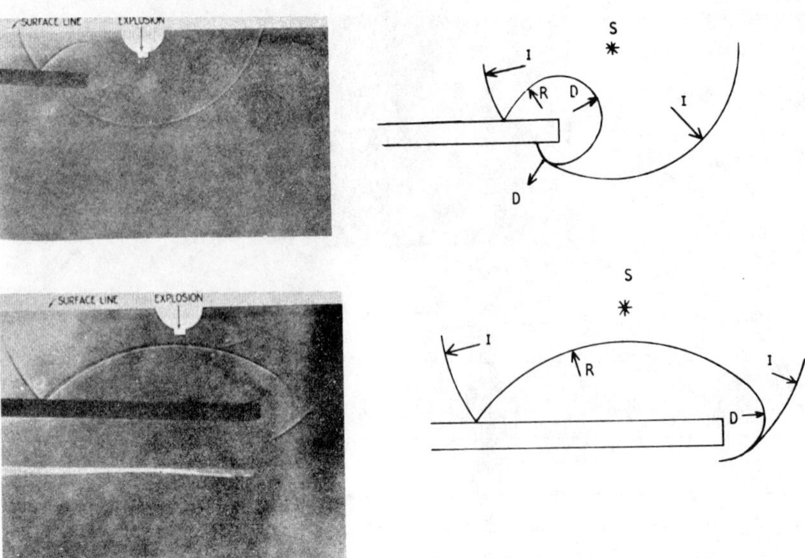

FIGURE 2.13. *Diffraction from the edge of a reflector. (From Rieber 1936, courtesy of Geophysics.)*

MODEL STUDIES

As far back as the 1930s, seismologists were concerned with records characterized by abnormally large moveout. They reasoned that, in areas containing faults or steeply folded strata, it would be necessary to go a step beyond the ordinary geometrical paths assumed by ray theory. Rieber (1936) made a study of diffractions from models of certain geological structures. He devised an experiment using shadow photography for observing wavefronts of a sound wave propagating in air and striking solid models of geological structures, as shown for the fault models in Figure 2.13. The incident wavefront I is circular because the medium has constant velocity, with the source at the center of the circle. The wavefront of the reflected wave R merges smoothly with the circular diffraction wavefront D generated at the upper edge of the barrier.

A fault model study by Angona (1960) shows the seismic data obtained by shooting directly over a vertical fault (Figure 2.14). The reflection from the high side, R_H, does not terminate abruptly at the fault, as ray theory would predict; rather, it continues beyond the fault, gradually fading in amplitude. Likewise, the reflection from the low side, R_L, continues beyond the fault. The diffraction moveout exceeds the reflection moveout, as noted on the time-distance plot of the R_H and R_L events. The reflection from the base of the model, R_B, is discontinuous because of the different thicknesses and velocities of the two layers; each branch has its associated diffraction, because a discontinuous reflection cannot end abruptly but must have a diffraction at the discontinuity.

Other events identified on the seismic section are the direct compressional (P) and shear (S) waves; the P-S conversions on both the high and low sides, which show the amplitude increase with offset distance in agreement with the curves in Figures 2.8(a) and 2.9(a); the discontinuous reflection from the base, R_B; and multiple reflections R_H(mult) and R_L(mult). The R_H(mult) is the surface multiple whose path is from surface to copper-top to surface to copper-top to surface, with reflectivity $R_H(-R_0)\,R_H$, where R_0 is the surface reflection coefficient and the negative sign indicates approach of the surface from below. This type of multiple is called a W-multiple. The R_L(mult) is an internal multiple whose path is from surface to base to copper-top to base to surface, with reflectivity $R_B(-R_L)R_B$.

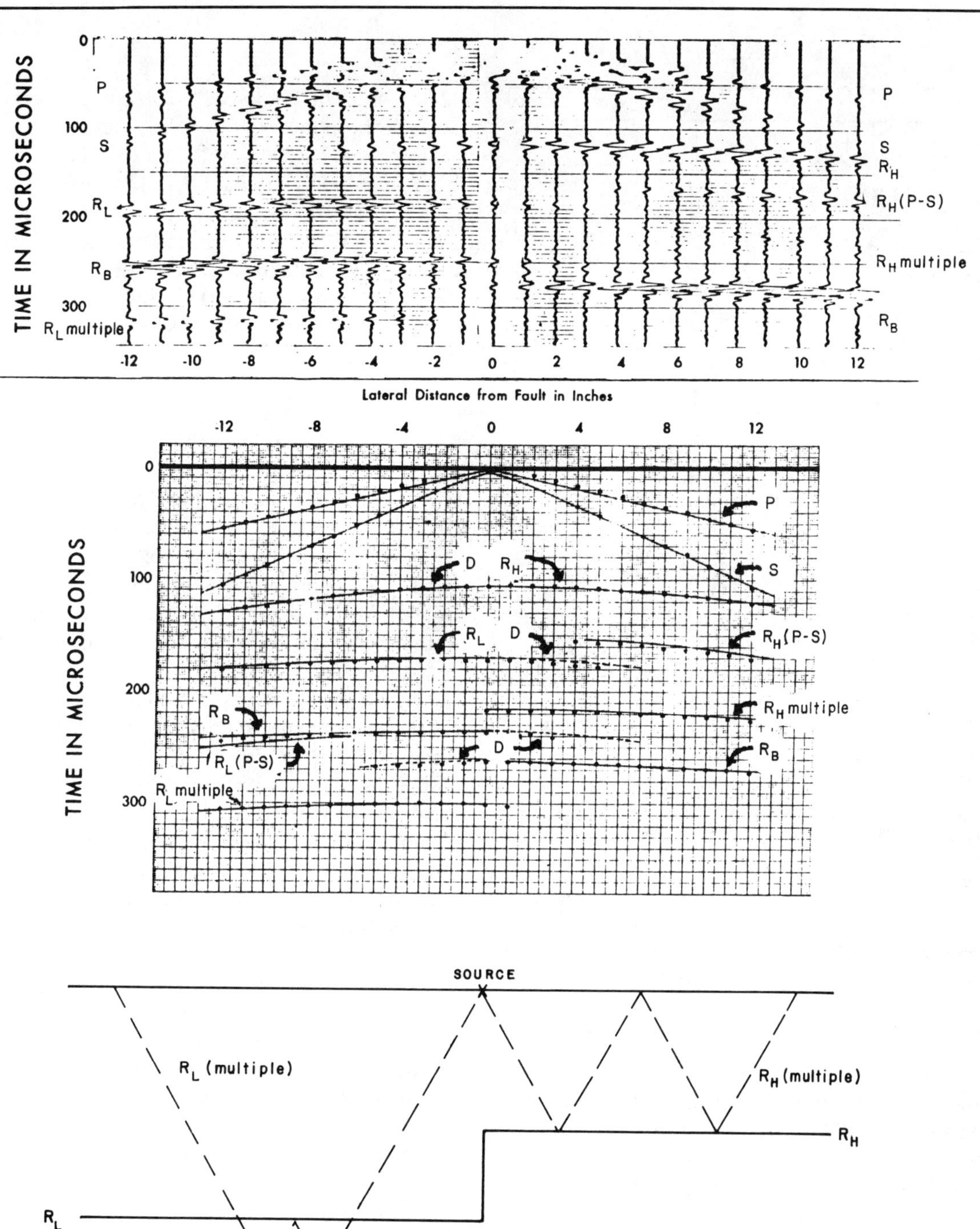

FIGURE 2.14. *Fault model study. (From Angona 1960, courtesy of Geophysics.)*

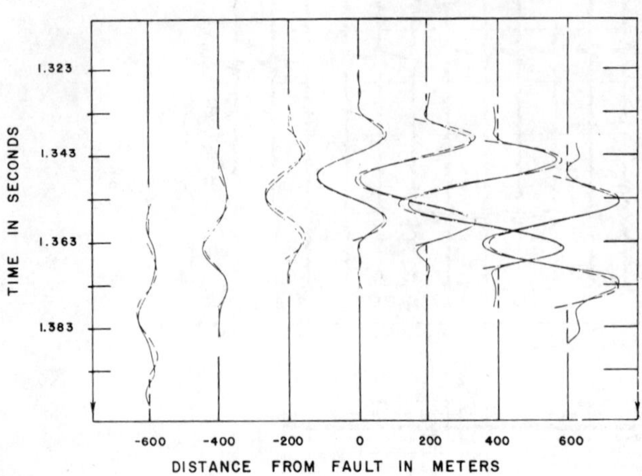

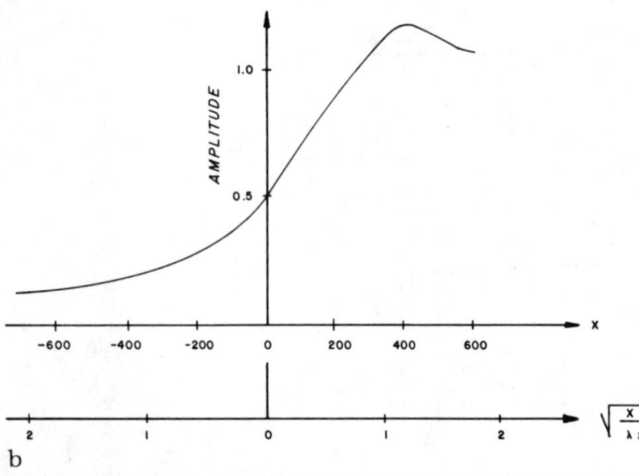

FIGURE 2.15. *Diffraction from a fault. (From Krey 1952, courtesy of Geophysics.)*

The model is aluminum over copper, with compressional velocities of 17.75 and 12.67 ft/ms, respectively. Because the first layer has higher velocity than the second, there is no critical angle for refraction; hence, there is no refraction from the copper layer. The compressional velocity of aluminum can be found by measurement of the moveout of the direct *P*-wave (57 μs to travel one foot), giving $V_p = 17.54$ ft/ms. The direct *S*-wave has velocity $V_s = 9.35$ ft/ms (107 μs to travel one foot). Using these measured velocities, Poisson's ratio is found to be 0.30, using equation (2.5). The compressional velocity can also be obtained from the normal moveout of the compressional reflection R_H and the diffraction moveout of its attendant diffraction. Substituting measured values $T_0 = 107$ and $T_x = 121$ μs at $x = 1$ ft in equation (1.8) gives $V = 17.70$ ft/ms from NMO measurements. Substituting $T_d = 130$ μs at $x = 1$ ft in equation (1.10) gives $V = 18.29$ ft/ms from diffraction measurements, which is within 4 percent of the velocity determined from NMO measurements.

DIFFRACTION AMPLITUDES

Krey (1952) determined the amplitude dependence of diffractions from a simple fault by applying the expressions developed by Fresnel and Sommerfeld for the diffraction of light to the seismic problem. The solution was obtained by considering a source located at its image point and an impervious sound barrier placed in the position of the reflector. This geometry is then analogous to that of a light source and a straight edge, as worked out by Fresnel.

Krey applied the steady-state expressions to the Fourier components of the pulse, which was taken as 1½ cycles of a 50 Hz cosine wave. Diffraction arrivals from a fault located at a depth of 2000 meters are shown in Figure 2.15(a). The amplitude as a function of horizontal distance is plotted in Figure 2.15(b). Notice that the amplitude of the diffraction is equal to half that of the normal reflection at the shotpoint and decreases as the detector is moved away from the fault into the geometric shadow zone.

The diffraction amplitude is a function of distance x, depth z, and pulse wavelength λ. By plotting the amplitude against the dimensionless parameter $\sqrt{x/\lambda z}$, the curve can be used to predict the diffraction amplitude for all values of x, z, and λ. Such a dimensionless scale has been placed on Figure 2.15(b) for the 60 m wavelength that Krey used in his example.

Berryhill (1977) developed the diffraction response for nonzero separation of source and receiver and came to the unexpected conclusion that the diffraction amplitudes are controlled almost exclusively by the location of the source-receiver midpoint. Since the data summed together in stacking all share a common shot-geophone midpoint, the diffraction amplitudes on the stacked trace should behave in good approximation to the zero-separation theory, which is a special case of nonzero separation. Thus, theoretical support is obtained for applying the zero-separation theory to stacked seismic data.

There is an excellent theory available relating subsurface reflector geometry to the amplitude properties of diffraction patterns in seismic sections (Trorey, 1970). This theory bears on the gross question of whether or not diffractions should be discernible in a particular set of circumstances, and it predicts the detailed amplitude characteristics to be expected in observable diffractions. An understanding of this theory offers a potential for extracting additional information from seismic data and for placing useful constraints on seismic interpretations.

A mathematically equivalent theory implicit in a paper by Hilterman (1970) shows that the theory correctly predicts the diffraction amplitudes observed in acoustic model experiments. A second paper by Hilterman (1975) develops additional aspects of the theory.

Berryhill (1977) shows that the diffraction amplitude at zero shot-geophone distance, caused by the termination of a planar reflector, can be calculated by convolving the reflection wavelet with a time-domain operator called the normalized diffraction response, $D_0(t)$, as shown in Figure 2.16. D_0 depends only on the onset time t_0 and the angle θ_0 between the raypath and the normal to the reflecting plane. D_0 and its Fourier transform evaluated at $t_0 = 2.0$ sec for four values of θ_0 are shown in Figure 2.17. The Fourier transforms have been normalized to show unit response at the origin, thus emphasizing the decrease in high-frequency response with increasing θ_0.

The theory leads to the following conclusions concerning diffraction amplitudes at zero separation resulting from a single reflection termination:

(1) Diffraction amplitudes are frequency-dependent. High-frequency pulses excite less diffraction response than low-frequency pulses, and the waveform changes shape along a diffraction hyperbola because high-frequency components die out at a faster rate.

(2) The maximum amplitude a diffraction can attain is half that of the associated reflection. The maximum amplitude occurs (on a seismic stacked section) where the diffraction meets the reflection, which is not at the apex of the diffraction hyperbola if the reflector is dipping.

(3) Diffraction hyperbolas are divided into two regions, in which the algebraic signs of the amplitudes are opposed. The part of the hyperbola off-end from the associated reflection has the same polarity as the reflection, while the part beneath the reflection (if visible) has the opposite polarity.

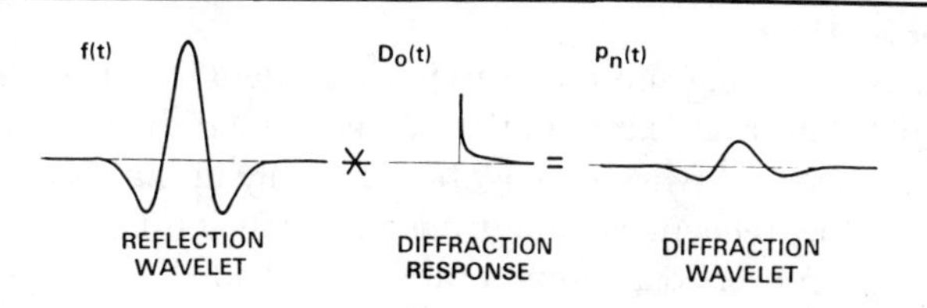

FIGURE 2.16. *Diffraction wavelet as convolution of reflection wavelet and the diffraction response. (From Berryhill 1977, courtesy of Geophysics.)*

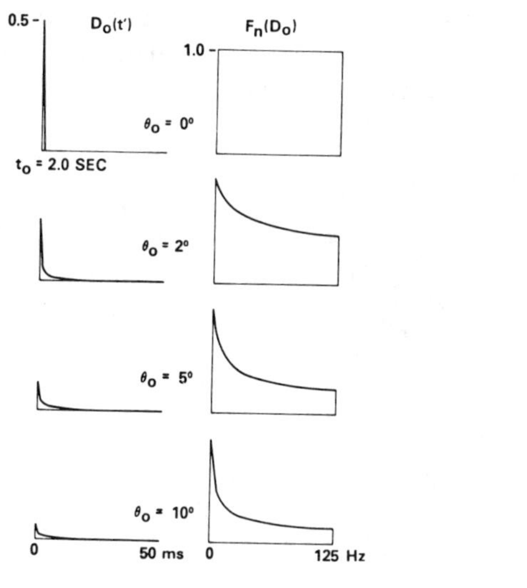

FIGURE 2.17. *Diffraction response as a function of angle θ_0. (From Berryhill 1977, courtesy of Geophysics.)*

SUMMARY

It is becoming more and more important for those involved in seismic exploration to understand the underlying theory of wave propagation, for only in that context can the reflectivity and transmissitivity of the earth, mode conversions from compressional to shear waves, and diffraction and migration be understood. The recent development of wave equation migration is the most significant triumph for the wave propagation model to date. A lesser-known triumph is Foster's development, which shows that there is no transmission loss in a continuous medium, as measured at the surface. Until that development, no one seemed to question the fallacy that sampling creates layers.

Chapter 3 Seismic Signatures

INTRODUCTION

The seismic method uses active seismic sources to generate elastic waves. There are two classes of sources: (1) impulsive sources, such as dynamite, weight drops, and gas and air guns, which generate a pulse of seismic energy whose time duration is very short compared to the length of the seismic record; and (2) sources whose signatures extend over time durations on the order of the length of the record and which rely on correlation procedures to effect a pulse compression down to the order of the duration of pulses from impulsive sources.

The pulse waveform in the vicinity of the source and the factors that influence the waveform as the pulse travels in the real earth are of interest in understanding the seismic method. In the immediate vicinity of the source, the pressure is such that the stress-strain relationship is nonlinear, and theoretical studies of wave propagation in this region are difficult at best. As the wave propagates outward from the source, its amplitude diminishes until, eventually, the pulse enters the region where the stress-strain relation is linear. Propagation of the seismic pulse that emerges from the relatively small, nonlinear region of ignorance into the linear region can be described by difficult but tractable mathematics.

In an ideal, nonattenuating earth, plane waves travel without change in amplitude or waveform. Spherical waves, although they do not exhibit waveform changes, decrease in amplitude with distance because the energy-density on the ever-increasing sphere must decrease. This is known as the *spherical spreading loss*. In the real earth,

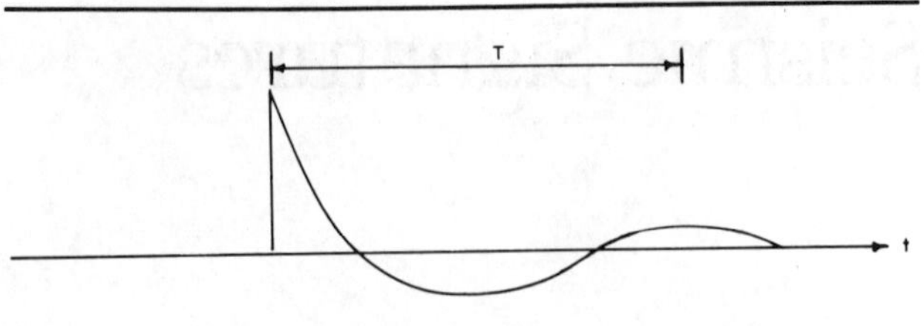

FIGURE 3.1. *Pressure wave near a dynamite source.*

in addition to the spreading loss, the wave loses amplitude because of attenuation. Because attenuation increases with frequency, the pulse waveform broadens with distance traveled.

LAND SOURCES

The predominant seismic source used in land prospecting is dynamite exploded in boreholes. Dynamite is a highly concentrated source of energy; it produces an intense impulsive signature and is simple to use. It is also expensive, and the requirement of drilled holes adds significantly to the overall exploration costs. It is also destructive and cannot be used near dwellings, water wells, and so forth. It is also out of tune with ecologists.

Various impulsive surface sources have been developed over the years, each of which has a realm of applicability. Thumpers generate energy by dropping a 10-ton weight from a 15 ft height (Peacock and Nash, 1962). Dinoseis (trademark of Atlantic-Richfield Company) explodes a propane-oxygen mixture in a chamber that is pressed against the earth's surface (Godfrey, Stewart, and Schweiger, 1968). Primacord buried in a long, shallow, plowed furrow is also used as a surface source. The air gun is becoming a popular impulsive surface source on land, primarily because it delivers a wideband spectrum at low cost. The wideband spectrum can be preserved when an array of guns is used because of precise control of firing, which insures optimal in-phase summation of the input signals from the array.

DYNAMITE SOURCES

An intense, short, transient pulse is generated in the immediate vicinity of dynamite sources. In this region, the stress-strain is nonlinear. Eventually, however, the amplitude diminishes as the wave propagates until the pulse enters the linear region. Although the extent of the nonlinear region is dependent on the initial pressure, this region is relatively small, even with large explosions. Measurements near a 1.5 kiloton explosion in Nevada showed that nonlinearity extended only 275 feet from the source.

The waveform of the signal that emerges from the nonlinear region is dependent on many factors, including size of charge and type of rock. Pressure measurements near the source in a typical sand-shale subweathering sequence have the waveform shown in Figure 3.1. The pressure wave has a very sharp onset, followed by an approximate exponential decay and overshoot. The time duration of the negative overshoot increases with charge size. Fourier analysis of waveforms from different charge sizes shows a spectral shift toward lower frequencies as the charge size increases. In one experiment, measurements 20 ft from the source show that a charge of 5/8 lb had its spectral peak near 400 Hz, while a 20 lb charge had its peak near 80 Hz. Thus, if one desires to enhance the high-frequency content of seismic data, the smallest possible charges should be used, in multiple holes, if necessary, to increase penetration and to improve the signal-to-noise ratio.

The waveform depends critically on the type of rock in which the charge is placed. In general, the frequency spectrum peaks higher in frequency as the rocks become more consolidated. The spectrum of the input signal in the least consolidated of all rocks, the weathering, peaks at the lowest frequency. The weathering is an inefficient medium for generating acoustic energy. Experiments show that spherically corrected amplitude of a downgoing pulse measured beneath the source is relatively constant for shots in the subweathering but is reduced by as much as a factor of 50 for shots in the weathering.

VIBROSEIS SOURCES

Vibroseis (trademark of Continental Oil Company) is the surface source that threatens to displace dynamite in seismic prospecting on land. Based on an entirely new concept for seismic sources, it had its origin in chirp radar developed by Bell Laboratories after World War II (Klauder et al., 1960). Until that development, radar had always increased its range by transmitting more intense

Parameters of input signals

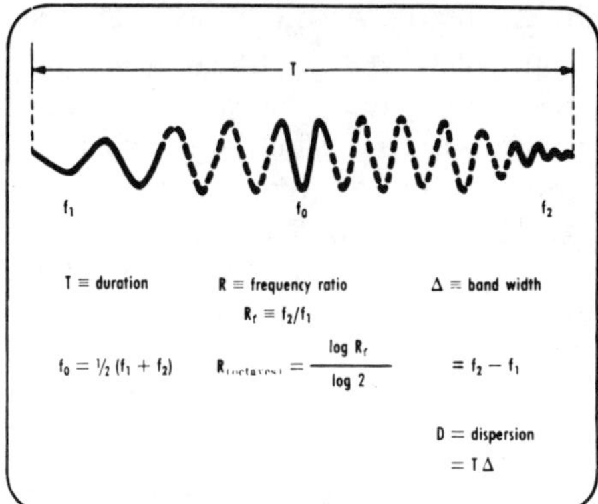

Parameters of Klauder wavelet

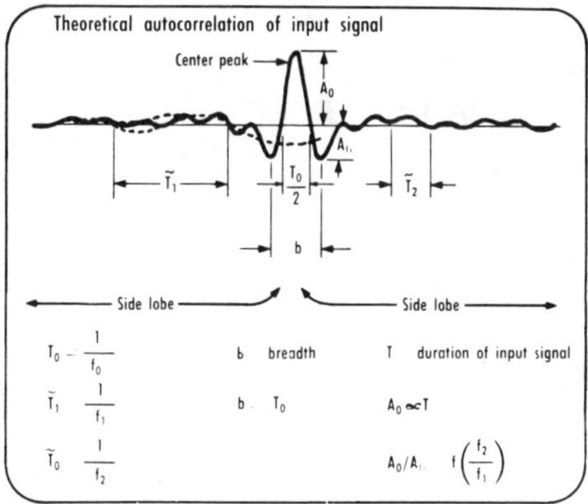

Klauder wavelet is theoretical autocorrelation of the 17-68 Hz input (Vibroseis pilot signal)

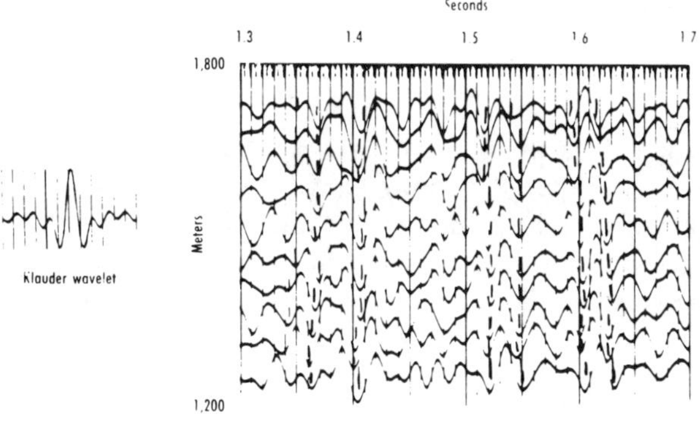

FIGURE 3.2. *Vibroseis characteristics. (From Geyer 1970, courtesy of Oil and Gas Journal.)*

impulses. Chirp radar increases energy by transmitting a long, oscillatory signature whose instantaneous frequency changes linearly with time. Each received echo is compressed to a pulse by correlating the received signal with the input signature. Conoco applied the chirp principle to seismic prospecting (Crawford, Doty, and Lee, 1960) (see Figure 3.2).

The source is a vibrating plate that is pressed against the ground by the jacked-up weight of a huge truck or tractor. The plate is vibrated through a hydraulic mechanism in synchronization with an electrical signal transferred from the recording truck to the Vibroseis unit. The input signal is a long train of oscillations whose instantaneous frequency changes linearly with time. The starting and ending frequencies f_1 and f_2, respectively, and the time duration of the sweep of frequencies can be selected by the operator. The amplitude spectrum of the input signal will be essentially constant between f_1 and f_2 and zero outside. The total input energy increases with the product of sweep length T and the bandwidth $\Delta f = f_2 - f_1$. The input signal is collapsed into a short wavelet (called the Klauder wavelet) by autocorrelation. The pulse width of the Klauder wavelet is inversely proportional to the bandwidth.

After the Vibroseis data are processed by cross-correlating the input signal with the received data, the result is very similar to seismic data obtained directly by shooting dynamite. The main advantage of Vibroseis over dynamite is that the source is nondestructive and can be used on roads and near dwellings, water wells, and so forth. It has been used in the midst of cities, such as Los Angeles and Chicago. There may also be economic advantages and possibly some improvement in data quality because of the ability to control the input spectrum. By using nonlinear sweep of instantaneous frequency, it is possible, for example, to increase the high-frequency content to compensate for increasing loss of high-frequency energy because of attenuation (Figure 3.3).

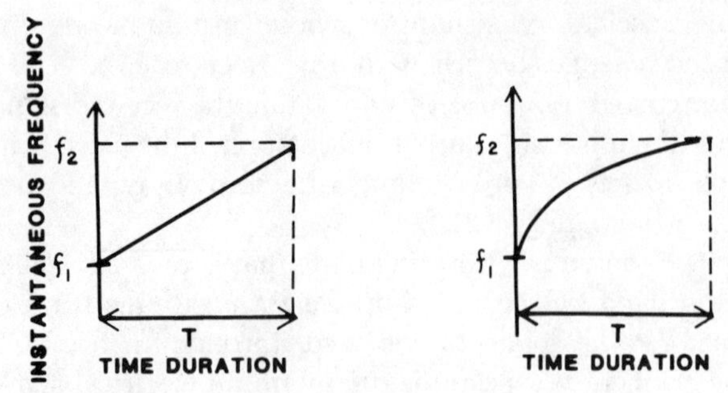

FIGURE 3.3. *Linear and nonlinear change in instantaneous sweep frequency.*

MARINE SOURCES

The first attempts at marine exploration used large dynamic charges as the seismic source. This source was virtually eliminated because of the inherent danger with large stores of dynamite, the expense of the two-boat operation necessary for using large dynamite charges, massive fish kills and the resulting irate complaints of ecologists, the introduction of gas gun sources, and, finally, the widespread use of horizontal stacking.

Gas guns were originally developed for shallow exploration, as shown in the cross section in Figure 3.4, from McClure, Nelson, and Huckabay (1958). The marine sonoprobe used a mixture of propane and oxygen; detonated at the top of a partially submerged section of pipe to create a compressional wave that traveled down the pipe and into the water. Cost factors and the unavailability of the mixture components in remote areas led to the replacement of oxygen with compressed air, and then to the use of compressed air alone. Air guns are now the most widely used source in marine acquisition, although the propane-oxygen source is still used for deep exploration by Western Geophysical Company (trademark Aquapulse).

One of the problems with marine sources is the presence of secondary pulses caused by oscillation of the gas bubble that forms in the water (Cole, 1948). The oscillation period decreases with depth of the bubble and increases with increased air volume and pressure. The equation for bubble oscillation period T is given by the Rayleigh-Willis equation,

$$T = \frac{k(PV)^{1/3}}{(D + 10)^{5/6}} \tag{3.1}$$

where P is pressure, V is volume, and D is depth in meters. Figure 3.5 shows air gun signatures for various pressures, with depth and volume held constant. Then, according to the Rayleigh-Willis equation, $T/P^{1/3}$ should be constant. Measurements show that this is substantially true:

$P =$ 20 bar	$T =$ 75 ms	$T/P^{1/3} = 27.6$
= 40	= 92	= 26.9
= 80	= 120	= 27.9
= 160	= 152	= 28.0

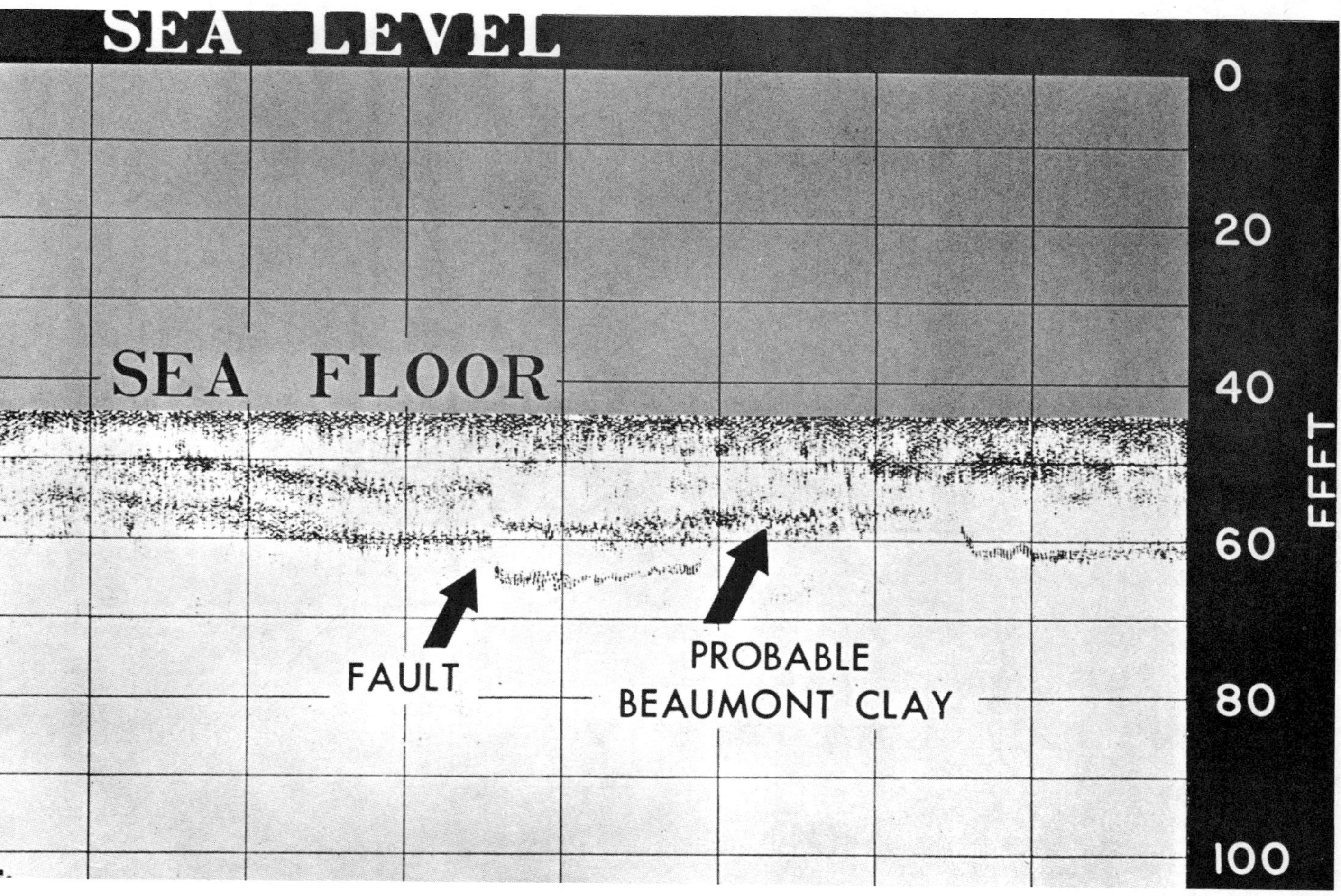

FIGURE 3.4. *Marine sonoprobe profile, High Island area, Texas, showing Pleistocene Beaumont clay outcropping on the sea floor. (From McClure, Nelson, and Huckabay 1958, courtesy of AAPG.)*

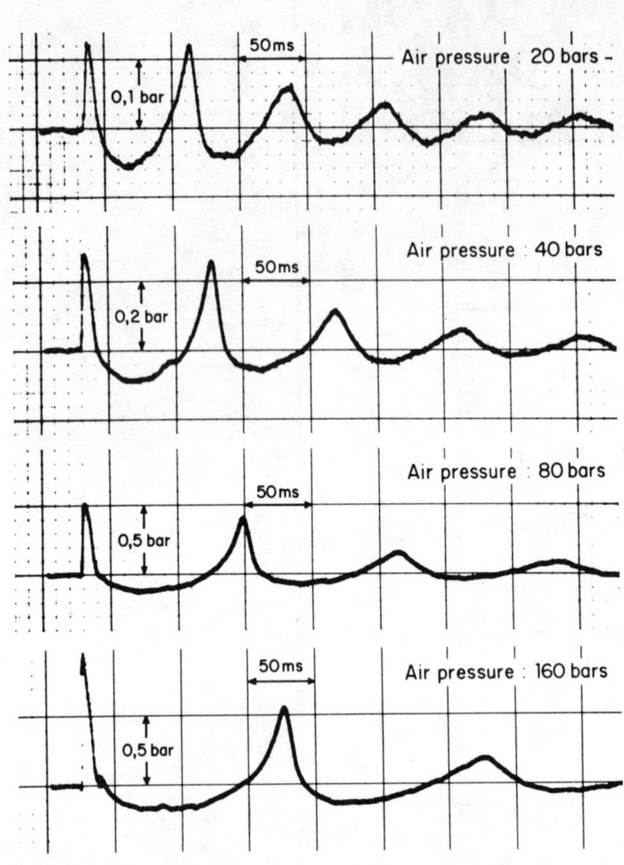

FIGURE 3.5. *Air gun signatures showing increase in period with increase in pressure. (From Farriol, Michon, Muntz, and Staron 1970, courtesy of SEG.)*

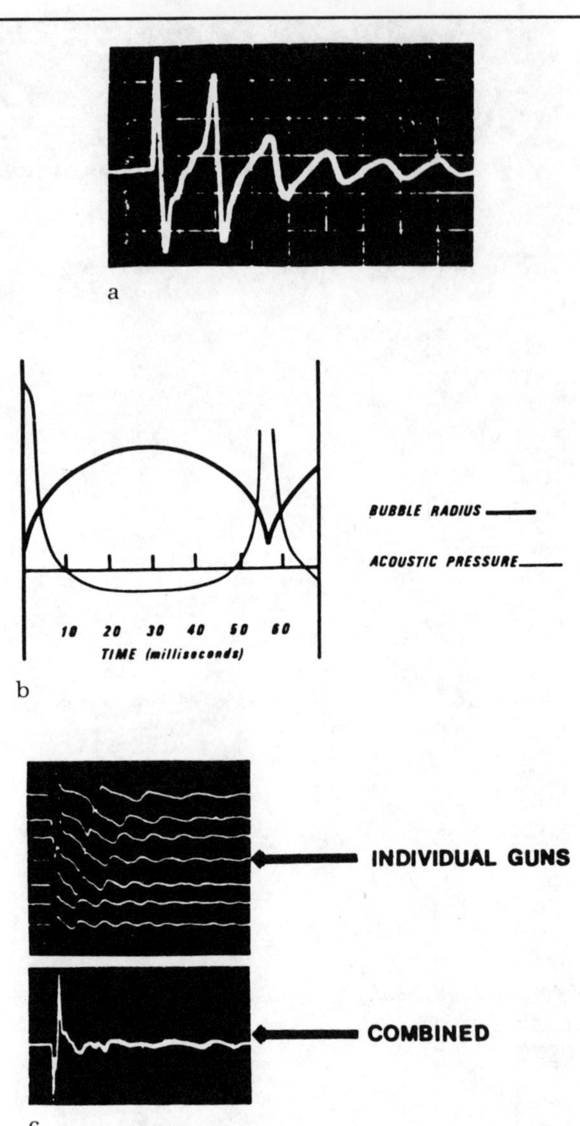

FIGURE 3.6. *Air gun array tuned to suppress bubble pulses. (a) One gun without wave-shaping kit. (b) Bubble radius and pressure. (c) Seven-gun tuned array, with wave-shaping kits. (From Kologinczak 1974, courtesy of OTC.)*

An array of guns of different sizes at the same depth will reinforce the initial compressional wave and spread out the effect of the bubbles. Tests by Kologinczak (1974), using a seven-gun array called the Stagaray, shows bubble suppression of 12:1 (Figure 3.6).

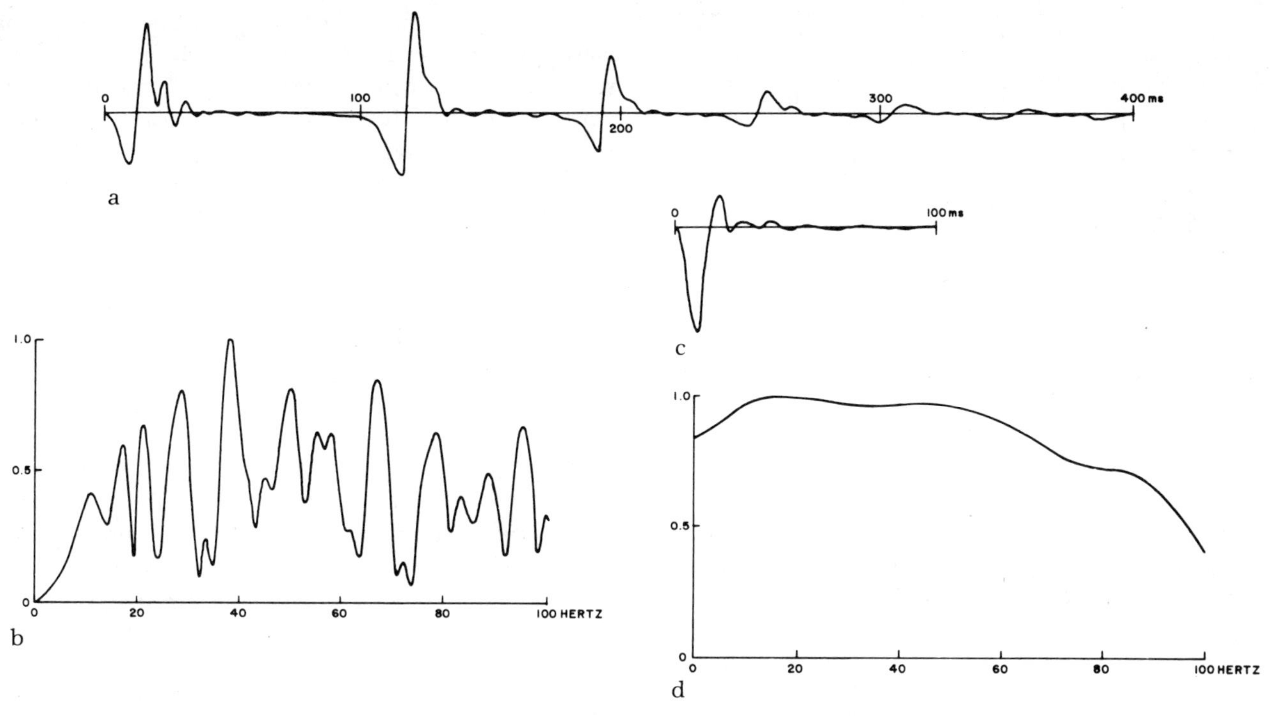

FIGURE 3.7. *Maxipulse before and after pulse compression. (a) Maxipulse: initial pulse and bubble-sequence pressure signature. (b) Maxipulse: initial pulse and bubble-sequence amplitude spectrum. (c) Output pulse pressure signature. (d) Output pulse amplitude spectrum. (Courtesy of Western Geophysical Company.)*

Small explosive sources on the order of 0.2 kg are being used by Compagnie Generale de Geophysique (trade name Flexotir) and Western Geophysical Company (trade name Maxipulse) (Figure 3.7). With Flexotir, the charge is detonated within a perforated cast-iron cage whose purpose is to damp out the effects of the bubble collapse by dissipating the energy of the collapsing bubble in turbulent flow through the perforations. With Maxipulse, no attempt is made to suppress bubble oscillations during recording. Thus, no energy is wasted in preventing, suppressing, or spreading out the effect of the bubbles. Because the input signal is nonminimum-phase, it cannot be collapsed into an impulse by minimum-phase (Wiener) deconvolution that is used to suppress reverberations present on marine data. Therefore, a preprocessing stage is required to convert the nonminimum-phase seismic data into minimum-phase data before applying Wiener deconvolution. Western Geophysical calls this stage *debubbling*. To do effective debubbling, it is necessary to measure the input signal free from interference from nearby reflectors, such as the ship and the water bottom, and then use this data to devise the debubbling operator.

Another source developed by CGG (trademark Vaporchoc) uses collapse of a steam bubble as the source (Figure 3.8). With this source, there is no bubble oscillation, because the steam condenses to water upon collapse of the bubble. The signature is nonminimum-phase and the recorded data require conversion to minimum-phase before applying Wiener deconvolution, as must be done with Maxipulse or any other nonminimum-phase sources. CGG calls their procedure Wapco; it is described in detail in Chapter 7.

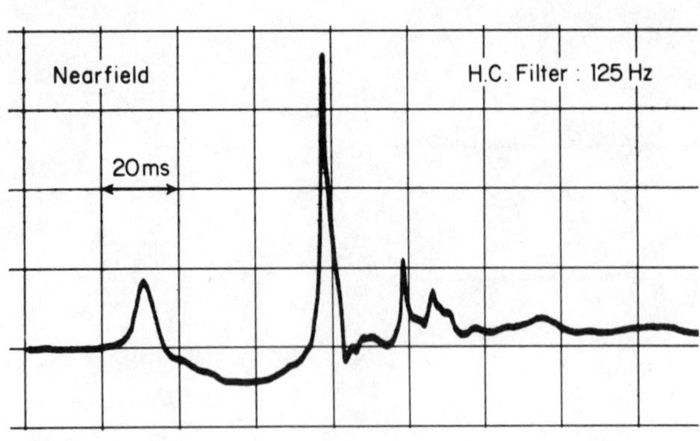

FIGURE 3.8. *Vaporchoc signature. (From Farriol, Michon, Muntz, and Staron 1970, courtesy of SEG.)*

Vibroseis was used successfully in marine operations, but the additional step of correlation in processing the large volume of data made the technique uneconomical compared to other marine sources, such as air guns. Also, the penetration was somewhat inferior to that of other sources.

For high-resolution work, electrical discharge energy sources are usually used because they are high-frequency sources. The penetration increases as the source frequency decreases. Various terms are applied to these sources, such as pinger, boomer, and sparker, in increasing order of penetration.

SHEAR SOURCES

It is more difficult to produce a relatively pure shear wave than it is to produce a relatively pure compressional wave. Conoco has developed a practical shear wave source using Vibroseis techniques (Cherry and Waters, 1968; Erickson, Miller, and Waters, 1968). This source has undergone extensive field tests and appears to be a satisfactory source of shear waves.

Shear waves from explosive sources have been discussed by White and Sengbush (1963). Explosive sources produce a mixture of shear and compressional waves, with the *P/S* ratio increasing with charge size. Such sources do not appear to have much applicability as a practical source of relatively pure shear waves.

SPHERICAL SPREADING

When a medium is disturbed at a point, the disturbance travels outward with a spherical wavefront, provided that the velocity is constant. Since the total energy on the ever-increasing sphere is constant, the energy density (energy per unit area) must decrease in proportion to the total area of the sphere. Thus, the energy of the pulse at any point on the sphere decreases as the square of the distance from the source. Since energy is proportional to the square of the amplitude, the amplitude decreases as the first power of the distance.

Geometric spreading of the energy in the ever-expanding wavefront causes a reduction in reflection amplitude with record time. Figure 3.9 shows a typical decay rate curve whereby amplitude is plotted as a function of record time. The amplitude can be corrected for spherical spreading by multiplying the amplitude by the distance traveled, or it can be corrected approximately by multiplying by the record time. In this example the amplitude decays by a factor of about 200 over a range of record time from 0.2 to 1.5 sec. A factor of 7.5 is due to spherical spreading. There are no transmission losses in a continuous medium, according to Foster (1975); therefore, the remaining factor of about 30 is largely due to attenuation, although some of the difference may result from a change in magnitude of the reflection coefficients with depth.

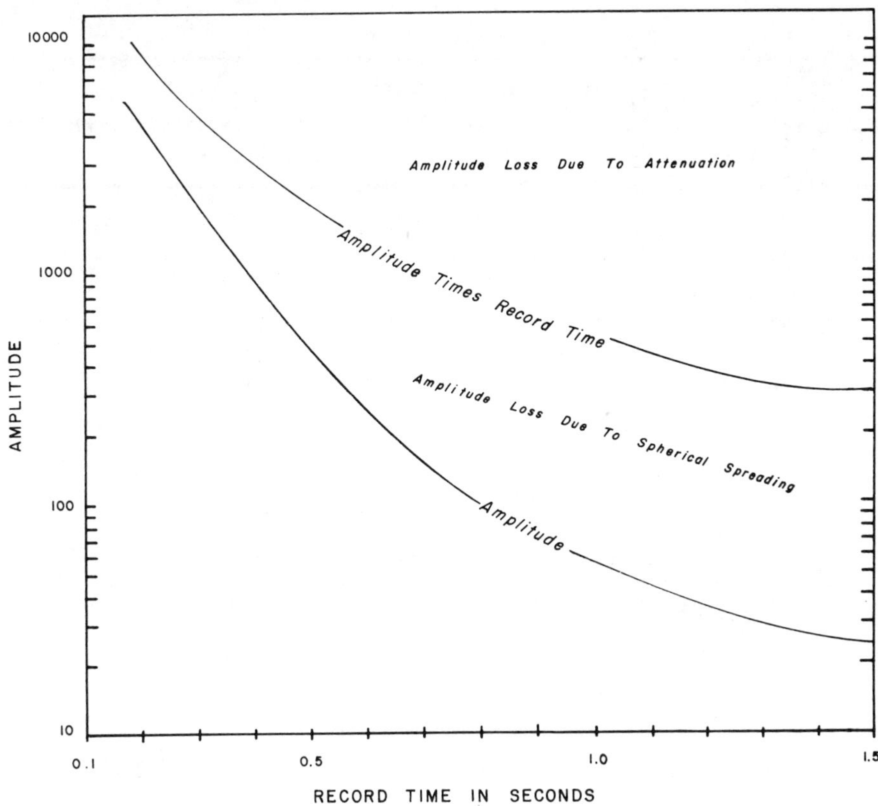

FIGURE 3.9. *Amplitude loss due to spherical spreading and attenuation.*

ATTENUATION

It is well known that reflections tend to have lower frequency as record time increases, indicating an increasing loss in high-frequency energy with distance traveled. Classical elastic theory does not admit to such a frequency-dependent energy loss, and spherical spreading is insufficient to account for the total energy loss. The mathematical model that encompasses the classical elastic theory evidently has assumptions that do not hold in the real earth. In the model, there is no mechanism for energy loss by conversion of elastic energy to heat. Of the various mechanisms that have been proposed to explain this energy loss, the one that most closely matches experimental results is the linear attenuation model. In this model, the attenuation is a linear function of distance traveled, and the amplitude loss is given by $\exp(-\gamma x)$, where the attenuation factor γ is a function of frequency and x is the distance traveled.

Field experiments in the Pierre shale near Limon, Colorado, are discussed by McDonal et al. (1958) (Figure 3.10). In the range of frequencies in the seismic band from 10–500 Hz, the attenuation increases as the first power of the frequency. The field procedure involved recording the downgoing shot pulse at various depths beneath the shot, analyzing the waveforms for their frequency spectra, and determining the attenuation for each frequency component. The spectra of the waveform generated by a 1 lb shot at 260 ft depth show that the peak in the spectrum shifts to lower frequency as the distance increases. The attenuation in decibels per 1000

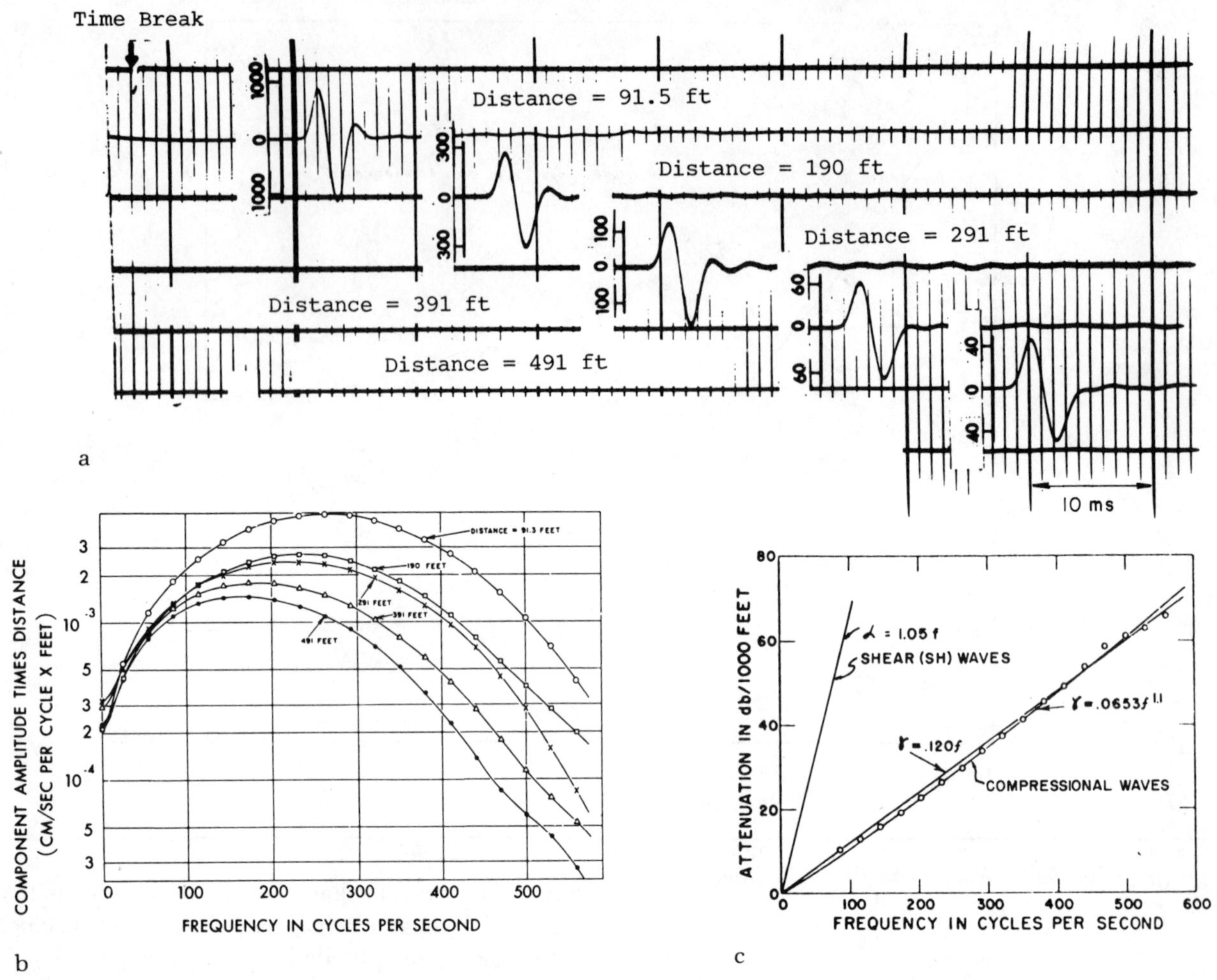

FIGURE 3.10. *Attenuation in the Pierre shale. (a) Vertically traveling compressional waves. (b) Spectra of vertically traveling compressional waves. (c) Attentuation of vertically traveling compressional and shear waves. (From McDonal, Angona, Mills, Sengbush, Van Nostrand, and White 1958, courtesy of Geophysics.)*

ft, plotted as a function of frequency, shows that it fits closely a linear dependence on frequency, $\gamma = 0.120f$.

Attenuation of shear waves was also measured at the same location in the Pierre shale. The functional dependent was linear in this case also, $\gamma = 1.05f$. The attenuation rate (the coefficient of f) for shear waves is about nine times greater than that for compressional waves. This indicates that shear waves are rapidly attenuated in comparison to compressional waves and suggests that, when thick shale sections are present, converted shear waves probably are not a serious source of seismic noise.

Similar experiments performed in the Ellenburger

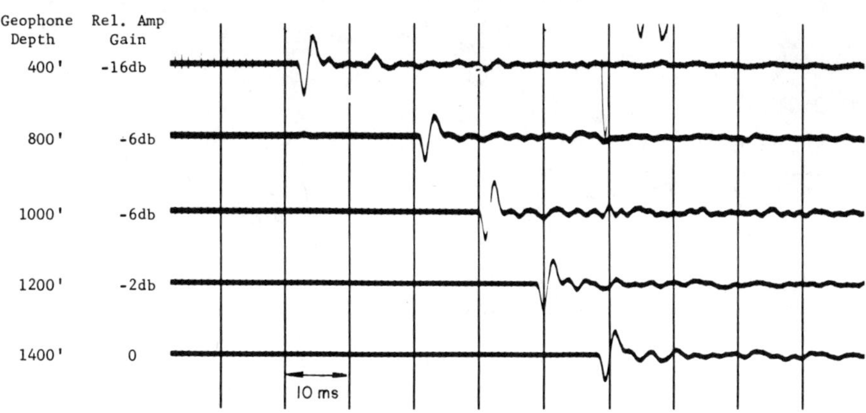

FIGURE 3.11. *Vertically traveling compressional wave in Ellenberger limestone. (From White 1978, courtesy of Pexcon Press.)*

limestone show that the propagating waveform remains unchanged in form in traveling from 400–1400 ft, Figure 3.11 (White, 1978). After applying the geometric spreading correction, the loss in the product (amplitude × distance) as a function of distance was about 4 dB/1000 ft, on the average. The apparent frequency of the signal is about 300 Hz. Therefore, assuming linear attenuation with frequency, the attenuation in the Ellenburger is given by $\gamma = 0.011f$, which is one-tenth the rate of attenuation found in the Pierre shale.

Measurements of attenuation in near-surface shale in East Texas show a sevenfold increase in attenuation in going from the subweathering to the weathering. The attenuation in dB/1000 ft is $0.3f$ in the subweathering, and $2.2f$ in the weathering.

In the McDonal et al. (1958) study, it was concluded that the measured linear attenuation was not accompanied by dispersion (that is, frequency-dependent velocity of propagation). In terms of filter theory, this implies that attenuation has zero-phase response. This condition is untenable in our causal world; causality demands dispersion if there is attenuation.

Wuenschel (1965) used the Pierre shale data to disprove the zero-phase character of attenuation by taking the waveform closest to the source and predicting the waveform measured farthest from the source under two assumptions, one zero-phase and the other minimum phase, both with the same attenuation (amplitude) response (Figure 3.12). The minimum phase assumption produced a far better prediction of the waveform measured farthest from the source, showing that anelastic wave propagation is causal. His careful study also showed a small change in phase velocity as a function of frequency:

f = 25 Hz	V = 6720 ft/sec
= 50	= 6770
= 100	= 6820
= 200	= 6840

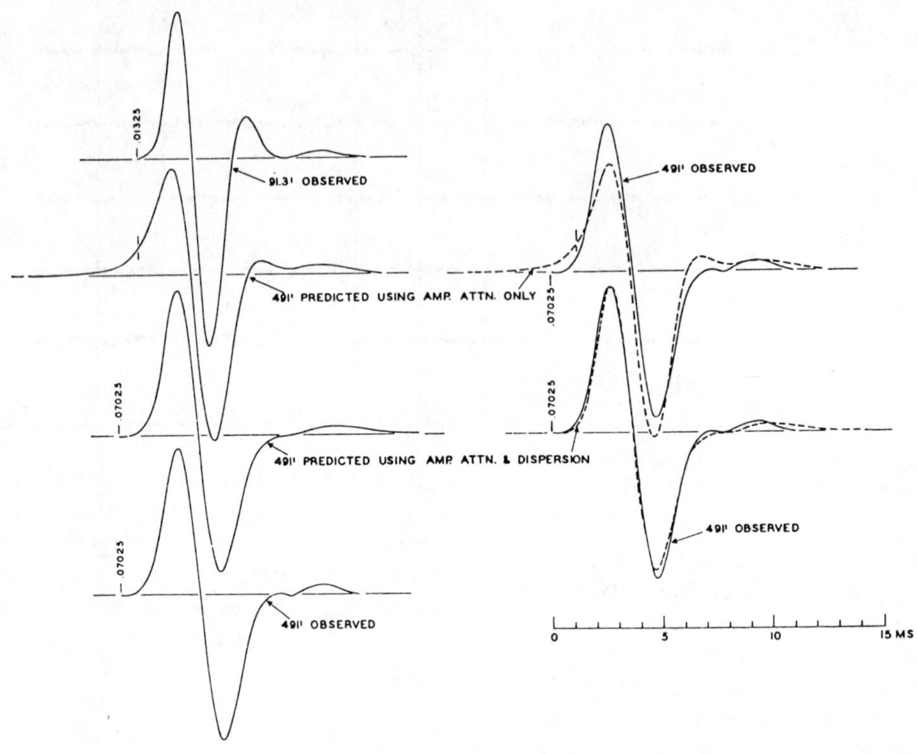

FIGURE 3.12. *Waveform prediction in the Pierre shale using zero-phase and minimum-phase assumptions. (From Wuenschel 1965, courtesy of Geophysics.)*

SUMMARY

Measurement of the seismic signature near the energy source and its change in character as it travels in the earth are vitally important in seismic exploration. The mechanics of attenuation have been studied widely and are still controversial; however, the most important aspect of attenuation with respect to seismic data processing is Wuenschel's demonstration that, for all intents and purposes, attenuation acts like a minimum-phase filter.

The importance of the phase of the recorded seismic data is now well known, because Wiener deconvolution to suppress reverberations and ghosts and reduce shot-point variable demands that the data be minimum-phase. It is apparent that the only significant departure from minimum-phase is introduced by the seismic signature itself; therefore, its effect must be removed from the seismic data by signature deconvolution so that subsequent application of Wiener deconvolution will be effective. This requires either measurement of the signature near the source or estimation of the seismic pulse from the reflection data.

Chapter 4

Analysis and Suppression of Seismic Noise

INTRODUCTION

The goal of seismic prospecting is to provide data that can be interpreted in terms of the subsurface geology. Seismic data consist of signal plus noise, where signal is that component of the data used in interpreting the subsurface and noise is any interference that tends to degrade interpretation. It is very important to reduce noise during acquisition in order to collect data that can be most effectively enhanced by further processing techniques. Before beginning a seismic survey, one should conduct experimental noise studies to optimize routine field procedures to be used in the production work. This optimization to suppress noise and enhance signal is always subject to economic constraints.

Source-independent noise includes the natural ground unrest due to wind (land) and cable motion, wave action, and boat noise (marine). These random noises usually constitute the quiescent noise level in seismic prospecting. Other independent noise sources include machinery, aircraft, pipelines, powerlines, and instrument noise. Source-independent noise is minimized by use of properly designed and operated equipment and good field procedures. Good geophone plants on land and acceleration-cancelling pressure detectors in water are basic ingredients of good data acquisition. Nulling devices at the input to seismic amplifiers usually can reduce electrical pickup to a satisfactory level. By placing the digitizer at the geophone rather than in the recording truck, crossfeed between channels and electrical pickup are greatly reduced. In areas of unusually high noise, special patterns or a Vibroseis source may be required.

Horizontally traveling noise generated by the source is suppressed by patterns of sources and detectors and by use of frequency and velocity filters. Noise in the reflection path—ghosts, reverberations, and multiples—is suppressed by special acquisition and processing techniques—notably, deconvolution for ghost and reverberation suppression, horizontal stacking for multiple suppression, and source-receiver patterns for long-period multiple suppression in marine operations. Random noise can be reduced only by increasing the number of elements—detectors, sources, fold, and trace mix.

SUPPRESSION OF RANDOM NOISE

Consider the output of the ith detector to be the sum of signal s_i plus noise n_i:

$$g_i(t) = s_i(t) + n_i(t). \tag{4.1}$$

The mean square output of the ith detector designated $[g_i^2]$ is the time average

$$[g_i^2] = \frac{1}{T}\int_0^T g_i^2(t)\,dt. \tag{4.2}$$

In terms of the components, $[g_i^2]$ is given by

$$[g_i^2] = [s_i^2] + [n_i^2] + 2[s_i n_i]. \tag{4.3}$$

Now consider the sum of the output of N detectors, denoted g_s:

$$g_s(t) = \sum_{i=1}^{N} g_i(t). \tag{4.4}$$

The mean square of the sum is then given by

$$[g_s^2] = \sum_{i,j}^{N} [s_i s_j] + \sum_{i,j}^{N} [n_i n_j] + 2\sum_{i,j}^{N} [s_i n_j]. \tag{4.5}$$

If the noise and signal are uncorrelated, then $[s_i n_j] = 0$ for all i, j, and the mean square of the sum is

$$[g_s^2] = \sum_{i,j}^{N} [s_i s_j] + \sum_{i,j}^{N} [n_i n_j]. \tag{4.6}$$

We make the following assumptions:

(1) The signal is identical in each output; $s_i = s$ for all i.
(2) The noise power is identical in each output; $[n_i^2] = n$ for all i.
(3) The noise is uncorrelated from output to output; $[n_i n_j] = 0$ for all $i \neq j$.

Then, the mean square of the sum is given by

$$[g_s^2] = N^2[s^2] + N[n^2]. \tag{4.7}$$

The signal-to-noise power ratio is improved by the factor N compared to the power ratio in any individual output. Thus, the signal-amplitude-to-rms-noise ratio is improved by $\sqrt{N}$.

Improvement in the signal-amplitude-to-rms-noise ratio as the square root of the number of elements applies more generally, under the foregoing assumptions, where the number of elements is given by the product of number of detectors in an array times the number of separate records that are vertically stacked, times the common depth point (CDP) fold, times the number of traces mixed.

If the signal power in each trace is identical but the cross correlation between signals is not perfect, then the cross signal power $[s_i s_j] = \phi_{ij}[s^2]$, where ϕ_{ij} is the correlation coefficient bounded by ± 1 and defined by

$$\phi_{ij} = \frac{[s_i s_j]}{\sqrt{[s_i^2][s_j^2]}}. \tag{4.8}$$

If the correlation coefficient $\phi_{ij} = \phi$ for all $i = j$, then

$$[g_s^2] = N[s^2] + (N^2 - N)\,\phi[s^2] + N[n^2] \tag{4.9}$$

$$= \{N^2\phi + N(1 - \phi)\}\,[s^2] + N[n^2]. \tag{4.10}$$

The signal-to-noise power ratio is then reduced from N to $N\phi + (1 - \phi)$, which is approximately equal to $N\phi$ for large N. For example, if $N = 2$ and $\phi = 0.5$, then the signal-to-noise power ratio is 1.5 instead of the maximal value of $N = 2$ for perfect signal correlation.

Now assume that the signal power is equal on two traces but the noise power is unequal, being equal to $[n^2]$ on one trace and $a^2[n^2]$ on the other. Simple addition of the two traces gives mean square output of

$$[g_s^2] = 4[s^2] + (1 + a^2)[n^2], \tag{4.11}$$

or a signal-to-noise power ratio of $4/(1 + a^2)$, compared to the theoretical ratio of $N = 2$ for equal noise power. Note that, for $a^2 = 1/3$, the signal-to-noise power ratio is 3, which is the same as the ratio for the second trace alone. For $a^2 < 1/3$, the ratio is better for the second trace alone than it is for the sum of the traces. For $a^2 = 3$, the ratio for the sum of traces is the same as for the first trace; and, for $a^2 > 3$, the ratio for the first trace is better than the ratio for the sum. For intermediate values of a^2 between $a^2 = 1/3$ and $a^2 = 3$, the ratio is better for the sum than for either trace individually.

When the signal-to-noise ratio differs from trace to trace, the optimal sum is a weighted sum, with the weights decreasing as the signal-to-noise ratio decreases (Foster and Sengbush, 1971).

In the case of separation of random signal from random noise, the optimal filter to separate signal from noise is the Wiener filter, given by

$$G(f) = \frac{\Phi_s(f)}{\Phi_s(f) + \Phi_n(f)}, \tag{4.12}$$

where Φ_s and Φ_n are the power spectra of signal and noise, respectively.

HORIZONTALLY TRAVELING NOISE

Horizontally traveling noise consists of waves that travel in the near surface at velocities that depend on the near-surface layering. The noise consists, in general, of three classes that tend to separate with increasing distance from the source because their characteristic velocities are different.

(1) Compressional refractions and their attendant multiple refractions are the first arrivals, because they have the highest velocity of the three classes.
(2) Shear refractions arrive later than the compressional refractions, because the shear velocity is always considerably less than the compression velocity.
(3) Surface waves, which are predominately of the Rayleigh wave type, arrive still later, because their velocities are always less than the shear velocity.

Figure 4.1 (from Dobrin, Lawrence, and Sengbush, 1954) shows a noise profile in the Delaware basin, Texas, with the various events plotted on a time-distance plot. The first arrivals are always compressional refractions, followed by the shear refractions, and finally by the long, leggy train of Rayleigh waves, often called ground roll by seismologists. These classes of horizontally traveling waves occupy distinct regions of t-x space because of their differences in propagation velocity.

MULTIPLE COMPRESSIONAL REFRACTIONS

The characteristics of multiple compressional refractions are determined by the near-surface compressional velocity layering. Consider a simple two-layer case with velocity V_1 in the first layer and V_2 in the second, with $V_2 > V_1$, as shown in Figure 4.2. Beyond the critical distance x_p, the primary refraction travels with velocity V_2. As the event moves along the profile, energy is continuously fed back into the first layer, creating a multiple refraction that is reversed in polarity with respect to the primary because of reflection from the free surface and is delayed behind the primary by $t_d = t_0/\cos a_p$, where t_0 is the two-way vertical time in the first layer and a_p is the critical angle. Beyond the distance $2x_p$, the multiple refraction travels with velocity V_2. Repeating this pattern gives rise to a succession of multiple refractions that have alternating polarities and delay times that are integral multiples of t_d. The kth multiple refraction has polarity $(-1)^k$, delay time kt_d, and critical distance $(k + 1)x_p$. The impulse response of the multiple compressional refraction mechanism indicates that the frequency component whose period is $T = 2t_d$ will be in phase. Thus, the succession of multiple refractions tends to produce a train of cycles with period $T = 2t_d$ and phase velocity V_2.

SURFACE WAVES

In 1885, Lord Rayleigh developed a theory of waves on the free surface of a semi-infinite, ideal elastic solid. His

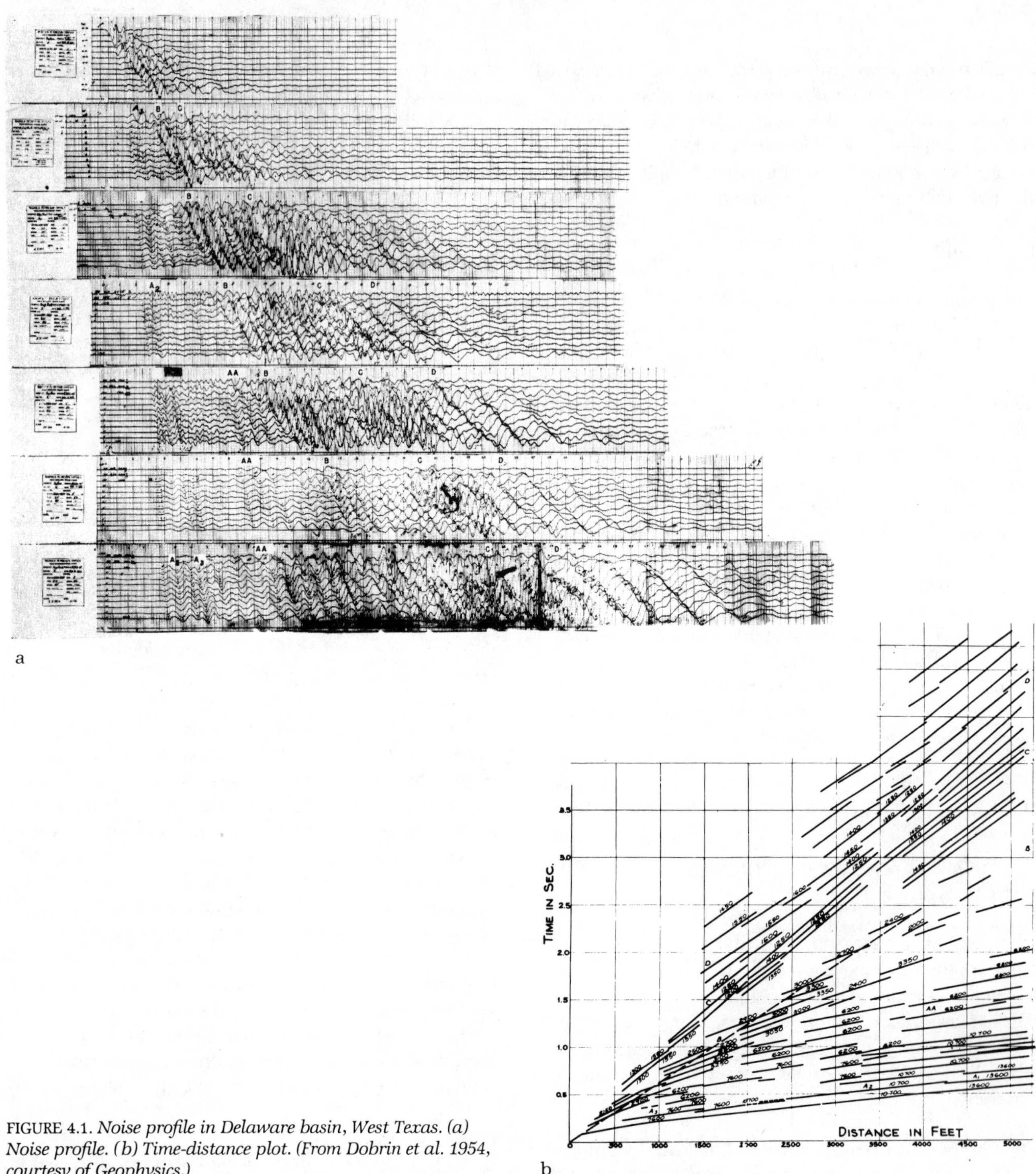

FIGURE 4.1. *Noise profile in Delaware basin, West Texas. (a) Noise profile. (b) Time-distance plot. (From Dobrin et al. 1954, courtesy of Geophysics.)*

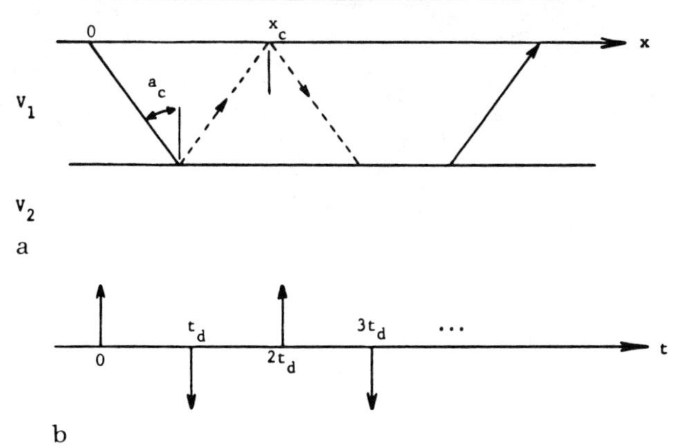

FIGURE 4.2. *Multiple compressional refraction mechanism. (a) Generation of multiple refractions. (b) Impulse response.*

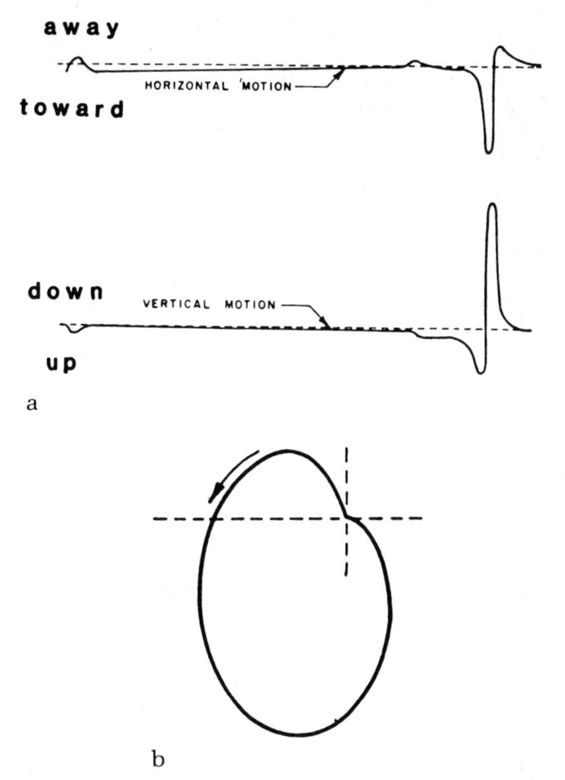

FIGURE 4.3. *Particle motion of Rayleigh waves. (a) Components of motion. (b) Particle motion. (From Sorge 1965, courtesy of Geophysics.)*

theory gave the following characteristics to the surface waves that bear his name:

(1) Rayleigh waves propagate with a velocity equal to 0.92 of the shear velocity.
(2) The particle motion at the free surface is elliptical retrograde, with the vertical displacement about 1.5 times the horizontal displacement.
(3) The vertical component of the particle motion decreases exponentially with depth, while the horizontal component decreases and becomes zero at a depth of $z = 0.19\ \lambda$, where λ is the wavelength. Below this depth, the particle motion reverses its direction and becomes forward elliptical, similar to the particle motion of waves on the surface of the water. Such motion is called hydrodynamic motion.
(4) The amplitude measured at the surface decreases exponentially with source depth.

Lamb (1904) predicted the Rayleigh wave motion that would be observed at the surface of a half-space from an impulsive source, as shown in Figure 4.3 (from Sorge, 1965). The trajectory shows the retrograde elliptical motion deduced by Rayleigh.

Experimental results substantiate Rayleigh's theory (Figure 4.4). Noise studies from the Delaware basin show that the particle motion at the free surface is a series of cycles having elliptical retrograde particle motion. Other experiments show that the particle motion changes from elliptical retrograde near the surface to forward elliptical at a depth of 40 ft, which agrees with the theoretical change of direction at 42 ft, based on the measured wavelength of 220 ft. Experimental results also show that the amplitude decreases exponentially with depth of receiver and with depth of source, provided the measurements are made in a relatively uniform medium.

The effect of charge depth on surface waves is shown on the noise profile in Figure 4.5. The noise profile has 20 ft trace spacing, with the first trace being offset 1980 ft from the source. At each shot depth, a 15 lb dynamite charge was used as the source. The surface waves weaken relative to primary reflections as the shot depth increases from 40 to 165 ft. The purpose of this study was

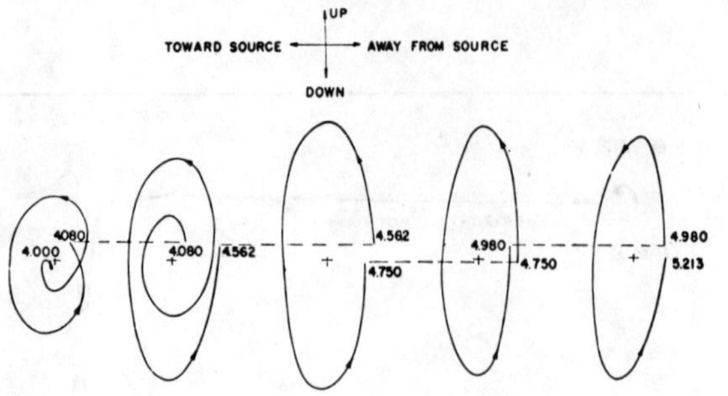

UP
TOWARD SOURCE
AWAY FROM SOURCE
DOWN
4.000
4080
4.080
4.080
4.562
4.562
4.750
4.980
4.750
4.980
5.213

a

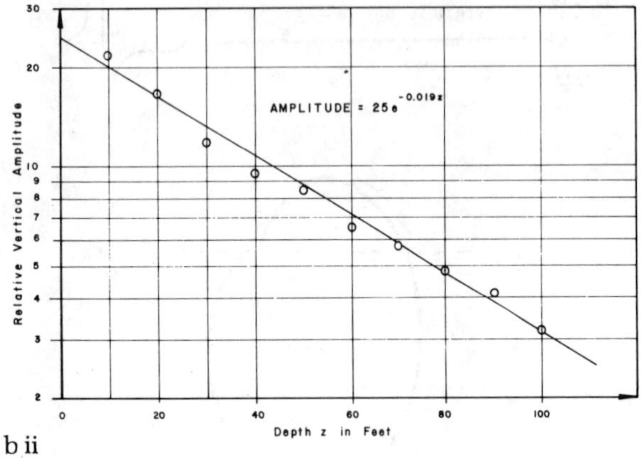

AMPLITUDE = 25e$^{-0.019z}$
Relative Vertical Amplitude
Depth z in Feet

b ii

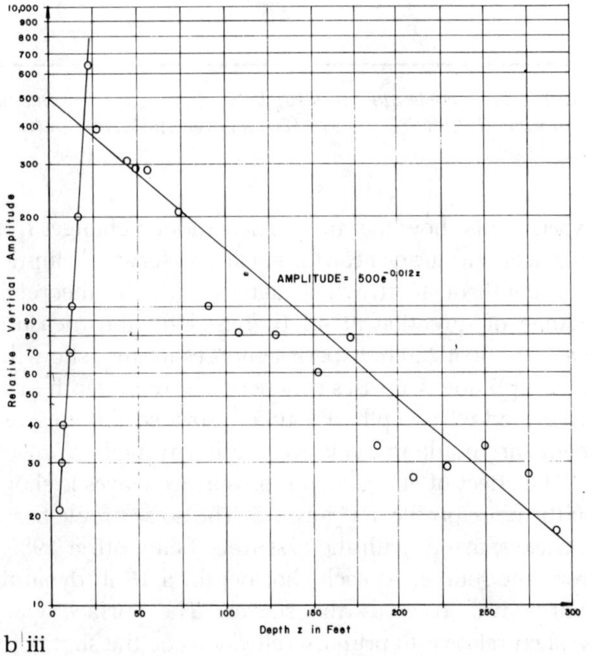

AMPLITUDE = 500e$^{-0.012z}$
Relative Vertical Amplitude
Depth z in Feet

b iii

UP
TOWARD SOURCE
AWAY FROM SOURCE
DOWN
Depth
Surface
10'
20'
30'
40'
50'
60'
70'
80'
90'
100'

b i

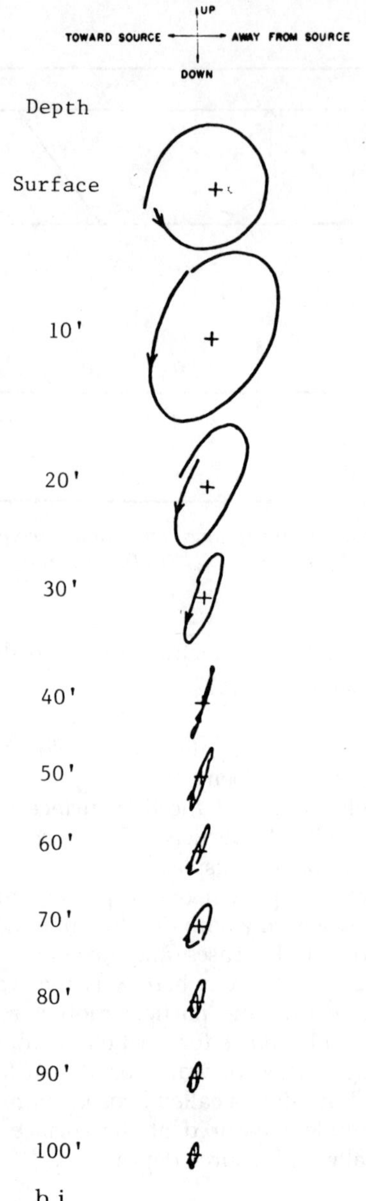

FIGURE 4.4(a). *Particle motion of Rayleigh waves at free surface, Delaware basin, Texas. (From Dobrin et al. 1954, courtesy of Geophysics.)*

FIGURE 4.4(b). *Experimental measurements of Rayleigh waves, Henderson County, Texas. (i) Particle motion as a function of depth. (ii) Amplitude vs. receiver depth. (iii) Amplitude vs. source depth. (From Dobrin et al. 1951, courtesy of AGU.)*

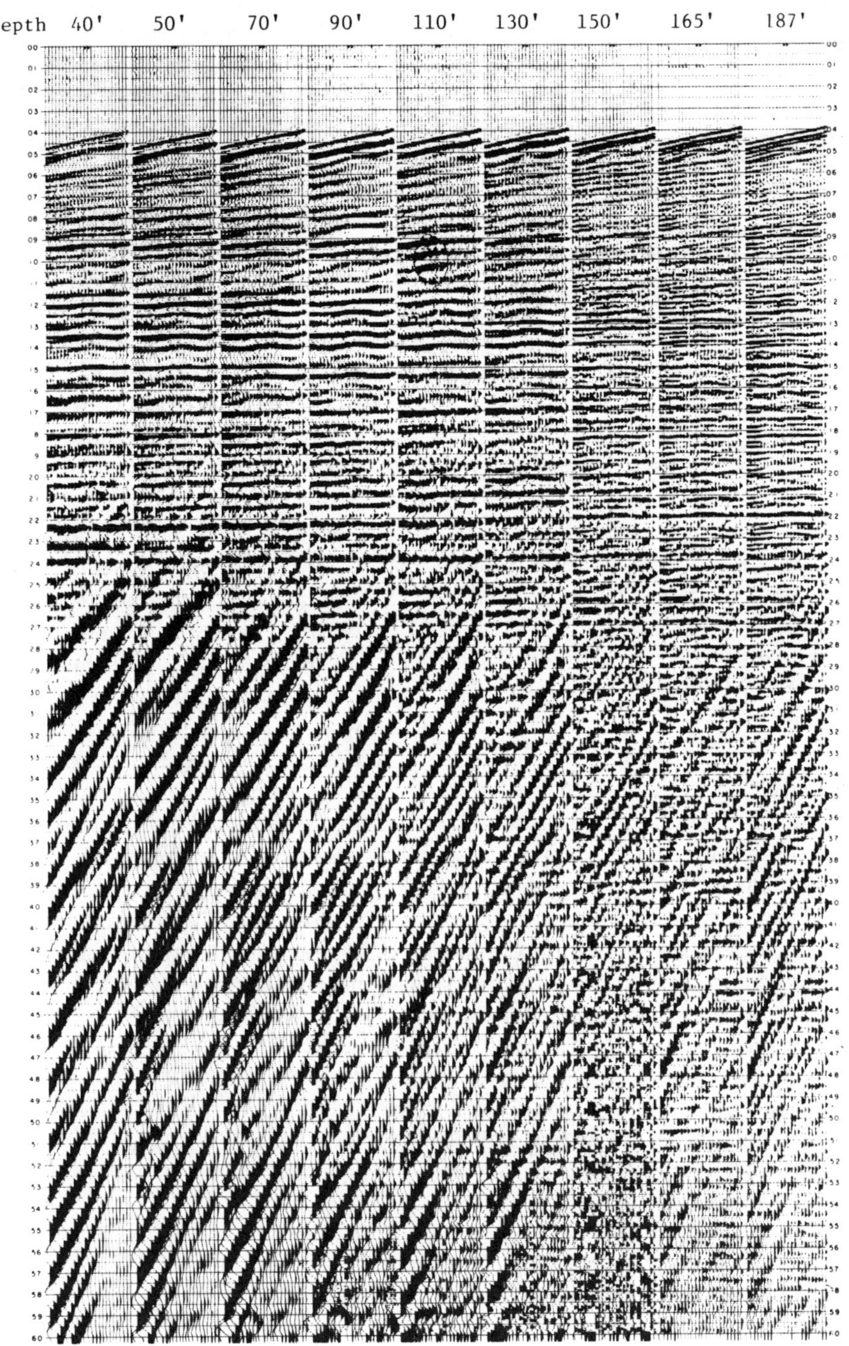

FIGURE 4.5. *Effect of shot depth on surface wave generation in southern Louisiana. (Courtesy of Harry Barbee.)*

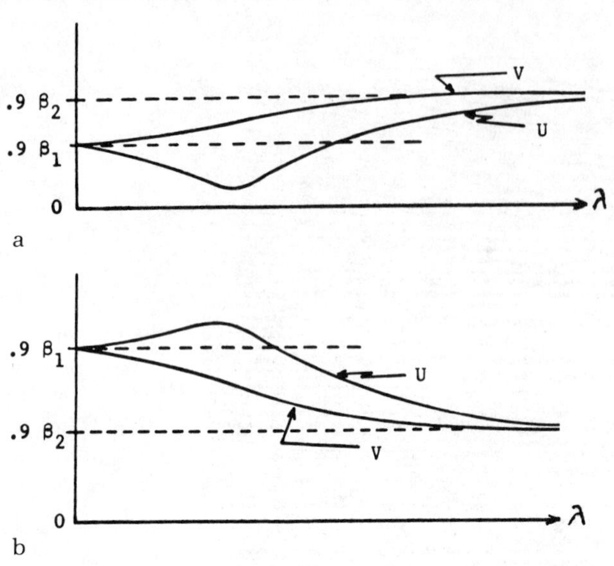

FIGURE 4.6. *Rayleigh wave dispersion curves. (a) Case I, normal dispersion. (b) Case II, inverse dispersion.*

to find the optimal shot depth from the standpoint of minimal surface wave noise and adequate high-frequency reflections, for high-resolution acquisition; 165 ft shot depth appears optimal from both standpoints.

In a layered medium, the Rayleigh wave velocity depends on wavelength. Such wave propagation is said to be dispersive. In a dispersive medium, two measures of velocity are used to describe the wave propagation, the phase velocity V and the group velocity U. The phase velocity is defined by $V = f/k$, where k is the wave number (reciprocal of wavelength), and U is defined by $U = \partial f/\partial k$. Taking the partial derivative gives the relationship between U and V,

$$U = V + k\, dV/dk. \tag{4.13}$$

Equation (4.13) can be written in terms of wavelength as follows:

$$U = V - \lambda\, dV/d\lambda. \tag{4.14}$$

The phase velocity is commonly called the moveout velocity, $V = dx/dt$. The group velocity is the distance from the source divided by the travel time, $U = x/t$.

The dispersive characteristics of a medium depend on the shear velocity, density, and thickness of the layering. Consider the simplest case of a single layer of thickness H overlying a semi-infinite half-space, with the shear velocity in the upper layer V_{s1} and the shear velocity in the half-space V_{s2}. Because the short wavelengths ($\lambda \ll H$) are not influenced appreciably by the deeply underlying half-space, both U and V approach $0.92V_{s1}$. On the other hand, the long wavelengths ($\lambda \gg H$) are not influenced appreciably by the thin veneer layer at the surface, and both U and V approach $0.92V_{s2}$. As the wavelength increases from zero to infinity, the phase velocity varies smoothly between the two limiting values.

There are two cases of dispersion (Figure 4.6), depending on the relative magnitudes of V_{s1} and V_{s2}: case I, normal velocity layering, $V_{s1} < V_{s2}$; and case II, velocity inversion, $V_{s1} > V_{s2}$. In the normal case, the phase velocity increases with wavelength, and its derivative with respect to wavelength is positive. Therefore, the group velocity is less than the phase velocity. The group velocity has a minimum that occurs in the range $\lambda = 2H$ to $4H$. In the inverted case, the phase velocity decreases with wavelength, its derivative is negative, and the group velocity is greater than the phase velocity.

Numerical calculation of the phase and group velocity at one point on a noise profile is shown in Figure 4.7(a). The peak at $t = 2400$ ms on the trace at $x = 2300$ ft has group velocity

$$U = x/t = 2400/2300 = 0.96 \text{ ft/ms}. \tag{4.15}$$

Over the distance $\Delta x = 200$ ft shown, the time delay Δt of this peak is 145 ms, giving its phase velocity

$$V = \Delta x/\Delta t = 200/145 = 1.38 \text{ ft/ms}. \tag{4.16}$$

The period T of this event is 150 ms; therefore, its wavelength is

$$\lambda = VT = (1.38)\,(150) = 207 \text{ ft}.. \tag{4.17}$$

The points $(U, \lambda) = (0.96, 207)$ and $(V, \lambda) = (1.38, 207)$ lie closed to the phase and group velocity curves in Figure 4.7(b) obtained by Fourier analysis of the data, indicating that the numerical approximation is reasonably accurate.

In the normal case, the surface waves tend to develop long, dispersive wave trains. The maximum energy tends to concentrate at the minimum group velocity, which is called the Airy phase. The portion of the dispersion curve that is observable on field records depends on the thickness of the upper layer and the bandwidth recorded. Thin layering, $H = 60$ ft in East Texas, shows observable surface waves that decrease in wavelength and period with record time (Figure 4.8), because the observable surface waves lie to the right of the Airy phase, where the group velocity increases as the wavelength increases. Thick layering, $H = 280$ ft in the Delaware basin, shows observable surface waves that increase in wavelength and period with record time (Figure 4.9), because the observable surface waves lie to the left of the Airy phase, where the group velocity decreases as the wavelength increases.

In the inverted case, dispersive wave trains do not develop because there are large radiation losses into the lower medium for wavelengths on the order of the layer thickness, and the observable surface waves tend to be short, transient pulses traveling at Rayleigh wave velocities. This is a leaky propagation mode. In Dallas County, Texas, a velocity inversion occurs because the Austin chalk with shear velocity of 3500 ft/sec overlies Eagleford shale that has shear velocity of 1250 ft/sec (White and Sengbush, 1953). The Austin chalk is 95 ft thick, and the Eagleford shale that underlies it is 400 ft thick, and that, in turn, overlies the Woodbine sand.

The event labeled R_1 in Figure 4.10 is established as a Rayleigh wave by its particle motion. Because it does not develop a dispersive wave train, this suggests that it results from the high-velocity Austin chalk overlying the low-velocity Eagleford shale. Its phase velocity is 3100 ft/sec, which is equal to 0.89 of the shear velocity in the Austin chalk. This is in substantial agreement with the theoretical value of 0.92. However, its phase velocity exceeds its group velocity, which is contrary to what should happen in the case of inverse dispersion.

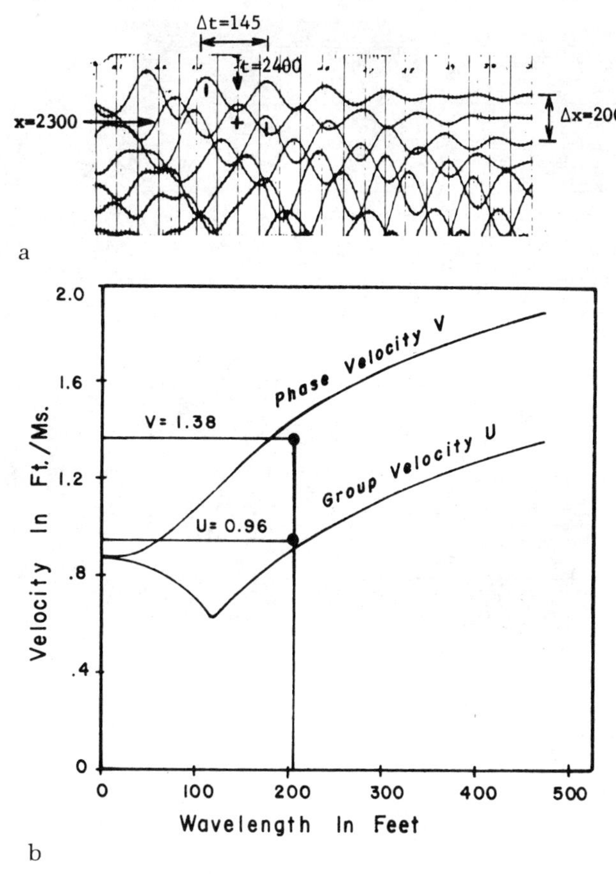

FIGURE 4.7. *Dispersion curves for a single layer over a half-space. (a) Numerical calculation of phase and group velocities. (b) Dispersion curves obtained by Fourier analysis.*

The event labeled R_2 corresponds to the Rayleigh wave that is obtained from the low-velocity Eagleford shale overlying the high-velocity Woodbine sand. Its wavelength is on the order of 300 to 600 ft, which in essence ignores the veneer of Austin chalk that overlies the Eagleford shale.

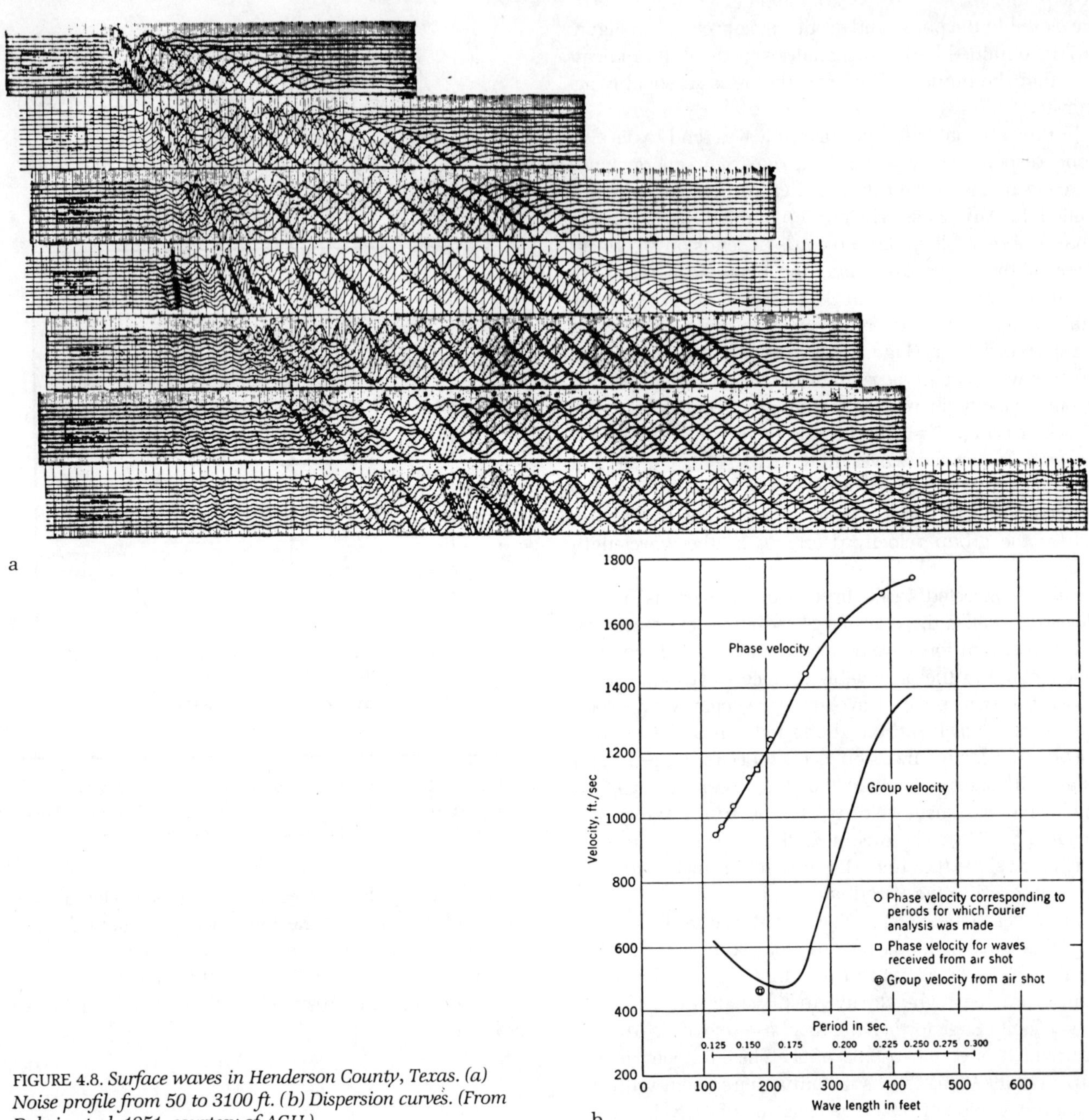

FIGURE 4.8. *Surface waves in Henderson County, Texas. (a) Noise profile from 50 to 3100 ft. (b) Dispersion curves. (From Dobrin et al. 1951, courtesy of AGU.)*

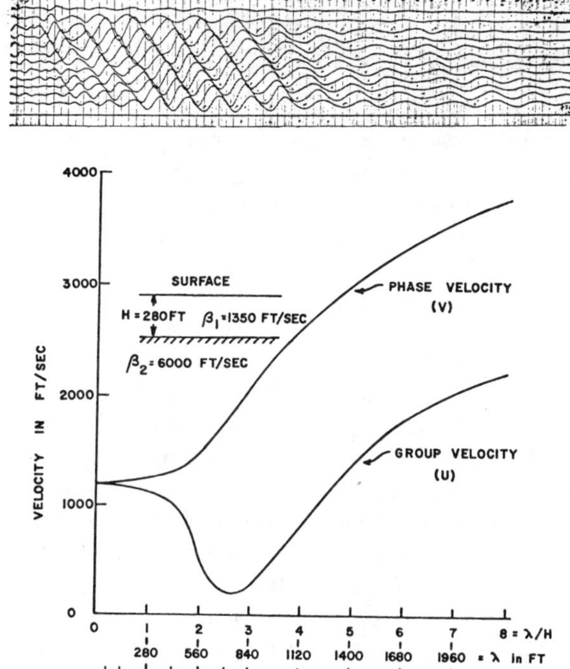

FIGURE 4.9. *Surface waves in the Delaware Basin, West Texas. (From Dobrin, Lawrence, and Sengbush 1954, courtesy of Geophysics.)*

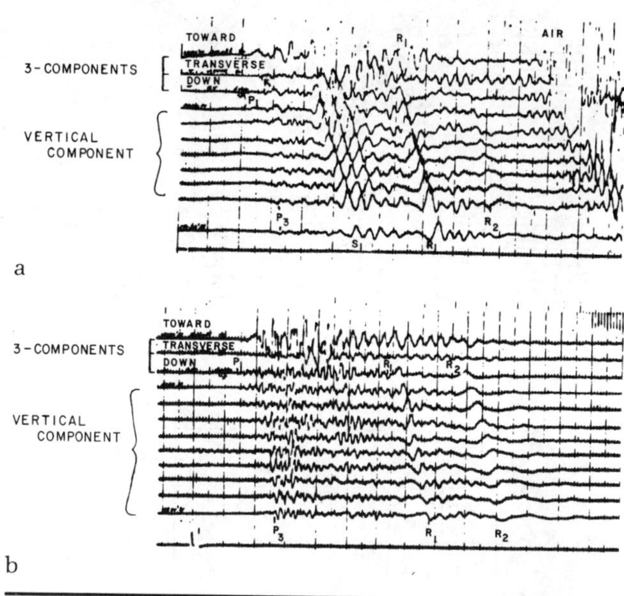

FIGURE 4.10. *Inverse dispersion of surface waves, Dallas County, Texas; Austin chalk overlying Eagleford shale. (a) Source in air. (b) Source in Eagleford shale at 100 ft depth. (From Press and Dobrin 1956, courtesy of Geophysics.)*

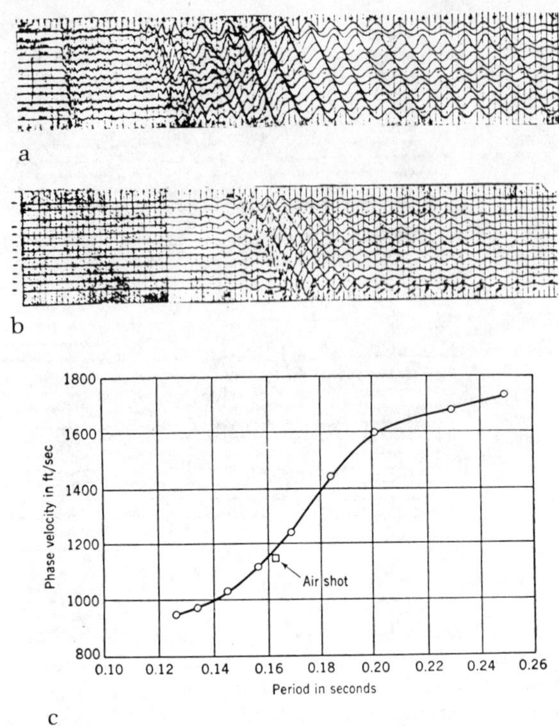

FIGURE 4.11. *Surface waves from sources in the subsurface and in the air. (a) Source in subsurface. (b) Source in air. (c) Dispersion curve. (From Elastic waves in layered media by Ewing, Jardetzky, and Press. Copyright © 1957, McGraw-Hill Book Co. Used with permission of McGraw-Hill Book Company.)*

Other identifiable events are P_1, the compressional refraction that travels within the high-velocity Austin chalk; P_3, the compressional refraction from the high-velocity Woodbine sand that underlies the Eagleford shale; and S_1, the shear refraction traveling through the Austin chalk. S_1 shows up most clearly on the air shot and is not generated by the shot at 100 ft, which is located in the Eagleford shale.

A horizontally traveling compressional wave in the air above the surface generates a constant-frequency Rayleigh wave that falls on the phase velocity dispersion curve obtained from data generated by subsurface shots, as shown in Figure 4.11 (from Ewing, Jardetzky, and Press, 1957).

COHERENT NOISE ANALYSIS

NOISE PROFILES

Noise profiles such as those shown in Figure 4.1 are analyzed to determine the apparent periods and wavelengths of horizontally traveling noise, so that procedures can be employed to suppress the effect of noise on field records. The dispersion characteristics of the noise per se are of secondary importance.

A noise profile is obtained using point sources and point receivers, with the source-to-receiver distance ranging over the distance to be used in the production shooting. It is also important to determine the effect of shot depth and charge size on the amplitude and frequency of the signal and noise.

The noise data may be displayed in time-distance (t-x) space or in its transform space, which is frequency-wave number (f-k) space. In time-distance space, the noise events are entered on the plot, as shown in Figure 4.1. The different classes of events occupy different regions of t-x space because of their different velocities. The second display, that in f-k space, is a more significant and useful display of a noise profile. The frequency and wavenumber of each event can be calculated from the period and phase velocity of the event as measured on the noise profile, and each event can be plotted as a point in f-k space. The slope of the line through this point and the origin is its phase velocity V. Its amplitude A plots in the third dimension above the f-k plane. This plot shows the regions in f-k space where the noise is most severe. This information is needed in order to design patterns (wavenumber filters) and to specify frequency bandwidth.

F-K SPACE

A noise profile is a two-dimensional function, $n(t, x)$, of time and distance that can be characterized in f-k space by its two-dimensional Fourier transform $\mathcal{F}[n(t,x)]$. The steady-state function $A \cos 2\pi(f_1 t - k_1 x)$ transforms into an impulse with weight A located at the point (f_1, k_1). The slope of the line through this point and the origin is its velocity V.

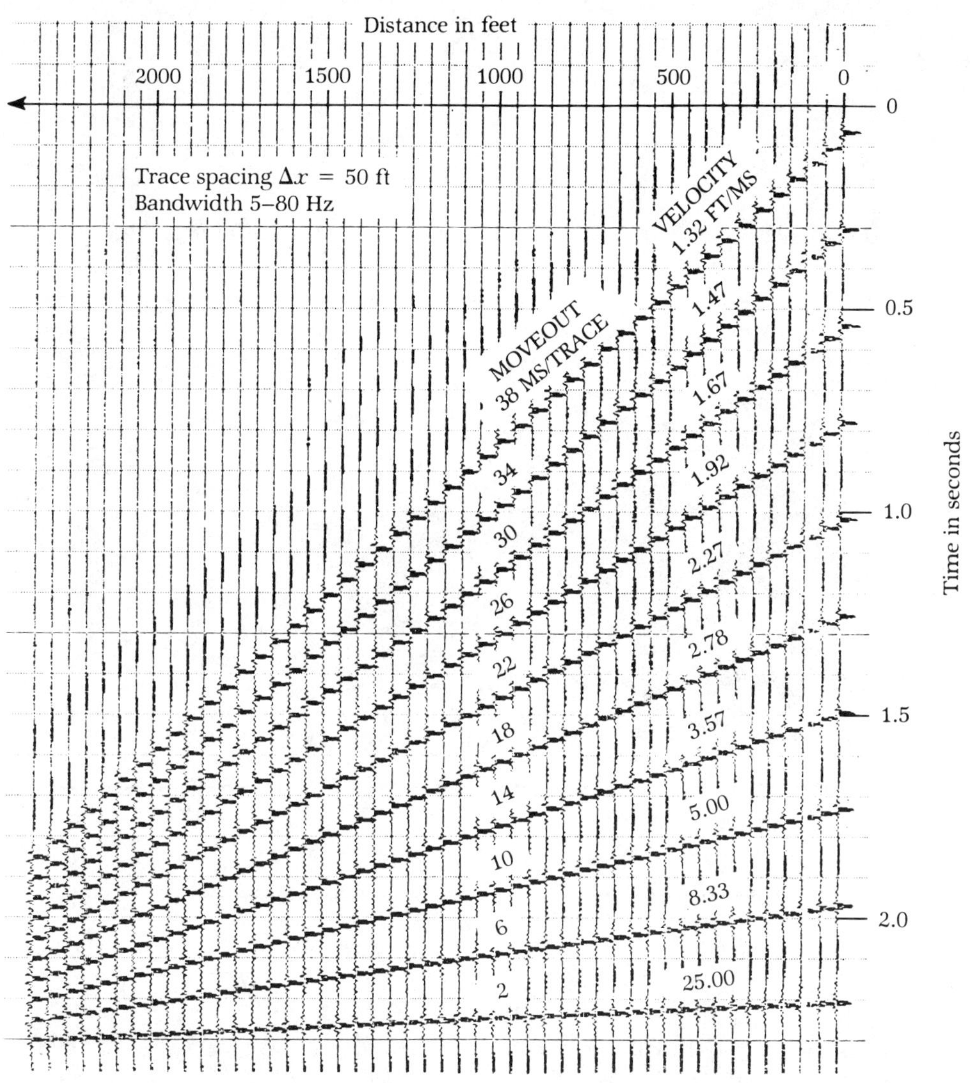

FIGURE 4.12(a). *A suite of broadband wavelets in* t-x *space, each traveling with constant-phase velocity.*

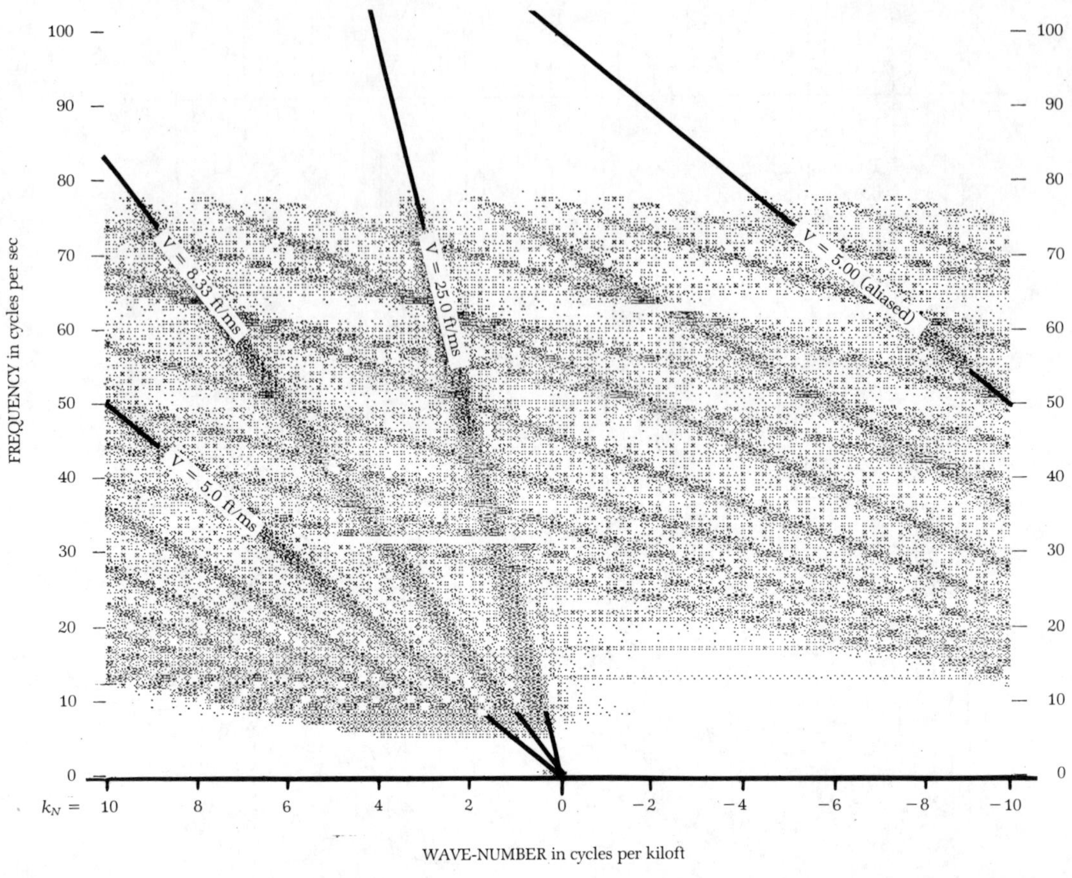

FIGURE 4.12(b). *Wavelets in* t-x *space (from Figure 4.12(a)) transformed into* f-k *space.*

A wavelet traveling with constant phase velocity has its spectral content concentrated along its velocity line in f-k space. Figure 4.12(a) shows a collection of broadband wavelets with moveouts ranging from 2 to 38 ms/trace, where the trace spacing $\Delta x = 50$ ft. In f-k space, their spectra are concentrated along the corresponding velocity lines, as shown in Figure 4.12(b). Aliasing occurs at the folding wavenumber $k_N = 1/2\Delta x = 10$ cycles per kilofoot whenever a wavelet has spectral content at frequencies that exceed $f = Vk_N$.

The noise profile in Figure 4.13 clearly shows the compressional refractions, dispersive Rayleigh waves, and reflections. A portion of the noise profile bounded by the (t, x)-axes and the lines $x = 2000$ ft and $t = 0.6$ sec has an f-k energy spectrum that shows the concentrations of refraction and surface wave energies. The f-k plot is bounded by the folding frequency $f_N = 1/2\Delta t$ and the folding wavenumber $k_N = 1/2\Delta x$, where the time-sampling interval $\Delta t = 2$ ms and the space-sampling interval (that is, the trace spacing) $\Delta x = 12.5$ ft. Therefore, $f_N = 250$ Hz and $k_N = 0.040$ cycles per foot. The refracted energy is concentrated to the right of the f-axis along a line with slope 9400 ft/sec. The reflections have apparent velocities on the order of 33,000 ft/sec and lie between the refraction velocity line and the f-axis. The refraction energy is much stronger than the reflection energy and dominates the region of the space where

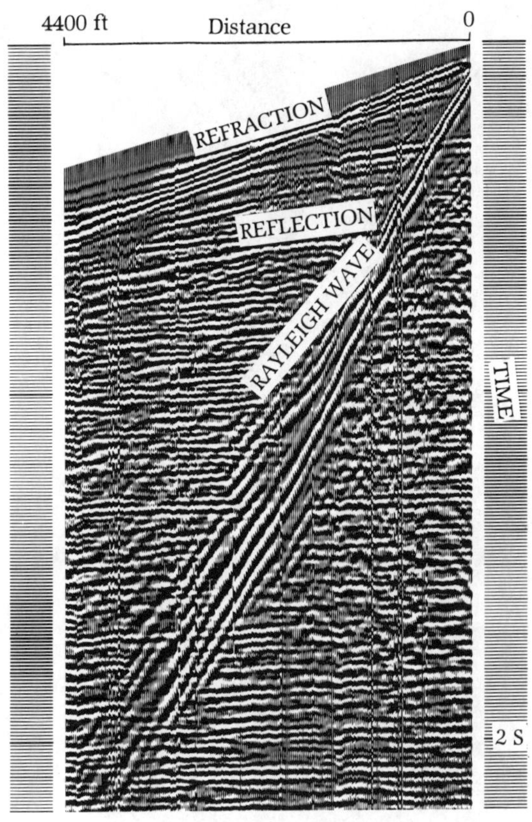

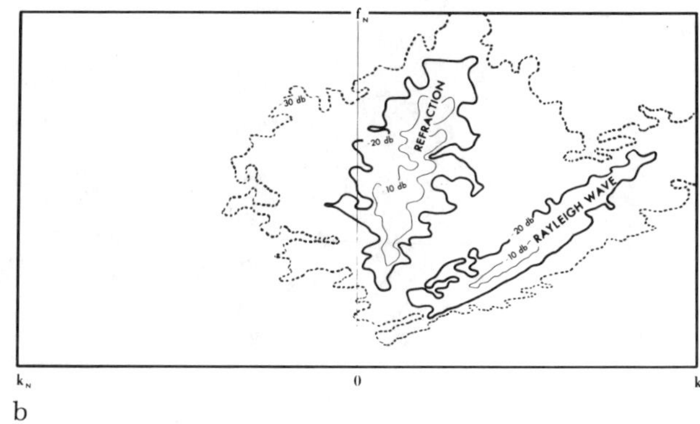

FIGURE 4.13. *A noise profile and its transform into* f-k *space. (a) Noise profile. (b) Energy spectrum of area bounded by 0–2000 ft and 0–0.6 sec. (From Sengbush and Foster 1972, copyright © 1972 IEEE; courtesy of IEEE.)*

these two types of events lie. The Rayleigh wave energy travels at a much lower velocity and is centered along a line whose slope is 2100 ft/sec.

EXAMPLE OF NOISE ANALYSIS

The two-dimensional *f-k* spectrum of a noise profile is the basis for pattern design. In the absence of this spectrum, a useful approximation can be obtained by measurements of the horizontal component of phase velocity V and the period T for selected events in the noise profile. The phase velocity is $V = \Delta x/\Delta t$, where Δx is some distance interval and Δt is the time for the event to travel across that interval. The reciprocal of the period T is frequency f and the wavenumber $k = 1/(VT)$. The frequency and wavenumber of a transient pulse as calculated earlier is located near the middle of its *f-k* spectrum. The outline of the *f-k* spectrum of the noise becomes better defined as more events are included in the analysis. Of importance are the highest and lowest velocity of each class of events, because the wedge bounded by these velocities will contain all events of the class. Consider the noise profile in Figure 4.14. It has 20-foot trace spacing and extends a distance of 5,760 feet from the seismic source. The refraction velocity near the source is a velocity of 5,650 ft/sec. This is the minimal refraction velocity, and all other refraction events fall between this velocity and the minimal reflection velocity line, which in this example is 9,000 ft/sec. The Rayleigh waves lie in the wedge bounded by 746 and 1,365 ft/sec velocity lines. The spectral analysis for these events is given in the following table, and these and other events are plotted in *f-k* space in Figure 4.15.

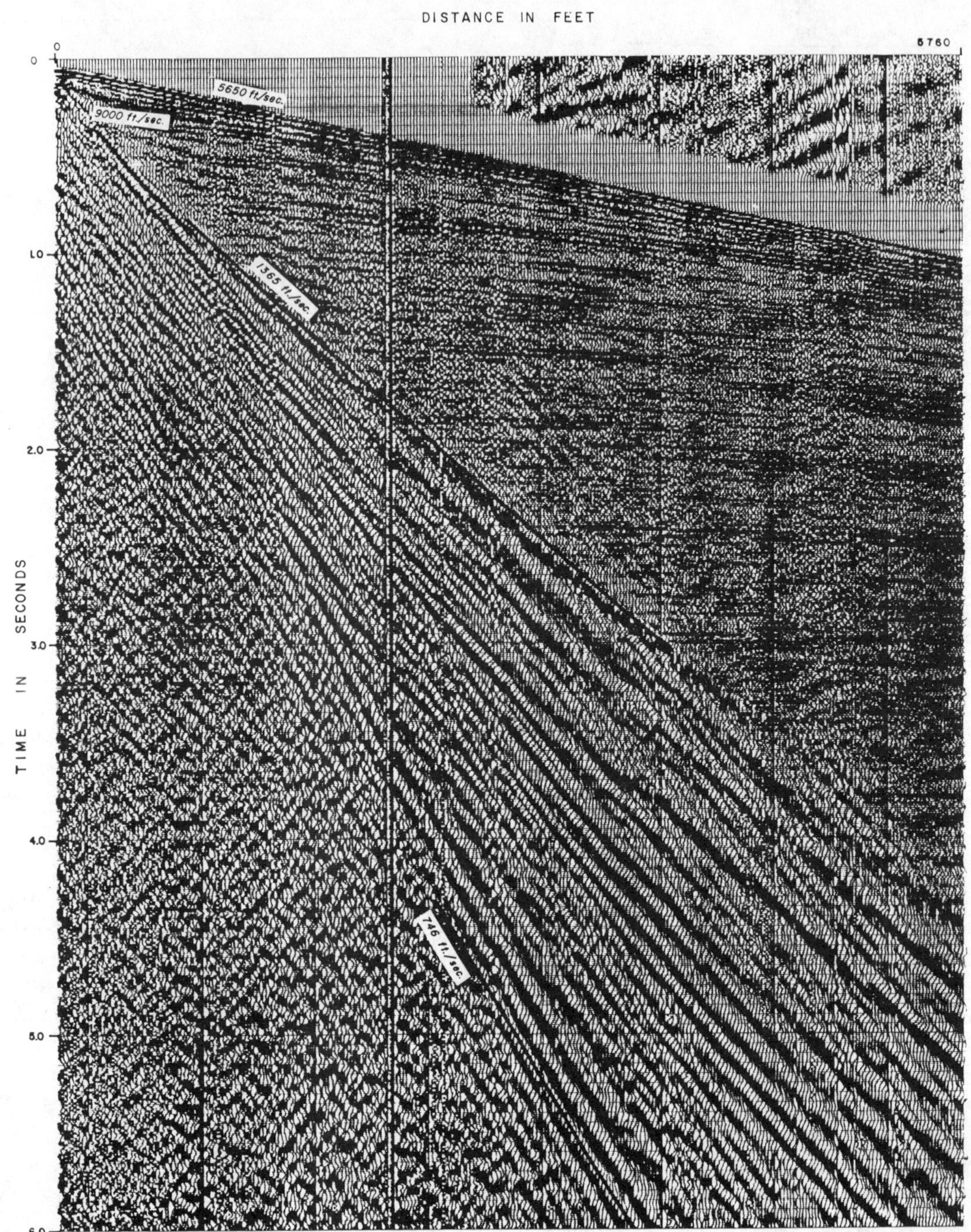

FIGURE 4.14. *Noise profile with spacing* $\Delta x = 20$ *ft. (Courtesy of Amoco Production Company.)*

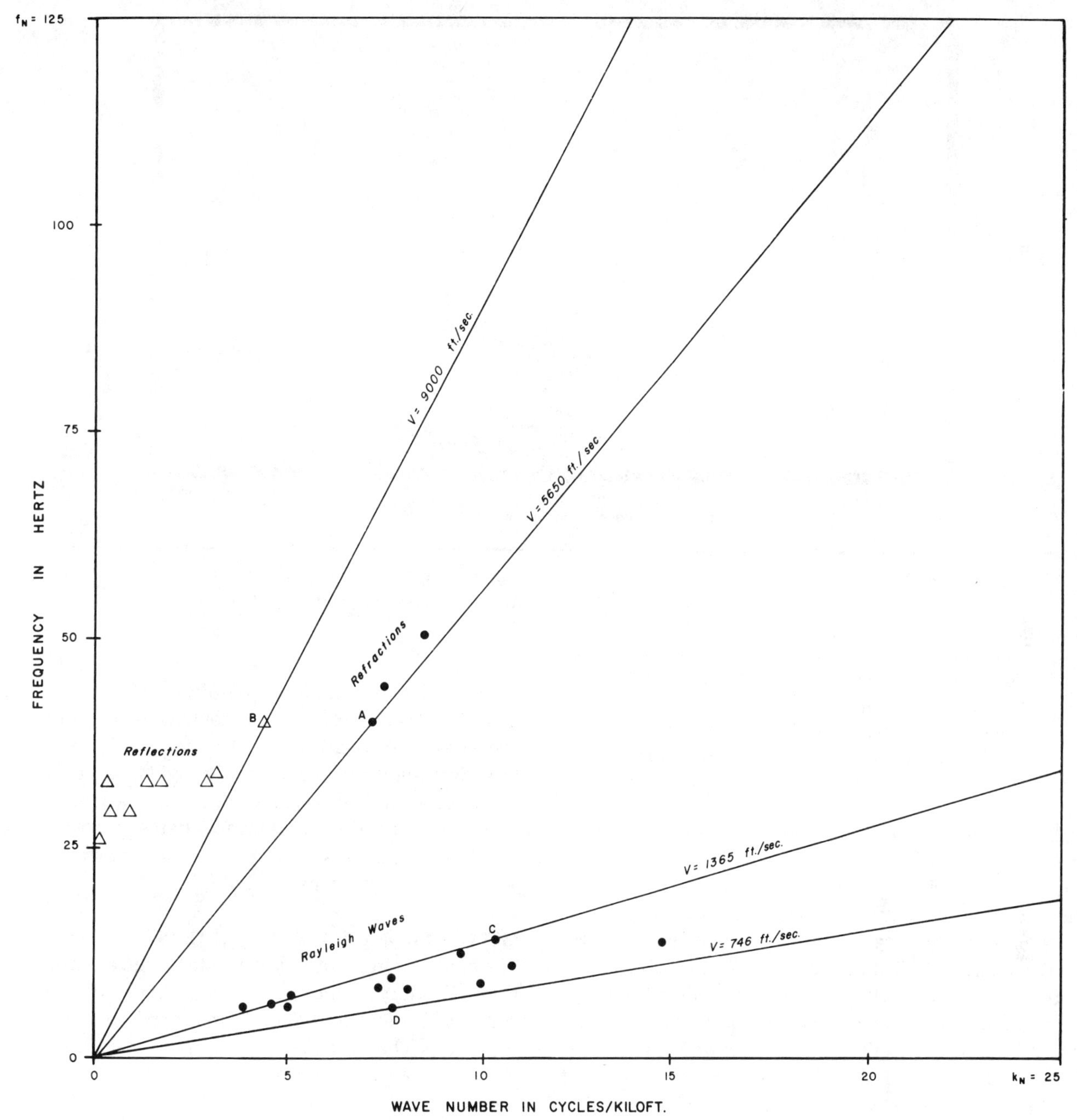

FIGURE 4.15. *Signal and noise events in* f-k *space from the noise profile in Figure 4.14.*

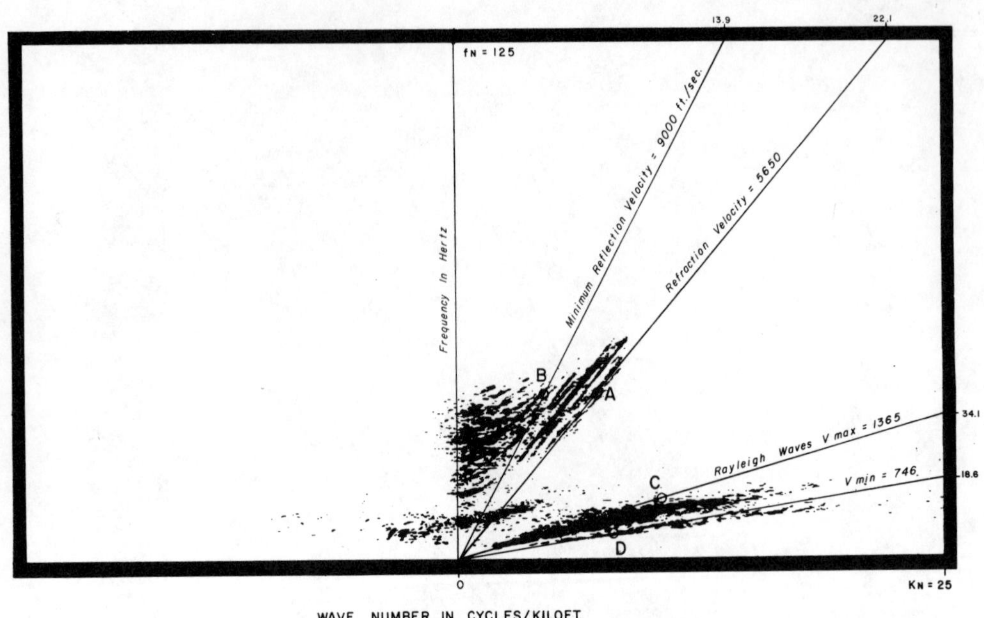

FIGURE 4.16. *Fourier transform of noise profile in Figure 4.14. (Courtesy of Amoco Production Company.)*

Event	V	T	f	λ	k
A	5650 ft/sec	25 ms	40 Hz	141 ft	$7.1*10^{-3}$ cpf
B	9000 ft/sec	25 ms	40 Hz	225 ft	$4.4*10^{-3}$ cpf
C	1365 ft/sec	70 ms	14.2 Hz	96 ft	$10.4*10^{-3}$ cpf
D	746 ft/sec	170 ms	5.8 Hz	129 ft	$7.8*10^{-3}$ cpf

The *f-k* transform of the noise profile is shown in Figure 4.16, with the events A through D and the velocity lines through those points superimposed. This figure shows that the spectral content of the reflections, refractions, and Rayleigh waves are largely confined to the velocity wedges specified by the events A through D.

It is desirable to devise a field procedure that will reject the noise (Rayleigh waves) without unduly affecting the signal (the reflections) by use of *k*-filters (patterns) and *f*-filters. These filters are separable and act only in their own domain, as shown in Figure 4.17, where the pattern rejects those wavelengths lying between λ_S = 50 ft and λ_L = 250 ft and the *f*-filter rejects those frequencies below 7 Hz. It is obvious that the noise wavelengths less than λ_S have little spectral content, but there is considerable spectral content in the noise with wavelengths greater than λ_L. An attempt to reject longer noise wavelengths with the pattern will be accompanied by rejection in the signal wedge; thus, a compromise must be struck. If the noise wavelengths greater than λ_L = 250 ft are bothersome, they may be suppressed by an *f*-filter with a low cutoff of 7 Hz.

PATTERNS: SPACE-DOMAIN FILTERS

The objective in pattern design is to design a spatial filter that will suppress the noise wavelengths without appreciably affecting the signal (primary reflections). A spatial filter is characterized by its impulse response $p(x,y)$ in two-dimensional x-y space, or by the corresponding Fourier transform $\mathcal{F}[p(x,y)] = P(k_x, k_y)$, where k_x and k_y are wavenumbers. In one-dimensional space, the impulse response $p(x)$ transforms into $P(k)$. The amplitude spectrum $|P|$ is called the pattern response.

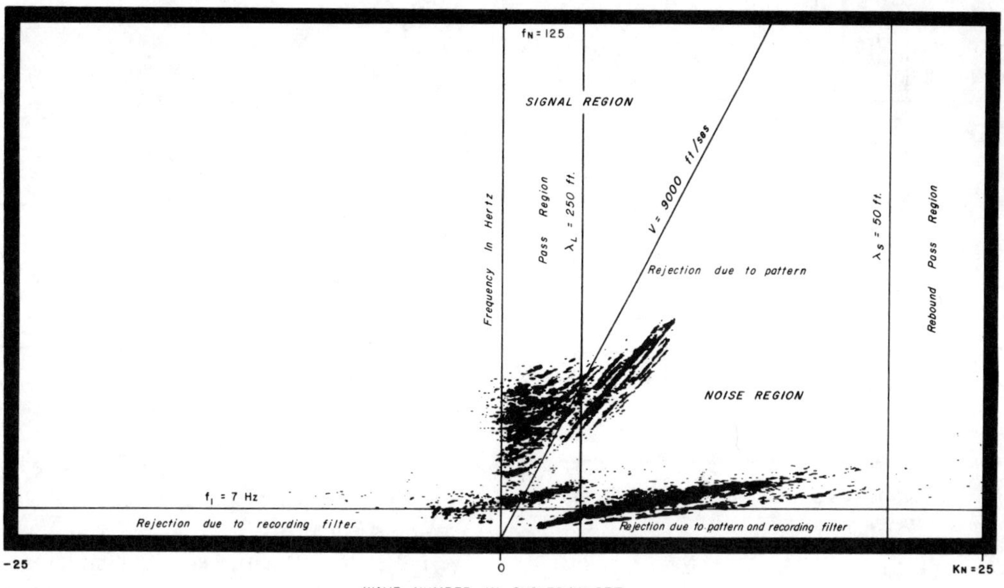

FIGURE 4.17. *Partition of* f-k *space by separable two-dimensional filters. (Courtesy of Amoco Production Company.)*

IMPULSE RESPONSE OF PATTERNS

Each element in a pattern can be represented by a weighted impulse located in x-y space, with its weight proportional to its sensitivity. The set of weighted, spaced impulses is the impulse response of the pattern in x-y space, designated $p(x,y)$. It is convenient to reduce the impulse response to one dimension by projecting all impulses onto a line perpendicular to a plane wavefront. The impulse response then depends on the direction of arrival of the wavefront.

STEADY-STATE RESPONSE OF PATTERNS

If $p(x)$ is the impulse response of a pattern produced by projection of all impulses onto the x-axis, then its steady-state response is the Fourier transform of $p(x)$, designated $P(k)$, and given by

$$P(k) = \int_{-\infty}^{+\infty} p(x) \exp(-j2\pi kx)\, dx. \tag{4.18}$$

The impulse response $p(x)$ of an n-element pattern is given by

$$p(x) = \sum_{l=1}^{n} p_l \delta(x - x_l). \tag{4.19}$$

By use of the sifting property of impulses,

$$P(k) = \sum_{l=1}^{n} P_l \exp(-j2\pi kx_l). \tag{4.20}$$

The amplitude spectrum $|P(k)|$ (the pattern response) is usually normalized by setting $\Sigma_l p_l = 1$. Then $|P(0)| = 1$.

When the impulse response is symmetrical about $x = 0$, the steady-state response is given by the cosine transform

$$P(k) = 2\int_0^{\infty} p(x) \cos 2\pi kx\, dx. \tag{4.21}$$

The impulse responses of symmetrical patterns with odd and even numbers of elements are shown in Figure 4.18. With an odd number of elements (n elements) and equal

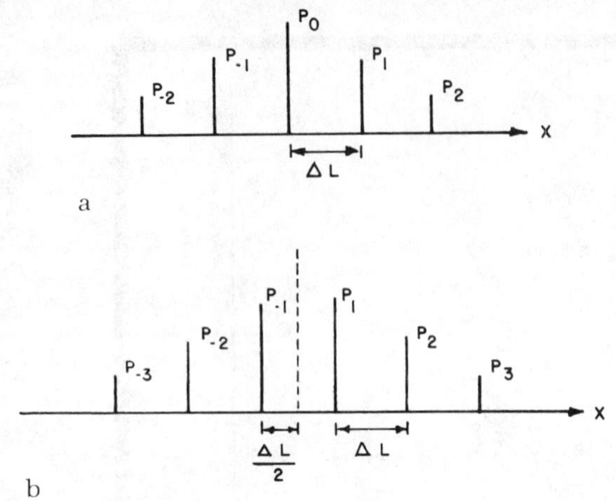

FIGURE 4.18. *Impulse responses of patterns. (a) Odd-number elements. (b) Even-number elements.*

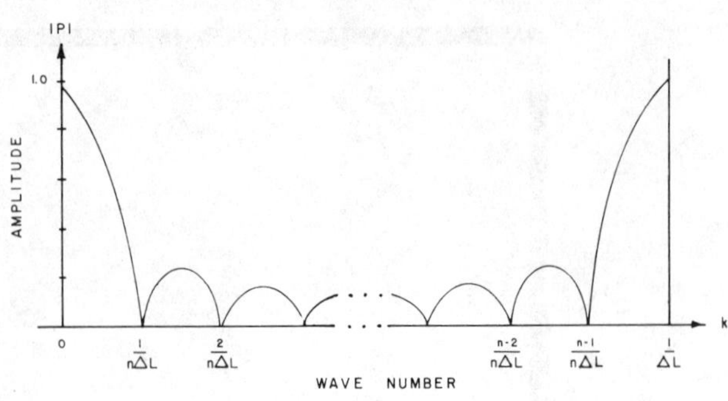

FIGURE 4.19. *Pattern response of an n-element uniform pattern.*

spacing ΔL, the impulse response of a symmetrical pattern is

$$p(x) = \sum_{l=-(n-1)/2}^{(n-1)/2} p_l \delta\,(x - l\Delta L). \tag{4.22}$$

Because of symmetry, $p_l = p_{-l}$, and each impulse pair is equally spaced about the origin. Thus, the steady-state response is given by the cosine transform

$$P(k) = p_0 + 2 \sum_{l=1}^{n-1/2} p_l \cos 2\pi kl\Delta L. \tag{4.23}$$

With an even number of elements and equal spacing, the impulse response and the steady-state response are given by

$$p(x) = \sum_{\substack{l=-n/2 \\ l\neq 0}}^{n/2} p_l \delta\left(x - \frac{2l-1}{2}\Delta L\right), \tag{4.24}$$

$$P(k) = 2\sum_{l=1}^{n/2} P_l \cos 2\pi k\left(\frac{2l-1}{2}\right)\Delta L. \tag{4.25}$$

With symmetrical patterns, the sine spectrum is zero and the amplitude spectrum is the absolute value of the cosine spectrum.

Patterns are sampled data systems, and when the sampling interval is constant, say ΔL, then the pattern response is periodic, with period $1/\Delta L$. Such patterns have a folding wavenumber, $k_N = 1/(2\Delta L)$.

UNIFORM LINE PATTERNS

Uniform line patterns have equal weight and equal spacing. The weight of each element normalizes to $1/n$, where n is the number of elements in the pattern and the spacing is ΔL. The k-spectrum rebounds at $k = 1/\Delta L$ to the value at the origin $P(0) = 1$ and equals zero at $k = 1/n\Delta L$, $2/n\Delta L$, $(n - 1)/n\Delta L$, as shown in Figure 4.19. The rejection band is defined as the band between $1/n\Delta L$ and $(n - 1)/n\Delta L$. The amplitudes of the minor peaks in the rejection band are given in Table 4.1.

NONUNIFORMLY WEIGHTED PATTERNS

Four methods can be used to obtain nonuniformly weighted patterns:

(1) In a uniformly-spaced line pattern, the sensitivities of the individual elements may be set to the sampled values of a nonuniform envelope.
(2) Equally weighted elements may be spaced nonuniformly in a line pattern to approximate equal areas under a nonuniform envelope.

TABLE 4.1. *Amplitudes of rejection band peaks, uniform line patterns*

n	1st peak	2nd peak	3rd peak	4th peak
3	.33	—	—	—
4	.27	.27	—	—
5	.25	.20	—	—
6	.24	.17	.17	—
7	.23	.16	.15	—
8	.23	.15	.12	.12
9	.22	.145	.118	.111
R	.212	.127	.091	.071

(3) Use a combination of source and receiver line patterns.
(4) Use an areal pattern.

These methods are not equivalent. With line patterns, changing the direction of arrival merely stretches or compresses the distance scale without changing the shape of the impulse response. With areal patterns, the shape of the impulse response depends on the direction of arrival of the noise events.

LINE PATTERNS WITH NONUNIFORM WEIGHTING

The ideal pattern response is one that has brickwall characteristics (Figure 4.20), that passes all positive wavenumbers less than k_1 without attenuation, and that rejects completely all wavenumbers outside the passband. The impulse response in the x-domain of a brickwall pattern response is a sinc function. This, of course, is an impractical pattern, because it is a continuous detector of infinite length. To be useful, the pattern length must be finite, which requires truncation of the impulse response to length L. Choosing L equal to the major lobe of the sinc function and sampling at intervals of ΔL gives a sinc-weighted pattern. This pattern is an approximation of the ideal pattern.

Holzman (1963) described Chebyshev weighted arrays. The pattern responses of such arrays have equal ripple in the reject band. The upper bound in the reject band is dependent on the number of elements; 27 dB attenuation can be achieved with 8 elements, 40 dB with 21 elements, and 60 dB with 30 elements.

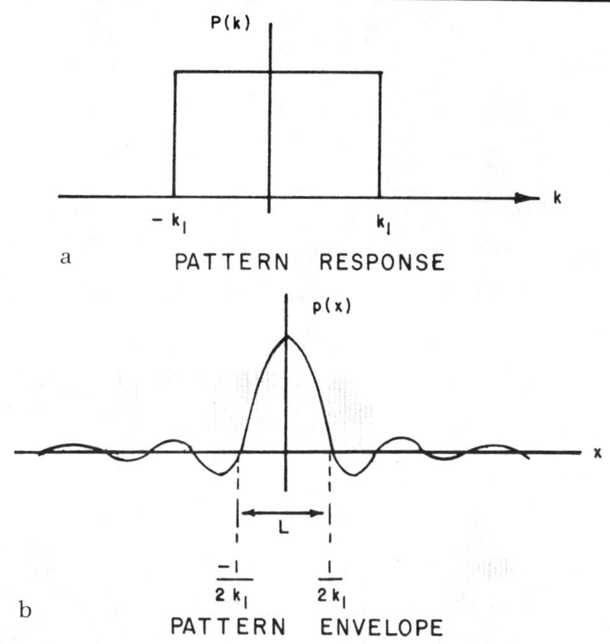

FIGURE 4.20. *Ideal brickwall pattern. (a) Pattern response. (b) Pattern envelope.*

Patterns with triangular, sinc, and Chebyshev weightings are compared with a uniform pattern with the same number of elements, $n = 8$, in Figure 4.21(a). The unnormalized weights for the patterns are as follows:

Triangular	3.5	2.5	1.5	0.5
Sinc	0.975	0.784	0.471	0.139
Chebyshev	1.0	0.831	0.562	0.330

The response curves show that the nonuniform patterns have better rejection over a more restricted bandwidth in comparison to the uniform pattern. In the reject band, the attenuation is greater than 12.8 dB with the uniform pattern, greater than 26.9 dB with the Chebyshev pattern, greater than 27.5 dB with the sinc pattern, and greater than 29.8 dB with the triangular pattern. The first notch in the response is at $k = 1/L$ with the uniform pattern, at $k = 1.4/L$ with the Chebyshev pattern, at $k = 1.6/L$ with the sinc pattern, and at $k = 2/L$ with the triangular pattern.

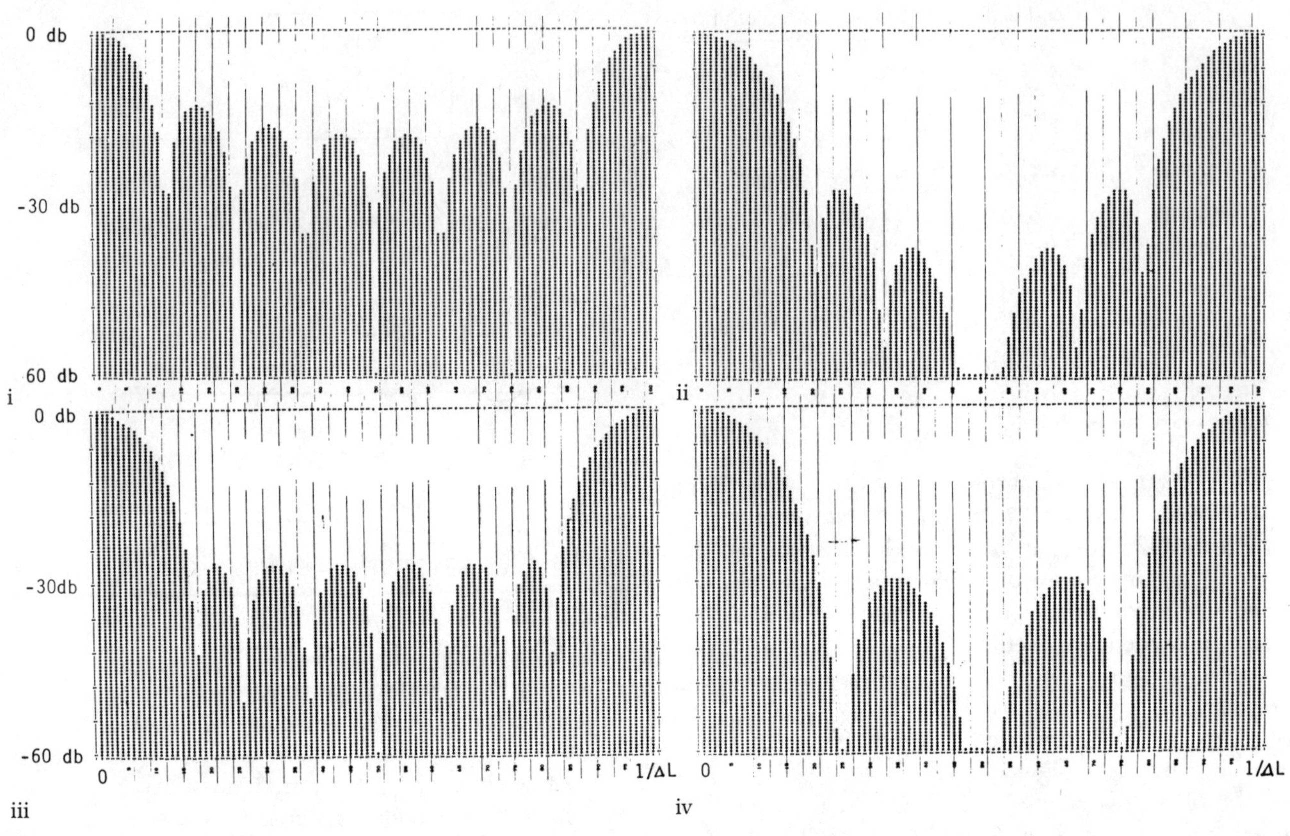

FIGURE 4.21(a). *Comparison of pattern responses of eight-element patterns. (i) Uniform pattern. (ii) Sinc pattern. (iii) Chebyshev pattern. (iv) Triangular pattern.*

For a nonuniform pattern to have the same rejection bandwidth as a uniform pattern, its length must be increased by a factor characteristic of the type of nonuniform pattern, and its spacing must remain the same as that used with the uniform pattern. As an example, a sinc pattern is compared with a uniform pattern of length L in Figure 4.21(b). The uniform pattern has its rejection band between the notches at $k = 1/L$ and $k = (n - 1)/L$. For a sinc pattern to have the same rejection band (the same notches), its length must be increased to $1.6L$. This increases the number of elements by the factor 1.6, which further increases the attenuation in the rejection band.

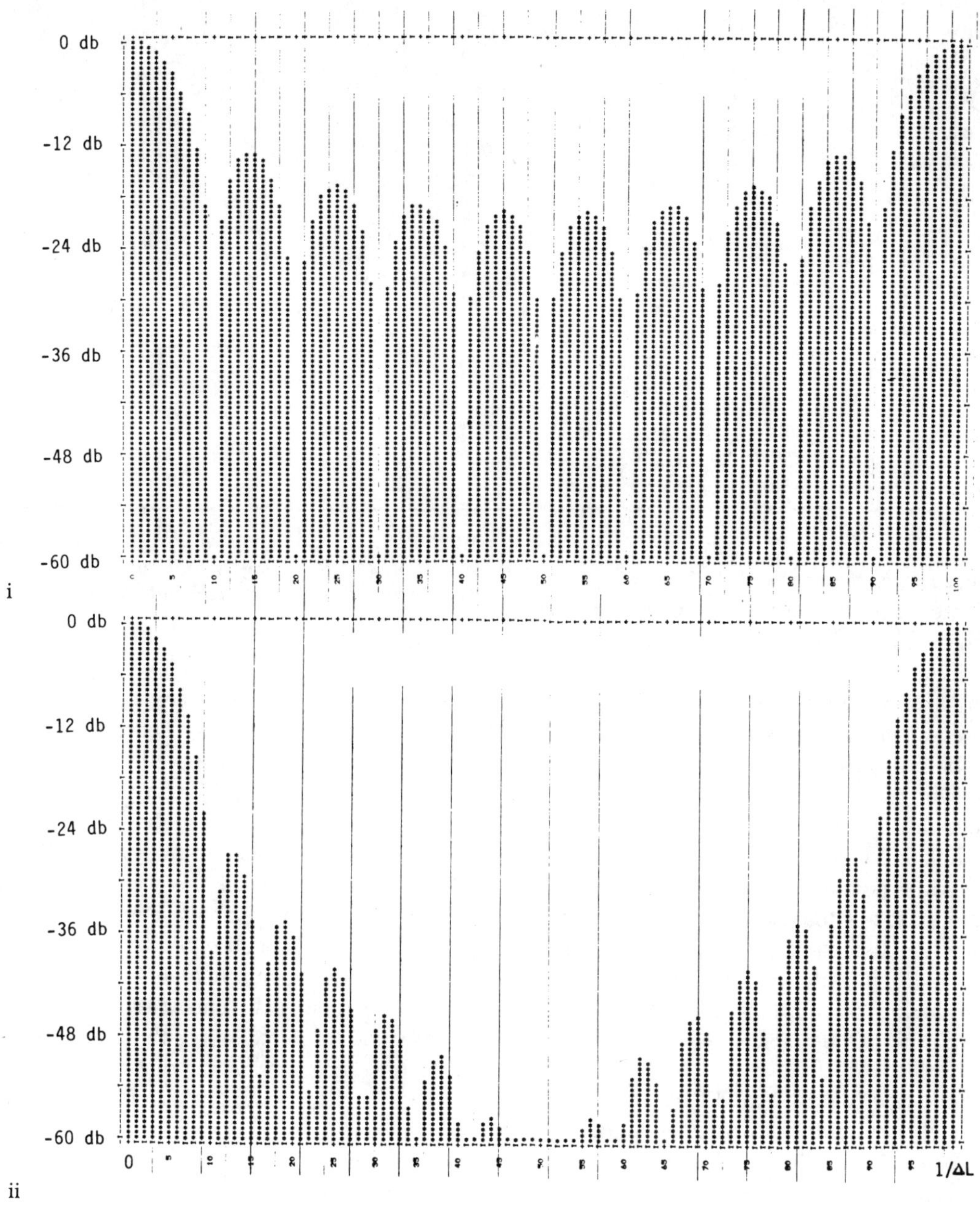

FIGURE 4.21(b). *Comparison of uniform and sinc patterns having the same rejection bandwidth. (i) 10-element uniform pattern of length* L. *(ii) 16-element sinc pattern of length 1.6*L.

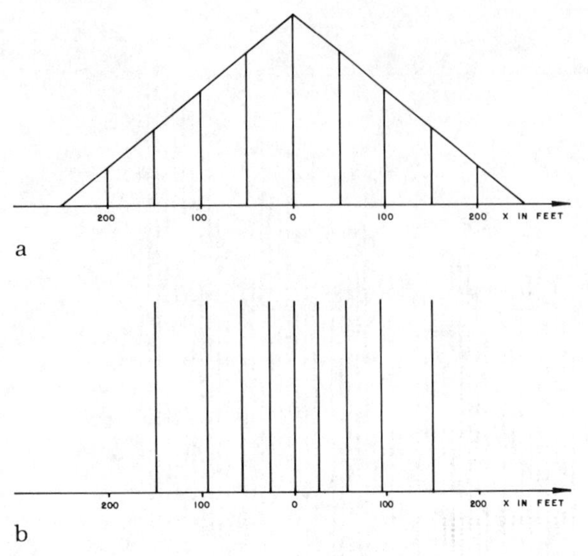

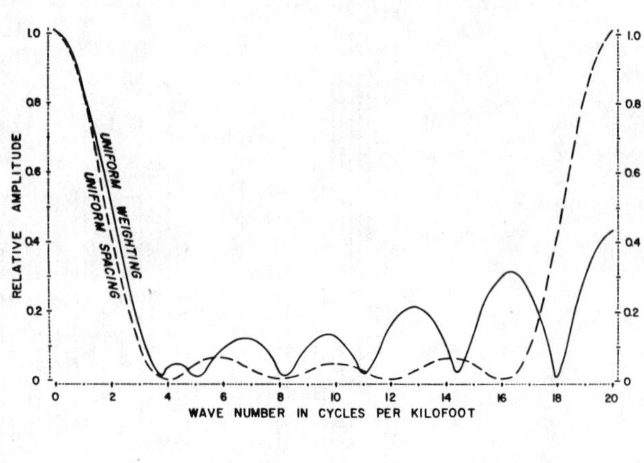

FIGURE 4.22. *Comparison of approximation to a triangular envelope by sampling, using (a) uniform spacing and nonuniform weights and (b) uniform weights and nonuniform spacing.*

LINE PATTERNS WITH NONUNIFORM SPACING

Nonuniform line patterns can be obtained by using nonuniform spacing of equally weighted elements. Given a continuous, nonuniform impulse response $p(x)$ in the space domain, the area under the curve can be divided into a set of n equal elementary areas, and each elementary area can be replaced by an impulse of weight $1/n$ located at its centroid. The resulting pattern will be a uniformly weighted, nonuniformly spaced array that is equivalent to the nonuniformly weighted, equally spaced array that would be obtained by sampling the continuous impulse response at equal spacing. As it is generally more convenient to use equally weighted elements to build an array, nonuniformly spaced arrays have practical merit in producing a nonuniformly weighted pattern, especially in marine acquisition, where the pattern is built into the cable.

Figure 4.22 shows a comparison of the responses of a uniformly spaced, nonuniformly weighted pattern (a) with a nonuniformly spaced, uniformly weighted pattern (b). The patterns are nine-element approximations to triangular weighting, with (a) uniform spacing of ΔL = 50 ft, with triangular weights of 1, 2, 3, 4, 5, 4, 3, 2, 1, and (b) uniform weights, with spacings 54.9, 37.8, 30.7, 26.6, 26.6, 30.7, 37.8, 54.9 ft. The pattern responses are almost identical in the passband to the first zero near k = 4 cycles per kilofoot. The differences within the rejection band can be attributed to the different numerical approximations used in sampling the triangular envelope and the coarseness of the sampling. Use of more elements (that is, finer sampling) would tend to reduce the differences between the responses in the rejection band out to the folding wavenumber $k_N = 1/2\Delta L$ of pattern (a). The response of pattern (a) is folded about k_N and rebounds to unity at $k = 1/\Delta L$. Pattern (b) does not have a folding wavenumber, and its response does not rebound to unity, as does the response of pattern (a). Therefore, there are major differences in response beyond the folding wavenumber of pattern (a).

COMBINATIONS OF SOURCE AND RECEIVER LINE PATTERNS

The combination of source and receiver line patterns produces a nonuniform line pattern whose impulse response is obtained by convolving the individual responses, $p(x) = p_1(x) * p_2(x)$. By properties of the

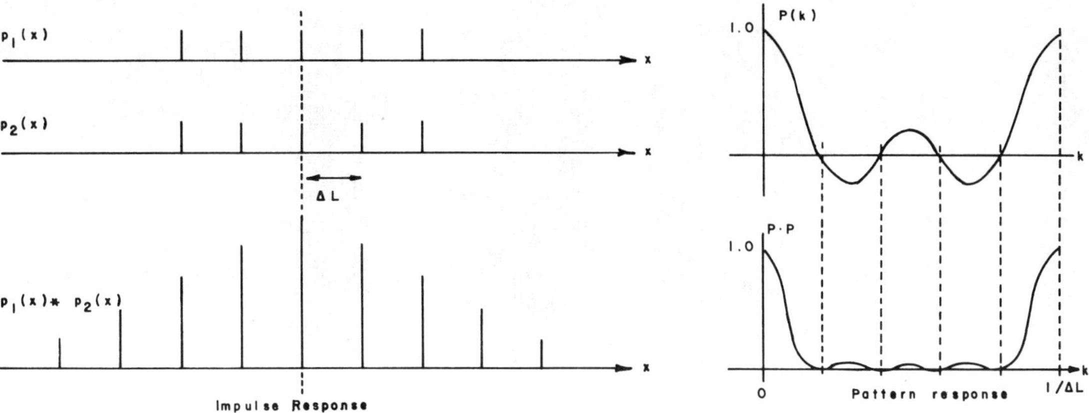

FIGURE 4.23. *Combination of two identical uniform patterns.*

Fourier transform, the pattern response of the combination is the product of the individual pattern responses, $P(k) = P_1(k)P_2(k)$. If each pattern is an n-element uniform pattern with the same spacing, the impulse response of the combination is triangular with $2n - 1$ elements and the pattern response of the combination is the square of the n-element uniform pattern response (Figure 4.23). If one of the uniform patterns has n elements and the other m elements, $m < n$, each with the same spacing, the resultant impulse response is trapezoidal, with $n + m - 1$ elements. The plateau level of the trapezoid has height m, and the number of elements along the plateau is $n - m + 1$. For example, if n is 16 and m is 9, the resultant pattern is a 24-element trapezoid with 8 elements along the plateau at a plateau level of 9. The pattern response of the individual patterns and the resulting trapezoidal pattern are shown in Figure 4.24. The individual responses have their notches at different values of k because the number of elements in individual patterns differs.

Source patterns are used with receiver patterns in areas where there is an extremely high amplitude surface noise that is confined within rather narrow wavelength limits. For example, Figure 4.25 shows the use of a 3-element, triangular-weighted source pattern in conjunction with a 12-element uniform receiver pattern. The receiver pattern has a spacing of 30 ft, and the source pattern has a spacing of 50 ft. The individual pattern responses are shown in Figure 4.25(a) and (b), and the pattern response of the combination is shown in Figure 4.25(c). Within the restricted wavelength limits of 88 to 120 feet, the combination pattern has rejection greater than 60 dB.

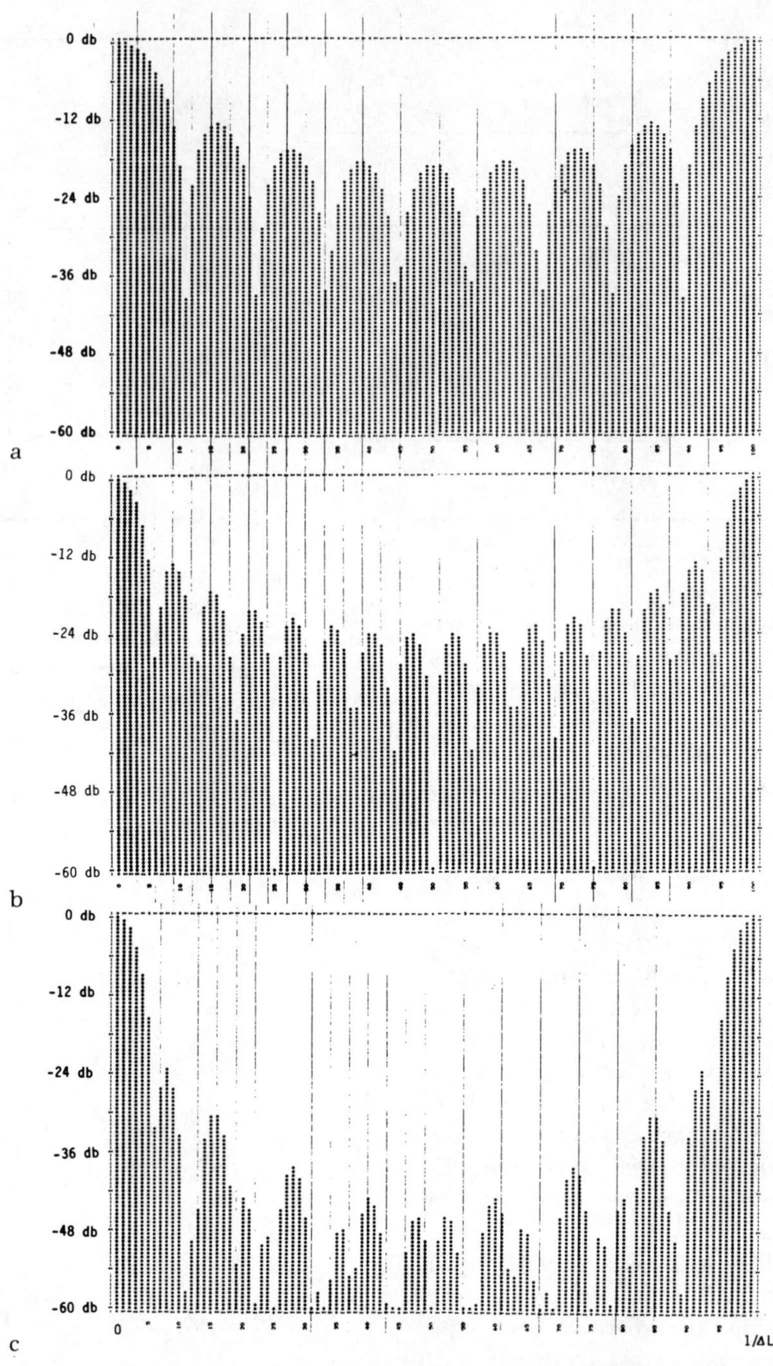

FIGURE 4.24. *A 24-element trapezoidal pattern produced by convolving a 9-element pattern with a 16-element pattern. (a) 9-element uniform pattern. (b) 16-element uniform pattern. (c) 24-element trapezoidal pattern.*

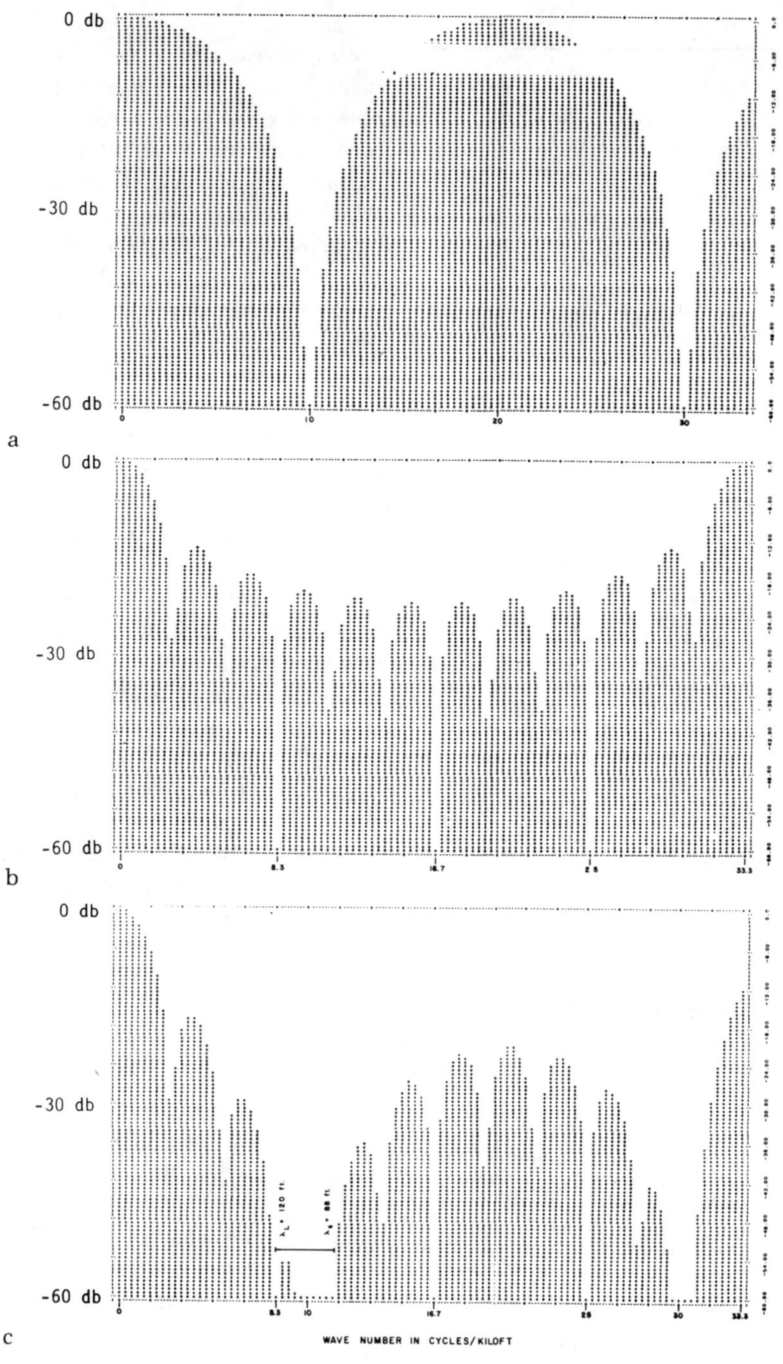

FIGURE 4.25. *Response of a 12-element uniform receiver pattern and a 3-element triangular source pattern. (a) 3-element triangular source pattern. (b) 12-element uniform receiver pattern. (c) Combination source/receiver pattern.*

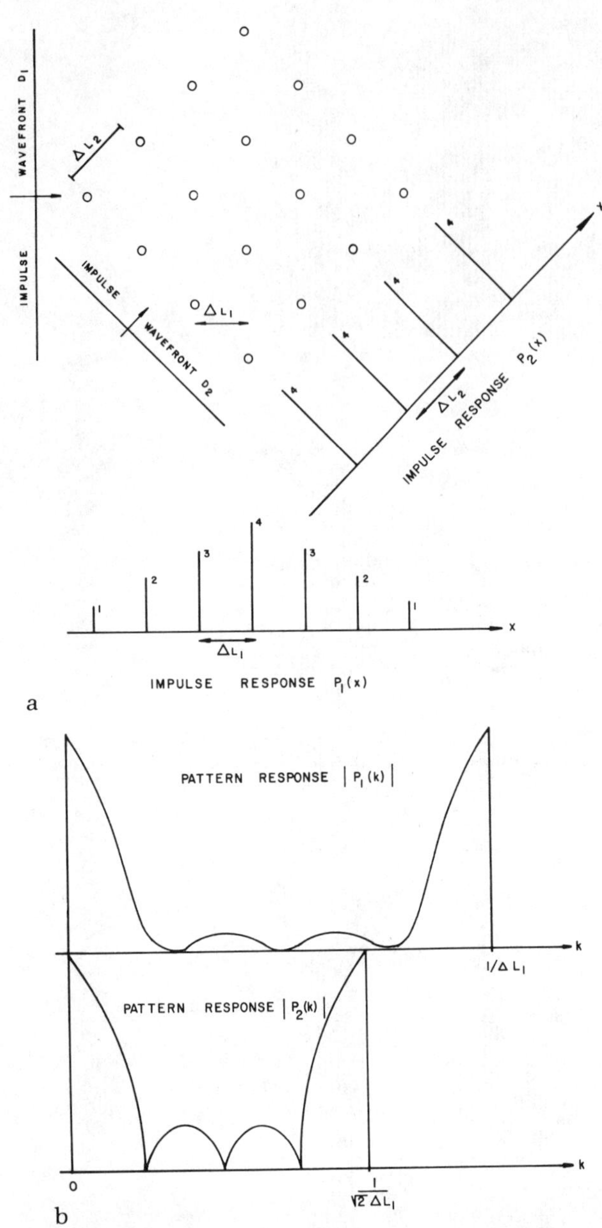

FIGURE 4.26. *Diamond areal pattern. (a) Space domain. (b) Wavenumber domain.*

AREAL PATTERNS

In regions of rough topography, horizontally traveling noise waves may arrive from many different directions because of reflections of the waves by canyon walls, mesas, sand dunes, and other types of rough topography. When this happens, it is usually necessary to use areal patterns. Both areal receiver patterns and areal source patterns may be required.

Areal patterns are usually analyzed by determining the impulse responses of the pattern to plane impulse wavefronts that travel across the pattern at various directions. Consider the 16-element diamond areal pattern shown in Figure 4.26. In direction of the diamond, D_1, the impulse response is a triangular pattern, with spacing ΔL_1. For an impulse-response wavefront arriving at 45° to the direction of the diamond, the impulse response is a four-element uniform pattern, with spacing ΔL_2 where ΔL_2 is equal to $\sqrt{2}\Delta L_1$. The pattern responses rebound at the reciprocal of the spacing—i.e., at $k = 1/\Delta L_1$ and $k = 1/\sqrt{2}\Delta L_1$.

Feather patterns are areal patterns that have trapezoidal or triangular impulse responses in the direction of the source. These impulse responses are good approximations of the major lobe of a sinc function. Such patterns are often used in areas where noise is severe. They are far superior to circular or star patterns because the impulse responses of circular and star patterns are too heavily weighted in the middle, which leads to definitely inferior pattern responses.

A typical 144-element half-feather pattern is shown in Figure 4.27. The pattern has 16 lines of elements, with each line of 9 elements placed at an angle ϕ of 45° with respect to the x axis. The line spacing equals ΔL and the element spacing along the lines equals $\Delta L/\cos\phi$. The areal impulse response projected onto the x axis has trapezoidal weighting that approximates the major lobe of a sinc function. The trapezoid is identical to that formed by the combination of line patterns with 9 and 16 elements each (Figure 4.24). For plane waves arriving at 45°, the feather pattern is a 16-element uniform pattern; for plane waves at 90°, it is a 9-element uniform pattern. Full-feather patterns are formed by two half-feathers that are symmetrical with respect to the x axis.

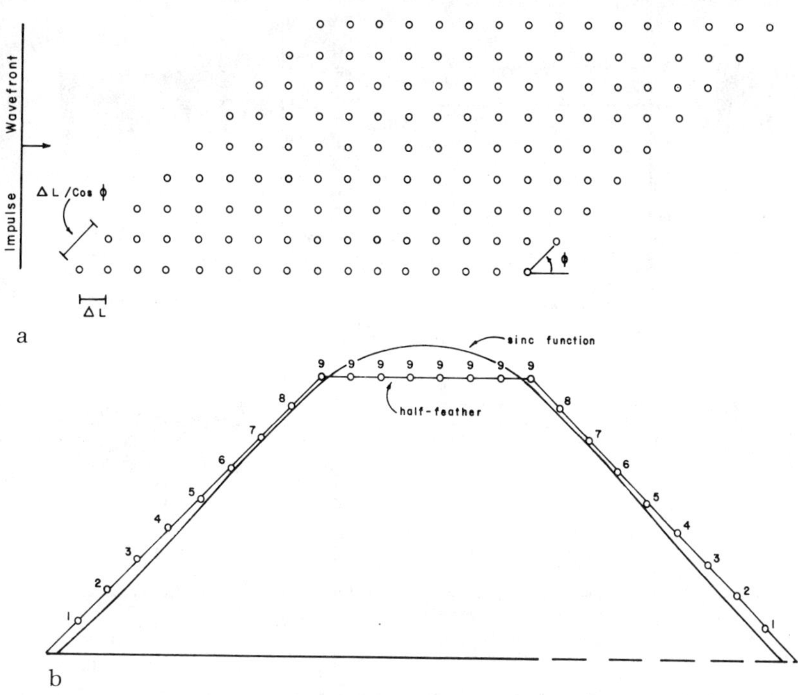

FIGURE 4.27. *Half-feather pattern: 16 lines, each with 9 elements. (a) Pattern geometry. (b) Impulse response.*

EFFECT OF RANDOM ERRORS ON PATTERN PERFORMANCE

In practice, patterns cannot be implemented with the same precision that is applied in design. The response actually achieved must be compromised by factors such as errors in position, errors in the sensitivities of the individual detectors, variations in ground coupling, and the effect of local irregularities in the near surface. Ultimately, the degree of pattern refinement that may be used effectively in any area is determined by the errors associated with the array environment. Complex arrays are expensive to operate; therefore, one should avoid overdesign that would not produce the desired results because of the aforementioned errors.

Newman and Mahoney (1973) discuss the effect of these errors and conclude that the more ambitious the pattern design, the less tolerant the design is to errors and the more care and precision in implementation is required in the field. The ensemble means from Monte Carlo experiments performed with Gaussian random errors in the weights of uniform, triangular, sinc, and Chebyshev arrays are shown in Figure 4.28. The standard deviation in the errors is 10 percent. The results indicate that the pattern response of the uniform array is the least affected by errors and that there probably will be little difference between the various tapered envelopes, such as triangular, trapezoidal, sinc, or Chebyshev, in actual field practice. Their passbands are not significantly changed, and the reject bands have essentially the same level of attenuation, about 30 dB. Errors in position produce results similar to those of errors in weight. Combining the errors in weight and position degrades the performance in the rejection band by about another 3 dB compared to errors in one or the other. In the concluding sentence of their paper, Newman and Mahoney state: “Theoretical response curves should, indeed, be taken with a pinch of salt.”

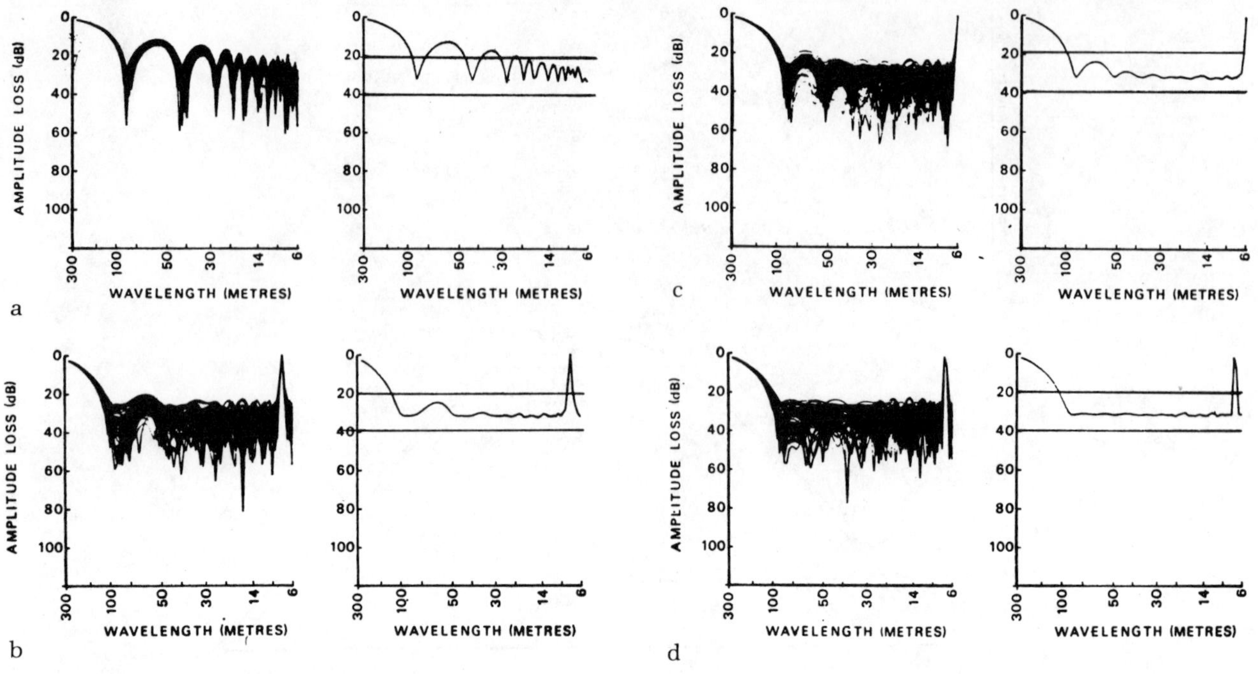

FIGURE 4.28. *Pattern responses with Gaussian random errors in the element weights (standard deviation 0.1):* (left) *ensemble of 50 runs;* (right) *ensemble mean. (a) Uniform pattern. (b) Triangular pattern. (c) Sinc pattern. (d) 40 dB Chebyshev pattern. (From Newman and Mahoney 1953, courtesy of Geophysical Prospecting; Blackwell Scientific Publications Limited.)*

COMBINATIONS OF PATTERNS AND FREQUENCY FILTERS

There is always a frequency filter in cascade with a pattern. At a minimum, the frequency filter consists of the seismic recording system (including the detector). The recording system is characterized by either its impulse response, $g(t)$, or its spectrum, $G(f)$. It discriminates against frequency without affecting wavenumber. The pattern is characterized by its pattern response, $P(k)$. It discriminates against wavenumber independently of frequency. The response of the combination pattern and recording system is the product $G(f)P(k)$, which is designated $\Psi(f,k)$. Because $G(f)$ and $P(k)$ operate on separate aspects of the data, and knowledge of $\Psi(f,k)$ on the f and k axes is sufficient to specify $\Psi(f,k)$ throughout f-k space, the two-dimensional Fourier transform is said to be separable. The amplitude response of a two-dimensional filter is shown as contours in f-k space. An example with a third-order Butterworth filter and a ten-element uniform pattern is shown in Figure 4.29.

Multiplying the two-dimensional spectrum of the system $\Psi(f,k)$ and the two-dimensional spectrum of the noise profile, $N(f,k)$, produces the output spectrum $O(f,k)$. The performance of the system is judged by how effective the system is in suppressing noise without appreciably affecting signal. Quantitative measures, such as the ratio of output to input rms noise level within a specified region of f-k space or the ratio of maximum noise to signal amplitude, are useful in deciding on system parameters. In the absence of such measures, judgment by observation is probably adequate.

PATTERN DESIGN

DESIGN CRITERIA

It is desirable in the design of patterns to choose a pattern in which the noise wavelengths will fall within the rejection band of the pattern and the signal wavelengths will fall in the passband of the pattern. In the case of

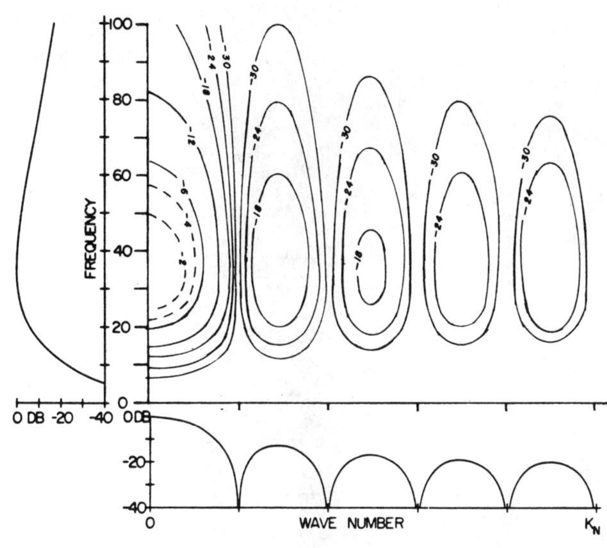

FIGURE 4.29. *Response of a ten-element uniform pattern and a third-order, 25–50 Hz Butterworth filter in* f-k *space.*

uniform patterns, the design involves choosing element spacing and pattern length such that the noise wavelengths fall into the rejection band between $k = 1/n\Delta L$ and $k = (n - 1)/n\Delta L$, as shown in Figure 4.30. The following principles apply:

(1) The spacing should be equal to or less than $(n - 1)/n$ times the shortest wavelength, λ_S, so that all the noise wavelengths will be to the left of the last zero before the rebound pass band:

$$\Delta L \leqslant \frac{n-1}{n} \lambda_S. \tag{4.26}$$

(2) The length should be equal to or greater than the longest noise wavelength, λ_L, so that all noise wavelengths will be to the right of the first zero:

$$L = n\Delta L \geqslant \lambda_L. \tag{4.27}$$

(3) The number of elements required to keep all noise wavelengths within the rejection band is

$$n \geqslant \frac{\lambda_L}{\lambda_S} + 1. \tag{4.28}$$

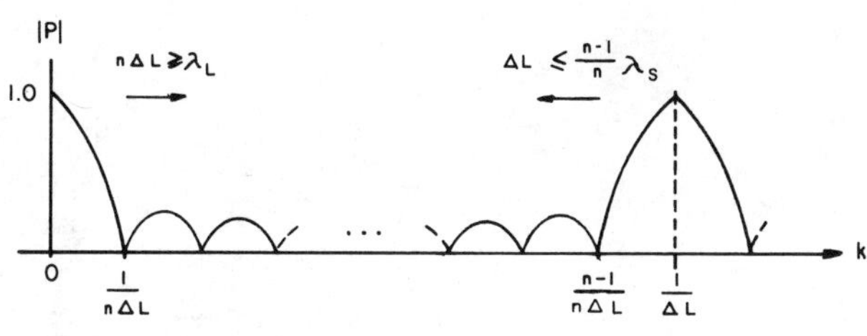

FIGURE 4.30. *Principles of uniform pattern design.*

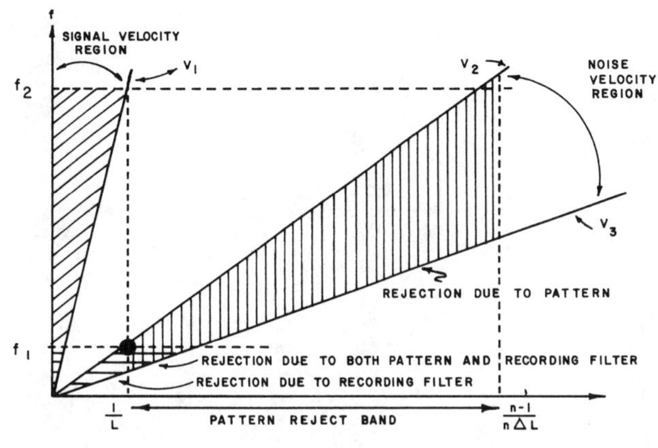

FIGURE 4.31. *Pattern design in* f-k *space when pattern length is restricted by signal considerations.*

Usually, a compromise must be made on pattern length based on the desire to suppress long noise wavelengths and the need to preserve high-frequency spectral characteristics of the reflections. Assuming that it is desired to preserve the frequency spectrum of the signal up to the frequency f_2, then, in the case of a uniform pattern, the first notch in the pattern should be located at $k = f_2/V_1$, where V_1 is the minimal signal velocity (Figure 4.31). Therefore, the maximum allowable pattern length based on signal considerations alone is

$$L = V_1/f_2. \tag{4.29}$$

With the pattern length restricted by signal considerations alone, the noise may spill over into the pattern

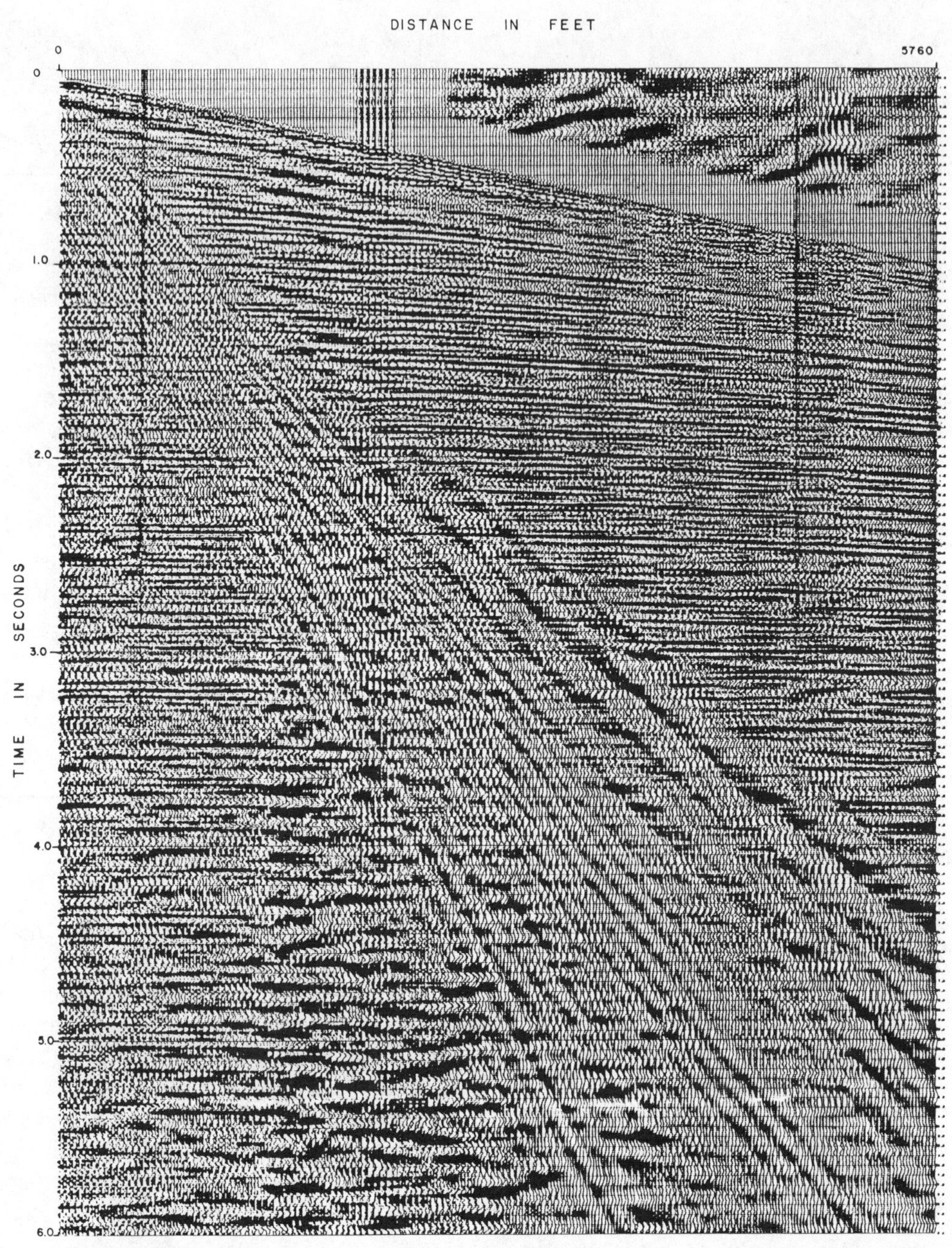

FIGURE 4.32(a). *Noise profile after filtering with six-element uniform pattern of length 240 ft. (Courtesy of Amoco Production Company.)*

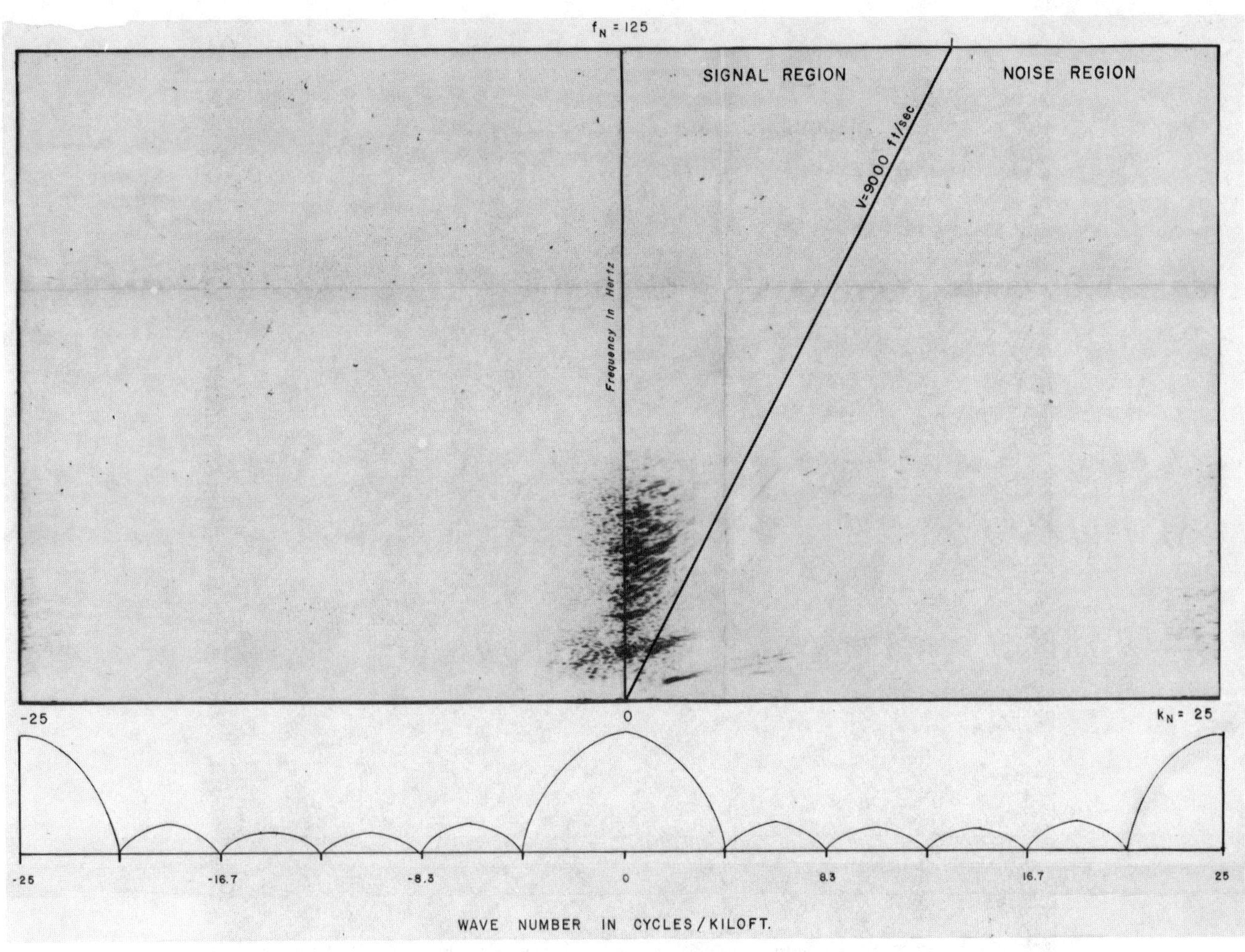

FIGURE 4.32(b). *Fourier transform of noise profile after filtering with six-element uniform pattern of length 240 ft. (Courtesy of Amoco Production Company.)*

passband, and it may be necessary to suppress the spill-over noise with a filter having its low cut-off frequency f_1 given by

$$f_1 = V_2/L = f_2 V_2/V_1, \tag{4.30}$$

where V_2 is the maximal noise velocity. For example, if $f_2 = 60$ Hz, $V_2 = 2000$ ft/sec, and $V_1 = 20{,}000$ ft/sec, then $f_1 = 6$ Hz.

EXAMPLE

Analysis of the noise profile in Figure 4.14 shows that the noise wavelengths fall between the range of 50 and 250 ft. The pattern designed to place these noise wavelengths in the rejection band is

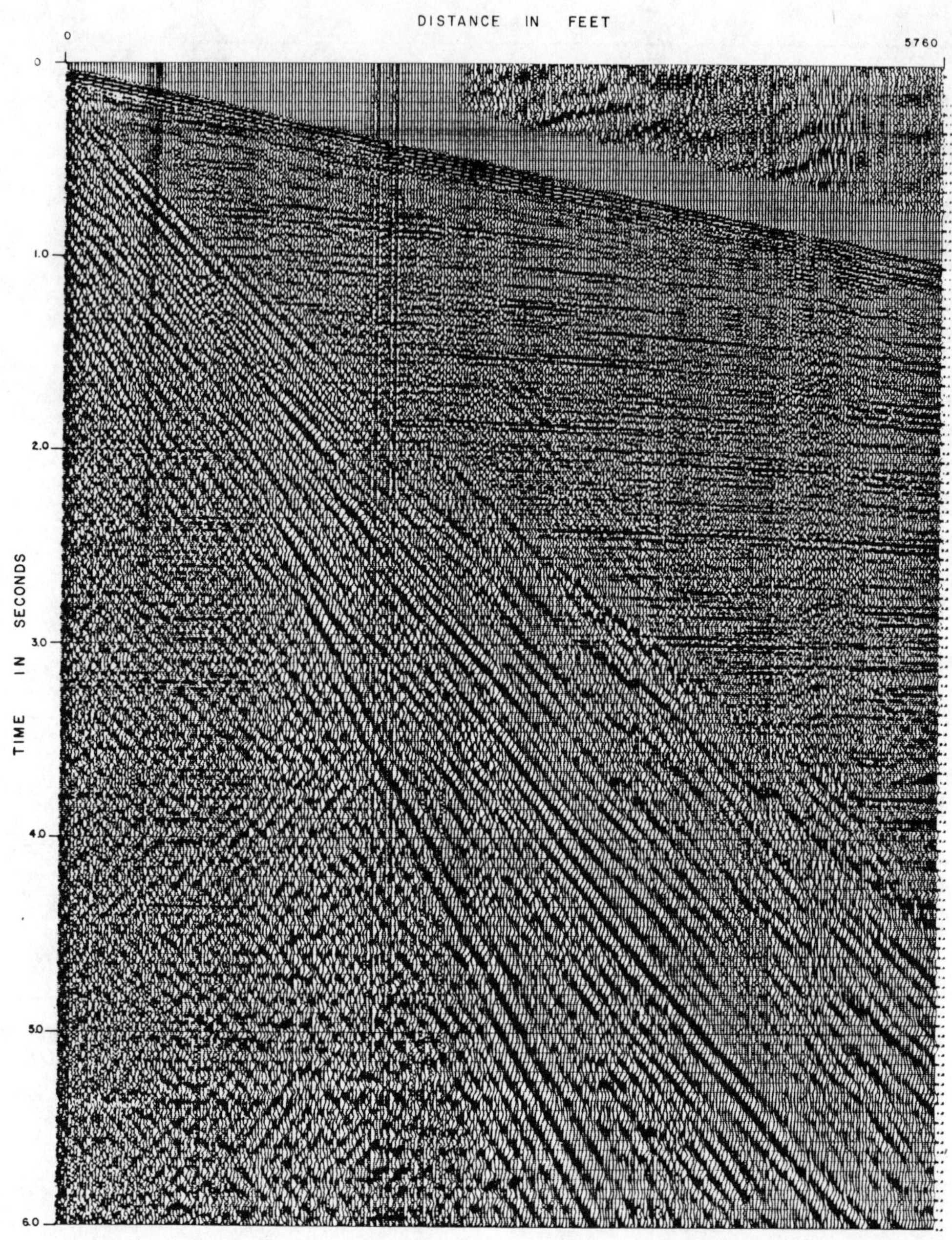

FIGURE 4.33(a). *Noise profile after filtering with two-element uniform pattern of length 240 ft. (Courtesy of Amoco Production Company.)*

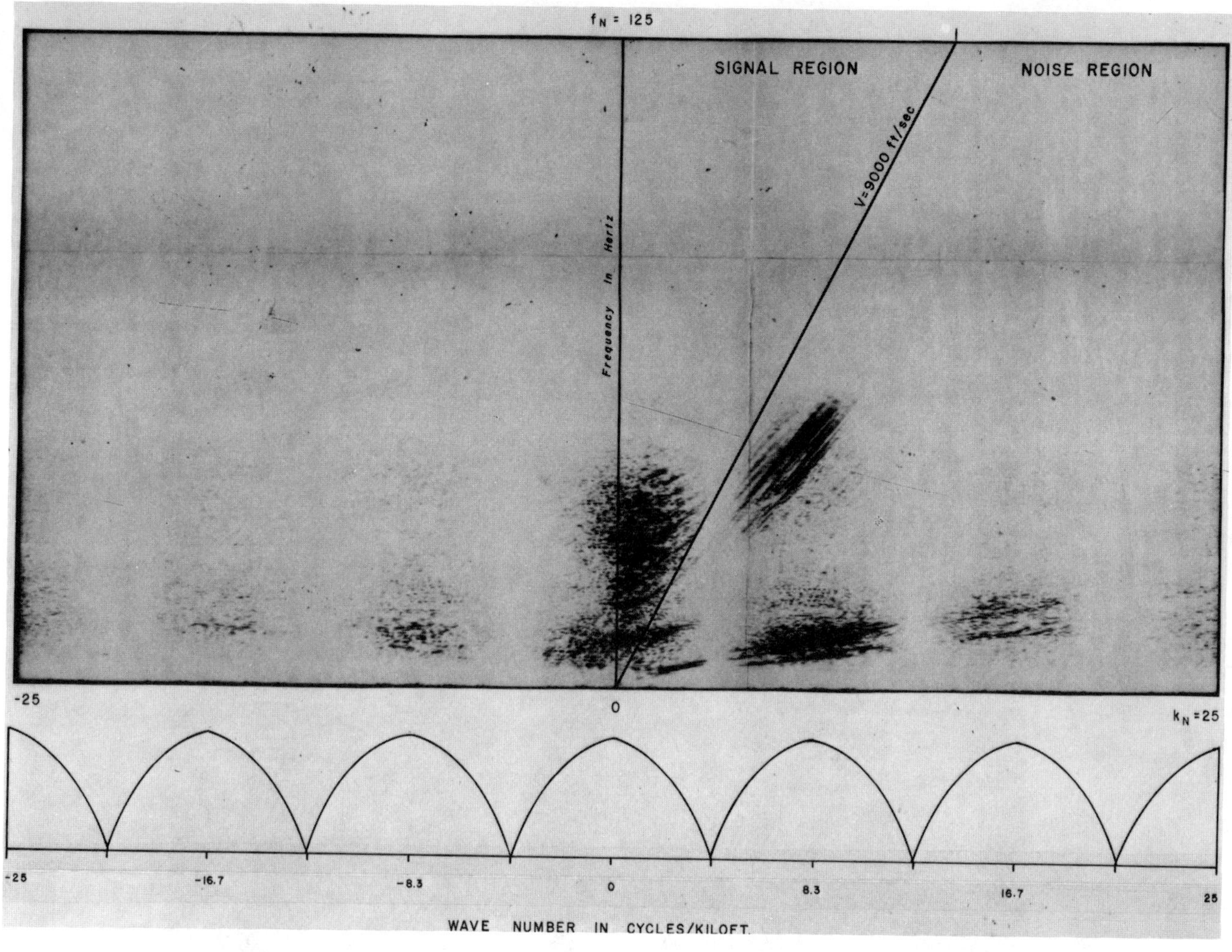

FIGURE 4.33(b). *Fourier transform of noise profile after filtering with two-element pattern of length 240 ft. (Courtesy of Amoco Production Company.)*

$$n = 250/50 + 1 = 6,$$

$$L = (5/6)(50) = 41.7 \text{ ft}, \tag{4.31}$$

$$L = n\Delta L = 250 \text{ ft}.$$

This design is minimal, as the longest and shortest wavelengths are placed at the very edge of the rejection band. These calculations are based on the noise being steady-state without spectral bandwidth. As the noise events are transients with appreciable bandwidth, it may be desirable to move the longest and shortest wavelengths away from the edges of the rejection band. This requires decreased spacing and increased length, with a corresponding increase in the number of elements. A compromise must always be reached on length to allow suppression of the noise without unduly affecting the reflections. Surface waves that fall in the passband because of this compromise may be suppressed by fre-

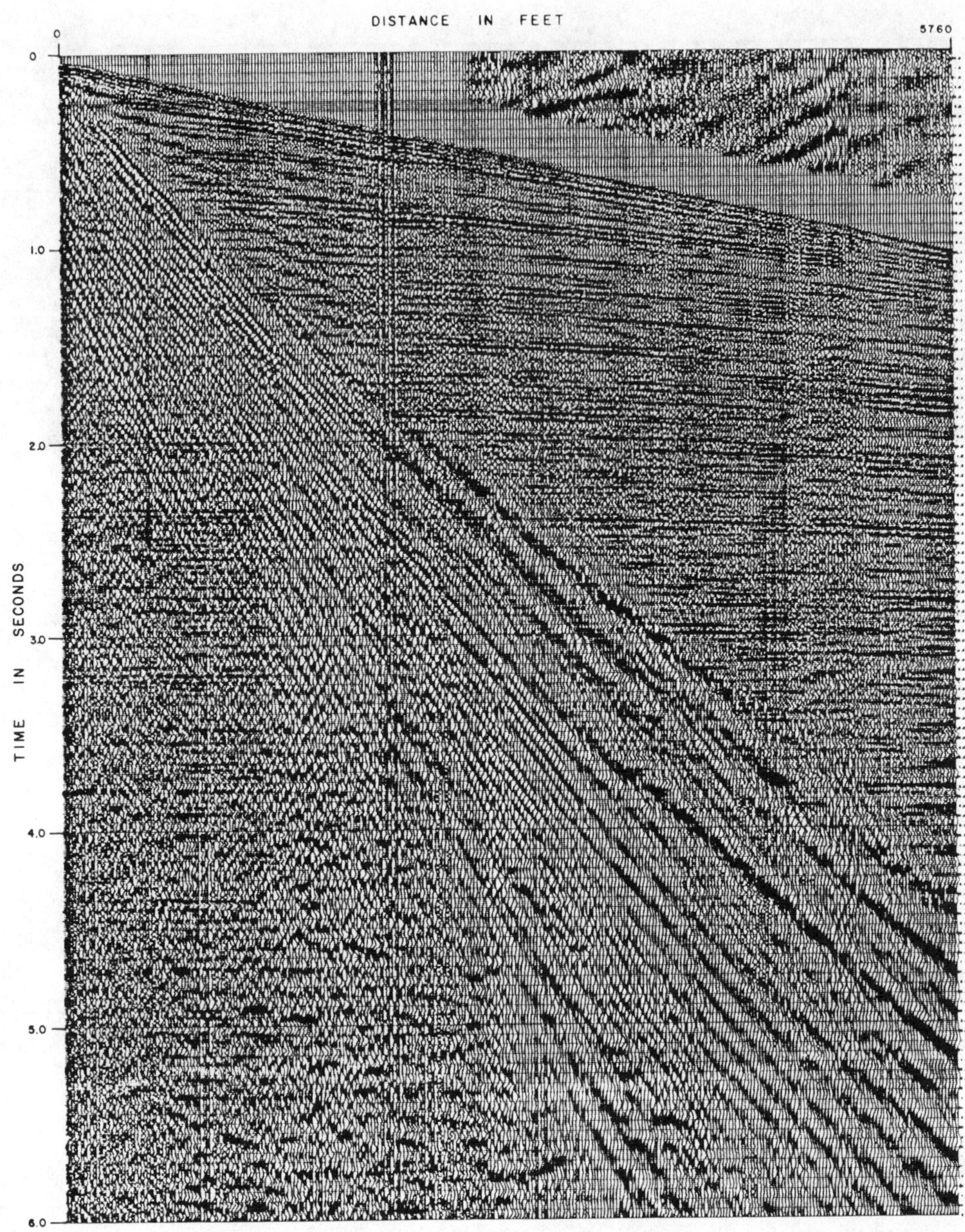

FIGURE 4.34(a). *Noise profile after applying two-element uniform pattern of length 120 ft. (Courtesy of Amoco Production Company.)*

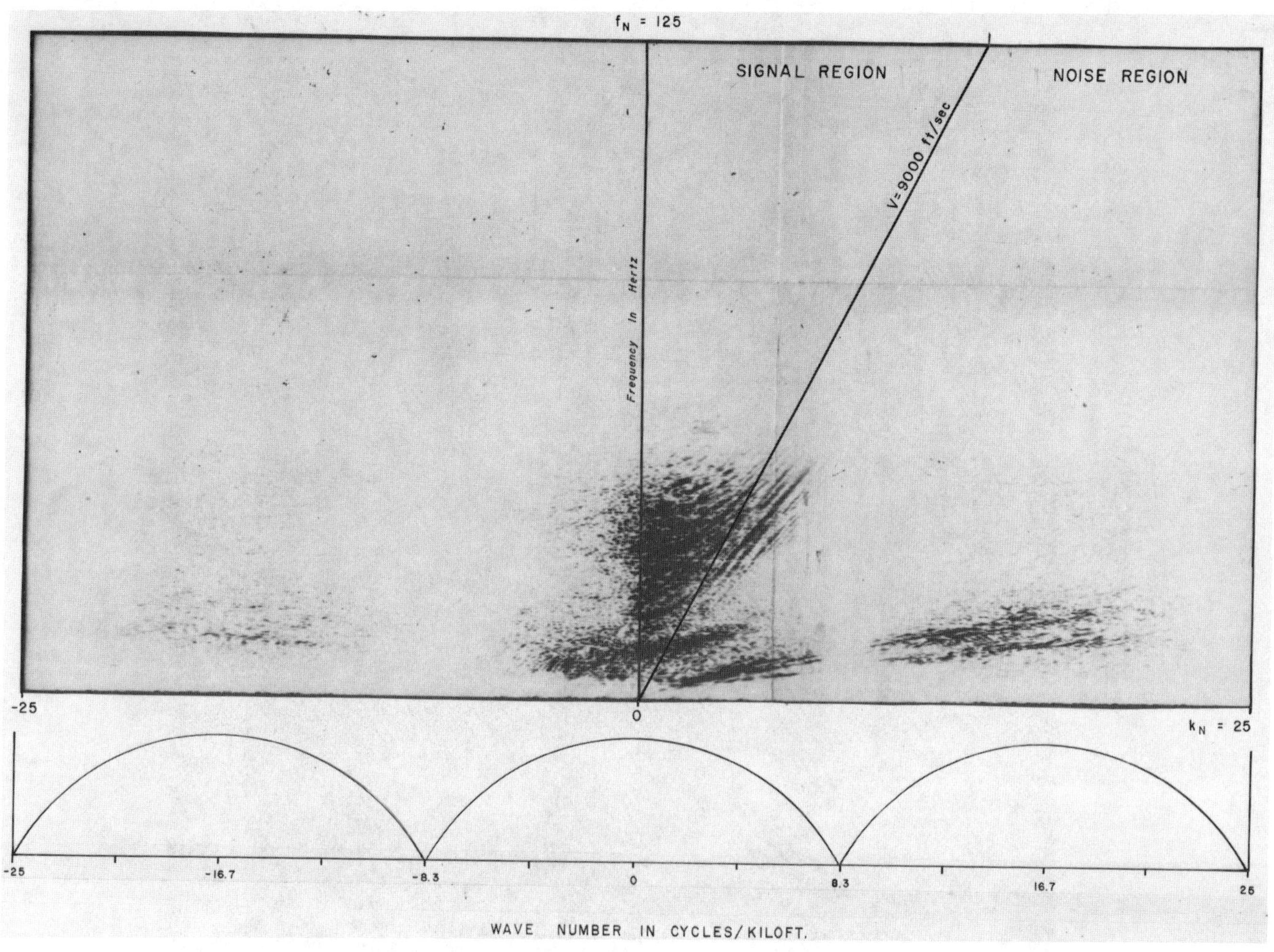

FIGURE 4.34(b). *Fourier transform of noise profile after filtering with two-element pattern of length 120 ft. (Courtesy of Amoco Production Company.)*

quency filters. In this example, a filter with low cut-off frequency of 7 Hz will suppress the surface waves that have spilled over into the passband of the pattern.

In Figure 4.32(a), the noise profile has been subjected to a simulated six-element uniform pattern with spacing $\Delta L = 40$ ft and length $L = 240$ ft. This has been done by summing six alternate traces to produce the first trace and then moving over one trace and summing the next set of six alternate traces, and so on, until the entire profile has been processed. The surface waves have been reduced considerably; however, the underlying reflections do not stand out because the surface waves have eaten up most of the dynamic range on the digital tape; and after the surface waves have been suppressed, there is not much dynamic range left for the underlying reflections. This, of course, is why one uses patterns in the field. In the field, the output of the pattern will have suppressed the surface waves before the data are re-

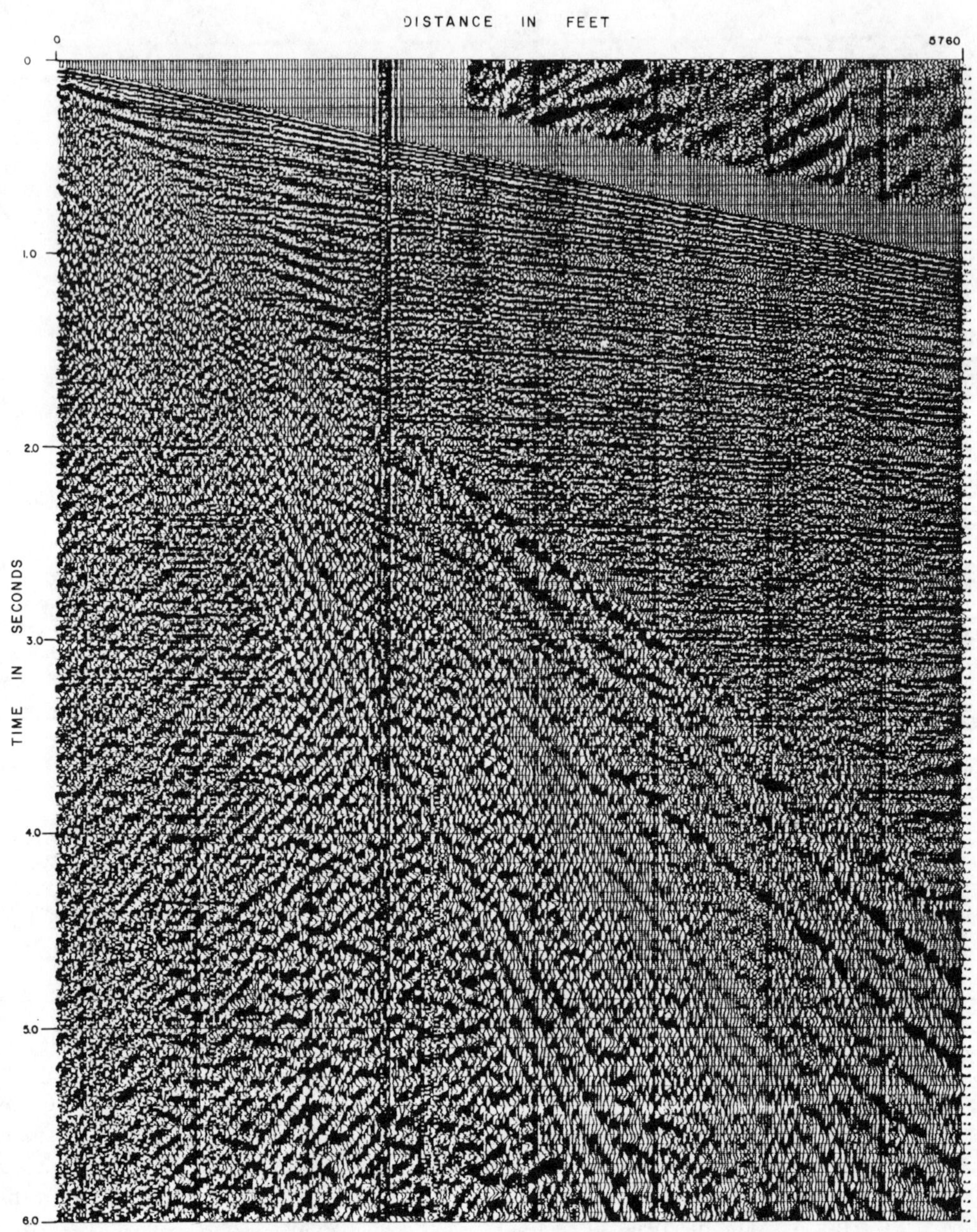

FIGURE 4.35(a). *Noise profile after applying velocity filter having rejection wedge between 672 and 1644 ft/sec. (Courtesy of Amoco Production Company.)*

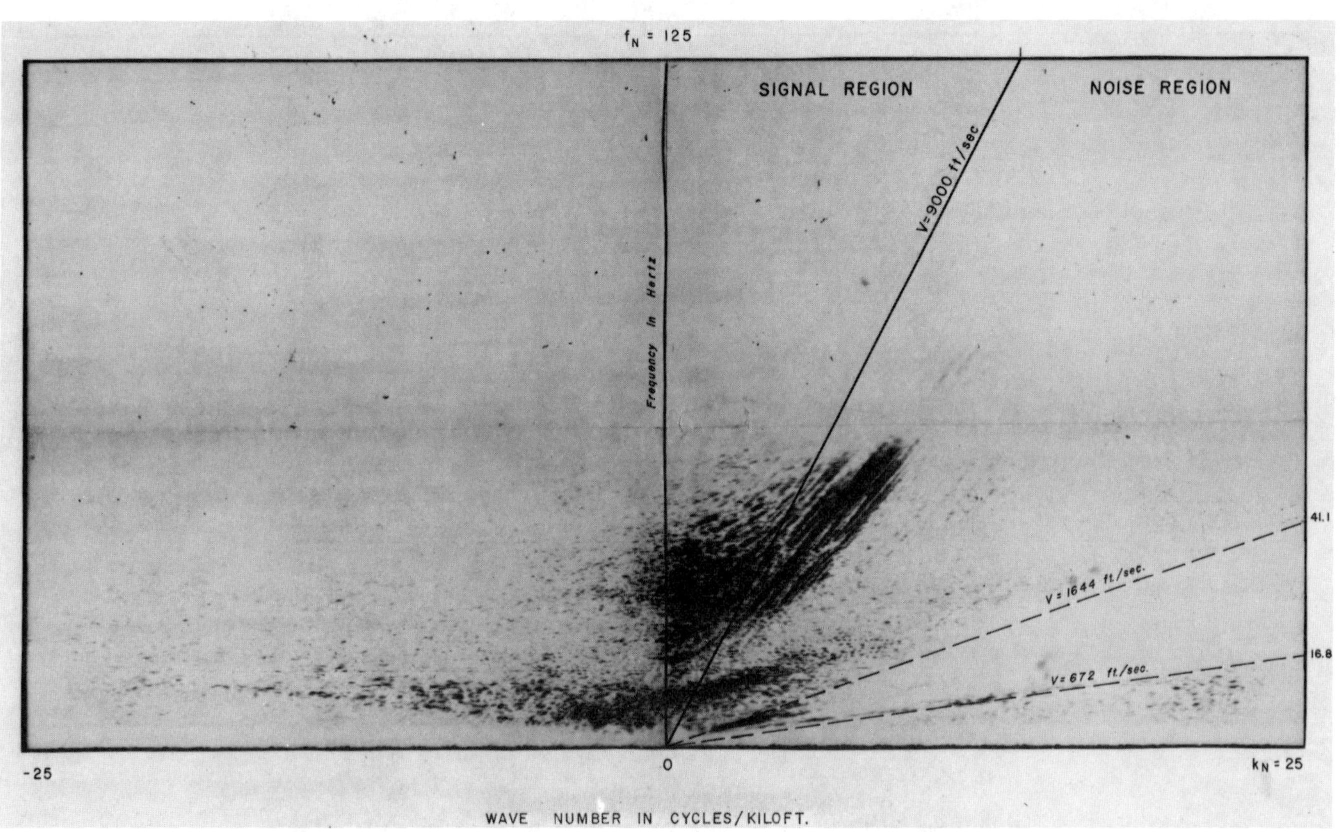

FIGURE 4.35(b). *Fourier transform of noise profile after applying velocity filter having rejection wedge between 672 and 1644 ft/sec. (Courtesy of Amoco Production Company.)*

corded, and the dynamic range of the tape will then be used to record reflections.

The *f-k* transform of noise profile of Figure 4.32(a) is shown in Figure 4.32(b), with the pattern response of the six-element pattern superimposed. The pattern response rebounds at $k = 1/\Delta L = 25$ cycles/kft, which is equal to the folding wavenumber of the data, $k_N = 1/2\Delta x$, where Δx, the trace spacing, equals 20 ft. The first notch in the pattern response is at $k = 1/L$. From the *f-k* spectrum, it is apparent that the noise has been suppressed without appreciably affecting the signal.

Figure 4.33(a) is the noise profile after processing with a two-element pattern with spacing $\Delta L = 120$ ft and length $L = 240$ ft, and Figure 4.33(b) is its *f-k* transform. The noise is not suppressed very well because much of its spectral content comes through the rebound passbands of the pattern. The primary passband is the same as with the six-element pattern, because the two patterns have

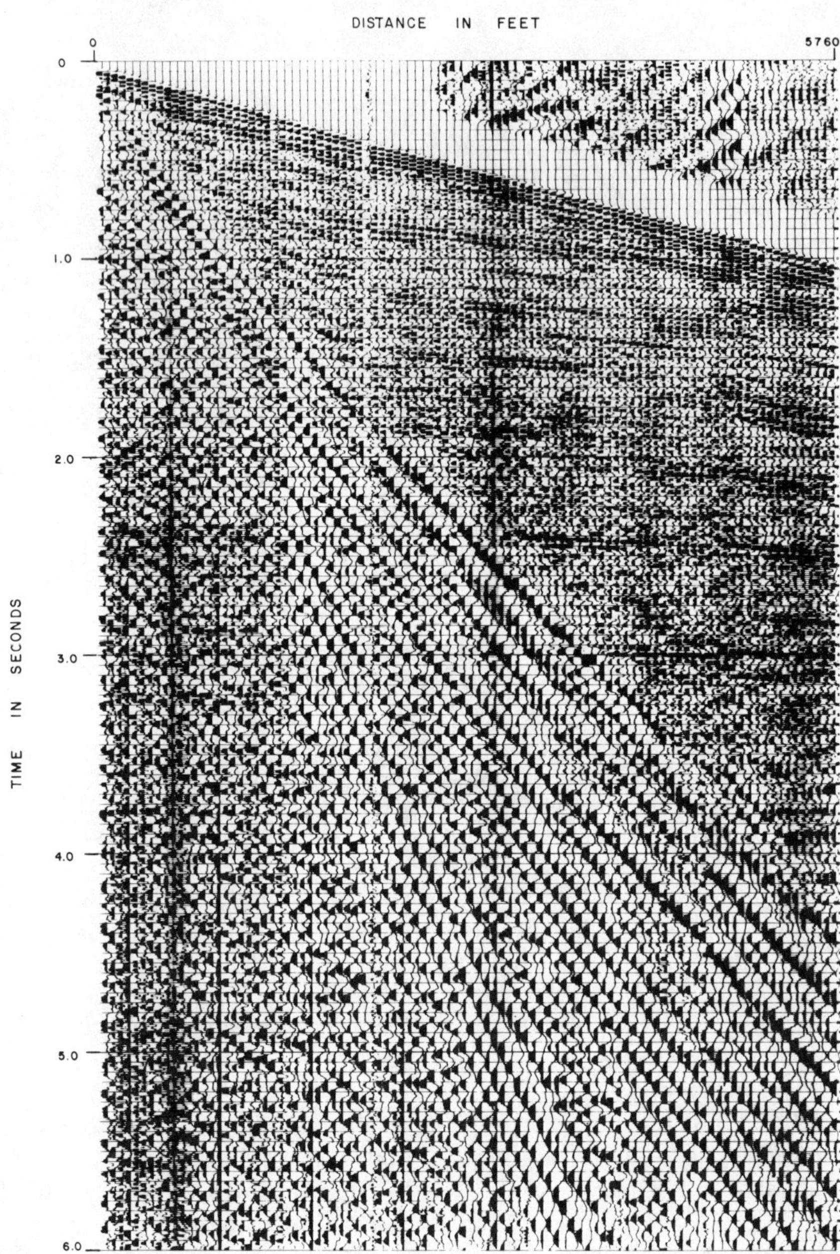

FIGURE 4.36(a). *Noise profile with spacing* $\Delta x = 60$ *ft. (Courtesy of Amoco Production Company.)*

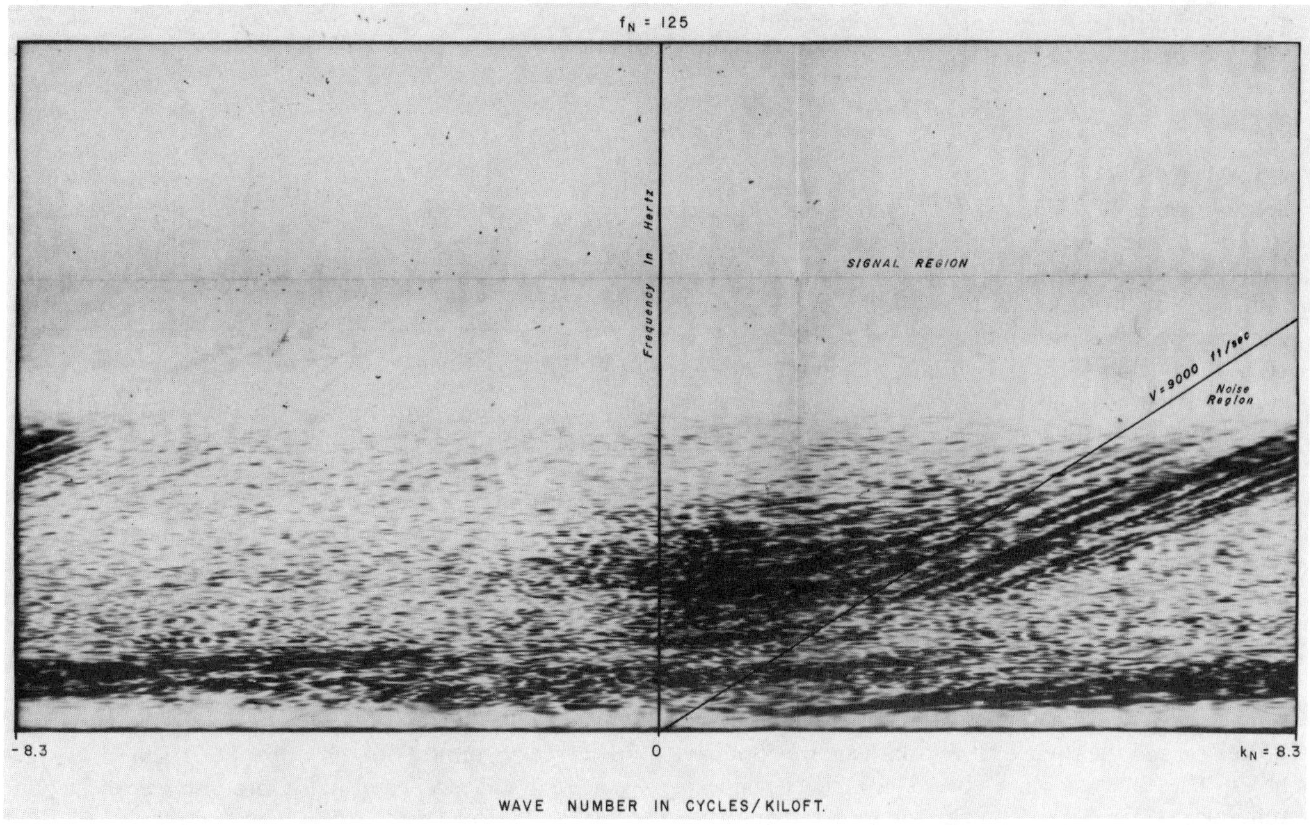

FIGURE 4.36(b). *Fourier transform of the noise profile in Figure 4.36(a), showing aliasing in* f-k *space. (Courtesy of Amoco Production Company.)*

the same length, so the signal is little affected by the pattern, as desired.

Figure 4.34(a) is the noise profile after processing with a two-element pattern having spacing ΔL = 60 ft and length L = 120 ft, and Figure 4.34(b) is its *f-k* transform. This pattern rejects only surface waves with wavelengths near the first notch at λ = 120 ft, and, of course, it is not very effective in suppressing the noise.

Figure 4.35(a) is the noise profile after processing with a filter that rejects noise lying in the velocity wedge bounded by V_1 = 672 ft/sec and V_2 = 1644 ft/sec. Filters like this one that discriminate on the basis of velocity are called velocity filters. They are discussed later in this chapter. The *f-k* transform after velocity filtering in Figure 4.35(b) shows good rejection of the noise spectra, but the filtered noise profile appears to have considerable noise that has phase velocities near the boundaries of the wedge.

The noise profile with spacing $\Delta x = 60$ ft (the intermediate traces have been deleted) in Figure 4.36(a) is included to demonstrate aliasing in the k-domain. Its f-k transform in Figure 4.36(b) with folding wavenumber $k_N = 8.3$ cycles/kft shows folding of the considerable spectral content above k_N back into the primary range between $\pm k_N$.

VELOCITY FILTERS

Because signal and noise occupy wedges in f-k space specified by their ranges of velocities, it is most appropriate to use two-dimensional f-k filters based on velocity rather than using separate filters in the frequency and wavenumber domains. Multichannel velocity filters have been thoroughly discussed in the literature, and are known variously as pie-slice (Embree, Burg, and Backus, 1963), fan (Fail and Grau, 1963), doublet (Foster, Sengbush, and Watson, 1964), laserscan (Dobrin, Ingalls, and Long, 1965), and optimal filters (Sengbush and Foster, 1968). Such two-dimensional filters are said to be nonseparable, because knowledge of their responses on the axes does not specify their response throughout the entire f-k space.

Optimal velocity filters are based on Wiener's theory of a stochastic model of the data-generating process. The signal, coherent noise, and incoherent noise are random processes, and their moveout velocities are random variables. The filter design is based on the principle of minimizing the Wiener mean-square-error measure. Two classes of optimal filters are considered; one passes a narrow band of velocities and rejects other velocities and the second rejects a narrow band and passes others. With the optimal filters, the coherent and incoherent noise-to-signal power ratios can be made frequency-dependent. This, together with control of the expected noise and signal moveout velocities, permits design of processing systems based completely on the signal and noise characteristics.

DESIGN OF OPTIMAL FILTERS

The objective in multichannel optimal filter theory is to find the set of filters $\{g_l\}$ to be applied to the input data $\{i_l\}$ such that the expected mean-square error between the signal estimate $\hat{s} = \Sigma_l g_l * i_l$ and the signal s is minimized; that is, choose the set of filters $\{g_l\}$ that minimizes $E(\hat{s} - s)^2$, as shown in the block diagram in Figure 4.37.

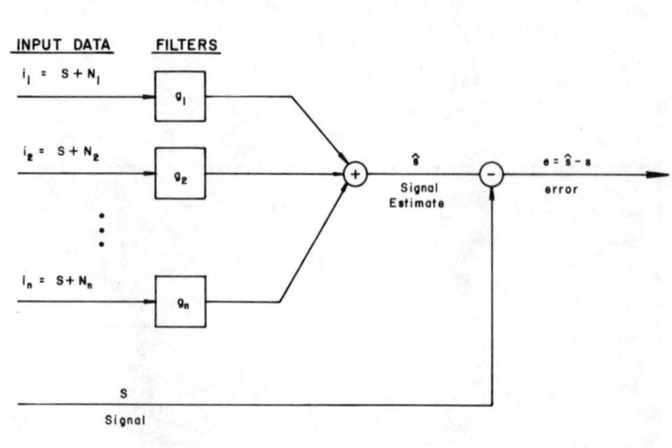

FIGURE 4.37. *Optimal filter design using Wiener criterion to find the set of filters* $g_1, g_2, \ldots, g_n$ *that will minimize the mean square error. (From Sengbush and Foster 1968, courtesy of Geophysics.)*

The statistical assumptions made about the signal and noise processes are as follows. The reflectivity is assumed to be a weakly stationary random process with power spectrum $P_r(f)$. With a time invariant source pulse $b(t)$, the signal is also weakly stationary, with power spectrum $P_s(f) = |B(f)|^2 P_r(f)$. Furthermore, the coherent and incoherent noises are also weakly stationary random processes, with power spectra $P_n(f)$ and $P_u(f)$, respectively, that are mutually uncorrelated and are uncorrelated with the signal.

The foregoing assumptions refer only to the statistical properties of the shape of the signal and noise. To specify the data-generation process completely, the probability distributions for the signal and noise moveouts must be specified. By assuming that the signal moveout per trace is equally likely between $\pm\tau_c$ and that the noise moveout per trace is equally likely in the intervals between $-\gamma\tau_c$ and $-\alpha\tau_c$ and between $\alpha\tau_c$ and $\gamma\tau_c$, a set of optimal pass filters is obtained. By reversing the roles of signal and noise, a set of optimal reject filters is obtained. As a consequence of these probability assumptions, the pass and reject regions in f-k space are defined as shown in Figure 4.38 for the optimal pass filter.

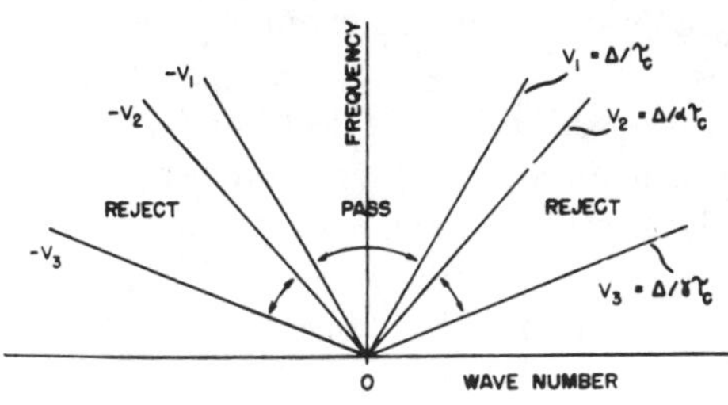

FIGURE 4.38. *Partition of* f-k *space into signal (pass) region and noise (reject) regions. (From Sengbush and Foster 1968, courtesy of Geophysics.)*

The optimal filters just specified have seven input parameters that can be varied: (1) number of channels, (2) location of the signal estimate with respect to the channel number, (3–5) limits of moveouts per trace τ_c, $\alpha\tau_c$, and $\gamma\tau_c$, (6) ratio of coherent noise to signal power $R_n(f) = P_n(f)/P_s(f)$, and (7) ratio of incoherent noise-to-signal power $R_u(f) = P_u(f)/P_s(f)$. All pertinent statistical information about the ensembles of signal and noise is contained in the power spectra. With optimal pass filters when γ is infinite, there is almost complete ignorance of the coherent noise moveout, the processor does not distinguish between coherent and incoherent noise, and the noise-to-signal power ratio $R(f) = R_n(f) + R_u(f)$ contains the pertinent statistical information about the signal and noise ensembles.

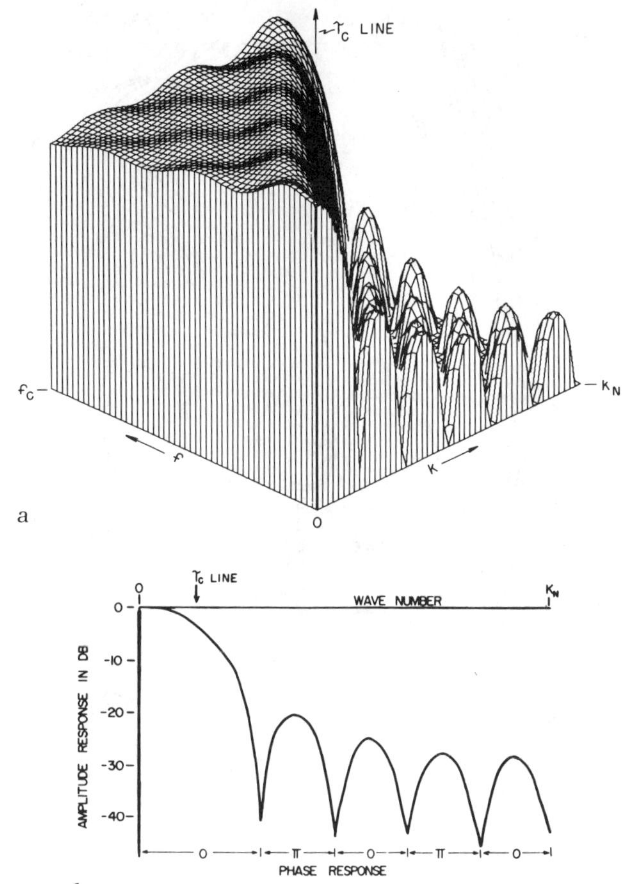

FIGURE 4.39. *Optional pass processor. (a) Isometric plot of amplitude response. (b) Slice through* f-k *space at* f = f$_c$/7. *(From Sengbush and Foster 1968, courtesy of Geophysics.)*

RESPONSE OF OPTIMAL FILTERS IN *F-K* SPACE

Consider the response of an optimal pass system having 12 channels, with center trace estimate, $\alpha = 1$, γ infinite, and $R(f)$ constant and equal to unity for all frequencies. An isometric representation in Figure 4.39 (a)of the amplitude response in decibels viewed down the τ_c line shows the signal region to the left of the τ_c line as a virtually flat plateau, with a steep escarpment upon entering the noise region, leading to a valley floor that has local hills trending parallel to the τ_c line. The local hills rise to about -20 to -30 dB. A constant-frequency slice through *f-k* space at $f_c/7$, where $f_c = 1/(2\tau_c)$, shows these features very well [Figure 4.39(b)]. Throughout the pass region, the phase shift is zero. The notches in the reject region separate regions where the phase shift changes by 180°. Thus, the hills alternate between zero and 180° phase shift.

For a given number of channels, the center trace estimate gives the best suppression of coherent noise. With $n = 12$, the end trace estimate has only about 10–15 dB rejection, compared to 30–40 dB rejection with the center trace estimate. Both the rejection rate of change at the edge of the passband and the rejection in the reject band increase as the number of channels increases. With cen-

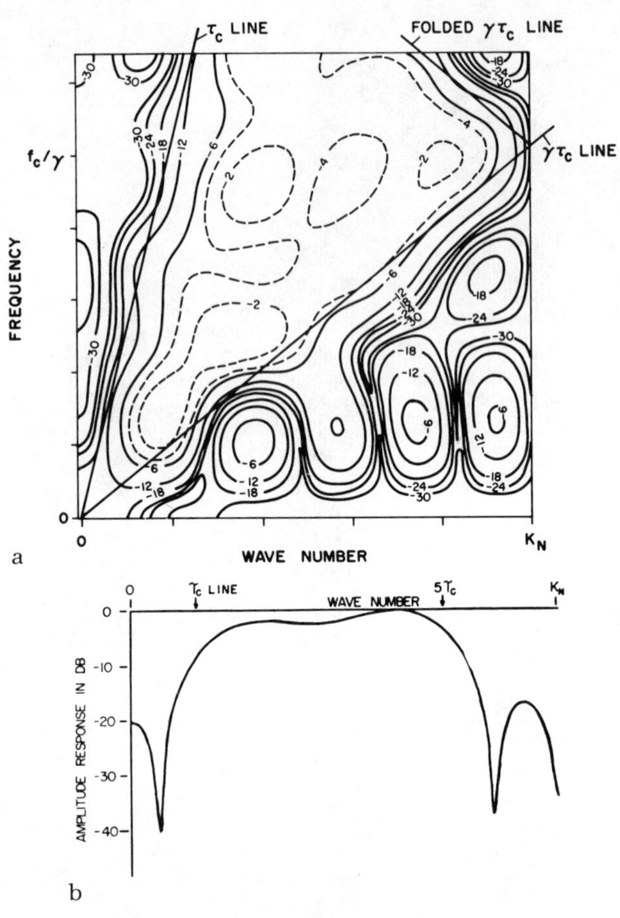

FIGURE 4.40. *Optional reject processor. (a) Amplitude response. (b) Slice through* f-k *space at* f = fc/7. *(From Sengbush and Foster 1968, courtesy of Geophysics.)*

ter trace estimates, a 12-channel processor has about 10 dB more attenuation in the reject band than a 6-channel processor. As γ increases, the width of the expected coherent noise moveout increases and there is less noise suppression over the wider band of expected noise moveouts. When the expected signal and coherent noise moveouts are separated, $\alpha > 1$, there is no coherent noise between τ_c and $\alpha\tau_c$, there is no need for sharp rejection rate near the τ_c line, and the processor has a more gradual rejection rate there.

The amplitude response in Figure 4.40(a) of an optimal reject system with 12 channels, center trace estimate, $\alpha = 1, \gamma = 5, R_n = 1.9$, and $R_u = 0.1$ shows (1) at least 20 dB rejection in the reject band, (2) sharp rejection rate near the τ_c line that separates signal and noise, and (3) little attenuation in the passband over a wide range of signal moveouts. A slice through f-k space at $f = f_c/7$ in Figure 4.40(b) shows these features.

TIME-DOMAIN RESPONSE

The performance of a multichannel filter can be observed in the time domain by feeding a set of traces that contain events with various moveouts into the multichannel filter and obtaining the signal estimate. Results are shown in Figure 4.41 for optimal pass and reject processors with 12 channels, center trace estimates, τ_c = 2 ms/trace, $\alpha = 1$, γ infinite (pass) and 5 (reject), R = 2.0 (pass), and $R_n = 1.9$ and $R_u = 0.1$ (reject). The signal estimate in each case shows that events falling within the signal passband are passed without waveform distortion and that events falling outside the signal passband are highly rejected. Events whose moveout per trace equals τ_c are passed with 6 dB attenuation but without waveform distortion.

APPLICATIONS OF OPTIMAL VELOCITY FILTERS

Optimal velocity filters are useful and improved replacements for any multitrace processing scheme that uses trace summation. The advantages over summing are due to undistorted signal estimates in the passband, sharper rejection rates at the edge of the signal passband, and increased rejection of the coherent noise.

Early applications of velocity filters in processing record sections were disappointing largely because the filters used were time- and space-invariant. However, by using dip search and time- and space-variant processors, velocity filters are useful in improving signal-to-noise ratio on record sections. Letton and Bush (1969) discuss such time-variant velocity filters.

Velocity filters also are useful and powerful replacements for summing methods currently used in horizontal stacking and patterns. Vastly improved noise suppression of horizontally traveling noise is attainable by recording subpatterns individually, possibly even recording each detector output separately, and combining the

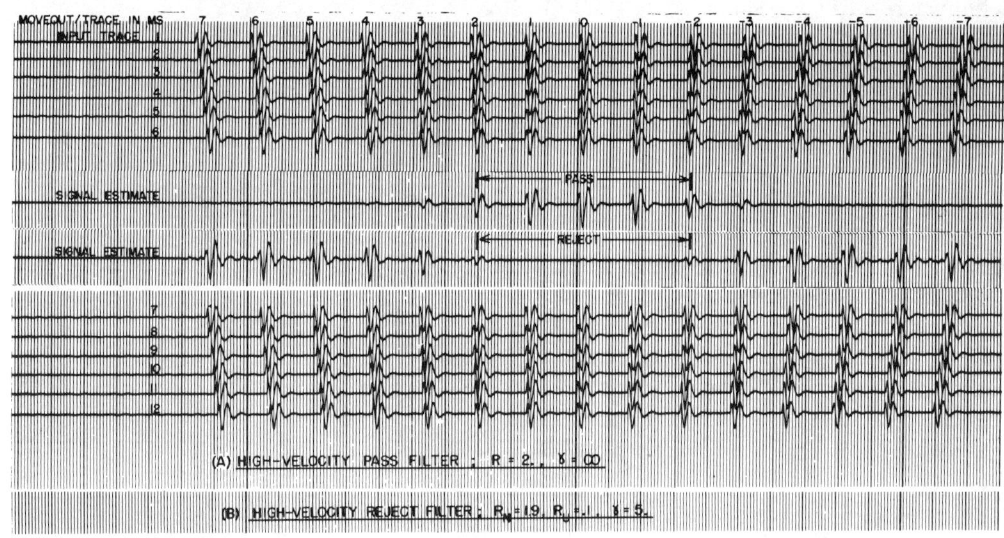

FIGURE 4.41. *Time-domain responses of optional pass and reject processors. (From Sengbush and Foster 1968, courtesy of Geophysics.)*

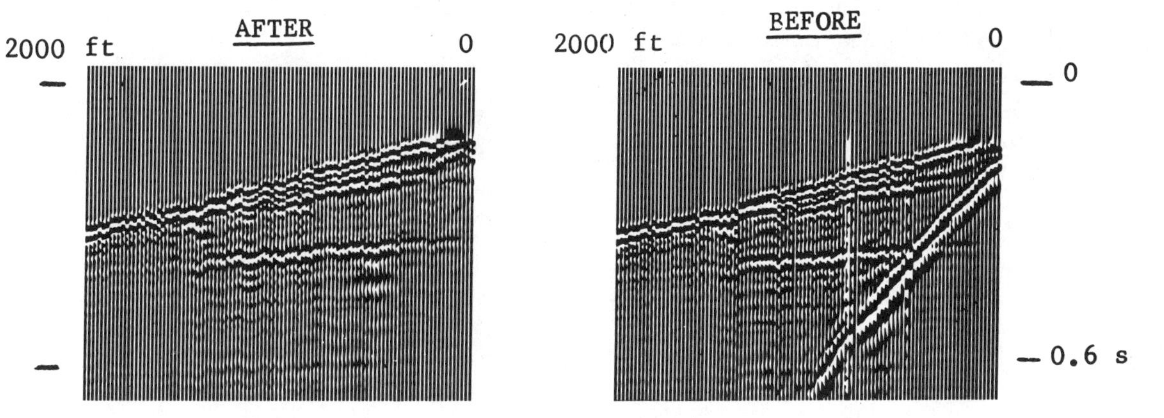

FIGURE 4.42. *Applying optimal pass processor to noise profile; processor designed to pass reflected and refracted energy and reject Rayleigh waves. (From Design and application of optimal velocity filters in seismic exploration by Sengbush and Foster, copyright © 1972 by IEEE; courtesy of IEEE.)*

result into a pattern using velocity filters instead of summation. Multichannel recording systems with 1000 or more channels soon will be available to make this feasible. Figure 4.42 shows the power of velocity filters in discriminating against low-velocity Rayleigh waves on a noise profile while allowing reflections and refractions to pass.

SUMMARY

Seismic prospects may be made or lost in the acquisition of data. No amount of reprocessing can rectify bad geophone plants, dead traces, reversed polarity within groups, cross-feed between channels, and other such multifarious causes of poor data. It should be apparent that acquisition must be carefully supervised; unfortunately, however, this aspect of seismic exploration is sometimes neglected.

Once the aforementioned routine problems are taken care of, it is time to optimize field procedures to suppress horizontally traveling noise. The prime consideration here is the design of source and receiver patterns to suppress the noise without appreciably affecting the reflections. To this end, the noise profile is indispensable. Its analysis in *f-k* space most clearly reveals the regions where signal and noise lie and allows for optimal pattern design. Probably the most important aspect of this chapter is the use of *f-k* space in noise analysis and in the design and application of patterns and velocity filters.

Chapter 5 The Seismic Signal Process

INTRODUCTION

In terms of linear filter theory, the seismic signal process is the convolution of the seismic pulse with the earth's reflectivity. This process is a mixture of determinism and randomness, because the seismic pulse has a deterministic waveform, while the reflectivity is a random function. This concept of the seismic signal process was developed by the Geophysical Analysis Group at MIT and is described in a number of publications, including Wadsworth et al. (1953) and Robinson (1957). They consider the reflectivity to be a set of weighted impulses whose weights and arrival times are mutually uncorrelated. The seismic pulse is deterministic because its characteristics obey basic physical laws.

Looking back, it is apparent that the invention of the continuous velocity log (CVL) by Summers and Broding (1952) was a turning point in seismic exploration. For the first time, a detailed measurement of velocity versus depth was available. Combining this measurement with later-developed continuous density measurements gives the acoustic impedance of the earth in fine detail. From the acoustic impedance, the reflectivity is easily derived. Peterson, Fillipone, and Coker (1955) synthesized seismograms from the CVL data, which, in many instances, closely matched the seismic field data obtained near the well. At last an adequate model of the seismic data-generation process was available. This marked the demise of the Age of Art and Mystery and ushered in the New Scientific Age.

THE LINEAR FILTER MODEL OF THE SEISMIC REFLECTION PROCESS

The linear filter model of the seismic reflection process has the two characteristics common to all mathematical models: it is precisely defined mathematically and it is only an approximation to reality. The primary purpose of a model is to answer some questions about the thing being modeled. The usefulness of models lies in their success in giving reasonably correct answers. The answers may be quantitatively inaccurate but qualitatively useful. Even when the answers are partially wrong, the model is useful, because it provides a framework for the description of deviations from ideality—and this may lead to a better model. Models are important from another standpoint—and this is very important in the case of the linear filter model of the seismic process—in that they force a certain precision of thought on the part of the problem formulator that had not been demanded by the relatively nebulous sort of thinking that preceded the model.

The linear filter model of the seismic process presented by Peterson, Fillipone, and Coker (1955) has the following properties:

(1) It is one-dimensional; the model earth is transversely isotropic, and the seismic pulse propagates in the z-direction as a plane wave striking the layers at normal incidence.
(2) It is noise-free; only signal (that is, primary) reflections produced from changes in acoustic impedance are allowed. Multiples, noise (ground roll, and the like), and distortions (ghosts, reverberations, and so on) are excluded from the model.
(3) It is time-invariant; the seismic pulse shape and amplitude are constant and do not change with travel time.

Peterson called the output of his model the synthetic seismogram. The fact that, in many instances, the synthetic is similar to the field record to a marked degree means that the model frequently gives correct answers, indicating that the model approaches reality. This is sufficient reason for a detailed study of the model.

The reflectivity of a layered earth is a set of weighted and delayed impulses,

$$r(t) = R_k \delta(t - t_k), \tag{5.1}$$

where R_k is the reflection coefficient at the kth interface and t_k is the two-way traveltime to that interface. Convolving $r(t)$ with the time-invariant seismic pulse $b(t)$ gives the set of primary reflections

$$p(t) = \sum_k R_k b(t - t_k). \tag{5.2}$$

Each reflection has the same waveform as the seismic pulse and has its own characteristic amplitude, polarity, and time delay.

The reflectivity of an earth with continuous acoustic impedance is given by

$$r(t) = \frac{1}{2} d(\ln Z(t)/dt. \tag{5.3}$$

Continuous functions can be converted to discrete reflectivity by bandlimiting before sampling.

The seismic pulse $b(t)$ is time-invariant in the model. In real life, however, its waveform changes as it travels through the earth. This is called earth filtering. Earth filtering causes an increasing loss in amplitude due to absorption and scattering as the frequency increases. Earth filtering occurs simultaneously with the reflection process and introduces time-variance into the reflection process in the real earth.

Synthetic seismograms are often made under the assumption that the acoustic impedance is adequately characterized by the velocity distribution obtained from a continuous velocity log. The fact that, in many instances, synthetics made using this approximation are similar to field records to a marked degree means that the linear filter model with $V(t)$ replacing $Z(t)$ is realistic.

The invention of the CVL made it possible for the first time to characterize the earth in all its complexity by a parameter that is directly related to reflections on seismic records. Before the CVL, the mathematical model used was the thick-layer model. It had the same properties as the Peterson model except that, in the absence of definite velocity information, the thick layers were characterized by uniform velocities within each layer. This model contained a number of interfaces, which gave rise to distinct reflections, with noise between reflections. Of course, there could be some overlapping of reflections.

The refinement introduced by Peterson—using the continuously varying velocity obtained from a CVL—brought the model closer to reality. It is an ideal model in that the synthetic seismogram contains nothing other than primary reflections. Other models have been introduced that depart from the ideal and become more realistic by introducing ghosts, reverberations, and other multiples.

Berryman, Goupillaud, and Waters (1958) and Wuenschel (1960) described models containing all multiples. The continuous velocity function from the CVL is sampled to give an n-layered earth, and all reflections—primary and multiple—are obtained by solving the one-dimensional wave equation, taking into account the reflection and transmission effects at each boundary. Foster (1975) showed that, in the continuous case, the transmission losses as observed at the surface are zero; hence, the method of sampling to produce a layered earth and then applying the wave equation leads to erroneous transmission effects. However, the sampled reflectivity functions give correct primary reflections, provided that the sampling is done in accordance with the time-domain sampling theorem.

Backus (1959) introduced water reverberations into the model by means of a filter in cascade with $b(t)$. Reverberations result from two unique factors: (1) a large reflection coefficient at the water-rock interface and a nearly perfect reflector at the water surface, and (2) very low absorption in the water layer. As a result of these factors, multiple reflections within the water layer produce highly distorted seismograms that are difficult to interpret.

Backus (1959) and Lindsey (1960) introduced ghosts into the model. Ghosts at the source result from energy initially traveling upward from the shot (Van Melle and Weatherburn, 1953). This up-traveling energy is reflected downward by reflectors at or near the surface and follows the initial downgoing signal into the reflection path. Thus, the ghost energy produces ghost reflections that are superimposed on the primary reflections. Ghosts are also produced by detectors located beneath the surface. The ghosting effect of buried sources and detectors can be described by an additional filter in cascade with $b(t)$.

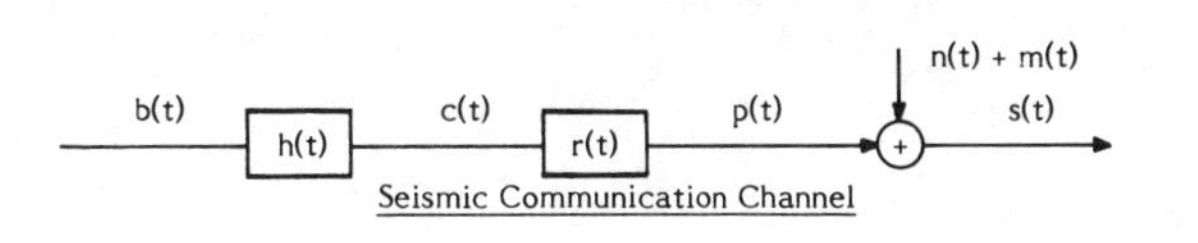

FIGURE 5.1. *Seismic communication channel.*

The additional filter may include the additional absorption in the ghost path as well as the acoustic layering above the buried sources and detectors.

It is convenient to consider ghosts and reverberations as distortions of the seismic pulse and to represent them as a filter $h(t)$ in cascade with $b(t)$. Convolution of $b(t)$ and $h(t)$ produces the distorted seismic pulse $c(t)$ that enters the reflection path to produce the primary seismic signal $p(t)$. Combining signal with additive noise $n(t)$ and multiples $m(t)$ gives the seismic data $s(t)$. This is the seismic method viewed as a communication channel (Figure 5.1).

SIMPLE ACOUSTIC IMPEDANCE FUNCTIONS

A single interface between two beds of different but uniform acoustic impedance layering has reflectivity $R\delta(t - t_1)$, where R is the reflection coefficient and t_1 is the two-way traveltime to the interface. Therefore, it produces a reflection that has the same waveform as the seismic pulse, with its amplitude and polarity given by the reflection coefficient R and its onset occurring at time t_1.

A three-layer case, with the first and third layers having the same impedance and the middle layer different, has a rectangular impedance function (Figure 5.2). The reflectivity function is a doublet given by

$$r(t) = R\,\delta(t - t_1) - R\,\delta(t - t_1 - \Delta), \tag{5.4}$$

where Δ is the thickness of the middle layer, expressed in two-way traveltime through the layer. Two reflections with opposite polarity are produced, one with onset t_1 and the other with onset $t_2 = t_1 + \Delta$. The amplitude spectrum of the reflectivity in this case is a rectified sine wave with notches at $f = n/\Delta$ and maxima at $f = (2n - 1)/2\Delta$ for $n =$ all integers.

The rectangular impedance function will be discussed

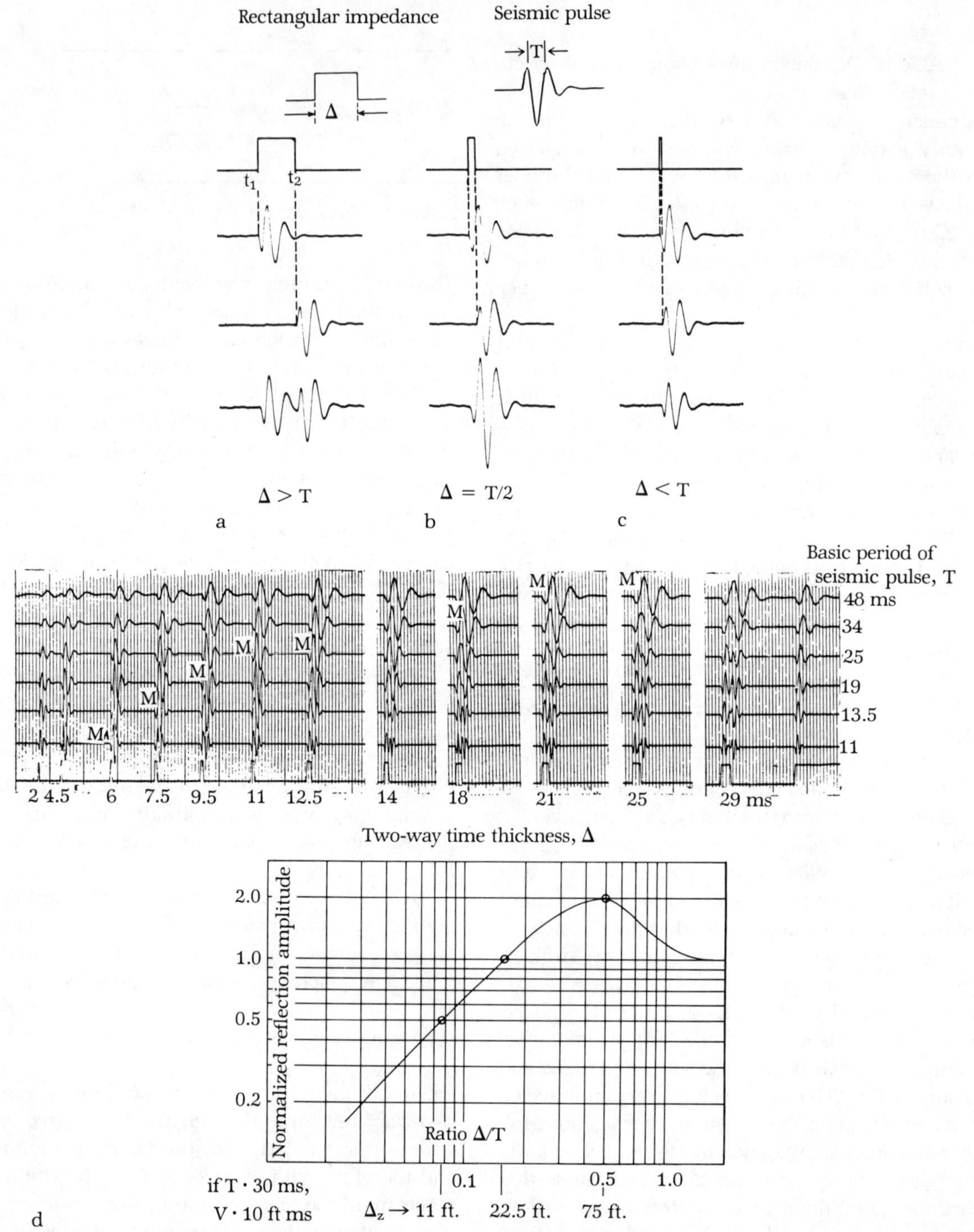

FIGURE 5.2. *Seismic response of rectangular impedance function. (a) Two distinct seismic pulses. (b) Tuning. (c) Differentiated seismic pulse. (d) Reflection amplitude vs. Δ/T. (From Sengbush et al. 1961, courtesy of Geophysics.)*

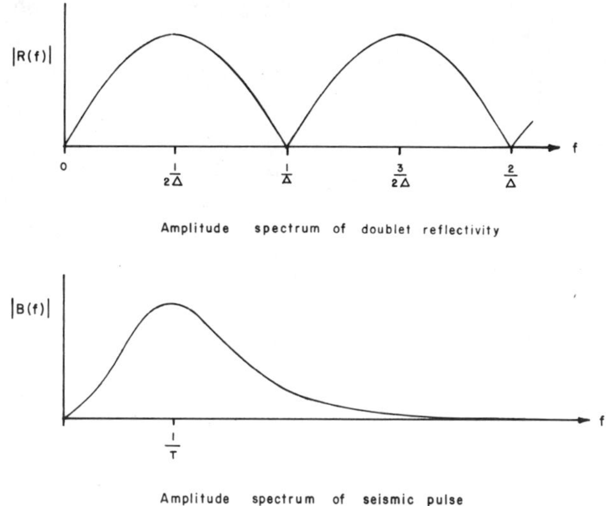

FIGURE 5.3. *Spectral analysis of tuned layering.*

by considering the ratio of Δ/T, where Δ is the bed thickness expressed in two-way traveltime and T is the basic period of the seismic pulse. When the bed thickness is large compared to the length of the pulse, the rectangular layering produces two isolated reflections of equal amplitude and opposite polarity. When $\Delta = T/2$, the composite reflection from the rectangular layering has maximum amplitude because of constructive interference between the top and bottom reflections. The composite reflection amplitude is approximately double the amplitude of either the top or bottom reflection. This amplitude effect is called tuning; that is, the seismic pulse is tuned to the reflectivity. The tuning results when the first maxima in the doublet spectrum at $f = 1/2\Delta$ coincides with the peak in the pulse spectrum at $f = 1/T$, that is, when $\Delta = T/2$ (Figure 5.3). As Δ/T decreases from one-half, the composite reflection amplitude decreases because of destructive interference between the two reflections. The doublet reflectivity acts like a numerical differentiator when Δ is much less than T. With the ratio Δ/T less than about 0.2, the composite reflection amplitude decreases at the rate of 6 dB/octave decrease in Δ/T. This is exactly the amplitude response of a differentiator.

The composite reflection amplitude in the rectangular impedance case is maximum at $\Delta/T = \frac{1}{2}$, which is equivalent to $\Delta z/\lambda = \frac{1}{4}$, where Δz is depth thickness and λ is the apparent wavelength of the reflection. As an example, with $T = 30$ ms and the velocity in the middle layer 10 ft/ms, the tuned layer thickness ($\Delta z = \lambda/4$) is 75 ft. A 22 ft layer ($\Delta z = \lambda/14$) produces a reflection as large as the reflection from an infinite layer. An 11 ft layer ($\Delta z = \lambda/27$) produces a reflection with half that amplitude. The significant seismic response of thin beds was first recognized by Widess (1957).

Tuning of two or more reflections results in increased amplitude due to constructive addition. Properly spaced additional interfaces will continue to increase the maximum amplitude until the number of equally spaced interfaces is the same as the number of half-cycles in the seismic pulse. Further increase in the number of equally spaced interfaces will cause no further increase in the maximum amplitude of the tuned reflection.

A linear increase in impedance with depth corresponds to a linear increase in the logarithm of impedance with time. This type of impedance function is called a ramp. The reflectivity of a ramp is a step-function; hence, the reflection from a ramp is the integrated seismic pulse. Its onset occurs at the change in

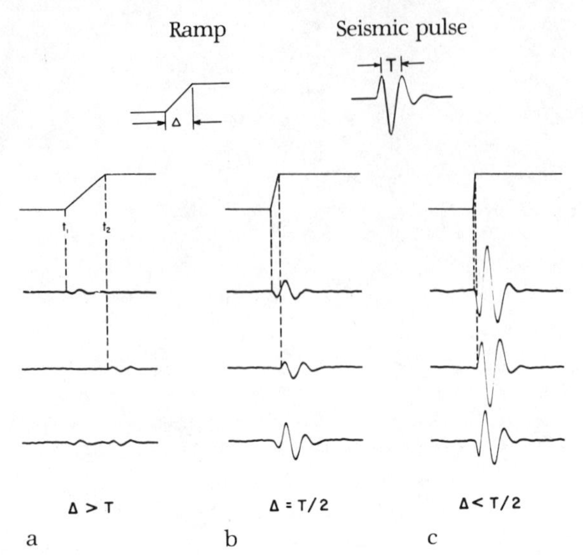

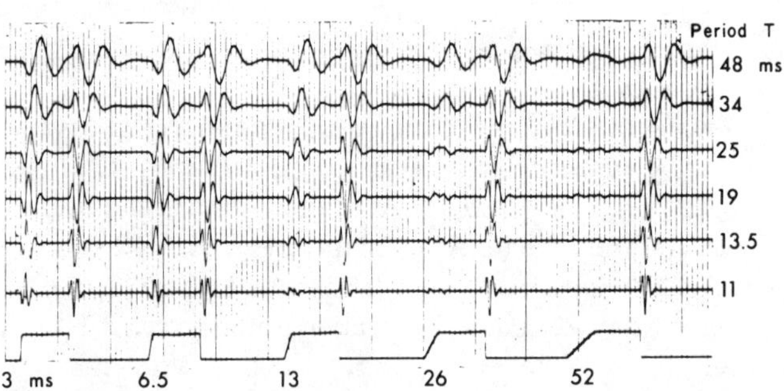

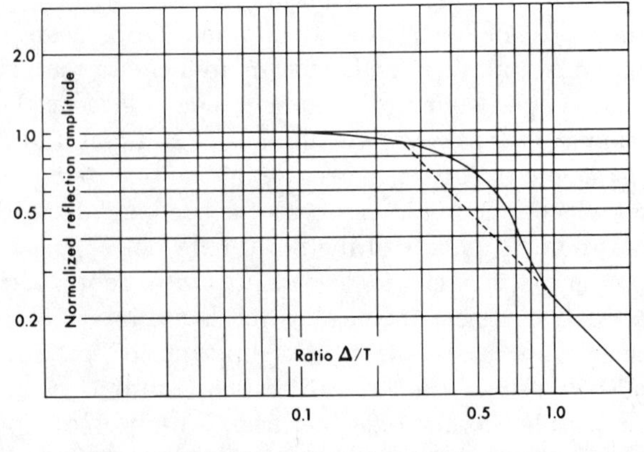

FIGURE 5.4. *Seismic response of ramp-transition impedance function. (a) Two distinct integrated seismic pulses. (b) Tuning of integrated seismic pulses. (c) Seismic pulse. (d) Reflection amplitude vs. Δ/T. (From Sengbush et al. 1961, courtesy of Geophysics.)*

slope, its amplitude is proportional to the change in slope, and its polarity has the sign of the change in slope.

A three-layer case, in which the middle layer has a linear increase in the logarithm of impedance and the layers on either side have constant impedance, has a ramp-transition impedance function and a rectangular reflectivity (Figure 5.4). The changes in slope at the beginning and end of the transition zone have equal magnitude but opposite sign. The resulting reflections are integrated seismic pulses of equal magnitude and opposite polarity, with the amplitude of each reflection proportional to the change in slope.

The ramp-transition zone will be discussed by considering the effect of the Δ/T ratio on the amplitude of the composite reflection. When Δ is large compared to T, the two reflections are distinct. Their amplitude decreases at the rate of 6 dB/octave increase in Δ/T. This is exactly the effect of an integrator. When Δ is small compared to T, the ramp-transition function approaches a step function, and the composite reflection waveform approaches that of the seismic pulse. When the Δ/T ratio is approximately one-half, there is a tuning effect due to the constructive addition of the two reflections. However, each reflection is reduced by the factor $1/\Delta$, with the net result that the maximum amplitude of the tuned reflection is always less than the amplitude from the step. The composite reflections from ramp-transition layering with varying thicknesses and seismic pulse periods shows a bulge near $\Delta/T = 1/2$ due to tuning.

CORRESPONDENCE BETWEEN DEPTH AND TIME

The preceding development shows that the reflection from an acoustic discontinuity at depth z_k has a unique onset time t_k on the reflection record. However, the discontinuity at z_k produces a response $b(t)$, beginning at the equivalent time point t_k and extending later in time for duration A equal to the total length of $b(t)$. Conversely, the entire depth zone of thickness A' above z_k (A' is equivalent to A in time) contributes to the reflec-

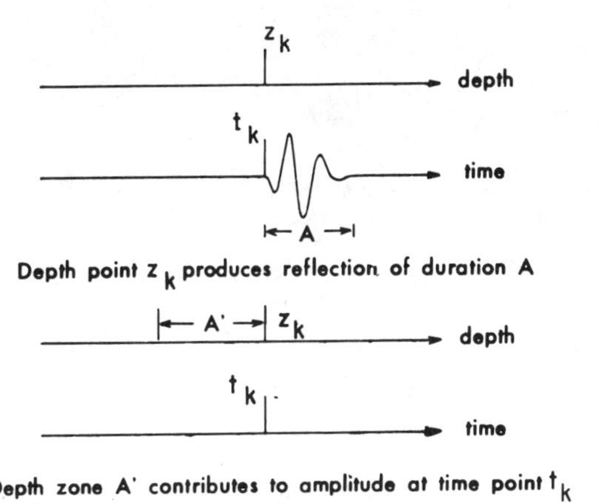

FIGURE 5.5. *Correspondence between depth and time. (From Sengbush et al. 1961, courtesy of Geophysics.)*

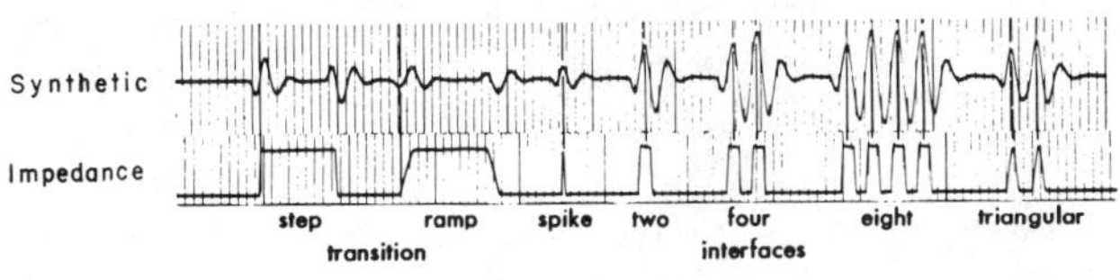

FIGURE 5.6. *Correlation of synthetic with impedance function delayed by 15 ms. (From Sengbush et al. 1961, courtesy of Geophysics.)*

tion amplitude at a time point t_k. Thus, a unique 1:1 correspondence does not exist between the seismic trace and its underlying reflectivity (Figure 5.5). Despite this lack of uniqueness, however, it is possible to specify a 1:1 correspondence, which, once established, must be honored throughout the entire trace when $b(t)$ is time-invariant. When $b(t)$ is time-variant and a 1:1 correspondence has been established at some point in time, then small shifts between the seismic trace and its reflectivity are to be expected at other times.

The filtering action of $b(t)$ shifts the field record later in time with respect to the reflectivity because $b(t)$ is realizable. The shift is called filter delay. The filter delay is constant when $b(t)$ is time-invariant and becomes progressively greater with record time in the time-variant case. Consider the example in Figure 5.6. The time-invariant seismic pulse $b(t)$ has essentially symmetric waveform consisting of a trough-peak-trough sequence, as shown by the reflection from the positive acoustic step. The strongest reflections are those tuned to the layering. The best correlation between the synthetic and the impedance function is reflection peaks to high-impedance layers, established by delaying the impedance function by 15 ms to compensate for filter delay. Another good correlation could have been made by shifting the impedance function 30 ms, in which case high-impedance layers would correspond to troughs on the synthetic. Thus, the synthetic does not have a unique correlation with the impedance function, although there is usually a best correlation.

In comparing seismic data processed by Wiener-Levinson spiking deconvolution followed by zero-phase bandlimiting operators, there should be no filter delay; that is, the processed data and the impedance function should have best match when there is no relative time-shift between them. This will be true for all zero-phase operators applied after spiking deconvolution, regardless of spectral content, and it is one of the main advantages of zero-phase seismic data. Vibroseis data, being essentially zero-phase, should also match the impedance function without shift.

Predictive deconvolution followed by zero-phase bandlimiting does not produce zero-phase seismic data, and filter delays that vary with prediction distance are to be expected.

COMPARING SYNTHETICS WITH FIELD RECORDS

In making synthetic seismograms for comparison with field records, it is necessary to estimate the effective seismic pulse in the field data. The reflectivity computed from the continuous velocity and density measurements is filtered with this estimate, and, if the estimate is good, the synthetic should compare favorably with field records taken in the vicinity of the well from which the impedance function is obtained. The pulse estimate can be considered correct only if the synthetic matches the field record.

Three criteria should be satisfied for a good match between synthetic and field records:

(1) The field and synthetic records should match in character. They should have the same interval time between large reflection events, and they should have the same dead zones.
(2) There should be no relative time-shift between the field and synthetic records on best character match; that is, the phase characteristics of the real seismic pulse in the field record and its estimate in the synthetic record should match.
(3) The polarities of the synthetic and field records should be consistent. If the polarity of the field record is established by making the first arrival break down, a step impedance will produce a reflection that initially breaks down on the field record. By placing such an isolated step on the impedance function and choosing the synthetic polarity that has an initially down-breaking reflection, proper polarity of the synthetic is assured.

There are many possible reasons for a poor match between the synthetic and field records. They fall into three classes:

(1) The filtering on the synthetic may not duplicate the filtering on the field record. In addition to the seismic pulse, the field data are filtered by a variety of filters external to the earth, such as detector response, detector coupling, amplifier filter, automatic gain control (AGC), and so forth.
(2) The assumptions made in formulating the model may not hold sufficiently well in the actual earth. Distortions such as ghosts and reverberations, multiples, and surface waves that appear on field data may not be included in the model from which synthetics are made. Synthetics made from velocity logs without accounting for density changes may be in serious error in formations containing salt and anhydrite and in porous zones containing gas.
(3) The measured acoustic impedance is subject to errors. The most serious errors occur in washed-out shale or salt sections and in gas-bearing sands. With one-receiver CVLs, the recorded velocity is too low in washouts. This results in exaggerated reflection amplitudes and in significant timing errors on the synthetics. With two-receiver CVLs, the timing errors are minor, but washouts may produce false character.

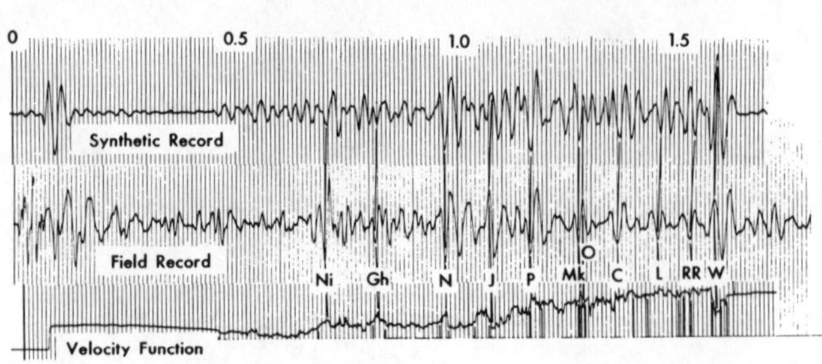

FIGURE 5.7. *Comparison of field and synthetic records in Williston Basin. (From Sengbush et al. 1961, courtesy of Geophysics.)*

Comparisons between analog field records and synthetics made from CVLs demonstrate the use of synthetics in identifying reflections on field data. With analog field data, the two most restrictive filters in the signal path are the seismic pulse and the field recording filter. Two bandpass analog filters are used in making the synthetic, one matching the recording filter used on the field data and the other simulating the source signature after it is modified by attenuation in the reflection path. The recording filter is known and the second filter is varied to give the best match between field and synthetic records. Although the seismic pulse producing the actual data is unknown, it has bandpass characteristics. In fact, it establishes the so-called seismic band.

EXAMPLE FROM WILLISTON BASIN

The objective in this study was to observe character changes in the Newcastle sand reflection caused by changes in its thickness; therefore, the time-invariant synthetic was generated to match the time-variant field data at Newcastle time (Figure 5.7). The synthetic was filtered with the 15–134 Hz field recording filter and a 30–43 Hz filter to simulate the attenuated source signature. This produced a good character match of the Newcastle reflection. The synthetic and field records are displayed with zero relative time shift and show that the second criterion is satisfied. The third criterion, consistent polarity between field and synthetic, is satisfied, as proved by the fact that both the first break on the field record and the reflection from a positive step in acoustic impedance on the synthetic show downward deflection initially.

One high-quality field trace was obtained by mixing several traces after static correcting for weathering and elevation. Column charges (Musgrave, Ehlert, and Nash, 1958) were used to reduce the strong ghosts that were present.

A 1:1 correspondence was established between the velocity function and the synthetic by equating the high-velocity peak in the Newcastle with the *N* trough on the synthetic. The velocity function has been shifted 23 ms to the right to line up these two points. The troughs marked Ni, Gh, P, Mk, C, L, and RR on Figure 5.6 correspond to high-velocity zones in the Niobrara, Greenhorn, Piper, Minnekahta, Charles, Lodgepole, and Red River, respectively. The peaks marked J, O, and W correspond to low-velocity zones in the Jurassic (Morrison), Opeche, and Winnipeg, respectively.

The synthetic and field records are timed below datum and are displayed with zero relative time shift between them. The *N* trough has an arrival time of 972 ms on both, showing that the criterion of no relative time shift upon best match is satisfied. The minor variations of a few milliseconds between corresponding events on the two traces are probably due to minor errors inherent in the time scale of the velocity function. The time scale of the field record is above reproach.

The reflection from the isolated step introduced at the left of the velocity function not only establishes the polarity but also shows the waveform of the seismic pulse. The basic period *T* of the pulse is 29 ms. From the value of *T*, the relative effectiveness of square and ramp-transition zones in producing reflections can be determined. For example, the Newcastle can be resolved into the sum of a ramp-transition zone whose thickness Δ is 22 ms and a negative step that is delayed 22 ms behind the onset of the transition zone. The Δ/T ratio of the ramp-transition zone equals 0.76, giving a normalized reflection amplitude of 0.35. By direct measurement on the velocity function, the ramp change is found to be only 0.71 as large as the negative step. Therefore, if the amplitude of the negative step reflection is considered unity, the amplitude of the ramp transition reflection will be $0.71 \times 0.35 = 0.25$. Thus, the negative step at the base of the Newcastle is four times more effective than the ramp-transition zone in producing a reflection.

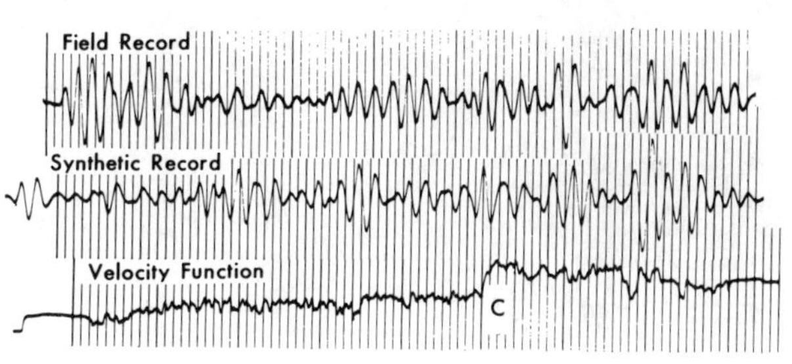

FIGURE 5.8. *Comparison of field and synthetic records in North-Central Texas. (From Sengbush et al. 1961, courtesy of Geophysics.)*

EXAMPLE FROM NORTH CENTRAL TEXAS

The example in Figure 5.8 illustrates the point that reflections from a step-velocity change in the earth are ordinarily smaller than reflections from tuned layers. The Canyon (C) limestone has a step-velocity change from about 11 to 17 ft/ms that is actually more abrupt than it appears in the figure because of the poor response of the recording galvanometer to step changes. The reflection from the step is considerably smaller on both the field and synthetic records than the later-arriving tuned reflections that are produced by alternating limestones and shales. The velocity function has been shifted later by about 22 ms to establish the correspondence between low-velocity zones and peaks on the synthetic. A relative shift between the field and synthetic records was allowed because the CVL was not checked by a well survey. Thus, the time scale on the velocity function was not anchored to the exact time scale of the field record.

EXAMPLE FROM SOUTH TEXAS

The strongest and most consistent reflection in this area is the tuned reflection produced by the cyclic variation in velocity in the alternating sands and shales, beginning with the Lopez (L) sand (Figure 5.9). The velocity function of these sands and shales has a basic period of about 20 ms, which is equivalent to a basic frequency of about

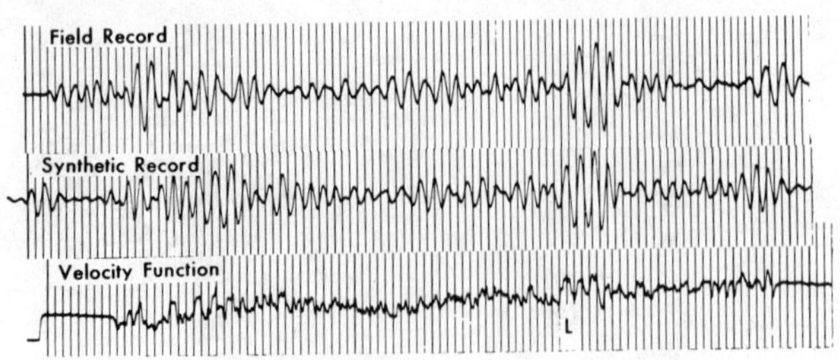

FIGURE 5.9. *Comparison of field and synthetic records in South Texas. (From Sengbush et al. 1961, courtesy of Geophysics.)*

50 Hz. This frequency falls within the passband of the seismic pulse, which was empirically determined to be 40–79 Hz. The seismic pulse filters out the thin bedding changes that are superimposed on the basic 20 ms layering. The velocity function has been time-shifted about 25 ms to the right to establish the correspondence between high-velocity zones and peaks on the field record.

The objective in this study was to identify the events corresponding to the sand-shale sequence and to study the effect of the known pinch-out of the Lopez sand on seismic data. The CVL was modified in accordance with the postulated change in stratigraphy, and the resulting synthetic shows how that change in stratigraphy will affect the field records (Figure 5.10). The pinch-out of the first sand in the alternating sand-shale sequence that is tuned to the seismic pulse results in the loss of amplitude of the first major leg of the resulting tuned reflection.

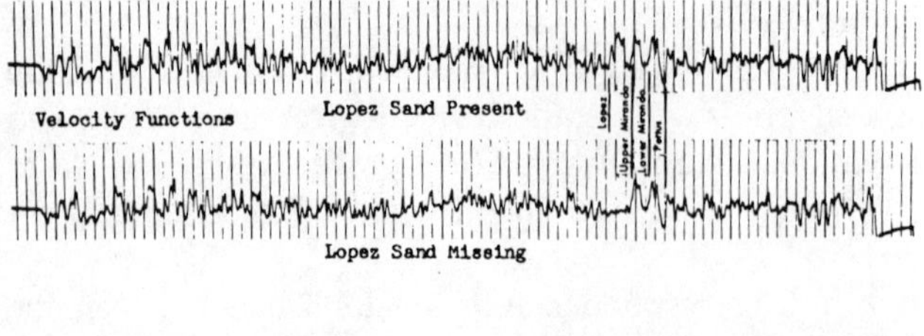

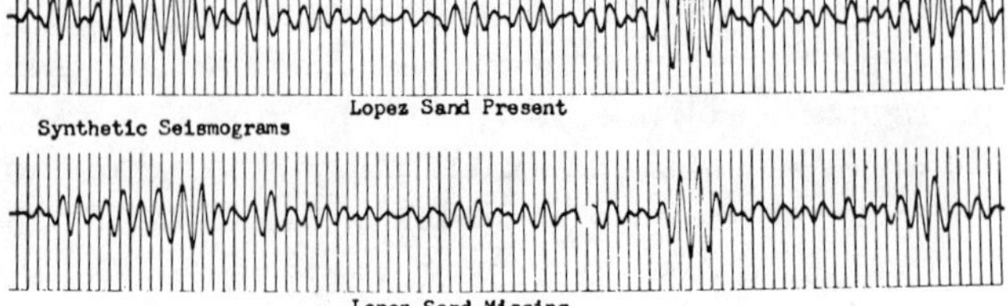

FIGURE 5.10. *Model studies using synthetics. (From Sengbush et al. 1961, courtesy of Geophysics.)*

EXAMPLE FROM BIG HORN BASIN

The Cottonwood Creek field, located in the eastern part of the Big Horn basin, produces from porous Phosphoria (Permian) dolomite at depths ranging from 5,000 to 10,000 ft. A facies change from marine dolomite to red shale and anhydrite in the Upper Phosphoria, designated the E zone, provides the trap for oil accumulation. Structurally, all beds are dipping steeply to the southwest at an average dip rate of 1,000 ft/mi. The Phosphoria has a total thickness ranging from 200 to 350 ft within the field, with the E zone thickness ranging from 40 to 90 ft. The zero porosity isopach is roughly coincident with the 40 ft thickness isopach. A thrust fault cuts across the field from the northwest to the southeast.

The map in Figure 5.11 shows the structural contours in the top of the Phosphoria through the Cottonwood Creek field area and shows the 40 ft Upper Phosphoria isopach that essentially delineates the edge of production. The two seismic lines that were shot in the area cross the facies change to the north and to the east and pass near four wells that had CVLs run in them. The CVLs in Figure 5.12 show that the E zone is bracketed by thin, low-velocity shale zones with two-way time thickness between the velocity minima ranging from 10 to 14 ms. The velocity of the Phosphoria dolomite is about 22,000 ft/sec when it is nonporous and about 18,000 ft/sec when it is porous.

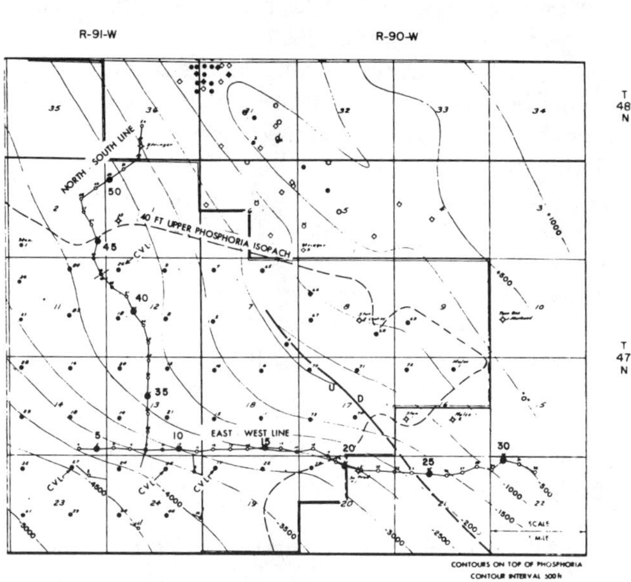

FIGURE 5.11. *Map of the Cottonwood Creek field, showing location of the seismic lines. (From Sengbush 1962, courtesy of Geophysics.)*

FIGURE 5.12. *CVLs in Cottonwood Creek field, showing effect of porosity on the velocity in the E zone. (From Sengbush 1962, courtesy of Geophysics.)*

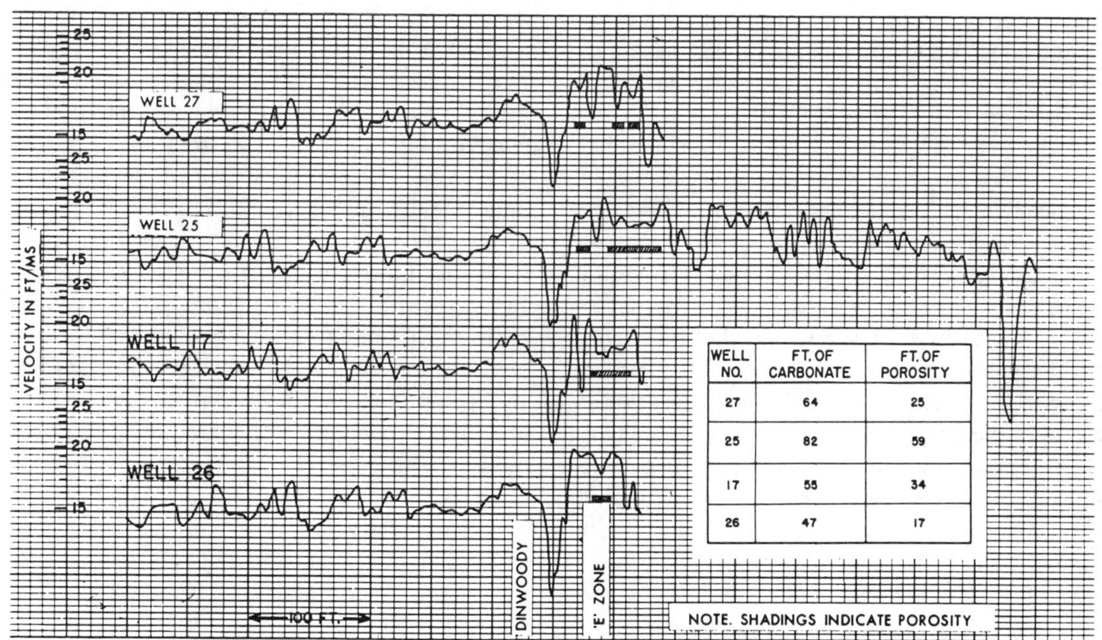

WELL NO.	FT. OF CARBONATE	FT. OF POROSITY
27	64	25
25	82	59
17	55	34
26	47	17

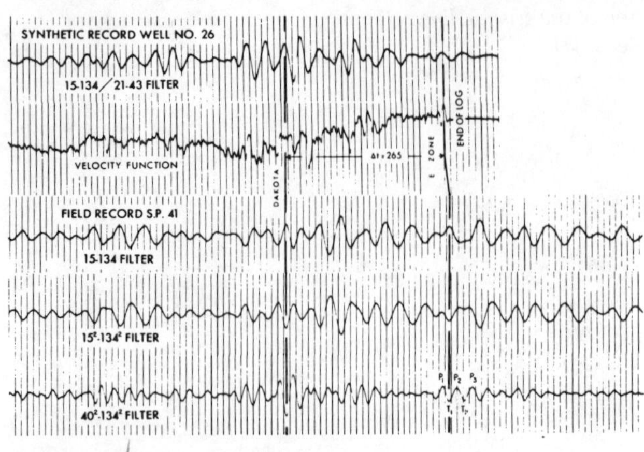

FIGURE 5.13. *Field and synthetic records used to identify reflections in the Big Horn Basin, Wyoming. (From Sengbush 1962, courtesy of Geophysics.)*

The identification of events of the field record was made by comparing synthetics made from the CVLs with field data nearby, as shown, for example, in Figure 5.13. The field data were recorded singlefold on analog magnetic tape and played back through a 15–134 Hz filter. The synthetic seismograms were filtered with two bandpass filters in series, a 15–134 Hz filter to match the response of the playback filter and a second filter that simulates the filtering action of the shot pulse that is effective in producing the reflections. The best comparison between field and synthetic records was obtained using a shot pulse filter of 21–43 Hz. This establishes the spectrum of the effective shot pulse and indicates that its spectrum peaks near 30 Hz. The effective shot pulse controls the character of the record, because the largest reflections arise from velocity zones that are tuned to the shot pulse. Here the pronounced variations in velocity in the Muddy-Dakota-Lakota sequence with basic periods in the range of 25–30 ms are tuned to the shot pulse.

A one-to-one correspondence is established between the high-velocity Dakota sandstone on the velocity function and the peak identified with a *D* on the synthetic and field record. On the velocity function, the high velocity in the E zone is delayed 265 ms behind the middle of the Dakota. This forces identification of the E zone reflection as an event with the same polarity as the D reflection and delayed behind it by about 265 ms.

Filtering the field record with a double-section 15–134 Hz filter has best correspondence of high-velocity Dakota to the indicated trough, with the resulting E zone trough delayed by about 265 ms behind the D trough. The change in correspondence between high-velocity to peak with the single-section filter to high-velocity to trough with the double-section filter is the result of change in impulse response in going from a single to a double section and has nothing to do with the absolute polarity of the trace, which has remained fixed. A double-section 40–134 Hz filter has the same correspondence as the double-section 15–134 filter. The E zone trough marked T_1 has peaks on either side, P_1 and P_2, that result from the velocity minima above and below the high-velocity E zone.

The E zone velocity layering tunes to seismic pulses ranging between 70 and 100 Hz. Because this is considerably higher than the peak frequency in the spectrum of the effective shot pulse, the E zone reflection is weak on the field records. To increase the E zone reflection amplitude, deep point sources were used in an effort to generate high-frequency input shot pulses; on playback, a 40^2–134^2 filter was used. The ideal prospecting method would differentiate between porous and nonporous dolomite, but the porous zones are too thin and the velocity changes due to porosity too small to affect the character of E zone reflections. This was beyond the resolving power of the seismic method in its state of development at the time. After all, these records were taken in 1958, before digital recording, before horizontal stacking, before deconvolution. Isopach mapping is not effective here because the maximum change in porosity from 75 ft to zero produces a change in two-way traveltime of only 1.5 ms. Because of the steep dip and rapid elevation and weathering changes in this area, it would be difficult to achieve the precision necessary to deduce porosity changes by isopach mapping.

The correspondence of high-velocity zones to reflection troughs on the 40^2–134^2 Hz filter establishes the correspondence on the field data. Several records in Figure 5.14 show the troughs identified as *D* and T_1. Troughs correspond to white on the seismic sections in Figure 5.15 (north-south line) and Figure 5.16 (east-west line).

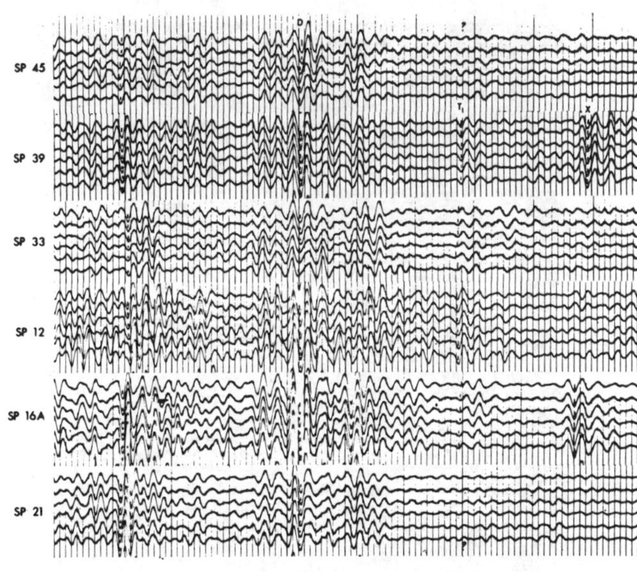

FIGURE 5.14. *Selected field records in Cottonwood Creek field. (From Sengbush 1962, courtesy of Geophysics.)*

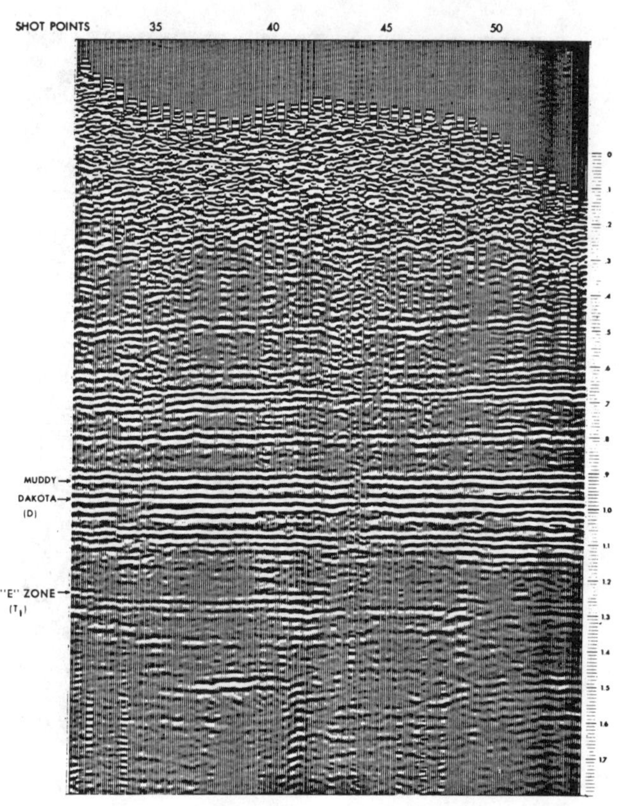

FIGURE 5.15. *North-south seismic line across Cottonwood Creek field, 40^2–134^2 Hz filter. (From Sengbush 1962, courtesy of Geophysics.)*

The data have been time-corrected using the Dakota reflection as datum. On the north-south line, the P_1–T_1–P_2 sequence fades out near SP 43. The reflection point for the Phosphoria at this shot point migrates to the vicinity of well 26 at the edge of the field. The well one-half mile to the north was nonproductive. The event P_3, which is deeper than the E zone, remains strong beyond SP 43. On the east-west line, P_1 is weak compared to P_2 and P_3. The entire sequence T_1–P_2–T_2–P_3 fades out near SP 20, located between a producer and a dry hole at the eastern edge of the field. The subsurface reflection point for the Phosphoria at SP 20 migrates updip to the vicinity of well 74, an edge well of the field. The thrust fault provides a complication on this line. The E zone reflection migrates to the fault face in the vicinity of SP 21 and SP 22, and the event P_3, which is deeper than the E zone, may fade out at SP 20 because of the presence of the fault.

The seismic sections clearly show the disappearance of the E zone reflection at shot points that are remarkably close to the edge of production. The loss of amplitude is too abrupt to be explained by thinning alone and probably results from a combination of thinning and facies change, with an assist from the fault on the east-west line. It is impossible to locate the high-frequency E zone reflection on the broadband cross section shown in Figure 5.17 because the E zone reflection is obscured by the preponderance of low-frequency reflections. This demonstrates the necessity of using filters that meet the interpretive objective.

In the conclusion of my paper (Sengbush, 1962) I state: This study, conducted in the summer of 1958 by the Socony Mobil field research laboratory, illustrates that seismic method can be used effectively to reveal stratigraphy through proper interpretation of amplitude and character changes on field records. Of

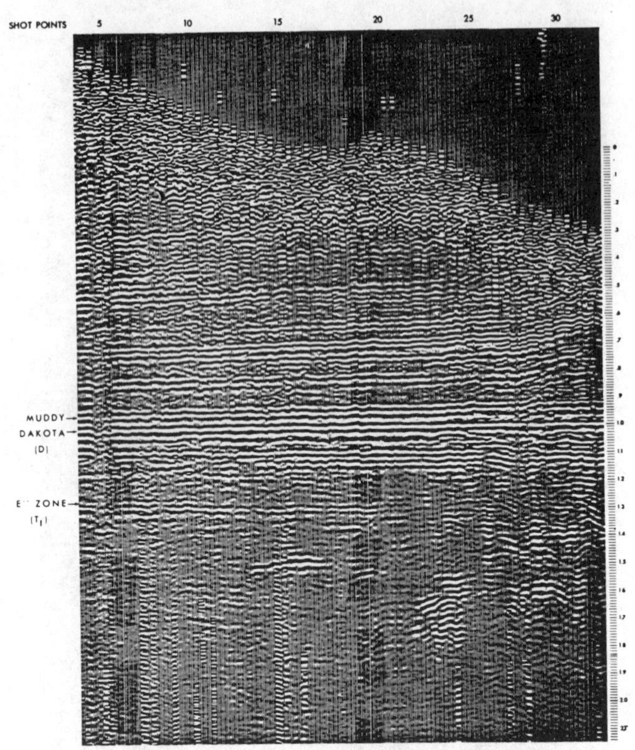

FIGURE 5.16. *East-west seismic line across Cottonwood Creek field, 40^2–134^2 Hz filter. (From Sengbush 1962, courtesy of Geophysics.)*

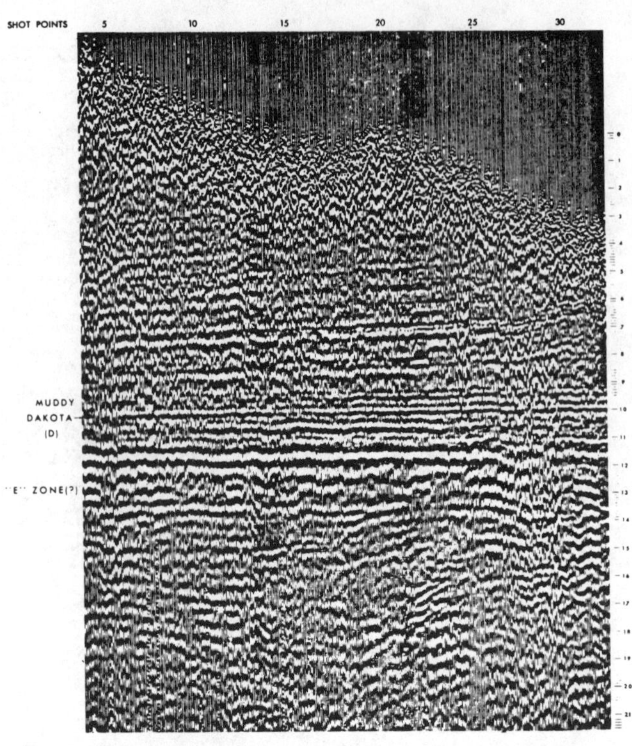

FIGURE 5.17. *East-west seismic line across Cottonwood Creek field, 13–134 Hz filter. (From Sengbush 1962, courtesy of Geophysics.)*

course, there are limitations to the seismic method, but its capabilities have been increased by modern techniques of magnetic tape recording and processing in conjunction with the filter theory approach to interpretation that was stimulated by the introduction of synthetic seismograms.

SUMMARY

Describing the seismic data-generating process as a communication system has brought an increased understanding of seismic data and has led to many advances in processing, the most significant being deconvolution. This model is limited because it is one-dimensional and does not allow for diffractions and other phenomena that can only be explained by use of the wave equation. Nevertheless, the communication theory model has been one of the major milestones in the history of seismic prospecting.

Chapter 6 Multiple Reflections

INTRODUCTION

Multiple reflections are seismic events that have undergone more than one reflection between source and detector. As a seismic wave travels through the earth, it is split into a reflected and a transmitted wave at each acoustic discontinuity. Each of these waves, upon encountering another discontinuity, splits into a pair of waves, and so on ad infinitum. Considering the complexity of multiple generation, it may seem remarkable that the seismic reflection method based on primary reflections works at all. The explanation seems to be that, although multiples exist whenever there are primaries, their amplitudes relative to primaries must be small. When this is not the case, multiple reflections are a serious problem. Most poor records probably result from severe multiple interference.

Multiple reflections have been recognized as a source of seismic noise since the beginnings of seismic prospecting. *Geophysics*, Volume 13, Number 1, 1948, was devoted to current knowledge of multiples, which consisted primarily of identification. Reproducible recording, which must necessarily precede techniques for multiple suppression, and the communication theory model of the seismic method, which gives deeper insight into multiple generation, were both yet to come.

GHOSTS AND REVERBERATIONS

In the early days of land exploration using dynamite in boreholes, one of the major problems was a phenomenon called shot-point variable. Two factors were involved.

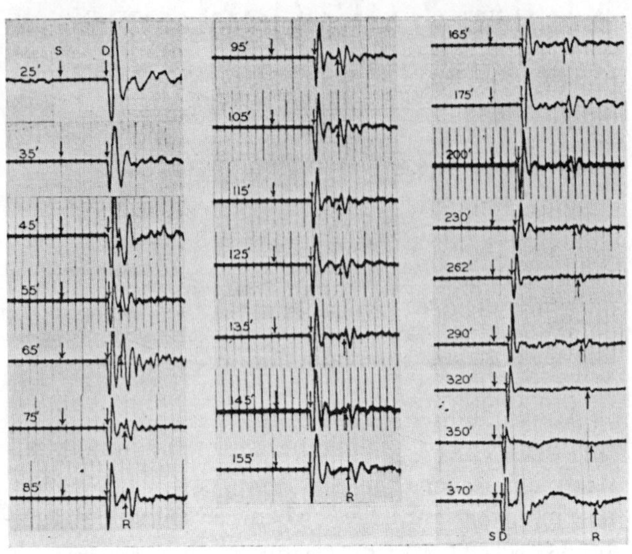

FIGURE 6.1. *Direct and ghost signals observed below shots. (From Van Melle and Weatherburn 1953, courtesy of Geophysics.)*

One was the effect of shooting in different formations, because the input pulse waveform was affected by the type of rock in which the charge was detonated. The second and more significant factor was the remarkable difference in the output seismic data caused by differences in shot depth. Van Melle and Weatherburn (1953) showed that the latter factor resulted from ghosts, that is, energy that traveled upward initially and then was reflected downward at the base of the weathering and other acoustic discontinuities that lay above the source. Figure 6.1, from Van Melle and Weatherburn (1953), shows that the input signal measured below the shot has both a direct and a ghost arrival. The ghost labeled *R* in this example is from the base of the weathering, as is the most often observed situation. Figure 6.2, from Musgrave, Ehlert, and Nash (1958), shows the changes in reflection character with shot depth in a West Texas example, where ghosts of the Devonian (marked *D*) interfere with the Ordovician reflections (marked *O*).

In the days before reproducible recording, ghosts were minimized by shooting just below the ghosting inter-

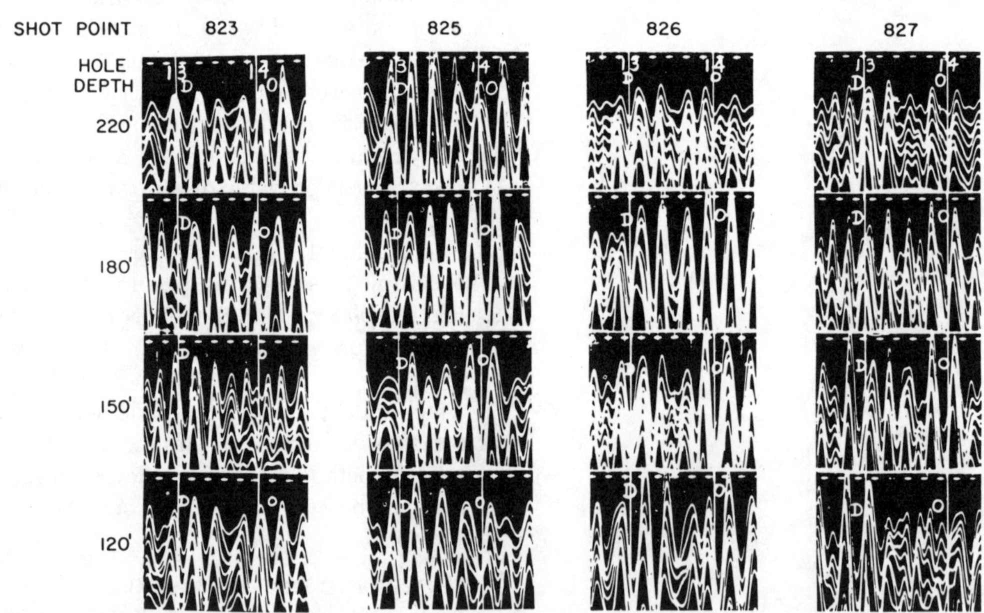

FIGURE 6.2. *Shot-point variable: interference of Devonian ghosts with Ordovician reflections, West Texas. (From Musgrave, Ehlert, and Nash 1958, courtesy of Geophysics.)*

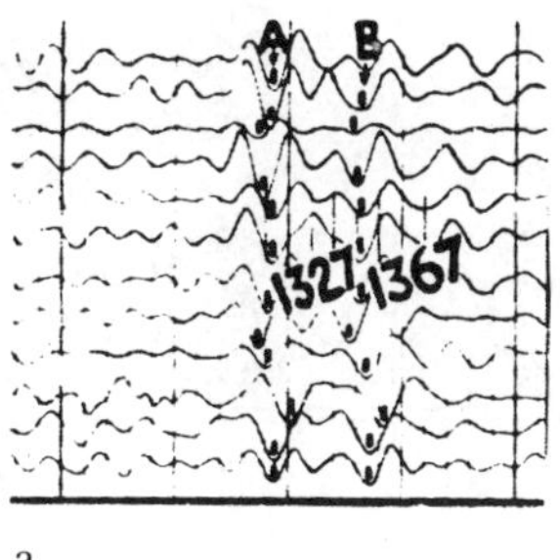

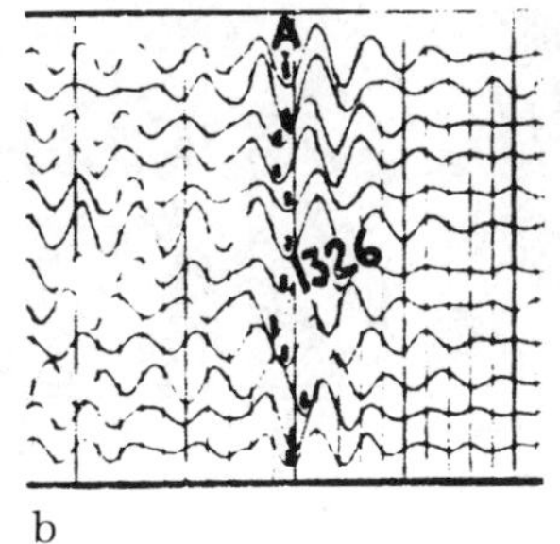

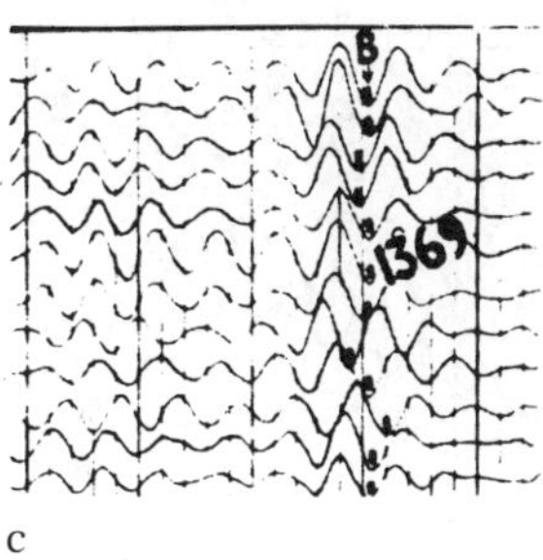

FIGURE 6.3. *Comparison of point and distributed sources. (a) Point source. (b) Distributed source detonated downward. (c) Distributed source detonated upward. (From Martner and Silverman 1962, courtesy of Geophysics.)*

face, usually at the base of the weathering, or they were suppressed by directional sources, such as column charges (Musgrave, Ehlert, and Nash, 1958) and broomstick charges (Martner and Silverman, 1962). The idea here is to have a long charge, on the order of 100 ft, whose detonation velocity matches the formation velocity. Then, by detonating the charge at the top, the downgoing signal is continuously enhanced and the upgoing energy is spread out in time and thus has reduced peak amplitude.

Martner and Silverman (1962) show the comparison of point and distributed sources on field data (Figure 6.3). The point source created strong reflections at times of 1.327 and 1.367 sec. The latter reflection is shown to be a ghost because it disappears when a distributed source is detonated downward. When a distributed source is detonated upward, the ghost at 1.367 sec is preserved and the primary at 1.327 sec is suppressed. In a field survey using point sources, a map on the second reflection would make geologic nonsense.

Directional sources are effective techniques for suppressing ghosts at the source, but they present many operational and economic problems. With the digital revolution, the problem of ghosts was solved almost completely by Wiener deconvolution.

The major problem in marine exploration is the reverberations caused by multiple reflections in the water layer. The almost perfect reflection at the water surface and the often strong reflection at the base of the water, coupled with low attenuation in the water layer, make multiple reflections within the water layer a severe problem. In water depths less than a few hundred feet, the seismic data are highly oscillatory, as shown in the example in Figure 6.4, from Lake Maracaibo (Levin, 1962).

The problem of ringing records in marine exploration was almost unsurmountable until Wiener deconvolution arrived on the scene, although the three-point operator had limited success in some areas (Backus, 1959). In deep water, long-period multiples are produced by the water layer. These are more difficult to suppress because Wiener deconvolution doesn't work in such situations. A variety of schemes for suppressing long-period multiples have been tried; super-long arrays, *f-k* processing, and velocity filters show the most promise.

MULTIPLE GENERATION WITHIN THE SEISMIC SECTION

Multiples within the seismic section have a wide range of coherence. Those easily identified on field data arise from multiple bounces between interfaces that have large reflection coefficients. Their raypaths can be deduced from the observed primary arrival times. A multiple that has twice the arrival time of a primary is called a *W*-type multiple. One whose arrival time is the sum of two different primary arrival times is called a peg-leg multiple.

Multiples such as the peg-leg and *W*-types that have three bounces in the path are called first-order multiples. Second-order multiples have five bounces; third-order, seven bounces; and so on. Because reflection coefficients have magnitude less than one, the multiple amplitudes tend to decrease as the order increases. Given an upper

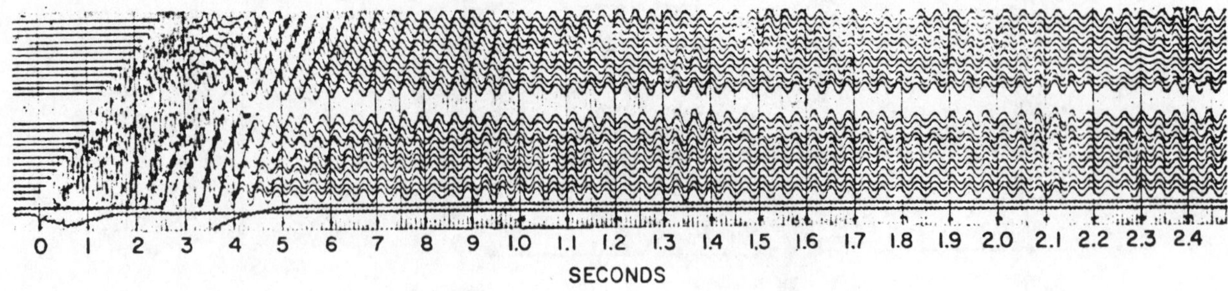

FIGURE 6.4. *Reverberated seismogram recorded in Lake Maracaibo. (From Levin 1962, courtesy of Geophysics.)*

bound on subsurface reflection coefficients of 0.3 and a surface coefficient near unity, it is apparent that first-order multiples have an upper bound of $(0.3)^3 = 0.027$ if all bounces are internal (all within the subsurface) and $(0.3)^2 = 0.09$ if the surface is the middle bounce. The latter type, called the first-order surface multiples (FOSM), is the most significant subset of first-order multiples. With second-order multiples, the upper bounds are $(0.3)^3 = 0.027$ for surface multiples and $(0.3)^5 = 0.00243$ for internal multiples. Higher-order multiples have still smaller upper bounds. It is apparent from considering the upper bounds that only first-order surface multiples, and possibly first-order internal and second-order surface multiples, can compete with primary reflections in regard to amplitude, and they will be the only ones recognizable as coherent events, except in special cases, such as in deep water when there is a large reflection coefficient at the water bottom or in areas where there is a strong shallow reflection. Then higher order surface multiples can be identified as coherent events. All other multiples besides FOSMs merely contribute to the underlying random noise in most cases. Internal multiples tend to smear the primary impedance function and obscure thin beds by increasing the random noise level and may create leggy reflections because of internal reverberations.

In Butte County, in the northern Sacramento Valley of California, a shallow basalt flow at 0.5 sec record time and a strong velocity contrast at the base of weathering create conditions conducive to the generation of surface multiples of high order. Johnson (1948) shows examples in which up to seven orders of multiples can be positively identified on the field data here through use of arrival times, increasing dip with increasing multiple order, and waveform correlation between primaries and multiples that shows alternating polarity with increasing multiple order because the base of the weathering interface has a negative reflection coefficient for upgoing waves (Figure 6.5). The base of the weathering has a large reflection coefficient, 0.67, and the basalt interface has coefficient 0.35, whereas other coefficients in the section are very small; therefore, the higher order multiples can override the primaries below the primary reflection from the basalt.

In considering the relative size of FOSMs compared to primaries, differential attenuation effects also should be included. In the normal case, where attenuation decreases with depth because rocks are increasingly more elastic, FOSMs will be reduced in amplitude more than primaries at a given reflection time, because FOSMs are making two round trips through the more highly attenuating near-surface layers, while primaries are only making one round trip. When the primary layering has large impedance changes at shallow depth because limestones are interspersed with shales, for example, the multiples tend to be severe because of both large reflectivity and low attenuation in the near surface. The primary-to-FOSM amplitude ratio will not be affected appreciably by other factors, such as spherical spreading and transmission losses, if any.

To derive the FOSM reflectivity function, consider the FOSM generated by the first interface R_1. Upon reflection from the surface, the primary from R_1 becomes a second-

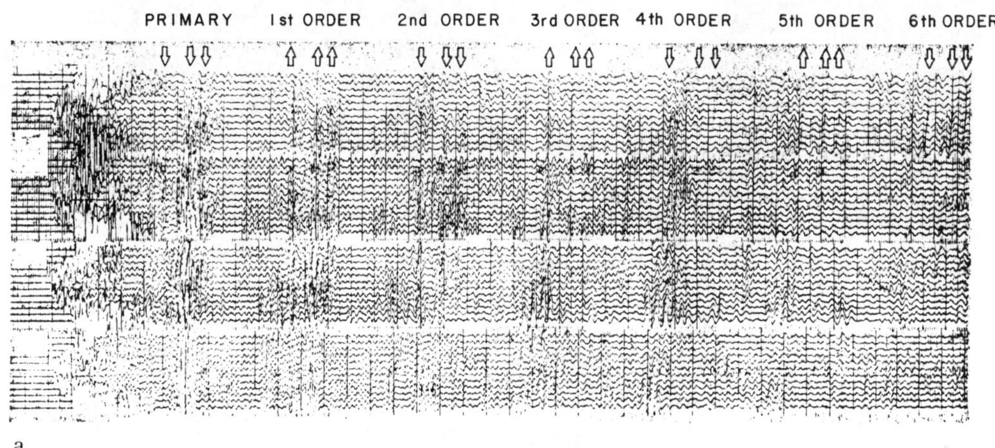

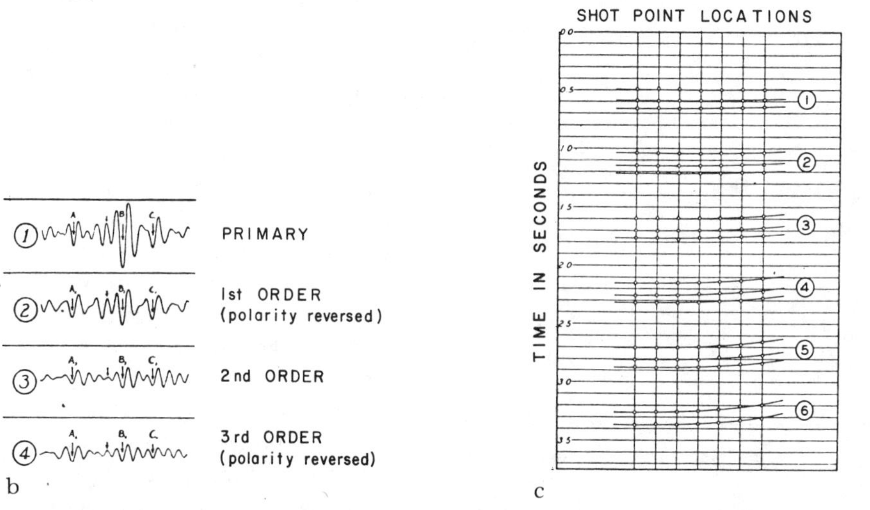

FIGURE 6.5. *Multiple reflections in the Sacramento Valley of California. (a) Records showing three separate reflections multiply reflected as a group. (b) Waveform correlation. (c) Time section of events A, B, and C. (From Johnson 1948, courtesy of Geophysics.)*

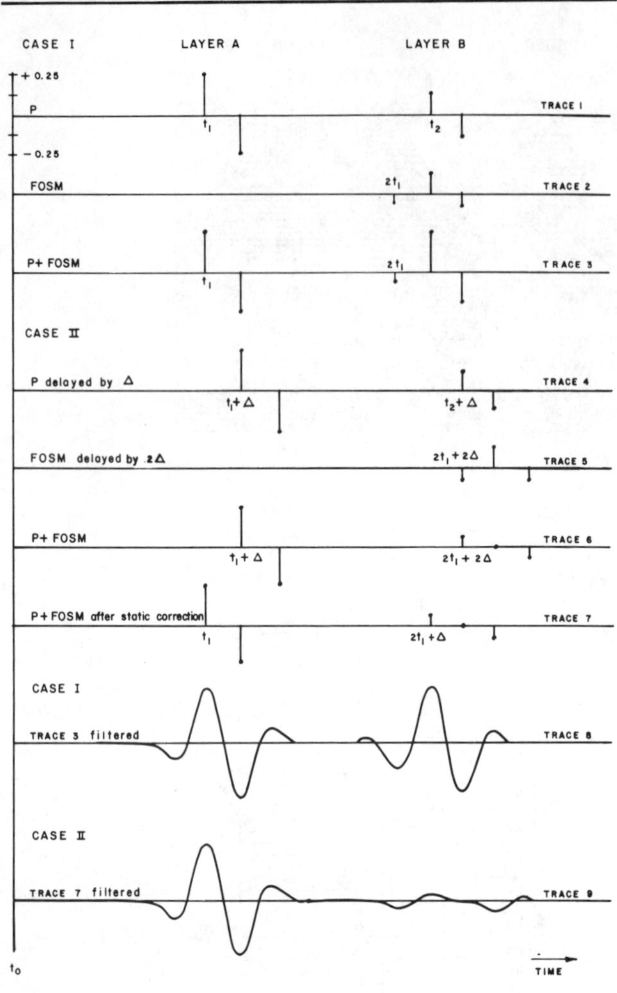

FIGURE 6.6. *Model study showing FOSM interference with primaries.*

ary source $-R_0R_1\delta(t - t_1)$, where $-R_0$ is the surface reflection coefficient as seen from below and t_1 is the two-way time of the primary from the first interface. This secondary impulse enters the subsurface and produces the FOSM set generated by R_1 by convolution:

$$-R_0R_1\delta(t - t_1) * r(t) = -R_0R_1r(t - t_1). \tag{6.1}$$

Likewise, each primary reflection $R_k\delta(t - t_k)$ acts like a delayed source upon reflection from the surface and produces the FOSM set generated by R_k. The totality of FOSMs generated by all primaries is the FOSM reflectivity function:

$$\text{FOSM}(t) = -R_0 \sum_k R_k r(t - t_k). \tag{6.2}$$

In the continuous case,

$$\text{FOSM}(t) = -R_0 r(t) * r(t). \tag{6.3}$$

In areas where FOSMs are significant, they play an important role in modifying the reflection amplitudes of the primary reflections they overlay. Consider the model in Figure 6.6, in which there are two tuned layers, A with top at t_1 and B with top at t_2, where t_2 is approximately equal to $2t_1$. This means that the FOSMs from layer A will superimpose on the primary from layer B. The reflectivity of A is a doublet with reflection coefficients of ¼ and −¼, and that of B is a doublet with coefficients ⅛ and −⅛. Two cases are shown. In Case I, the FOSM interferes most constructively with the B primary; in Case II, the interference is most destructive. The primary reflectivity in Case II is delayed by Δ compared to Case I, and this causes the FOSMs to be delayed by 2Δ. The result is that the maximum peak-to-trough amplitude of the composite of the B primary and the A FOSMs is 0.91 as large as the amplitude of the A primary in Case I, and 0.13 in Case II. When there is no FOSM interference, the B primary is 0.5 as large as the A primary. Thus, a differential shift between primary and FOSM reflectivity, arising because of differences in weathering, elevation, and dip between locations, can cause drastic changes in amplitude of the interfered-with primary. These changes may be erroneously ascribed to changes in bed thickness or in reflectivity, which also cause the reflection amplitude to change.

An example from southern Mississippi shows primaries, primaries plus first-order surface multiples, and primaries plus all multiples (Figure 6.7). The FOSM reflectivity function produced by autoconvolution of the primary reflectivity is added to the primary reflectivity to produce primaries plus first-order surface multiples. The

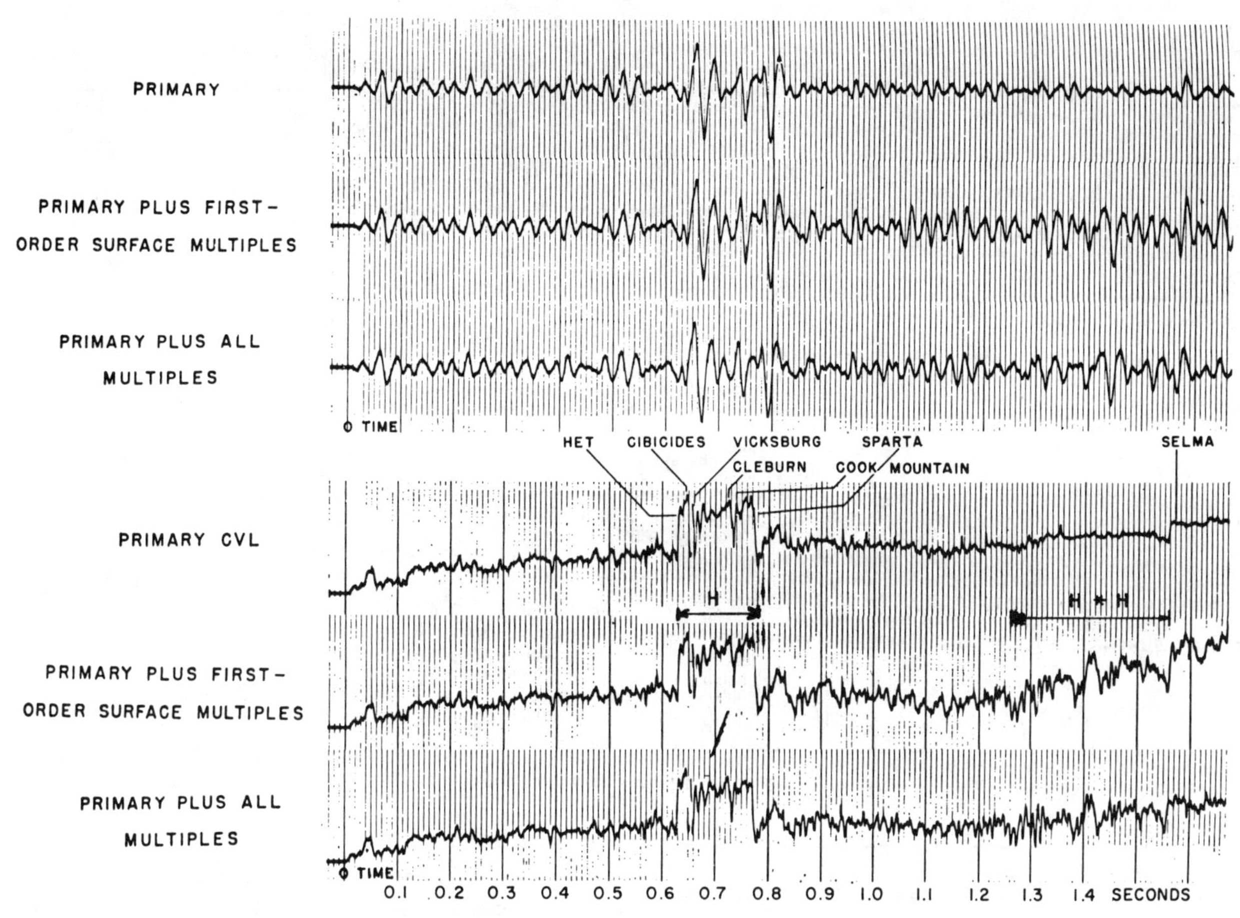

FIGURE 6.7. *Primary and multiple synthetic seismograms from southern Mississippi. (From Watson 1965, courtesy of Geophysics.)*

reflectivity functions are displayed as impedance functions, which are obtained by integrating the reflectivity functions. The layering is more easily comprehended by displaying impedance rather than reflectivity. The first significant series of reflections comes from the zone labeled H, a zone 150 ms thick that begins at the top of the Het limestone at 0.630 sec. This zone includes the Het, the Cibicides, the Vicksburg, the Cleburn, and the Cook Mountain. FOSMs from the H zone are spread out over a zone marked H*H, of length 300 ms, beginning with the *W*-multiple from the top of the Het limestone at 1.260 sec. Despite the large impedance change at the top of the Het, its *W*-multiple is small compared to the pegleg multiples generated within the H zone. The largest multiple contributions in the H*H zone occur in the middle of the zone, where many significant peglegs obtained in the autoconvolution process are superimposed. So, between the top of the Sparta and the top of the Selma, the most significant reflections are first-order surface multiples, the largest event being the multiple with a major trough at 1.445 sec. The base of the H*H zone has about the same arrival time as the top of the Selma. Downdip, as the H zone deepens, the H*H zone will spread out across the Selma chalk interface and will cause consider-

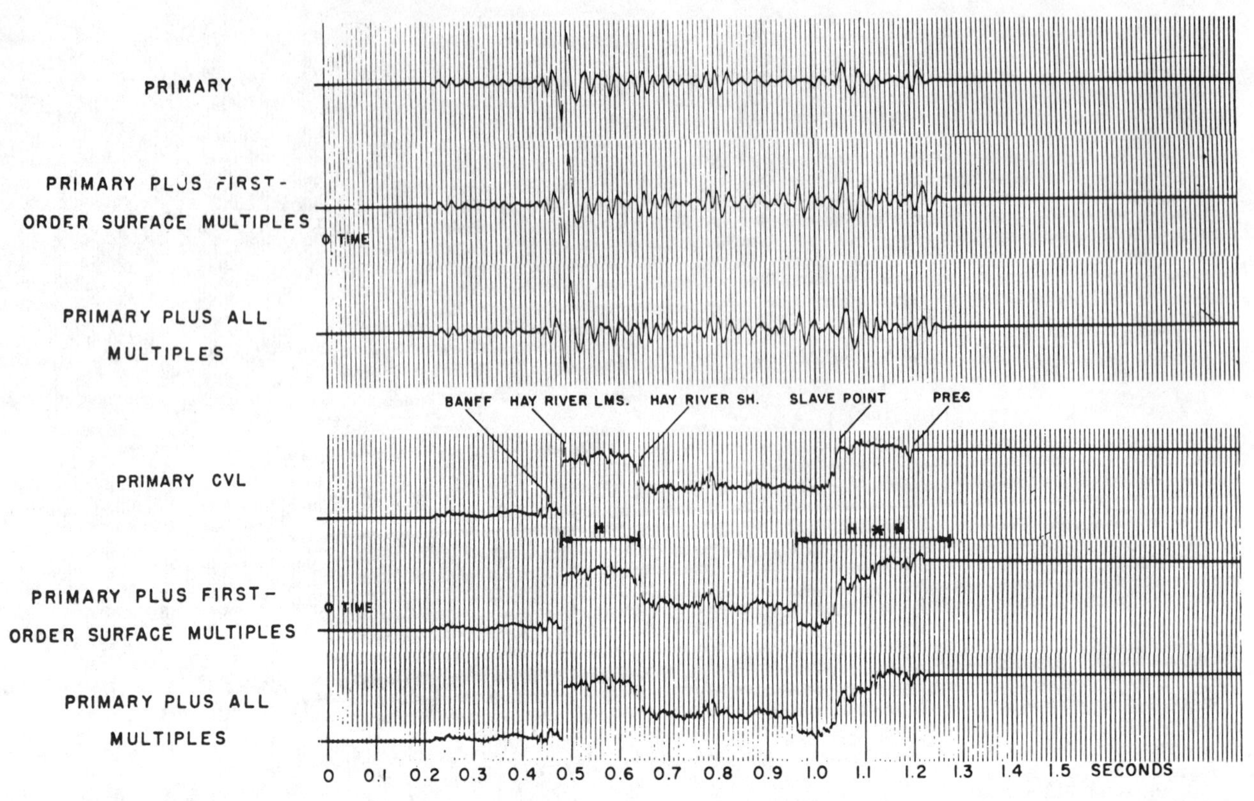

FIGURE 6.8. *Synthetic seismograms, including multiples, Alberta, Canada. (From Watson 1965, courtesy of Geophysics.)*

able interference with the Selma reflection and make it difficult to map. Two problems exist here:

(1) The proliferation of FOSM reflections below the base of the Cook Mountain; and
(2) The interference of the FOSMs on legitimate reflections, such as the Selma chalk.

Further addition of the rest of the multiples makes little contribution to the overall picture. The main difference is the appearance of additive random noise in the impedance function. The synthetic seismogram of primary plus all multiples is very similar to the synthetic from primaries plus FOSMs. Second-order surface multiples from the H zone will not begin to appear until 1.890 sec, which is beyond the length of the record shown here.

In Alberta, Canada, the Hay River limestone is a notorious multiple generator. In Figure 6.8, the top of the limestone has a step-discontinuity in velocity at 0.480 sec that produces a very strong primary reflection and is the principal source of multiples. The *W*-multiple from the top of the Hay River is clearly evident at 0.960 sec. A map on this multiple will give a good inverse map on the surface elevation. The base of the Hay River at about 0.640 sec is a ramp transition zone whose reflection amplitude is only about one-quarter the size of the reflection from the step. The H*H zone extends from 0.960 to 1.280 sec, which includes and overlaps the Slave Point. The H*H zone does not significantly modify the impedance through the Slave Point because the uniformity of velocity in the Hay River limestone means that peglegs

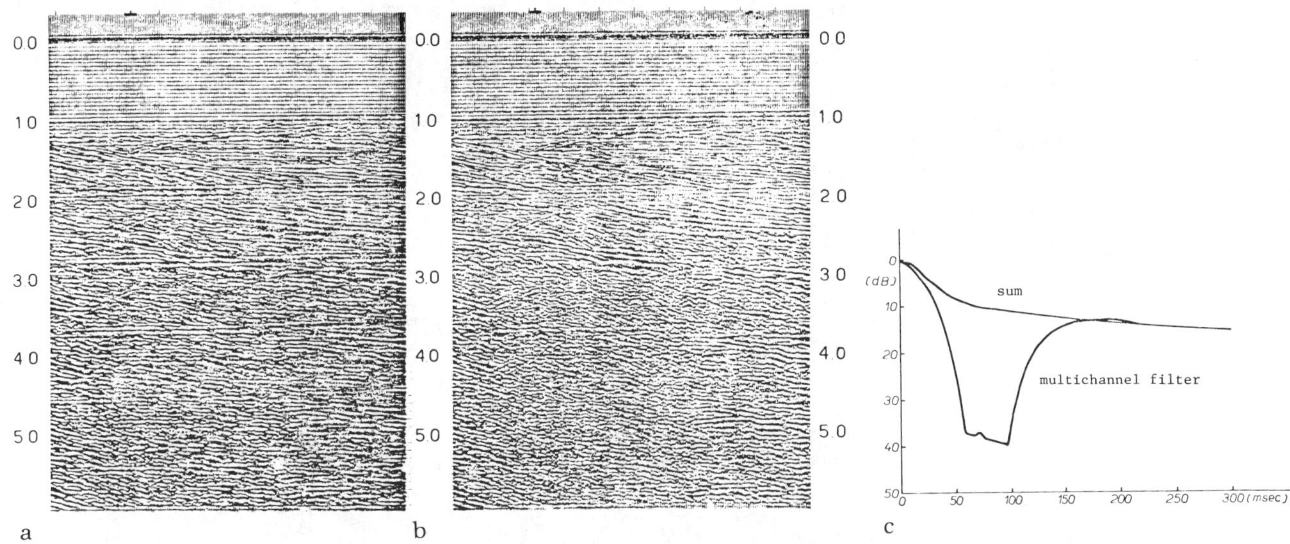

FIGURE 6.9. *Optimal multichannel filter for multiple suppression. (a) 48-fold stacked data. (b) Use of multichannel filter in place of stacking. (c) Typical multichannel filter response to moveouts in 50–100 ms rejection band. (From Cassano and Rocca 1973, courtesy of Geophysics.)*

generated by the Hay River are not significant. The internal multiple between the top and bottom of the Hay River is weak because of the ramp transition zone at the base and is not detectable on the synthetic. The addition of all of the other multiples has virtually no effect on the synthetic except for the appearance of random noise. When the Hay River is somewhat deeper, the *W*-multiple from the top of the Hay River will interfere with the Slave Point reflection.

MULTIPLE SUPPRESSION

HORIZONTAL STACKING

Horizontal stacking (Mayne, 1962) is the only multiple suppression technique with wide usage. Its effectiveness depends on the differential movement between primary and multiple reflections that occurs in the normal case, where velocity increases with depth. Then, a primary event will see a larger average velocity than the multiple events at any given record time. Hence, the primaries have less moveout than the multiples. Dynamic time shifting of the primaries to line them up for all record time at various source-detector separations will force the multiples to lie along their differential moveout curves. Summing the set of traces will then augment primary reflections because they are in phase and will suppress multiples because they are not in phase.

Simple stacking (summing) is not the most effective use of the potential power of the pattern response of an n-fold stack. The pattern response has nonuniform spacing because the differential moveout curve is hyperbolic. Simple stacking applies uniform weighting; hence, the pattern is overweighted where the spacing is small. To make the pattern have uniform sensitivity, the weight should be divided by the spacing. This simple procedure will produce an improved stack.

Application of velocity filters to the corrected gathers is superior to stacking, as shown by Cassano and Rocca (1973), who improved the suppression of long-period water bottom multiples with velocity filters (Figure 6.9).

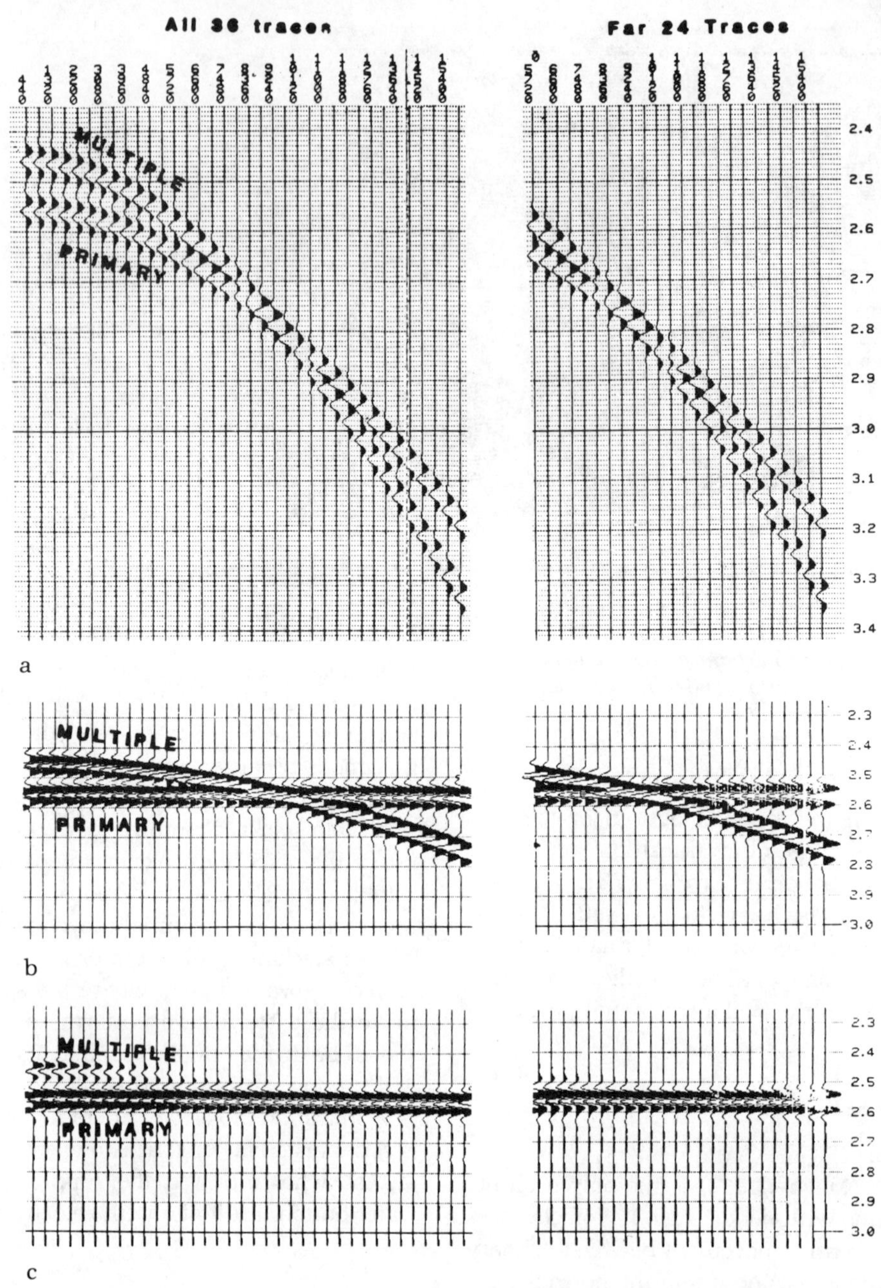

FIGURE 6.10. *Model of seismic gathered traces with a primary reflection and an overlapping multiple. (a) Before NMO correction. (b) After primary NMO correction and before* f-k *filter. (c) After* f-k *filter.*

F-K PROCESSING

Because primaries and overlapping multiples have different moveouts, it is possible to discriminate selectively against multiples by using *f-k* processing of the CDP-gathers, using a processing procedure developed by Charles R. Saxon and the author in 1977.

The model in Figure 6.10 shows the power of *f-k* filtering in suppressing multiples while leaving primaries unaffected. Figure 6.10(a) shows a gather of 36 traces at 400 ft spacing that has a primary reflection and an overlapping multiple. Figure 6.10(b) shows the input data after NMO correction to line up primaries before *f-k* filtering, and Figure 6.10(c) shows the result after *f-k* filtering. It is readily apparent that stacking of the primary moveout corrected traces to produce the one trace associated with the gather will show a very significant primary-to-multiple improvement after *f-k* filtering compared to ordinary stacking—that is, without *f-k* filtering. After *f-k* filtering, only the near traces show any significant multiple amplitude, so it is advantageous to increase the offset to further improve the primary-to-multiple ratio on the stacked trace, as can be seen by considering using only the far 24 traces, with near trace at 5720 feet, as shown on the right-hand side of Figure 6.10.

The *f-k* spectrum of the primary after NMO in Figure 6.11(a) shows that its spectral content is clustered around the frequency axis because the primary has zero moveout (infinite-phase velocity). The spectrum of the multiple blends into that of the primary when all traces are used, because the multiple-phase velocity ranges from infinity on the near traces to 28.9 ft/ms on the far traces, as shown in Figure 6.11(b). When the first 12 traces are deleted from the gather, the multiple and primary spectral components show a distinct separation, as shown in Figure 6.11(c).

The *f-k* filtering is accomplished by zeroing out the part of *f-k* space in which the multiples reside. Then, upon transformation back into t-x space, the multiples are suppressed on the filtered gathers, and they are still further suppressed by subsequent stacking of the gathers.

In practice, *f-k* filtering uses some variation of the above. When the primaries are overridden by multiples, it is best to moveout-correct on multiples and then zero out the entire right-half plane, or a little more, to suppress the multiples. The art in this technique is knowing how to optimize multiple suppression by manipulating the NMO corrections and selecting the region in *f-k* space that should be zeroed out.

Figure 6.12 shows the velocity spectra before and after *f-k* filtering on a field example from Louisiana. Below the strong Het limestone reflection at 1.75 seconds, the velocity spectrum before *f-k* filtering shows that its largest amplitudes lie to the left of the 8.0 ft/ms velocity line at about the stacking velocity of the Het (7.0 ft/ms). This is strong evidence that almost all significant seismic events below the Het are multiples generated by Het. After *f-k* filtering, the multiples below the Het are severely suppressed, and the largest amplitudes on the velocity spectrum are primaries.

SUPER-LONG SOURCE ARRAYS

Long-period multiples on marine data have been one of the major problems on marine data. These multiples are generated by multiple bounces in the water layer and have periodicities that are too long for Wiener deconvolution to be effective. The water bottom multiples are obvious on the record, but what is not obvious is that each multiple generates its own weighted and delayed seismogram, because each is a weighted and delayed seismic pulse. One technique for suppressing long-period multiples is through use of pattern theory via a super-long source array and trace mix. This technique works because the apparent wavelengths of multiples are less than those of primaries throughout all $T_x - x$ (moveout) space, and the pattern responses can be adjusted to suppress multiples without appreciably affecting primaries throughout a large portion of that space (Lofthouse and Bennett, 1978).

An example from the English Channel, in Figure 6.13, shows the comparison between a single air gun array and Shell's super-long array that has five arrays of five guns each spread out over a length of 350 m. The long-period multiples generated with the single array seriously obscure the underlying structure. The extended source suppresses the long-period multiples effectively and allows the underlying structure to become visible.

The design of the system to suppress long-period water bottom multiples without appreciably affecting primaries consists of the two steps that always enter into

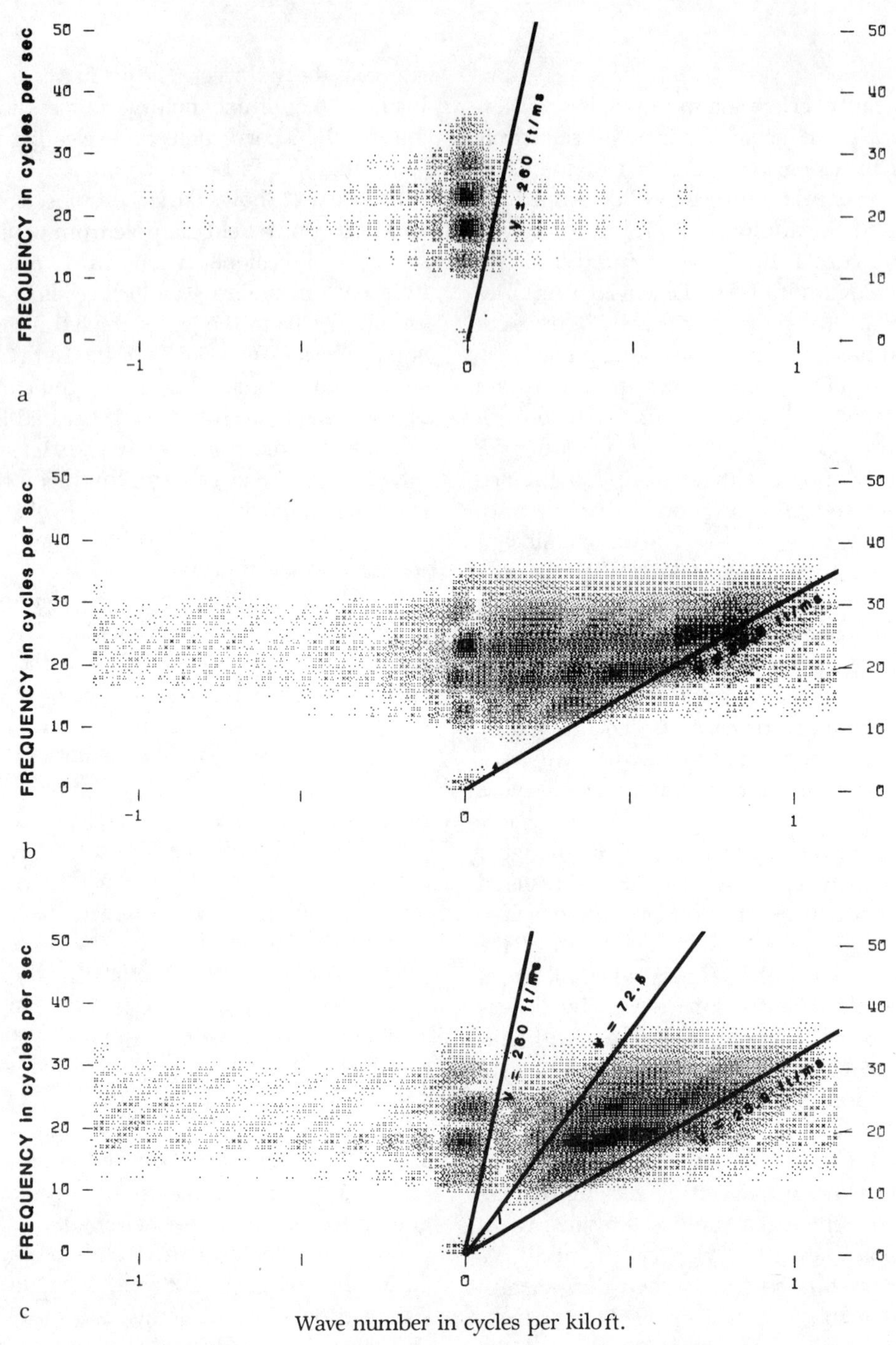

FIGURE 6.11. f-k *transform of seismic model after application of primary NMO function. (a) Primary reflection only. (b) Primary and multiple, all 36 traces. (c) Primary and multiple, far 24 traces.*

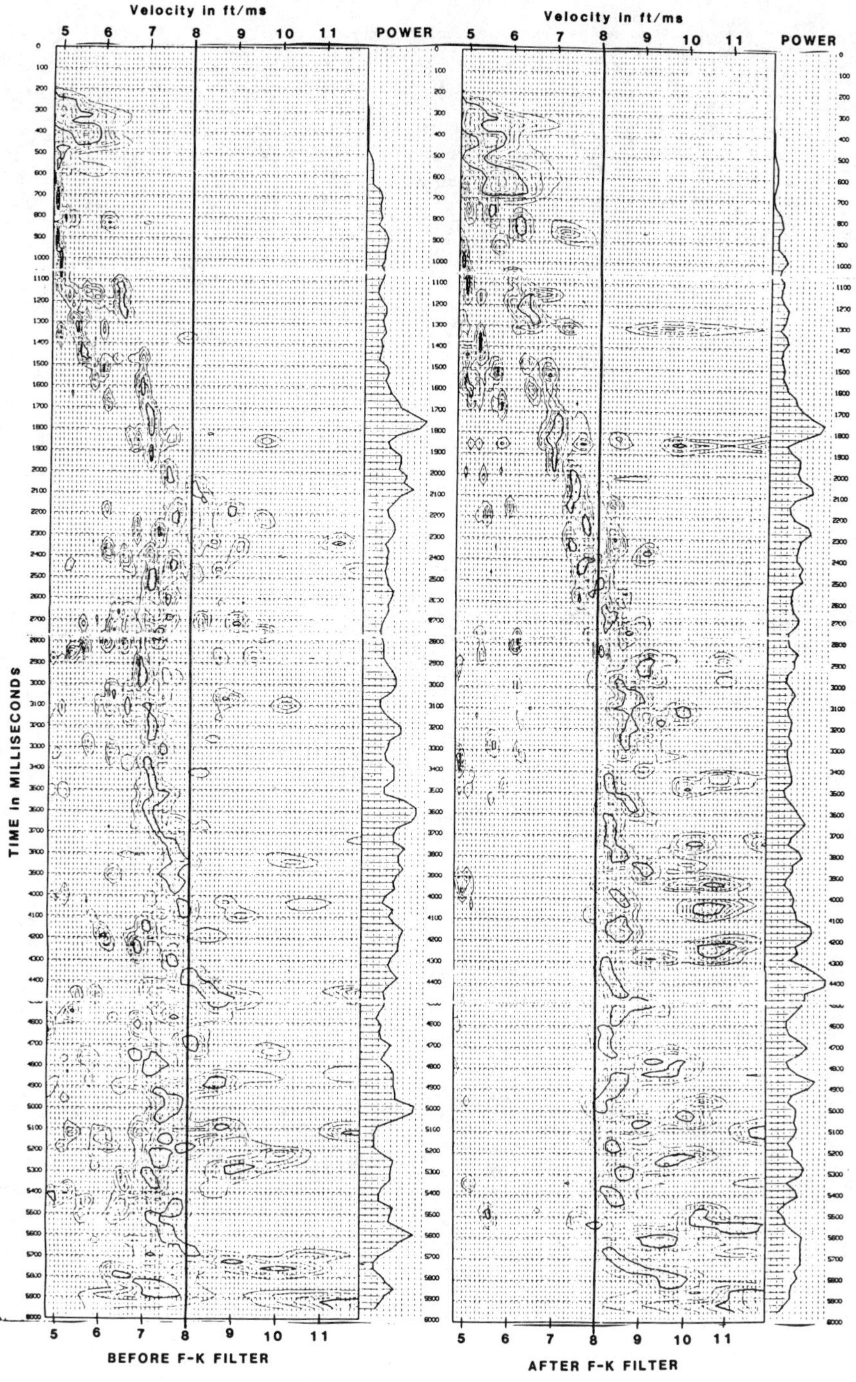

FIGURE 6.12. *Velocity spectrum before and after* f-k *processing.*

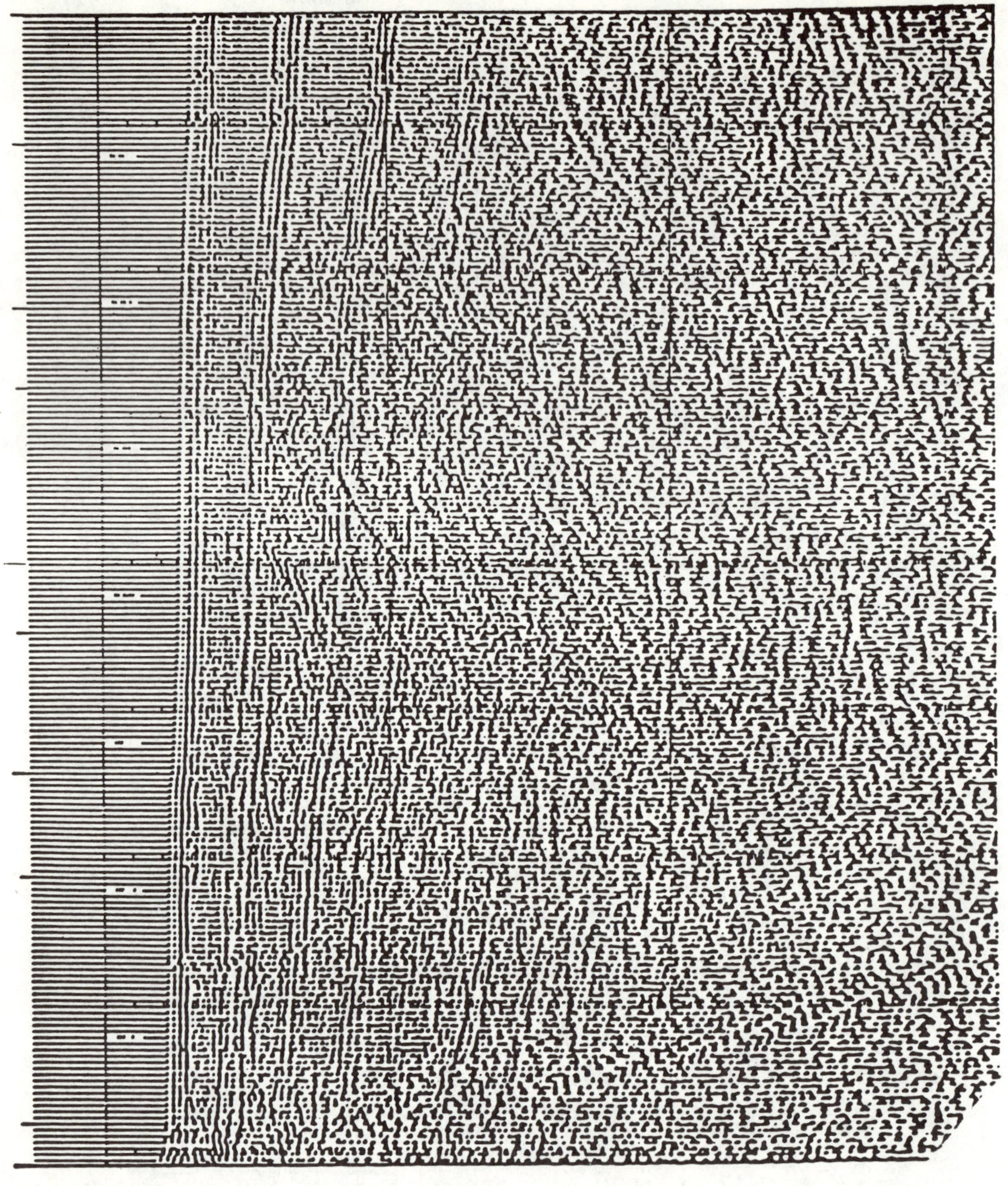

FIGURE 6.13(a). *Seismic section from English Channel produced by a single five-gun array, showing several orders of long-period multiples.*

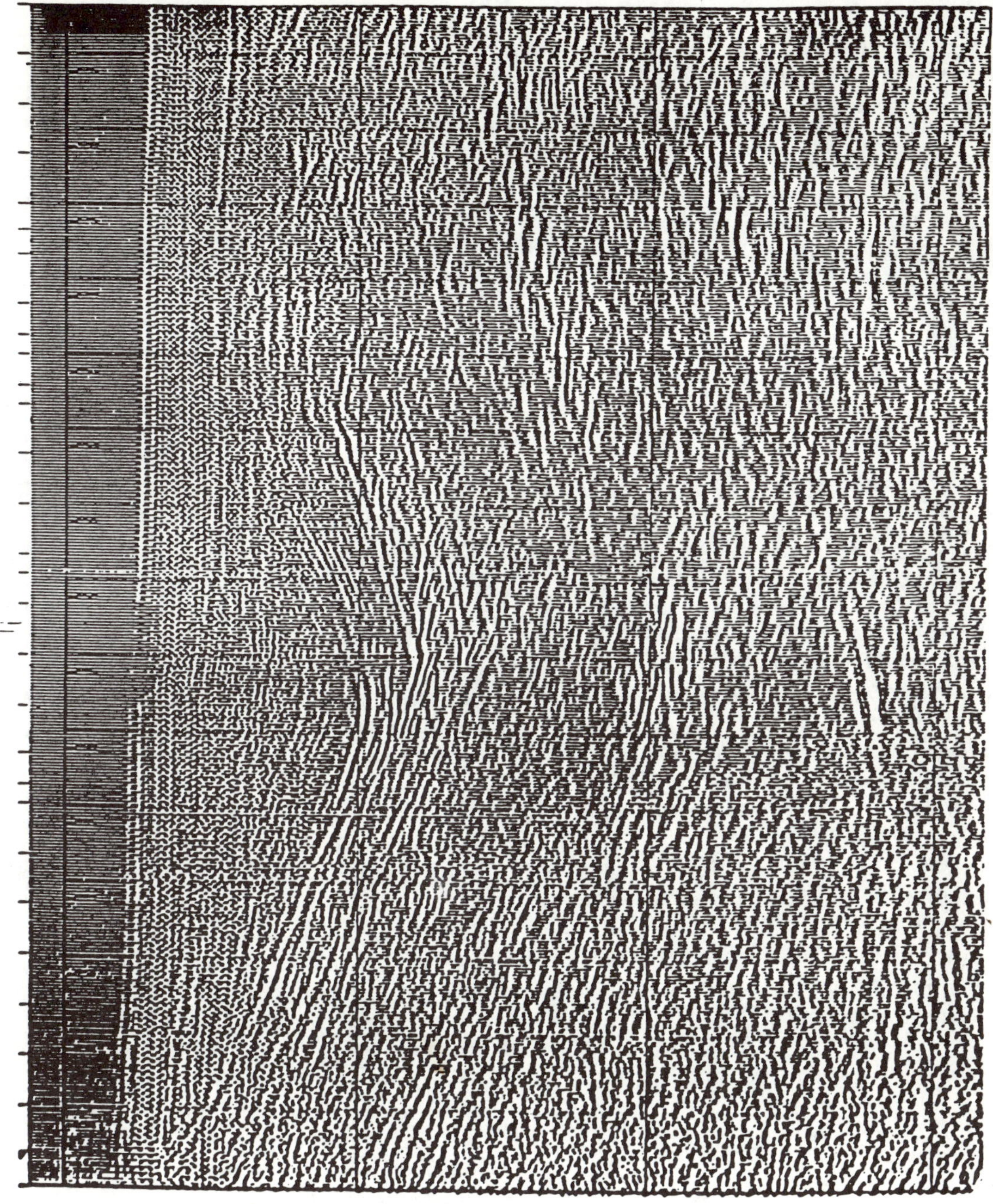

FIGURE 6.13(b). *Comparison section to (a) produced by Shell's super-long array consisting of five arrays, each with five air guns, extending over 350 m. The long-period multiples have been suppressed by the source pattern.*

pattern design: (1) determining the wavelengths of the signal and noise and (2) designing a pattern response that will suppress the noise without appreciably affecting the signal.

The slope of an NMO curve at offset x and time T_x is dT_x/dx, which is the reciprocal of the apparent velocity V_{ap} at that point. The derivative can be obtained by differentiating the NMO function with respect to x:

$$T_x^2 = T_0^2 + \left(\frac{x}{V}\right)^2, \tag{6.4}$$

$$\frac{dT_x}{dx} = \frac{x}{T_x V^2} = \frac{1}{V_{\text{ap}}}. \tag{6.5}$$

The apparent wavelength is then given by

$$\lambda_{\text{ap}} = \frac{V_{\text{ap}}}{f} = \frac{T_x V^2}{xf}, \tag{6.6}$$

where f is the frequency and V is the stacking velocity. For a given x, T_x, and f, the apparent wavelength is proportional to the stacking-velocity-squared of the event, be it primary or multiple. Because the multiples at a given (x, T_x) point have lower stacking velocity than a primary at that point in the normal case of increasing velocity with depth, as shown in Figure 6.14, the multiples will have a shorter wavelength.

In step (1), the apparent wavelengths of the signal (primaries) and noise (water bottom multiples) are calculated over moveout space using the stacking velocities for primaries and multiples and the average frequency of the events. Consider an example from Ursin (1978). The stacking velocity for primaries is

$T_0 = 0.25$ sec	$V = 1480$ m/sec
$= 1.0$	$= 1800$
$= 1.5$	$= 2000$
$= 2.0$	$= 2600$
$= 3.0$	$= 3200$
$= 4.0$	$= 3900$
$= 6.0$	$= 4500$

The stacking velocity for the water bottom multiples is the velocity of sea water, 1480 m/sec. Consider the average frequency to be 20 Hz over all of moveout space for

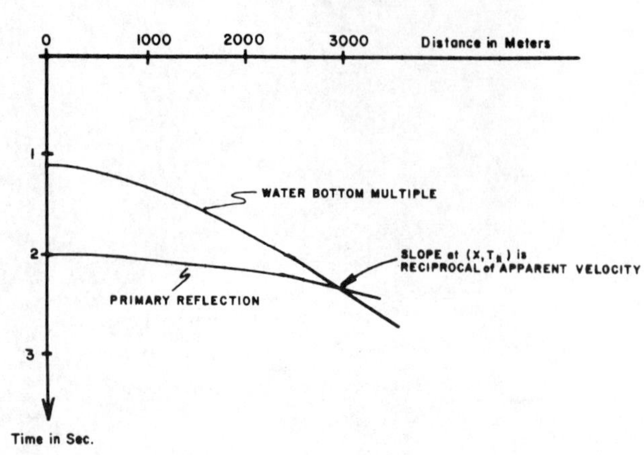

FIGURE 6.14. *Hyperbolic moveout functions in* t-x *(moveout) space.*

both primaries and multiples. Figure 6.15 shows the contours in moveout space for constant λ_{ap} for water bottom multiples and primaries. The contours are linear for the multiples because the stacking velocity is constant. The portion of moveout space between the 500 m wavelength contours for multiples and the 500 m contours for primaries is the region where significant improvement in signal-to-noise ratio can be achieved using super-long source patterns and weighted trace mix.

Ursin (1978) uses a five-element uniform source array with spacing equal to 54 meters and a five-trace mix with 50 m spacing and weights ½, 1, 1, 1, ½. The uniform source pattern response has notches at λ equal to the length of the pattern $L = (5)(54) = 270$ m and at $\lambda =$ 135, 90, and 67.5 m. The response rebounds to unity at λ equal to the spacing, 54 m. The five-trace mix is a uniform pattern of length $L = (4)(50)$, because the end elements are half of the interior elements (remember that the weights are weights of impulses, that is, areas under the curve being sampled), and has notches at λ equal to the length of the pattern $L = (4)(50) = 200$ m and at $\lambda = 100$ and 66⅔ meters, with a rebound at the spacing, 50 m.

The overall response of the source array and trace mix is shown in Figure 6.16. This response curve shows that wavelengths in the range from 100–300 m will be sup-

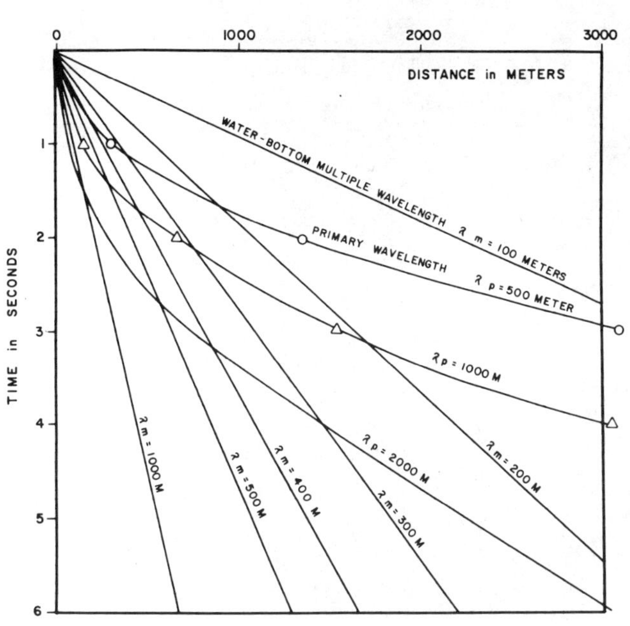

FIGURE 6.15. *Wavelengths of primary reflections and water bottom multiples in* t-x *(moveout) space. (From Ursin 1978, courtesy of Geophysical Prospecting; Blackwell Scientific Publications Limited.)*

pressed by 32 dB or more, and wavelengths greater than 500 m will be suppressed by 7 dB or less. This means that, in the region in moveout space between multiple wavelengths of 300 m and primary wavelengths of 500 m (see Figure 6.15), the primary-to-multiple ratio is improved by 25 dB or more. As the multiple wavelengths increase beyond 300 m, the primary-to-multiple improvement lessens until, at multiple wavelengths of 500 m, the improvement is 7 dB or less. As the multiple wavelengths continue to increase beyond 500 m, the improvement continues to decrease from 7 dB toward zero decibels.

Figure 6.17, from Ursin (1978), compares a conventional source and super-long source with trace mix along a profile in the Barents Sea. The super-long source and trace mix is that discussed previously in this section. The effectiveness of this technique in suppressing long-period water bottom multiples and in improving the continuity of the primary reflections is readily apparent.

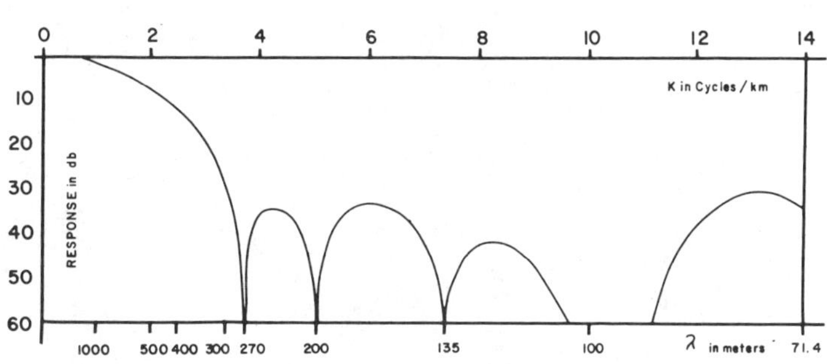

FIGURE 6.16. *Response of the combination of a five-element uniform source array with 54 m spacing and a five-trace mix with 50 m spacing and ½, 1, 1, 1, ½ weighting. (From Ursin 1978, courtesy of Geophysical Prospecting; Blackwell Scientific Publications Limited.)*

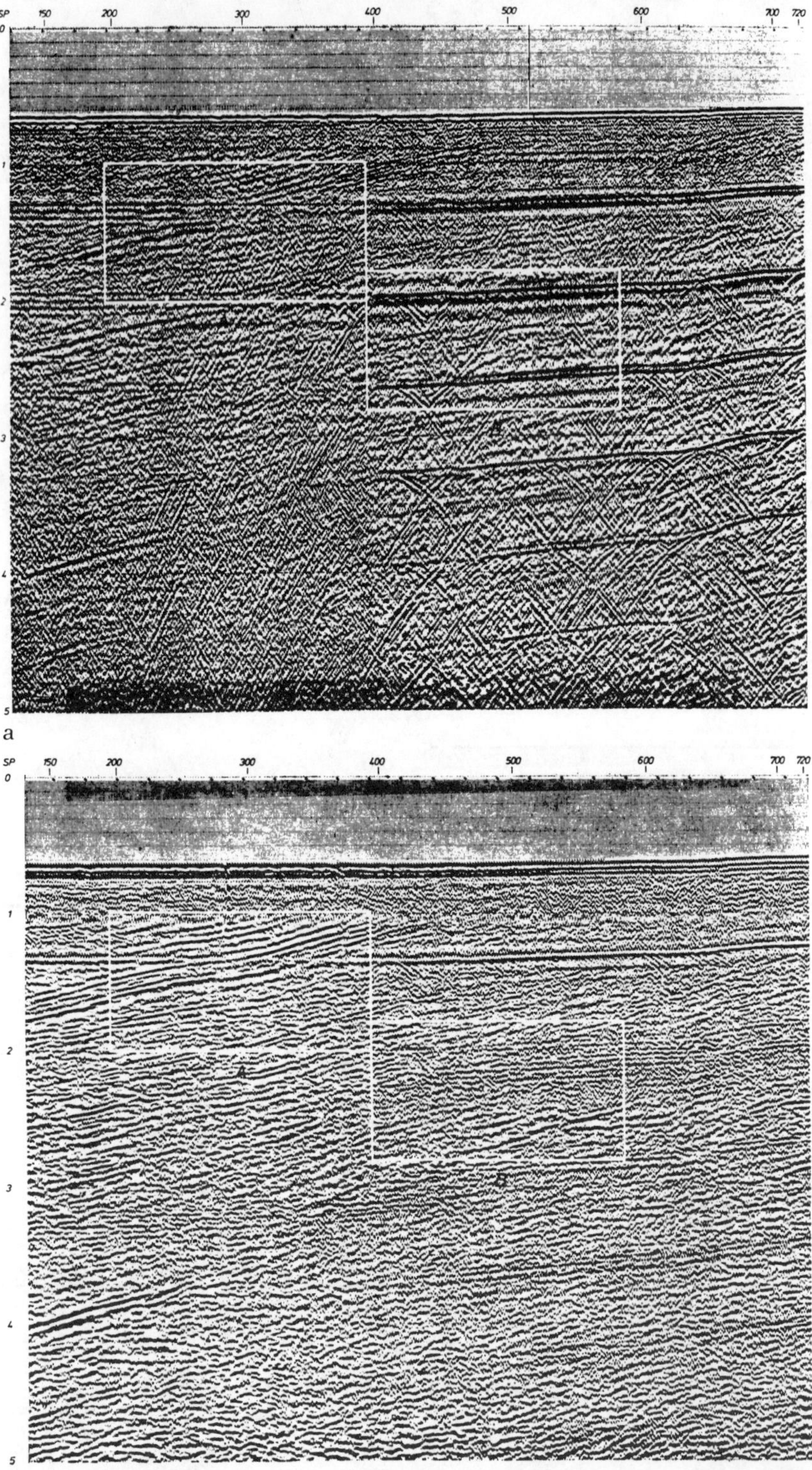

a

b

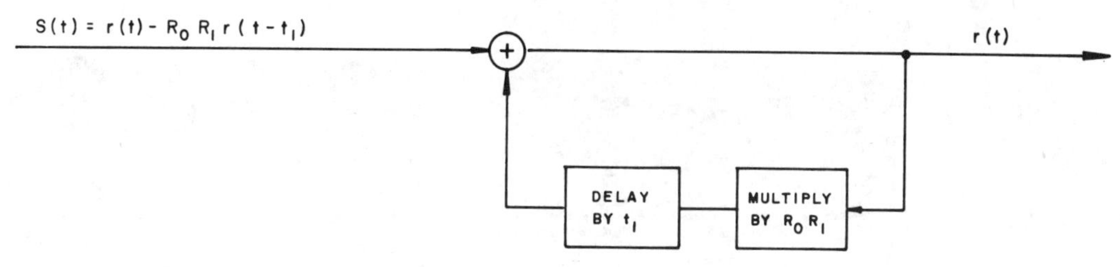

FIGURE 6.18. *Block diagram of Watson's feedback method of multiple suppression.*

FEEDBACK METHOD

Watson (1965) used feedback methods to strip out first-order surface multiples from the field data. The technique is similar to the one described by Lindsey (1960) for suppressing ghosts. The success of Watson's method depends on the presence of only a few strong multiple-generating interfaces.

Consider the case of one such interface, with reflection coefficient R_1 and two-way traveltime t_1. Surface multiples generated by R_1 will not begin to arrive until after t_1; therefore, the first t_1 seconds of the data will be essentially free of multiples. The FOSMs generated by R_1 are given by

$$\begin{aligned} m_1(t) &= -R_0R_1\delta(t - t_1) * r(t) \\ &= -R_0R_1r(t - t_1). \end{aligned} \tag{6.7}$$

The data $s(t)$ are the sum of primaries plus m_1,

$$s(t) = r(t) - R_0R_1r(t - t_1). \tag{6.8}$$

To remove the multiples m_1, the data $s(t)$ are delayed by time t_1, modified in amplitude by R_0R_1, and fed back additively to the data. By this method, the multiples are continuously being stripped out of the data at the summing junction, leaving primary $r(t)$ to be fed back at each instant of time, as shown in the block diagram in Figure 6.18.

FIGURE 6.17. *Comparison of superlong array with conventional source in Barents Sea. (a) Conventional section. (b) Superlong section. (From Ursin 1978, courtesy of Geophysical Prospecting; Blackwell Scientific Publications Limited.)*

Each interface that produces strong FOSMs must have its own feedback circuit. Problems involved include estimation of the parameters R_k and t_k for each feedback circuit. Because this method is a wavelet-canceling technique, the delay times must be known very precisely. Autocorrelation of a data window that includes the primaries and W-multiples from each multiple-generating interface can be used to obtain accurate estimates of t_k. Another problem is caused by additional absorption in the multiple paths, as compared to primary paths. Because primary wavelets are canceling multiple wavelets, filters simulating the differential absorption should be included in the feedback paths.

The feedback technique was applied by Watson to an area in Alberta, Canada, that had one strong multiplying interface—the Hay River limestone—whose W-multiple was a significant event that was not conformable with the underlying Slave Point reflection (Figure 6.19). A well had been drilled on what was mistakenly thought to be a reef development above the Slave Point but, in truth, was an anticlinal buildup on the Hay River W-multiple that resulted from a gentle surface depression. A single feedback circuit was all that was necessary, and its application to the data resulted in obliteration of the Hay River multiple, leaving the underlying Slave Point reflection intact.

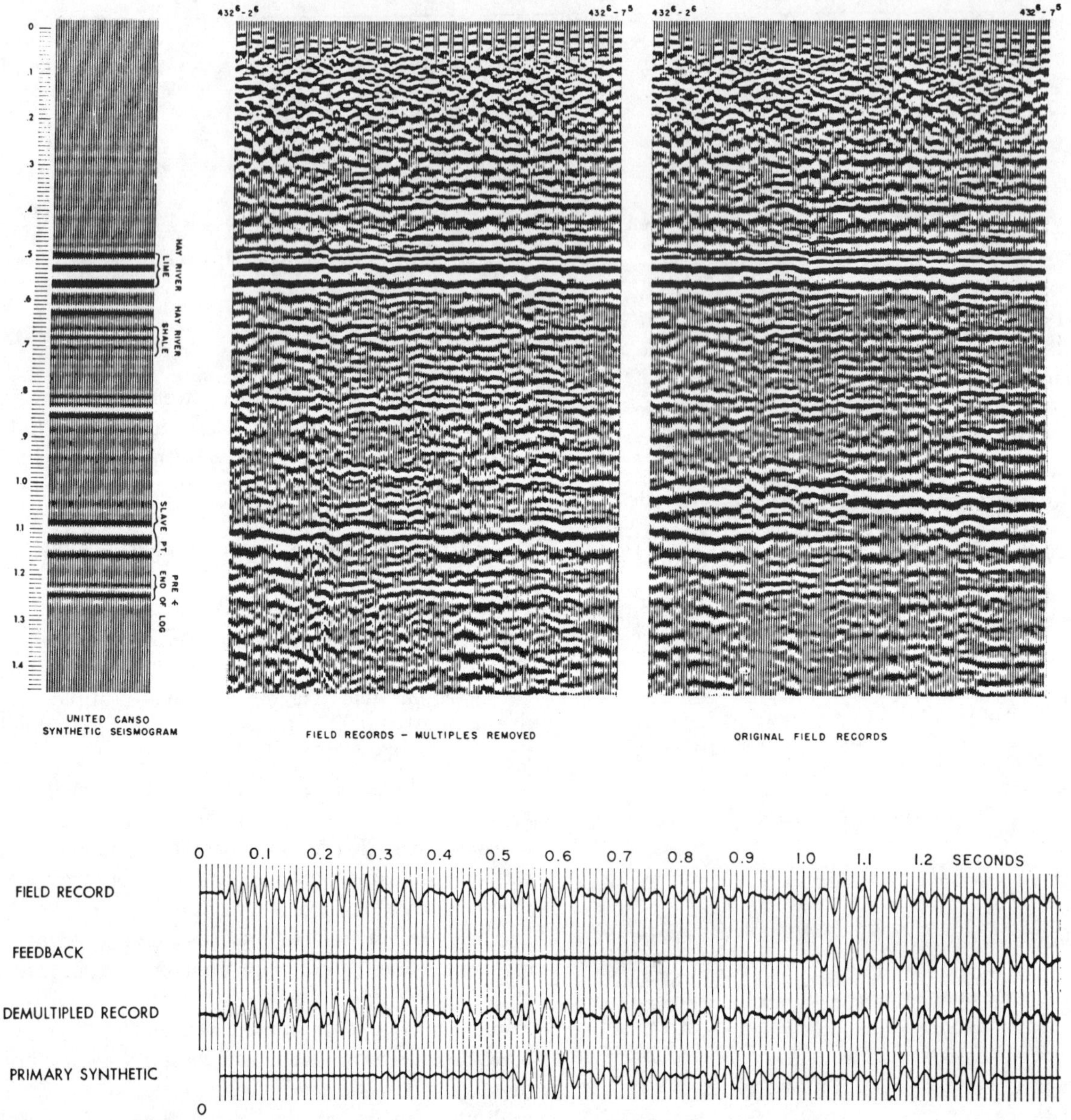

HAY RIVER LIME
HAY RIVER SHALE
SLAVE PT.
PRE € END OF LOG

UNITED CANSO SYNTHETIC SEISMOGRAM

FIELD RECORDS – MULTIPLES REMOVED

ORIGINAL FIELD RECORDS

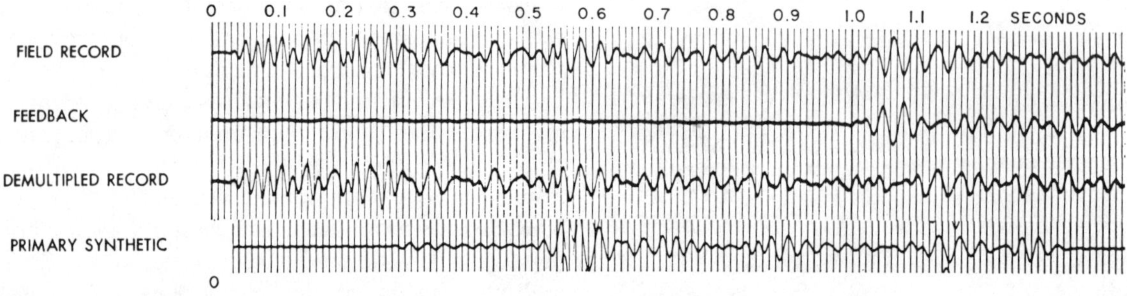

FIGURE 6.19. *Suppression of FOSMs generated by Hay River limestone, Alberta, Canada. (From Watson 1965, courtesy of Geophysics.)*

SUMMARY

Multiple reflections, including ghosts and reverberations, have become better understood in the context of the communication theory model of the seismic data-generation process, and many of the techniques for suppression of multiples have been developed through use of that model. In the next chapter, reverberations and ghosts are considered to be additional filters in the signal path. The filters are called distortion operators, and their suppression is then an exercise in inverse filtering, or deconvolution, as it is usually called.

Super-long arrays have the virtue of suppressing multiples before recording. This may be important in very high noise areas, where the dynamic range of the recording is used up by the noise and where, after noise suppression by deconvolution or other techniques, there is little or no signal left in the data.

The only general multiple suppression scheme is horizontal stacking, and the power of the stack in suppressing multiples can be improved dramatically by applying *f-k* filtering to the gathers before stacking.

Chapter 7 Distortion Operators and Deconvolution

INTRODUCTION

Seismic field records are the basic data from which the geological layering is inferred. In the reflection method, a seismic pulse is generated near the surface of the earth. This pulse, distorted by ghosts and reverberations, travels into the earth. Its waveform, continuously changing because of attenuation, is reflected by changes in the acoustic impedance of the subsurface layering. The field records contain additive noise and multiples in addition to the distorted reflections. As a further complication, the basic seismic signature may vary from place to place, making deduction of the geological layering still more difficult.

Theoretical development of an adequate mathematical model of the seismic data-generation process—a linear filter model—has led to a better understanding of the seismic process. Introduction of this model has led to inverse filter techniques to suppress the distortion present in seismic data. The most versatile technique is based on Wiener's (1949) optimum filter theory. The complicated processing involved in this technique is feasible only with digital processing, which, in turn, requires reproducible recording on either analog or, preferably, digital magnetic tape. Other technological developments that have contributed to improve seismic interpretation are horizontal stacking, migration, and presentation of final results as a corrected record section, preferably as a wiggle-variable-area display. The conciseness and clarity of such displays allows one to observe both stratigraphic and structural changes.

As a historical footnote, Professor Wiener applied statistical prediction to fire control during World War II. His classified document describing the technique became known as the "Yellow Peril" because of its yellow cover and erudite mathematics, which was completely beyond the average engineer and scientist of that era. After the war, this document was published in 1949 as "Extrapolation, Interpolation and Smoothing of Stationary Time Series." In the appendix, Professor Norman Levinson (1947) gave the discrete computational procedure that is used in deconvolution of seismic data today. Robinson's thesis of 1954, published in *Geophysics* in 1957, is the basis for the application of Wiener theory to seismic data processing. Foster, Hicks, and Nipper (1962) applied the theory to shortening the spacing of acoustic logs.

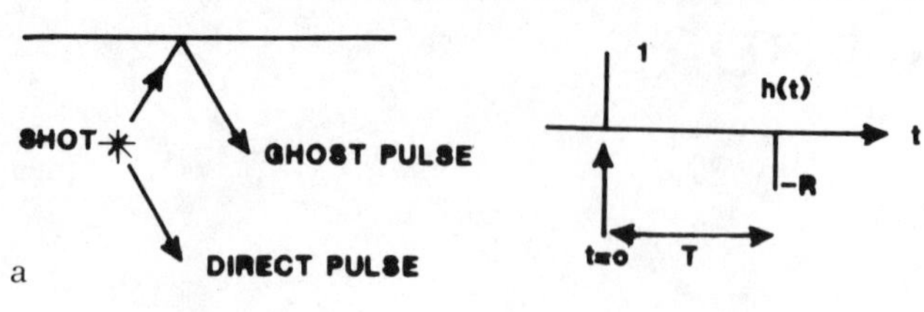

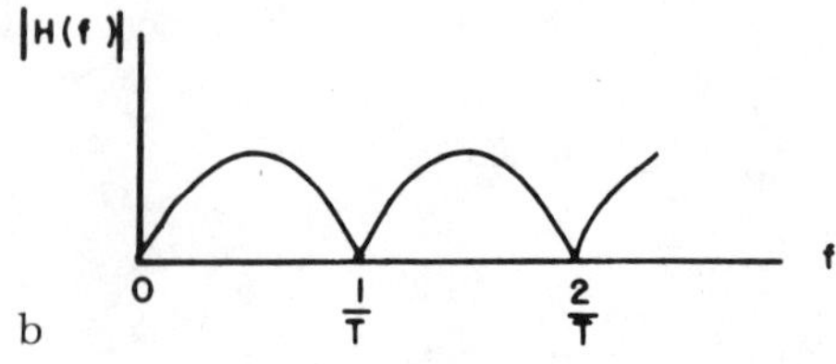

FIGURE 7.1. *Simple ghost distortion operator and its spectrum. (a) Impulse response. (b) Amplitude spectrum,* R = *1.*

DISTORTION OPERATORS AND THEIR DETERMINISTIC INVERSE OPERATORS

Distortions such as ghosts and reverberations can be considered additional filtering in the reflection path. This section discusses the pertinent characteristics of such distortion operators in simple cases.

GHOST DISTORTION OPERATOR

The ghost distortion operator accounts for energy traveling upward initially from a buried source (Figure 7.1). In the simple case of one reflecting interface above the source and no additional absorption in the ghost path, the operator is the sum of two impulses—the direct unit impulse and the ghost impulse weighted by the reflection coefficient $-R$ of the ghosting interface (where the negative sign accounts for the fact that the interface is being approached from below) and delayed by the two-way time T to the ghosting interface. The distortion operator is then

$$h(t) = \delta(t) - R\delta(t - T). \tag{7.1}$$

Using the sifting property of impulses, its Fourier transform is

$$H(f) = 1 - R\exp(-j2\pi fT). \tag{7.2}$$

Its amplitude spectrum is

$$|H(f)| = \sqrt{1 - 2R\cos 2\pi fT + R^2}. \tag{7.3}$$

The amplitude spectrum is periodic, with notches at $f = k/T$ and peaks at $f = (2k - 1)/2T$, for all integers k. The peak amplitude is $1 + R$, and the notch amplitude is $1 - R$. In the case of a perfect reflector ($R = 1$), the amplitude spectrum is a rectified sine wave,

$$|H(f)| = |2\sin\pi fT|. \tag{7.4}$$

The Z-transform of the simple ghost distortion operator is

$$H(z) = 1 - Rz. \tag{7.5}$$

Hence, its deterministic inverse is

$$H^{-1}(z) = 1 + Rz + R^2z^2 + R^3z^3 + \cdots. \tag{7.6}$$

The fact that R is less than unity shows that the inverse operator is stable; hence, the ghost operator is minimum-phase. The deterministic inverse is perfect and will completely eliminate the ghosts if the model is a good approximation to reality.

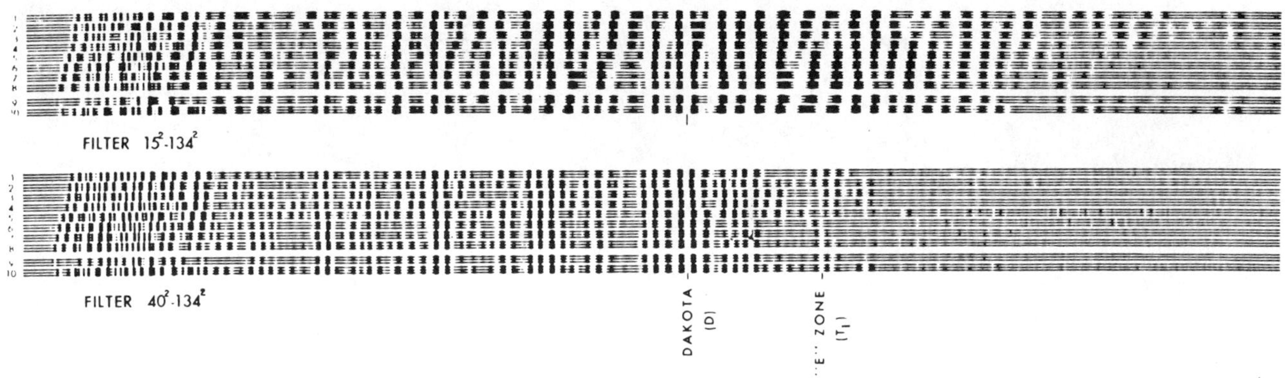

FIGURE 7.2. *Ghost identification in the time domain. (From Sengbush 1962, courtesy of Geophysics.)*

In the more general case, with n reflecting interfaces above the shot and additional absorption in the ghost path, the ghost distortion operator is

$$h(t) = \delta(t) - \sum_{k} R_k a_k(t - T_k), \tag{7.7}$$

where R_k is the reflection coefficient of the kth ghosting interface and a_k represents the additional absorption in the kth ghost path.

Ghosts may be identified on seismic data in either the time or frequency domain. In the time domain, the traces from a series of shot depths are shifted in time to line up primary events. The ghosts move out across the array of traces because they arrive progressively later as the shot depth increases. Behind each primary event a ghost follows, causing considerable interference on the data. If data from only one shot depth are available, it may be difficult to tell if ghosts are present; furthermore, it may be difficult to identify which events are primaries and which are ghosts. In the frequency domain, it is possible to identify ghosts from data from a single shot depth by spectral analysis of several windows from a field trace. If the spectra from the several windows show consistent notch separation, this is strong evidence that the notches are caused by ghosts, because the notch separation Δf is related to the ghost delay time T by $\Delta f = 1/T$.

An example of ghost identification in the time domain is shown in Figure 7.2 (from Sengbush, 1962). The data were obtained in the Big Horn basin of Wyoming during the course of a stratigraphic trap study in the area. The first eight traces of the display are the seismic data received at a given offset distance from the shot point, with the depths of the shot ranging from 240 ft on the first trace to 100 ft on the eighth trace and the shot spacing being 20 ft. The data are shifted to line up the primary reflection from the Dakota sandstone. The ninth trace of the display is the sum of the records from all shot depths. The primaries, because they are lined up, will be enhanced, and the ghosts, because they move out across the display, will be suppressed. The tenth trace of the display is a column charge of length 96 ft, with the top of the charge at 92 ft. The column charge is detonated at the top and has a detonation velocity approximating the formation velocity; therefore, the downgoing signal is continuously enhanced, and the upgoing signal, the ghost, is spread out in time and reduced in amplitude. The display is shown with a broadband filter, two stages of a third-order Butterworth filter whose bandwidth is 15–134 Hz. The second display is the same data through a highpass filter, which is also two stages of a third-order Butterworth filter whose bandwidth is 40–134.

Spectral analysis for ghost identification is usually performed by autocorrelating windows of length W and truncating them to length L, which is longer than the expected ghost delay time T but considerably shorter than W, and then taking the Fourier transforms of the truncated autocorrelation functions. By truncating the

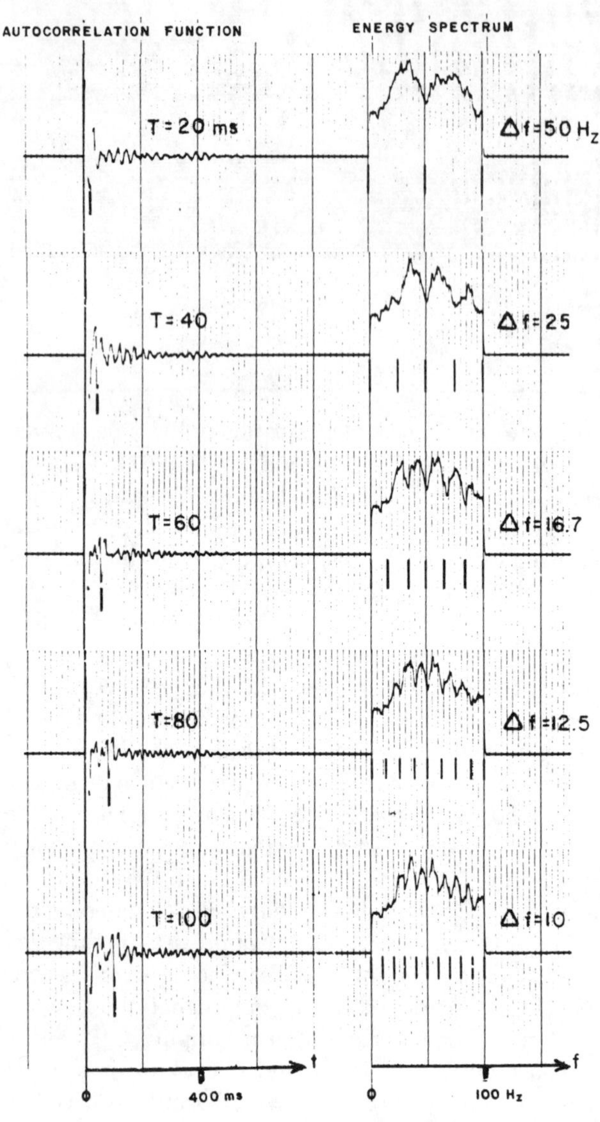

FIGURE 7.3. *Ghost identification in the frequency domain. (From McDonal and Sengbush 1966.)*

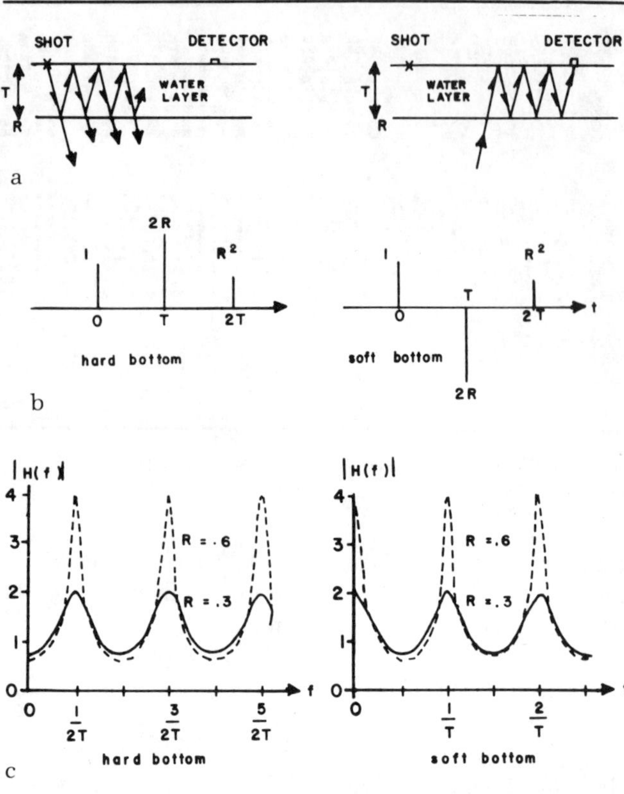

FIGURE 7.4. *Reverberation distortion operator. (a) Reverberation-generating mechanism. (b) Backus three-point dereverberation operator. (c) Amplitude spectrum of reverberation distortion operator.*

autocorrelation functions with the constraints $T < L <$ W, the spectral contributions of the reflectivity in the windows are smoothed out and the contributions due to ghosts are preserved. An example of ghost identification in the frequency domain is shown in Figure 7.3. The data are ghosted play data, with the ghosts' delay time ranging from 20 to 100 ms. The ghost pulses can be seen in the truncated autocorrelation functions. The energy spectra show ghost notches with separation from 50 Hz to 10 Hz.

REVERBERATION DISTORTION OPERATOR

The reverberation distortion operator is derived from the multiple reflections occurring in the water layer at both the source and the detector (Figure 7.4). Consider the simple case of a water layer with two-way traveltime thickness T and reflection coefficient R at the base. A unit impulse generated at the free surface travels downward in the water and is reflected at the base of the water layer. The upgoing reflection of size R arrives at the free surface at time T, where it becomes a secondary source upon reflection downward. The free surface has reflection coefficient (-1) as seen from below, so the secondary impulse is $-R\delta(t - T)$. This secondary impulse undergoes the same fate as the initial impulse, producing another secondary source with its impulse $R^2\delta(t - 2T)$. This repeats ad infinitum, producing a train of secondary impulses $\Sigma_k(-R)^k\delta\,(t - kT)$. Upon passing into the subsurface, each of the impulses is modified by the transmission coefficient $(1 - R)$. The source distortion operator is then given by

$$h_1(t) = (1 - R)\sum_{k=0}^{\infty}(-R)^k\delta(t - kT). \tag{7.8}$$

At the detector, a unit impulse entering the water layer from the subsurface has transmission coefficient $(1 + R)$ applied. This impulse is multiply reflected within the water layer, producing the detector distortion operator h_2, given by

$$h_2(t) = (1 + R)\sum_{k=0}^{\infty}(-R)^k\delta(t - kT). \tag{7.9}$$

The continual feeding of seismic energy into the subsurface at the source and the continual feeding of energy into the water layer by reflections from the subsurface are the mechanism for reverberation generation.

The reverberated seismogram is given by

$$s = b * h_1 * r * h_2 + n, \tag{7.10}$$

where b is the seismic pulse, r the reflectivity, and n the additive noise. To cancel the reverberations, it is necessary merely to apply the inverse of the distortion operator, $h^{-1} = h_1{}^{-1} * h_2^{-2}$, to the data; then,

$$s * h^{-1} = b * r + n * h^{-1}. \tag{7.11}$$

The inverse operators $h_1{}^{-1}$ and h_2^{-2} are given by

$$h_1^{-1} = (1 - R)^{-1}[\delta(t) + R\delta(t - T)], \tag{7.12}$$

$$h_2^{-2} = (1 + R)^{-1}[\delta(t) + R\delta(t - T)], \tag{7.13}$$

as can be seen most readily by multiplying the Z-transforms of the distortion operator and its inverse and observing that the product equals one:

$$\begin{aligned}(1 - R)^{-1}\,H_1(z) &= 1 - Rz + R^2z^2 - R^3z^3 + \cdots,\\ (1 - R)H_1^{-1}(z) &= 1 + Rz\\ \hline H_1(z)H_1^{-1}(z) &= 1\end{aligned}. \tag{7.14}$$

The inverse operator h^{-1} has Z-transform

$$\begin{aligned}H^{-1}(z) &= H_1^{-1}(z)H_2^{-1}(z) = (1 - R^2)^{-1}\,(1 + Rz)^2\\ &= (1 - R^2)^{-1}\,(1 + 2Rz + R^2z^2).\end{aligned} \tag{7.15}$$

In the time domain, the inverse distortion operator becomes

$$h^{-1}(t) = (1 - R^2)^{-1}[\delta(t) + 2R\delta(t - T) + R^2\delta(t - 2T)]. \tag{7.16}$$

Because the inverse is stable, the simple reverberation operator is minimum-phase.

This perfect, deterministic three-point inverse operator, derived by Backus (1959), cancels reverberations exactly in the simple reverberation model. To apply it to field data, it is only necessary to estimate two parameters—the reflection coefficient R at the base of the water and the two-way traveltime T in the water layer. In the usual case, where the acoustic impedance of the water is less than that of the underlying earth, the reflection coefficient at the water base is positive. This is called the hard-bottom case. In unusual circumstances, where the sea bottom is gaseous mud, as in Lake Maracaibo (Levin, 1962), the water has larger acoustic impedance than the mud, and R is negative. This is called the soft-bottom case. In the three-point operator, the polarity of R affects only the second point.

The frequency spectrum of the reverberation distortion operator can be obtained most easily from the inverse operator. By the sifting property of impulses and the convolution theorem, the Fourier transform of the inverse operator is derived as follows:

$$\mathcal{F}[h_1^{-1}(t)] = (1 - R)^{-1}[1 + R\exp(j2\pi fT)], \quad (7.17)$$

$$\mathcal{F}[h_2^{-1}(t)] = (1 + R)^{-1}[1 + R\exp(j2\pi fT)], \quad (7.18)$$

$$\mathcal{F}[h^{-1} = h_1^{-1} * h_2^{-1}]$$

$$= (1 - R^2)^{-1}[1 + R\exp(j2\pi fT)]^2. \quad (7.19)$$

The Fourier spectrum of the distortion operator (which is the reciprocal of the spectrum of its inverse) and its amplitude spectrum are as follows:

$$\mathcal{F}[h(t)] = H(f) = \frac{1 - R^2}{[1 + R\exp(j2\pi fT)]^2}, \quad (7.20)$$

$$|H(f)| = \frac{1 - R^2}{1 + 2R\cos 2\pi fT + R^2}. \quad (7.21)$$

The amplitude spectrum is periodic, with resonant peaks occurring when the term $2R\cos 2\pi fT$ becomes most negative. In the hard-bottom case, this term becomes most negative when $\cos 2\pi fT = -1$, which occurs when $fT = 1/2, 3/2, 5/2, \ldots$. In the soft-bottom case, R is negative, so this term becomes most negative when $\cos 2\pi fT = 1$, which occurs when $fT = 0, 1, 2, \ldots$. The separation between resonant peaks in either case is $1/T$. The magnitudes of the resonant peaks are

$$|H(f)|_{max} = \frac{1 - R^2}{1 - 2R + R^2} = \frac{1 - R^2}{(1 - R)^2}. \quad (7.22)$$

Intermediate to the resonant peaks, the amplitude spectra have minimum values of

$$|H(f)|_{min} = \frac{1 - R^2}{(1 + R)^2}. \quad (7.23)$$

The maximum and minimum values are tabulated as follows as a function of R:

R	H_{max}	H_{min}
.1	1.22	.81
.2	1.50	.67
.3	1.86	.54
.4	2.33	.43
.5	3.00	.33
.6	4.00	.25
.7	5.67	.18
.8	9.00	.11
.9	19.00	.05
1.0	∞	0

Examples of the use of the three-point operator on field data in both the hard-bottom case (Persian Gulf) and the soft-bottom case (Lake Maracaibo) are shown in Figure 7.5. In each case, the three-point operator produces significant suppression of reverberations. Spectral analysis of the Persian Gulf data shows that the bottom must be hard and that the water depth is about 200 ft. This is apparent by noting that the separation between notches is 12 Hz, with reverberation peaks at 42 and 54 Hz. This fits the hard-bottom case because, with 12 Hz notch separation, the notches in the spectrum should occur at 6, 18, 30, 42, 54, 66, . . . , Hz. The water depth from the spectral analysis is obtained from $T = 1/f = 2z/V$. Substituting $f = 12$ Hz and $V = 4800$ ft/sec gives $z = 200$ ft, in agreement with the known facts.

SOURCE AND RECEIVER DEPTH DISTORTION OPERATORS

The reverberated seismogram in water-covered areas is the result of the combined filtering of the source pulse, the water layer, the reflectivity, the source depth, and the detector depth. The filtering effect of source depth consists of a two-point time-domain operator $h_s(t)$ given below:

$$\text{Source:} \quad h_s(t) = \delta(t) - \delta(t - T_s), \quad (7.24)$$

where T_s is the two-way time from the source to the surface and the surface reflection coefficient is -1, as seen from below.

The filtering effect of receiver depth depends on the type of detector used. At a free surface, boundary conditions dictate zero pressure and double amplitude of the velocity particle motion; hence, a positive-pressure im-

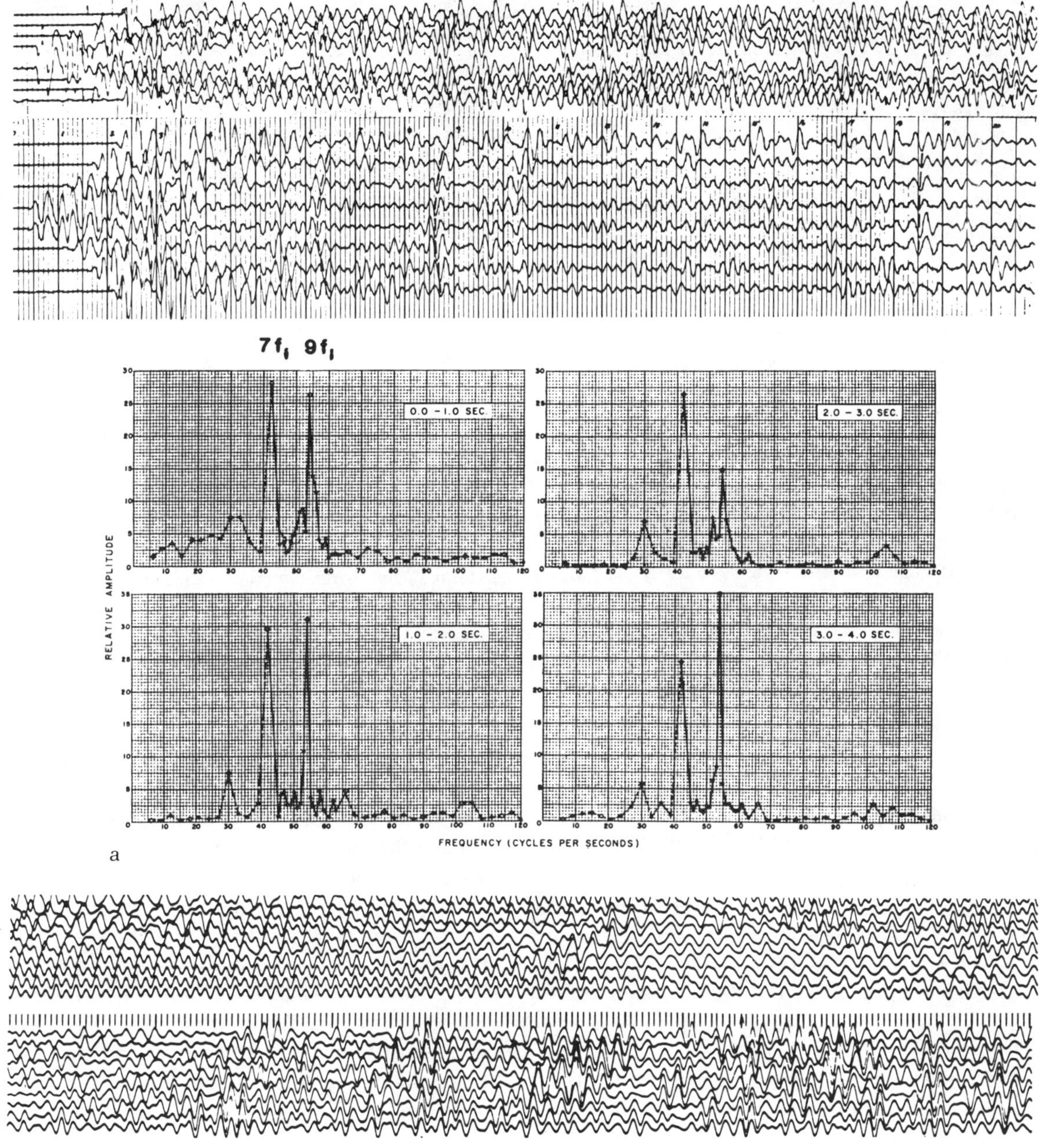

FIGURE 7.5. *Dereverberation using the three-point operator. (a) Persian Gulf, hard bottom, water depth 200 ft. (b) Lake Maracaibo, soft bottom, water depth 80 ft. (From Backus 1959, courtesy of Geophysics.)*

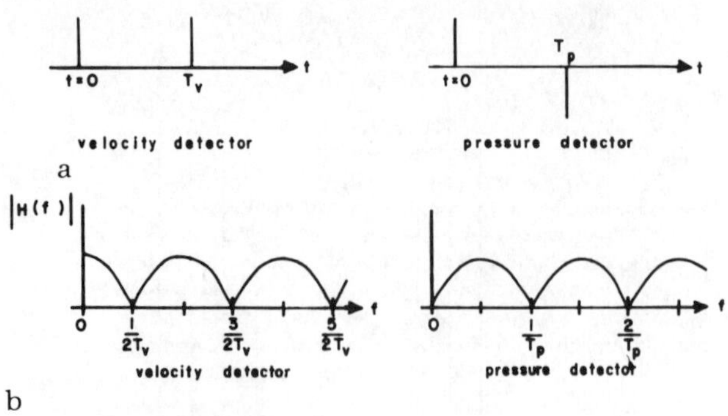

FIGURE 7.6. *Detector depth distortion operator. (a) Detector depth distortion operator. (b) Amplitude spectrum of detector depth distortion operator.*

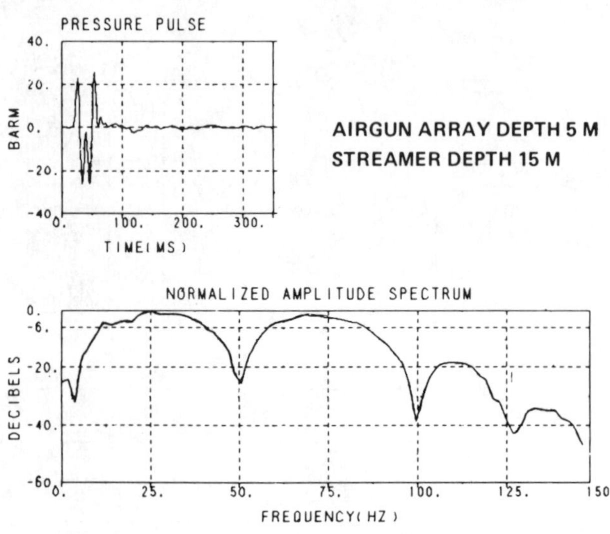

FIGURE 7.7. *Air gun signature with source at 5 m and detector at 15 m. (From Brandsaeter, Farestveit, and Ursin 1979, courtesy of Geophysics.)*

pulse traveling upward becomes a negative-pressure impulse upon reflection at the free surface, and a positive-velocity impulse is reflected as a positive-velocity impulse. The impulse response of a detector at depth is then a two-point operator with impulses separated by the two-way time between the detector and free surface. The polarity of the second impulse depends on detector type:

Velocity-type detector: $$h_v(t) = \delta(t) + \delta(t - T_v), \quad (7.25)$$

Pressure-type detector: $$h_p(t) = \delta(t) - \delta(t - T_p). \quad (7.26)$$

The impulse responses and amplitude spectra of source and receiver depth distortion operators are shown in Figure 7.6. The amplitude spectra of source depth and pressure detector depth filters are rectified sine functions, with notches at $f = k/T_s$ and $f = k/T_p$, for k = any integer. The velocity detector depth filter amplitude response is a rectified cosine function, with notches at $f = k/2T_v$, for k = any odd integer. Source and detector depths should be chosen such that the first notch in the spectrum is above the seismic band of interest. Preservation of the seismic band to 125 Hz requires source and pressure detector depths of 20 ft or less.

In shallow water, where detectors are allowed to rest on the water bottom, it is advantageous to use velocity detectors in the hard-bottom case and pressure detectors in the soft-bottom case. Then, notches in the receiver depth spectrum occur at reverberation peaks and suppress reverberations before recording (Sengbush, 1967).

Measurements of the signature of an air gun source at 5 m depth and pressure detector at 15 m depth are shown in Figure 7.7 (from Brandsaeter, Farestveit, and Ursin, 1979). The detector at 15 m produces notches in the spectrum at $f = nV/2z$, where $V = 1500$ m/sec. Therefore, the notches occur at $f = 750\ n/z = 50, 100, 150, \ldots,$ Hz. The source at 5 m produces notches at $f = 150, 300, \ldots,$ Hz. The notch at $f = 125$ Hz cannot be explained by either source or detector depth but would be explained by a source depth of 6 m. This is possibly the answer.

WIENER INVERSE FILTER

Since the introduction of the concept that the seismic process can be modeled by a filter process, with the signal in the seismic data being the result of filtering the reflectivity of the earth with the seismic pulse, geophysicists have been interested in inverse filtering of the data to recover the reflectivity. Backus's three-point operator for suppressing reverberations and the inverse ghost operator are perfect inverse filters for the simple models assumed. Each requires estimation of only two parameters. Although these deterministic inverse filters achieved limited success, they failed generally because the models are not sufficiently close to reality. Attempts to increase the complexity of these deterministic operators lead to an increase in the number of parameters to estimate. With the coming of the digital revolution, the deterministic approach was doomed to oblivion by optimal Wiener processing. The complications of multibottoms, multighosting interfaces, added attenuation in the ghost path, and all the associated parameter estimation problems are of no consequence in the Wiener process.

The Wiener process suppresses the distortion and compensates for the variations in seismic pulse waveform, thereby producing records that have more consistent character. The spectrum is broadened and smoothed in the seismic band. The inverse of the unknown distorted and attenuated seismic signal is obtained from the data, without any presumptions of its character, and then applied to the data to suppress the distortions and reduce variations in the signature waveform, as shown in the block diagram in Figure 7.8.

Wiener processing is constrained by the requirement that the data must be generated by a minimum-phase system. The only possible part of the overall system that can contribute significant nonminimum-phase characteristics to the data is the source signature itself, because attenuation, distortions caused by ghosts and reverberations, detectors, and recording instruments are minimum-phase. Sources that are nonminimum-phase require preprocessing by nonminimum-phase inverses of the signatures before application of Wiener processors. This preprocessing is known as signature deconvolution.

The mathematical model of the data-generating process is described by

$$s(t) = b(t) * h(t) * r(t) + n(t) + u(t). \tag{7.27}$$

The noise is split into two components—coherent noise $n(t)$ and random noise $u(t)$. Ground roll and refractions are typical coherent events that can be followed

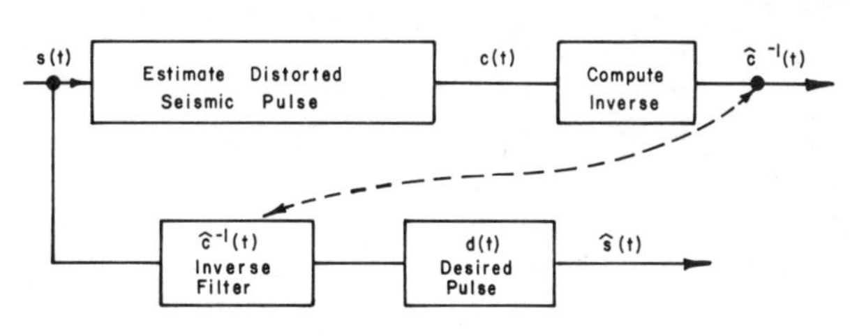

FIGURE 7.8. *Block diagram of optimal Wiener processor.*

from trace to trace but have not traveled in the reflection path. Random noise is the component that is uncorrelated from trace to trace. Every effort should be made to suppress additive noise by proper field procedures, because noise reduces the effectiveness of the Wiener processor.

Because the distortion operator $h(t)$ takes on many varied and generally unknown forms, and the seismic pulse $b(t)$ has unknown form, they are combined to give the distorted seismic pulse $c(t) = b(t) * h(t)$.

In deconvolution, the unknown distorted seismic pulse $c(t)$ is replaced with a desired pulse $d(t)$ that has special desirable properties, such as a flat amplitude spectrum and zero phase over a broad band. The process consists of two basic steps:

(1) The inverse of the unknown distorted seismic pulse $c(t)$ is estimated from the data, using the Wiener-Levinson algorithm. The estimate of the inverse is designated $\hat{c}^{-1}(t)$.
(2) The seismic trace from which the estimate is obtained is filtered by $\hat{c}^{-1}(t)$ to remove the effects of the distorted pulse and by $d(t)$ to produce the band-limited signal estimate $\hat{s}(t)$.

STEP 1: ANALYSIS OF THE DATA. The analysis is clarified by using the frequency domain, although deconvolution is usually performed in the time domain. A window of length W is selected from the seismic trace. The energy spectrum of the window is given by

$$\Phi_s = \Phi_c \cdot \Phi_r + \Phi_n + \Phi_u, \tag{7.28}$$

where

Φ_s = energy spectrum of window
Φ_c = energy spectrum of distorted seismic pulse
Φ_r = energy spectrum of reflectivity
Φ_n = energy spectrum of coherent noise
Φ_u = energy spectrum of random noise

The reflectivity and random noise are mutually uncorrelated random processes, and the spectral components Φ_r and Φ_u are estimates of their true power spectra. The window of length W must be sufficiently long to allow the random characteristics in the random processes to be developed, but it cannot be too long because of time-variance of the distorted seismic pulse $c(t)$. If the coherent noise is low, $\Phi_n \cong 0$, the energy spectrum of the window becomes

$$\Phi_s = \Phi_c \cdot \Phi_r + \Phi_u. \tag{7.29}$$

At this point, a separation of Φ_c from Φ_r is effected by using a priori knowledge about the spectral differences between the distorted pulse and the random processes. The spectrum of the distorted pulse is inherently smooth in contrast to the ragged spectral variations caused by the random processes, so the energy spectrum of the window is smoothed to produce the separation. The smoothed spectrum is an estimate of the energy spectrum of the distorted pulse.

Spectral smoothing is effected in the time domain by truncating the autocorrelation of the window to a length L that is much less than the window length W. To get effective separation of the deterministic distorted seismic pulse from the random reflectivity, the ratio L/W should be on the order of 1/10 and certainly not less than 1/5. Truncation is multiplication of time functions, which is equivalent to convolution in the frequency domain. The energy spectrum Φ_s is convolved with the spectrum of the truncator. Consider a rectangular truncator with truncation points $(-L, L)$. Its Fourier transform is proportional to sinc $2Lf$, which has a major lobe of frequency width $\Delta f = 1/L$. The spectral variations of Φ_s within this width are smoothed by convolution with the sinc function. This separates the spectral components in the reflectivity random process $r(t)$ from the components in the distorted seismic pulse $c(t)$. This is a crucial and sometimes unappreciated step in the deconvolution process.

Most often, before computing the Wiener operator, a small quantity is added to the autocorrelation function ϕ_s at the origin. This is called prewhitening, because it is exactly the effect that would be produced on ϕ_s if the seismic data contained a white random noise component. This small quantity is added to insure that the Wiener operator is stable. It serves no other purpose. The effectiveness of the deconvolution process is always reduced by prewhitening (Shugart, 1973), so as small a quantity as possible should be used; 0.1 percent of $\phi_s(0)$ should be sufficient.

STEP 2: PROCESSING THE DATA. In step 1 (analysis), the inverse of the distorted pulse $c(t)$ is estimated from the input data. The estimate $\hat{c}^{-1}(t)$ is obtained by numerical solution of the Wiener-Levinson algorithm. In step 2 (processing), the seismic data $s(t)$ are filtered by $\hat{c}^{-1}(t)$ and the bandlimited desired pulse $d(t)$ to produce the signal estimate $\hat{s}(t)$:

$$\hat{s}(t) = s(t) * \hat{c}^{-1}(t) * d(t). \tag{7.30}$$

Substituting the seismic data equation into the signal estimate equation gives

$$\hat{s} = c * r * \hat{c}^{-1} * d + (n + u) * \hat{c}^{-1} * d. \tag{7.31}$$

If the estimate of the inverse of the distorted pulse is reasonably good, $c * \hat{c}^{-1} \cong \delta(t)$, and the signal estimate is given by

$$\hat{s} \cong r * d + (n + u) * \hat{c}^{-1} * d. \tag{7.32}$$

If the noise terms are small,

$$\hat{s} \cong r * d. \tag{7.33}$$

The unknown and distorted pulse c has been replaced by the desired pulse d, which has known and desirable characteristics. The d-filter is necessary to restrict the bandwidth of the signal estimate, because $\hat{c}^{-1}$ is not a good estimate at low and high frequencies, where the signal-to-noise ratio is low. Thus, the Wiener process broadens and flattens the spectrum of the distorted pulse within the passband, where the signal-to-noise ratio is sufficiently good for $\hat{c}^{-1}$ to be a good estimate.

WIENER PROCESSING OF PLAY DATA

Wiener deconvolution is dissected and explained by processing play data, in which all components of the data-generation process are known.

INVERSE FILTERS

Given a function $b(t)$, its inverse is defined as the function $b^{-1}(t)$ that produces an impulse when it is convolved with $b(t)$. In the frequency domain, the spectrum of the inverse is the reciprocal of the spectrum of $b(t)$, so that the product of the spectra equals the integer 1, which is the spectrum of an impulse. This means that the amplitude spectrum of the inverse is the reciprocal of the amplitude spectrum of $b(t)$, and the phase spectrum of the inverse is the negative of the phase spectrum of $b(t)$. With sampled data, the Z-transform of the inverse $B^{-1}(z)$ satisfies the equation $B(z)B^{-1}(z) = 1$. $B^{-1}(z)$ is the exact inverse of $B(z)$ and is of infinite length if $B(z)$ is of finite length. Truncating the inverse results in an error between the product and the desired output, as shown in Appendix C in the section on correlation of time series.

The Wiener-Levinson algorithm produces the best finite length estimate $\hat{b}^{-1}(t)$ of the inverse of a function $b(t)$ in the sense that the mean square error between the output $b * \hat{b}^{-1}$ and the desired output (a spike at the origin) is minimal. Consider the example in Figure 7.9. The input is a finite-length minimum-phase sampled impulse response of a 25–50 Hz Butterworth filter. The window length used is the same length as the pulse, and its autocorrelation function is not truncated. No prewhitening factor is used, because there is no problem with stability in calculating the inverse. Convolving the Wiener operator $\hat{b}^{-1}(t)$ with the input $b(t)$ produces a spike at the origin and a residual rms error of 0.028. The Wiener process guarantees that no other estimate of the inverse with the same length will produce less rms error.

Incidentally, the analytical expression for the impulse response of a Butterworth filter is known to be minimum-phase, but truncating and sampling to produce a sampled function of finite length does not preserve the minimum-phase property. Taking the inverse

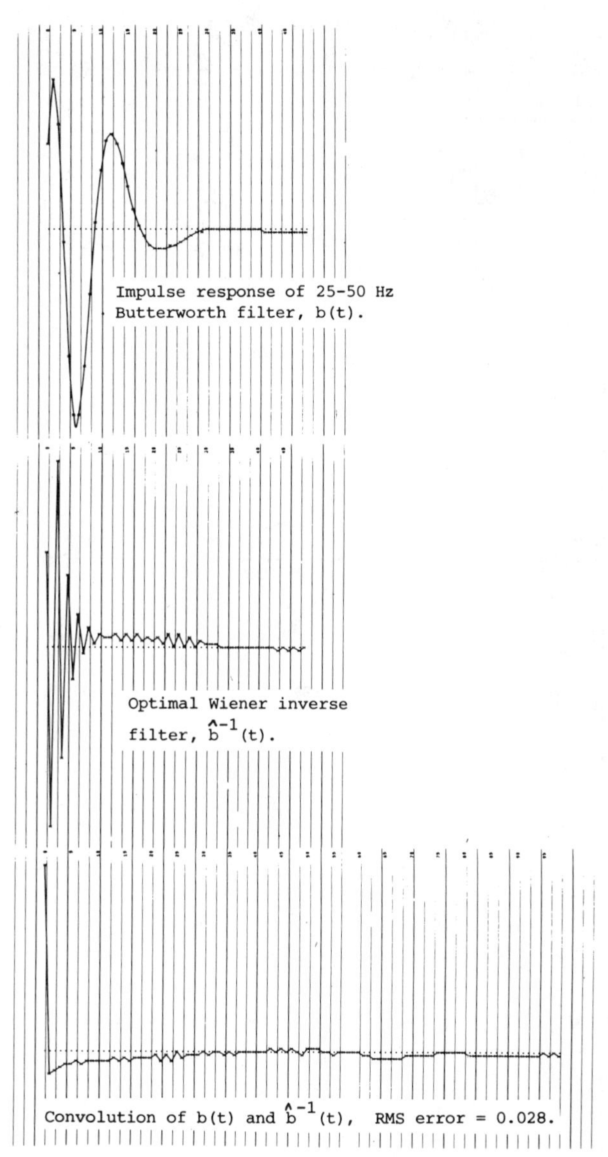

FIGURE 7.9. *Wiener processing of a minimum-phase function.*

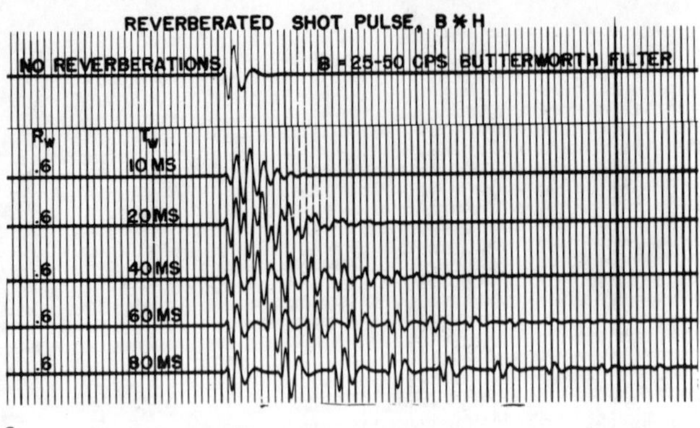

a

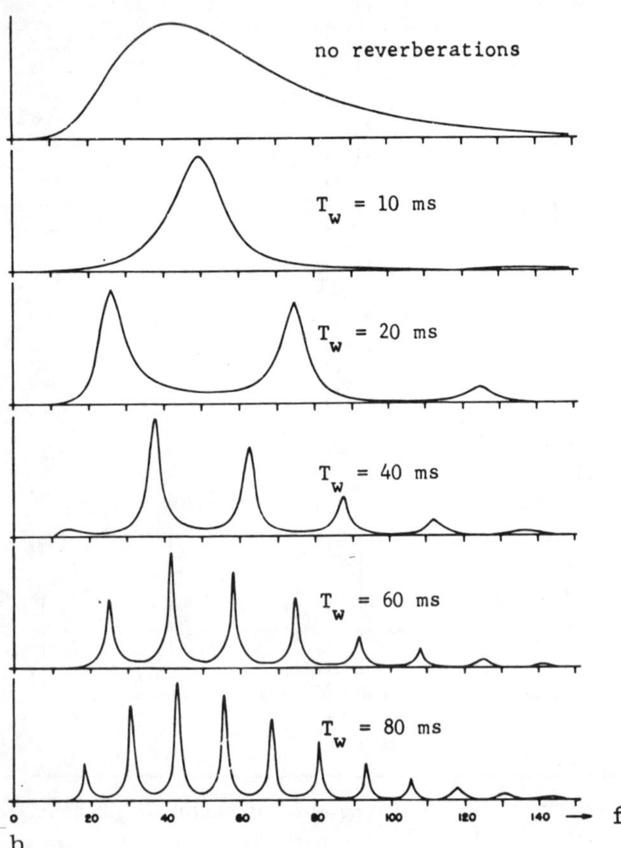

b

FIGURE 7.10. *Seismic pulses distorted with reverberations and receiver ghosts, and their spectra. (a) Time domain. (b) Frequency domain. (From McDonal and Sengbush 1966.)*

of the inverse of such a function, using the Wiener-Levinson algorithm twice, guarantees that the resulting function is minimum-phase. This was done to produce the $b(t)$ used in Figure 7.9.

REVERBERATED PLAY DATA

The simple reverberation play data generated for this test contain reverberations from a single layer and are free of coherent and incoherent noise. The data are assumed to be generated by a surface source and recorded with a pressure detector with ghost delay time $T_p = 6$ ms. The data $s(t)$ are described mathematically by

$$s(t) = b(t) * h(t) * r(t). \tag{7.34}$$

The seismic pulse $b(t)$ is the impulse response of a 25–50 Hz Butterworth filter, with 18 dB/octave rejection rates. The reflectivity function $r(t)$ was obtained from a continuous velocity log. The distortion operator $h(t)$ is the combination of the reverberation layer and the receiver ghost. The reverberating layer has a hard bottom, with reflection coefficient $R = 0.6$, and two-way time thickness T, ranging from 10–80 ms. The waveforms of the reverberated seismic pulses, including receiver ghosts, and the corresponding spectra are shown in Figure 7.10. The fundamental reverberation frequency is 50, 25, 12.5, 8.33, and 6.25 Hz on the successive traces.

In contrast to the inherently smooth spectrum of the distorted pulse, the energy spectrum of the reflectivity window, Figure 7.11, has erratic fluctuations characteristic of random processes. The smoothed spectrum is essentially flat, indicating a white noise process.

The spectrum of the input data is the product of the distorted pulse spectrum and the reflectivity spectrum, as shown by the energy spectrum in Figure 7.12, trace 1, for the case where $R = 0.6$ and $T = 60$ ms. The smoothed spectrum on trace 2 is a good estimate of the spectrum of the distorted pulse. The smoothed spectrum is obtained by spectral analysis of the truncated autocorrelation of the window. The spectrum of the bandlimited inverse of the distorted pulse is shown on trace 3. The bandlimiting has been done with a brickwall d-filter whose bandwidth is 20–65 Hz. Within this frequency range, the spectrum on trace 3 is inverse to the spectrum on trace 2. Outside this range, the spectral estimates of

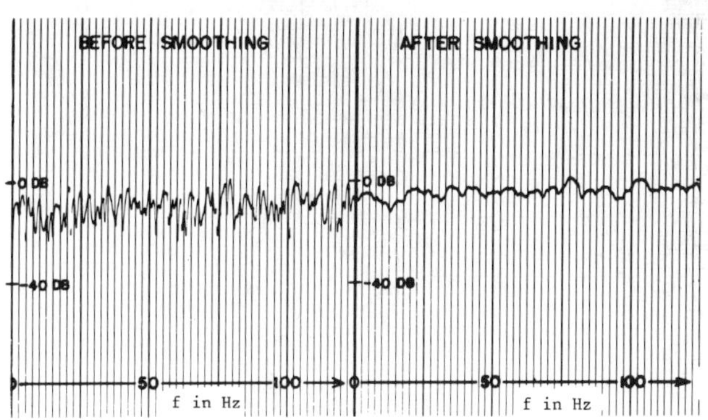

FIGURE 7.11. *Energy spectrum of reflectivity window. (From McDonal and Sengbush 1966.)*

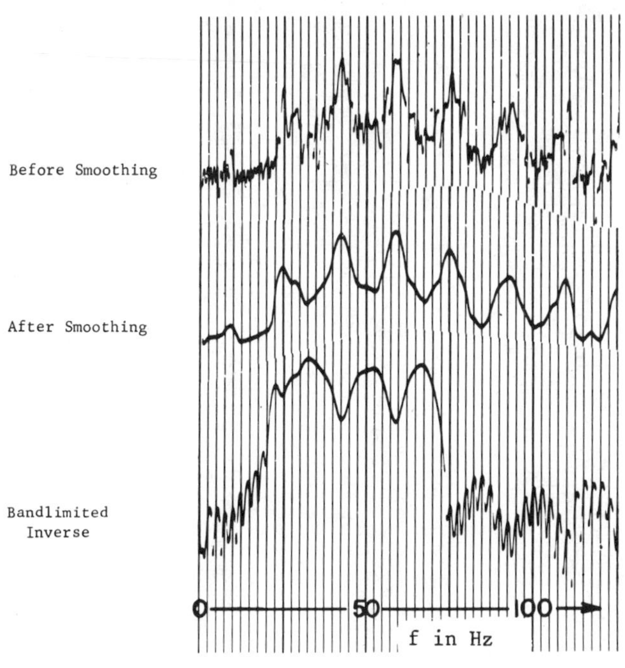

FIGURE 7.12. *Spectral smoothing to separate random reflectivity from deterministic distorted seismic pulse, in reverberation case where* R = *0.6 and* T = *60 ms. (From McDonal and Sengbush 1966.)*

the distorted pulse are not very accurate; hence, the bandwidth over which its inverse is obtained is restricted with a d-filter. Of equal importance, the phase spectrum of the inverse must be the negative of the phase spectrum of the distorted pulse for the inverse to collapse the distorted pulse into an impulse. This is assured by the Wiener process if the distorted pulse is minimum-phase, as it is in this example.

The reverberated play data before and after Wiener processing are shown in Figure 7.13. The deconvolution operator for each trace was obtained from the autocorrelation of a window of 1 sec length that was truncated at 250 ms and was bandlimited with a 20–65 Hz brickwall filter. The impulse response of this zero-phase filter is the desired pulse $d(t)$. The reverberations are highly suppressed by the bandlimited inverse operators. Applying these operators, which are generated from the seismic data, to the reverberated source pulses reduces the reverberated pulses to relatively simple pulses.

The bandlimited inverse operator $\hat{c}^{-1} * d$ that is applied to the data in the case where $R = 0.6$ and $T = 80$ ms, and its energy spectrum, are shown in Figure 7.14. The exact deterministic inverse of the reverberation filter in this example is the Backus three-point operator $(1, 2R, R^2) = (1, 1.2, 0.36)$, and it can be seen in the bandlimited inverse waveform. The three large pulses marked on the waveform are separated by T, and their relative magnitudes are in reasonable agreement with the values of the Backus operator. Exact agreement should not be expected, because the Wiener operator is an estimate obtained from the data and, in addition to suppressing reverberations and receiver ghosts, also changes the waveform of $b(t)$ to that of $d(t)$.

Applying the Backus operator would have completely suppressed the reverberations in this simple case. However, reverberations on actual field records often are complicated by multilayers, irregular bottom conditions, and so forth, and application of the three-point operator is not a satisfactory solution. In these cases, the Wiener processor does a better job, because the inverse is estimated from the field data itself, and its degree of complication is of no consequence to the Wiener processor, so long as the Wiener operator is of sufficient length to contain the significant information in the inverse of $c(t)$. The length of the Wiener operator is identical to the

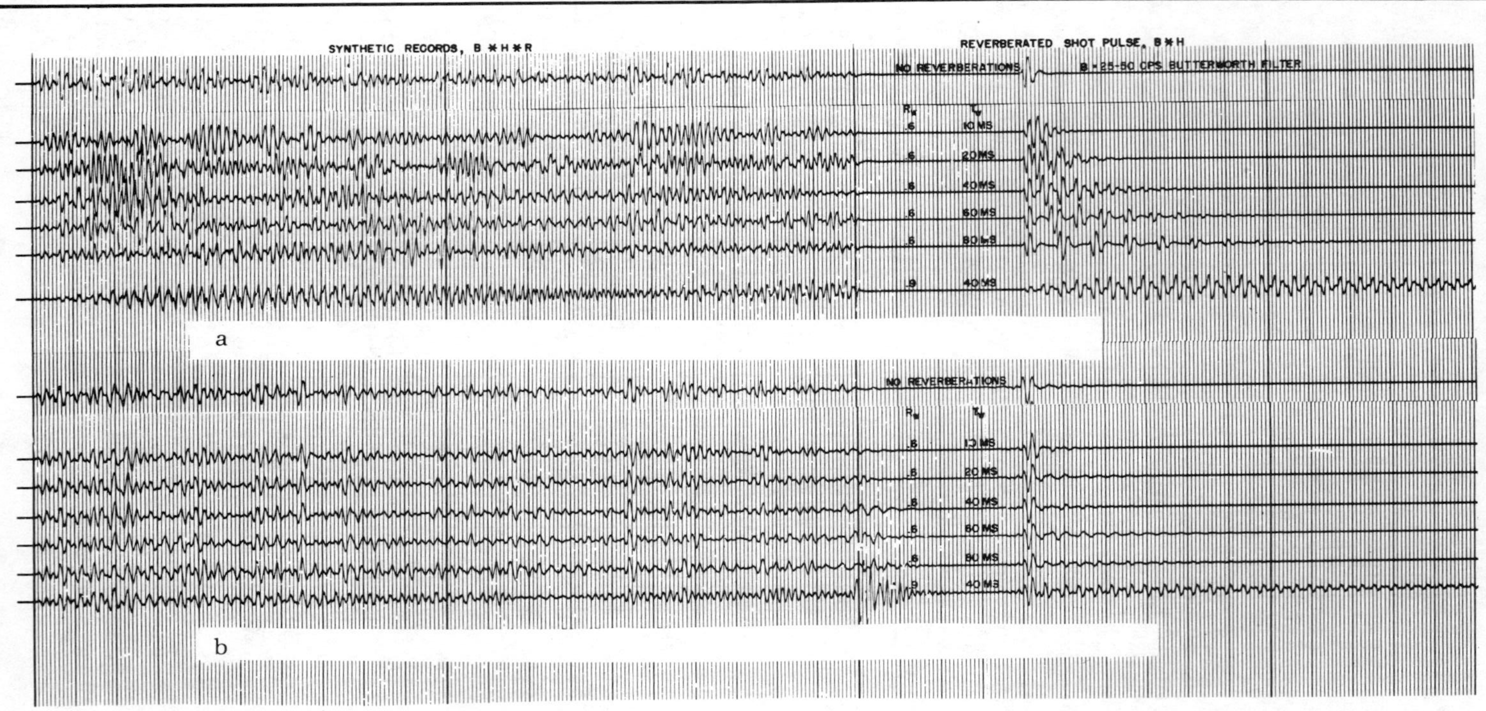

FIGURE 7.13. *Wiener optimal filtering of reverberated play data. (a) Reverberated records, surface source, pressure detector at* $T_p = 6$ *ms. (b) After bandlimited Wiener inverse filtering,* d(t) = *20–65 Hz brickwall filter. (From McDonal and Sengbush 1966.)*

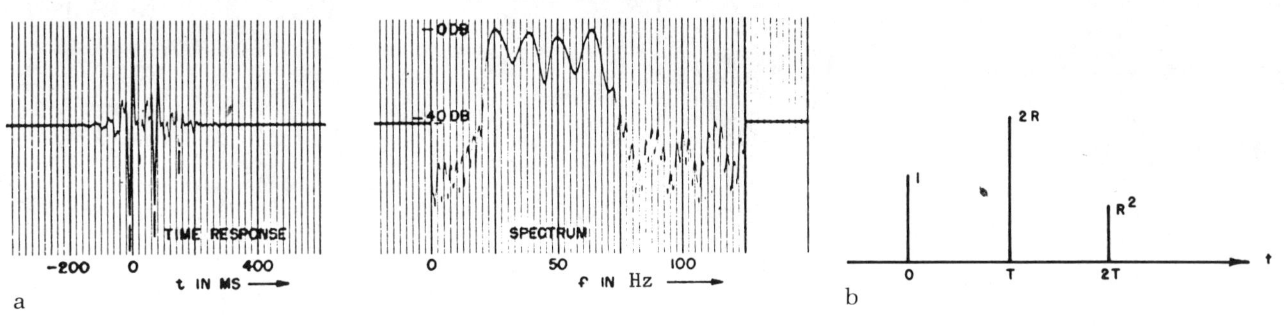

FIGURE 7.14. *Comparison of bandlimited Wiener inverse operator with exact deterministic inverse operator in reverberation case where* R = *0.6 and* T = *80 ms. (a) Bandlimited Wiener inverse operator and its energy spectrum. (b) Exact inverse of reverberation distortion operator. (From McDonal and Sengbush 1966).*

length of the truncator. To be effective in dereverberation, this operator must be of sufficient length to allow the three-point inverse to be developed. This requires an operator length L greater than $3T$. This points out the necessity of selecting parameter L and, hence, W in accordance with the distortions to be expected on the data.

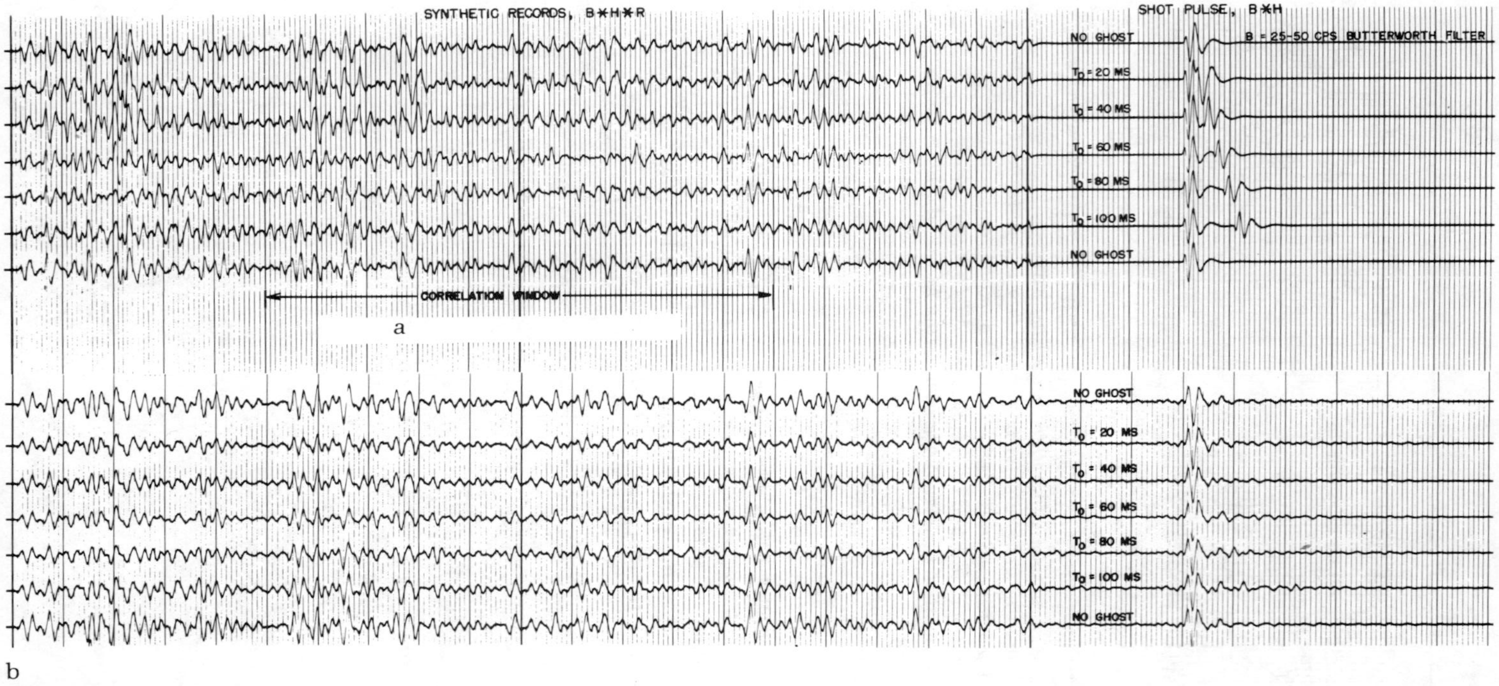

FIGURE 7.15. *Wiener optimal filtering of ghosted play data. (a) Ghosted records. (b) After bandlimited Wiener inverse filtering. (From McDonal and Sengbush 1966.)*

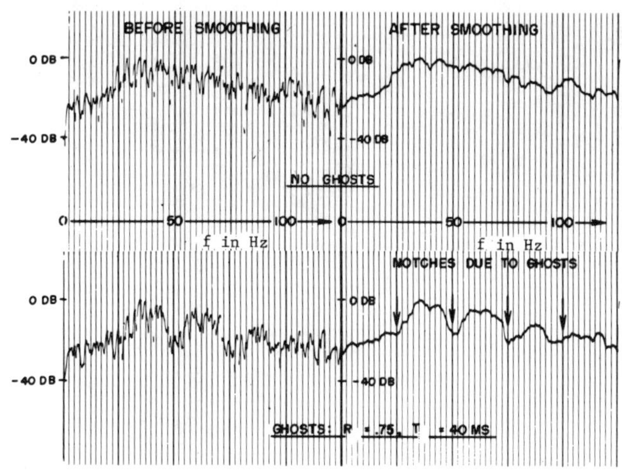

FIGURE 7.16. *Energy spectra for ghost and no-ghost cases. (From McDonal and Sengbush 1966.)*

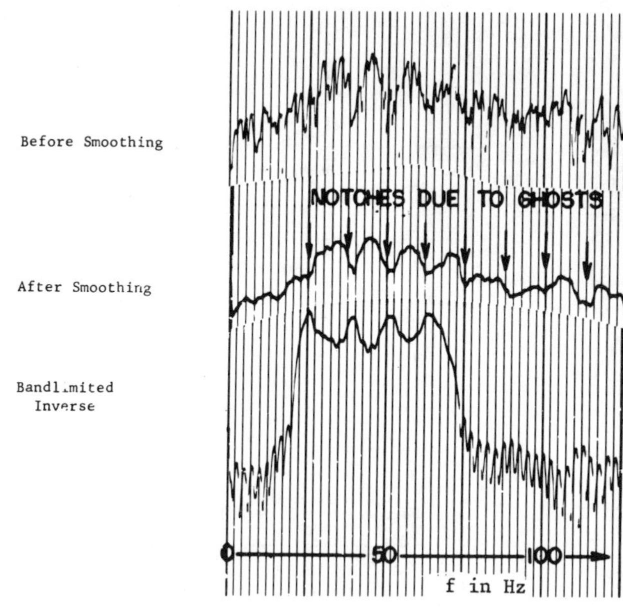

FIGURE 7.17. *Spectral smoothing to separate random reflectivity from deterministic distorted seismic pulse in ghost case where* R *= 0.75 and* T *= 80 ms. (From McDonal and Sengbush 1966.)*

GHOSTED PLAY DATA

The data contain distortion from simple ghosts and is noise-free. The distortion operators are two-point operators, with reflection coefficient $R = 0.75$ and two-way delay times T varying from 20–100 ms. The play data before and after Wiener processing are shown in Figure 7.15. The performance of the process is shown most clearly by the results of applying the bandlimited inverses obtained from the data to the distorted pulses. The ghost pulse is reduced by a factor of at least 5:1 in each case.

The spectra before and after spectral smoothing in the no-ghost case and in the ghost case where $R = 0.75$ and $T = 40$ ms are shown in Figure 7.16. The spectrum of the ghost distortion operator has notches at frequencies that are integral multiples of $1/T$, which in this case occur at integer multiples of 25 Hz. Spectral smoothing accomplished by truncating the autocorrelation of the window suppresses the rapid fluctuations in the energy spectrum of the window and gives a good estimate of the distorted pulse spectrum.

The bandlimited inverse spectrum has peaks where the distorted pulse spectrum has notches, and the product of the two results in a flat spectrum within the bandlimit allowed by the d-filter, as shown in Figure 7.17 for the case where $T = 80$ ms.

The bandlimited inverse in the case where $T = 60$ ms, shown in Figure 7.18, has a series of pulses separated by T and decreasing in amplitude. These pulses arise from the exact deterministic inverse of the ghost distortion operator. To get an effective Wiener inverse operator, the operator length L should be on the order of $3T$ to get the estimates of the first three impulses in the deterministic inverse. The succeeding impulses are usually small enough to ignore (the fourth impulse is of size R^3, and the succeeding impulses are still smaller).

As the ghost delay time increases, the Wiener processor becomes progressively less effective, because the notches in the spectrum caused by ghosts become closer together as T increases, and it becomes more and more difficult to get a clear separation between the spectral components of the reflectivity and the distorted pulse. With ghosts, the length L of the Wiener operator should

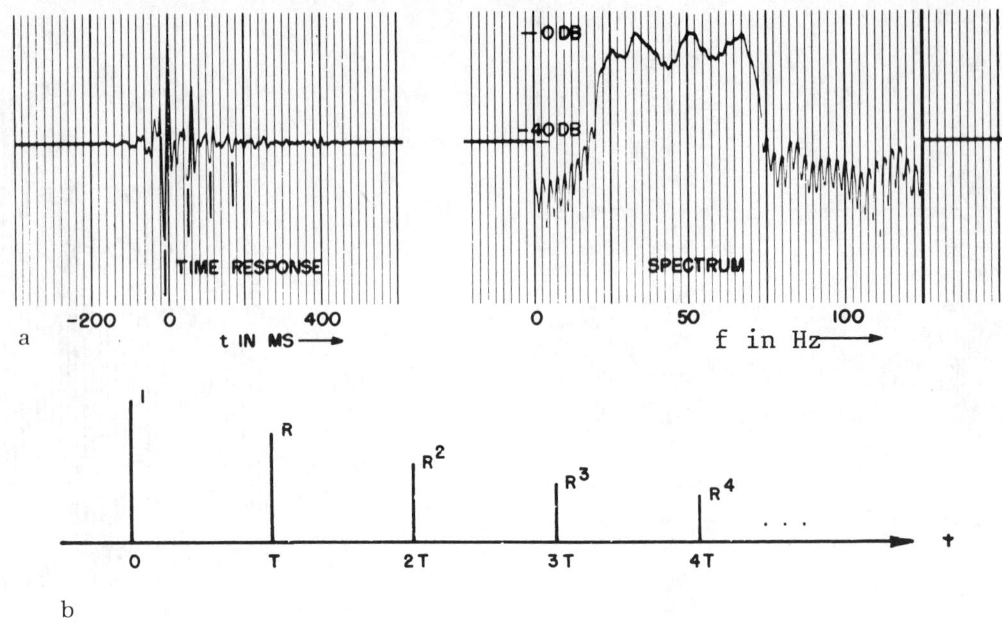

FIGURE 7.18. *Comparison of bandlimited Wiener inverse operator with exact deterministic inverse operator in ghost case where* R = *0.75 and* T = *60 ms. (a) Bandlimited Wiener inverse operator and its energy spectrum. (b) Exact inverse of ghost distortion operator. (From McDonal and Sengbush 1966.)*

be on the order of $3T$, and W should be on the order of $5L$ or more. Because real seismic data are time-variant and of finite length, it eventually becomes impossible to satisfy the foregoing constraints. Studies of play data suggest that the two-trace processor (Foster, Sengbush, and Watson, 1964) should be used rather than the Wiener processor if the ghost delay time exceeds 100 ms. This restriction is of little or no consequence in practice, because it would be rather uncommon to have field conditions such that the source would be placed that far below the ghosting interface.

The foregoing constraints limit the usefulness of Wiener processing in suppressing any type of long-period multiple, which means that other techniques, such as those described in Chapter 6, must be used to suppress such multiples.

WIENER PROCESSING OF FIELD DATA

REVERBERATED DATA

An example from the North Sea (from Rockwell, 1967) shows the improvement after deconvolution (Figure 7.19). Reflections that are completely obscured by reverberations become observable. This example shows the virtually impossible task faced by seismic interpreters in mapping structures on marine data before Wiener processing.

The autocorrelation function of the input data before deconvolution (Figure 7.20) is highly oscillatory, with period $T = 27$ ms, corresponding to reverberations with apparent frequency $f = 1/T = 37$ Hz. The input spectrum peaks at that frequency, as expected. The reverberation peaks in the spectrum have average separation Δf between peaks of 14.4 Hz. Therefore, this area must have a hard bottom (which is not surprising, as it is the North Sea), because the spectral peaks should be at 7.2, 21.6, 36.0, 50.4, 64.8, and 79.2 Hz, which is in good agreement with those observed on the input spectrum. The water depth z, based on the spectral peak separation $\Delta f =$ 14.4, calculates to be

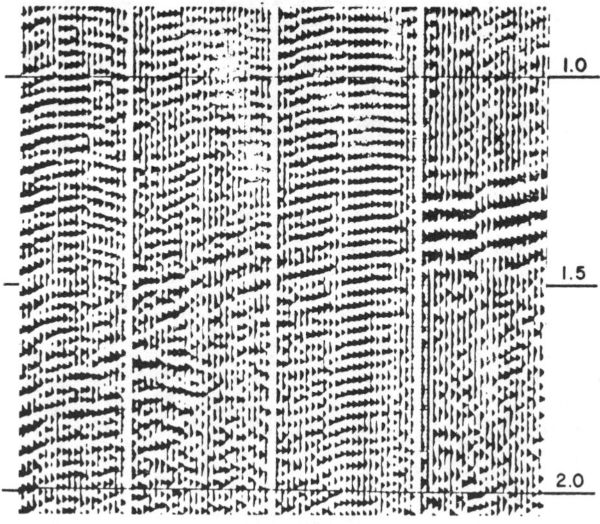

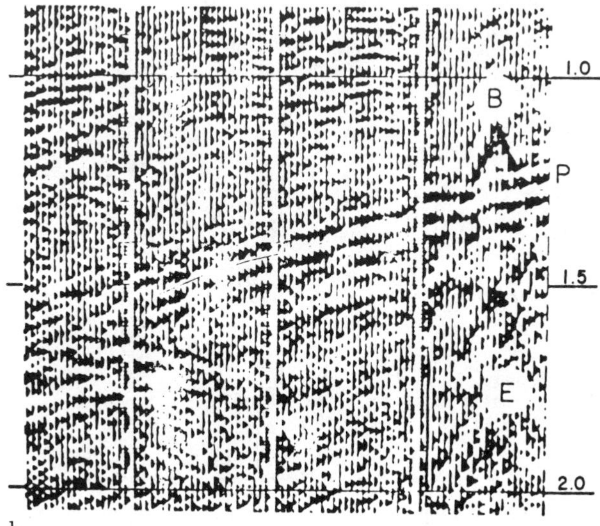

FIGURE 7.19. *Wiener deconvolution of North Sea data. (a) Record section (100 percent coverage) typical of English coastal waters with high-velocity sediments; true amplitude recovery, normal moveout, and statics applied. (b) After deconvolution. (From Rockwell 1967, courtesy of Geophysics.)*

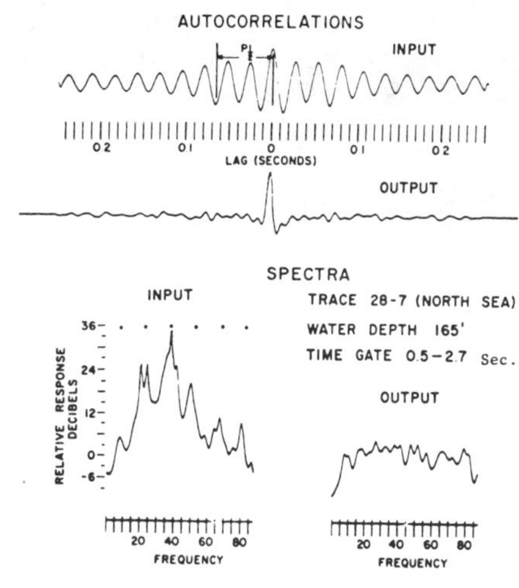

FIGURE 7.20. *Autocorrelation functions and power spectra of North Sea data before and after Wiener deconvolution. (From Rockwell 1967, courtesy of Geophysics.)*

$$z = \frac{VT}{2} = \frac{V}{2\Delta f} = 174 \text{ ft}, \tag{7.35}$$

which is in good agreement with the measured depth of 165 ft.

Autocorrelation after deconvolution shows that the reverberations are eliminated and the seismic pulse width is decreased, indicating spectral broadening. The output spectrum shows that deconvolution smooths the peaks and notches caused by the reverberations, broadens the bandwidth of data, and preserves the spectral content of the reflectivity.

Field comparisons invariably show Wiener processing to be superior to the three-point inverse. Because the three-point operator is the exact inverse in the one-layer reverberation model, it must be assumed that this model is only an approximation to reality. Extension of the exact deterministic inverse method to the multilayer case is impractical, because it requires knowledge of the reflection coefficient at each interface and traveltimes in each layer. Wiener processing circumvents the problem of parameter estimation.

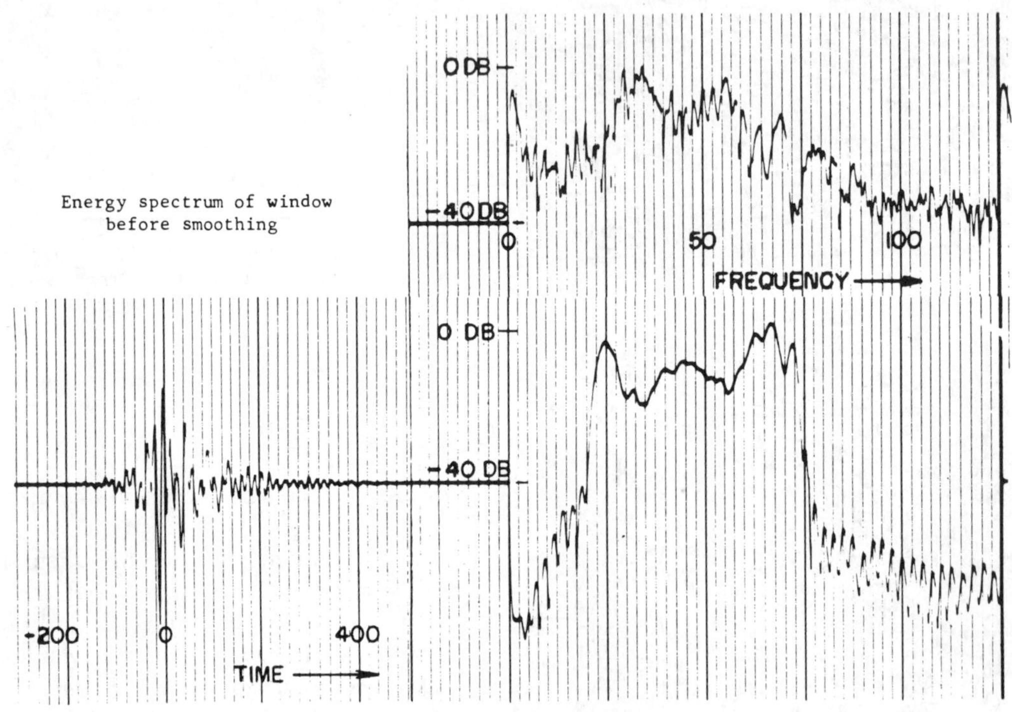

FIGURE 7.21. *Wiener processing of ghosted field data. (a) Ghost suppression using Wiener inverse operator. (b) Bandlimited inverse operator and its spectrum. (From McDonal and Sengbush 1966.)*

GHOSTED DATA

Wiener processing was applied to a set of field traces using the procedures described previously for the play data processing. Lumped charges at depths from 25–308 ft show ghosts on the variable-area display in Figure 7.21. After Wiener processing, the ghosts are insignificant within the window. Because of time variance in seismic data, the Wiener operator is generally dependent on the location of the window. Hence, a different operator should be used for ghost suppression beyond 1.5 sec.

Details of Wiener processing of a field data trace are also shown in Figure 7.21. The energy spectrum before smoothing shows the rapid fluctuations of the underlying reflectivity function. The distorted pulse is unknown, but its inverse is estimated by the Wiener processor. The bandlimited inverse operator has pulses with about 40 ms separation, which indicates that the distortion operator includes a simple ghost that has about 40 ms delay time. This produces notches in the spectrum of the input data at integral multiples of 25 Hz, and the spectrum of the bandlimited inverse shows peaks at about these frequencies, as expected.

Deghosting by Wiener processing is not subject to any severe restrictions on source depth; however, the Wiener method becomes less effective as the ghost delay time increases beyond 100 ms. The major advantage of Wiener processing over other deghosting schemes is that variations in the input source signature from record to record are compensated for and more consistent data are produced over an area.

WIENER DECONVOLUTION AND NONMINIMUM-PHASE DATA

Wiener deconvolution attempts to compress the distorted seismic pulse waveform $c(t)$ into an impulse and then is required to bandlimit the data with the desired seismic pulse $d(t)$ because the inversion is inaccurate in the low- and high-frequency regions as a result of poor signal-to-noise ratio in these regions.

Compressing c into an impulse means that the inverse must have an amplitude spectrum approximating the inverse of the amplitude spectrum of c and a phase spectrum approximating the negative of the phase spectrum of c. Because all information about c comes from an autocorrelation function, the amplitude spectrum is preserved in the estimate of c, and the phase is lost. However, the Wiener processor assumes that c is minimum-phase; hence, knowledge of the amplitude spectrum is tantamount to knowledge of the phase as well. If c is actually minimum-phase, then the inverse $\hat{c}^{-1}$ has the correct phase response to compress c into an impulse. If c is not minimum-phase, then the phase response of $\hat{c}^{-1}$ is not correct, and compression to an impulse is not achieved. Wiener processing in such cases may, in fact, render the seismic data useless.

The problems arising from applying Wiener deconvolution to nonminimum-phase data are shown in Figure 7.22 (from Shugart, 1973). Applying the Wiener inverse operator to the minimum-phase signature in that figure results in an impulse. However, applying the Wiener operator to a signature that has the same amplitude spectrum but has a constant-phase spectrum results in an oscillatory train, rather than an impulse. This example illustrates the havoc that would be created if Wiener processing were applied to nonminimum-phase data.

Sources such as Maxipulse (Western Geophysical), Vaporchoc (CGG), and air guns without adequate bubble suppression are not minimum-phase, and special deconvolution treatment must be applied. It is necessary to convert the seismic data into minimum-phase data by a preprocessing stage before applying Wiener deconvolution to suppress distortions. This requires measurement of the input source signature and computation of an operator (called the signature deconvolution operator) that will convert this signature to an impulse or some other minimum-phase signature. Then, the signature deconvolution operator is applied to the seismic data to make the data minimum-phase before applying Wiener deconvolution to suppress distortions.

In actual practice, Vaporchoc data are processed by a technique that CGG calls Wapco, as described by Fourmann (1974). Let $b(t)$ be the signature and let $b_m(t)$ be its minimum-phase counterpart. Their amplitude spectra are equal, $|B| = |B_m|$. It is desired to find the inverse of b, call it x, such that $b * x = \delta$, or, in the frequency domain $X = 1/B$. Multiplying numerator and denominator by the conjugate spectrum $\bar{B}$ gives

$$X = \frac{\bar{B}}{B\bar{B}}. \qquad (7.36)$$

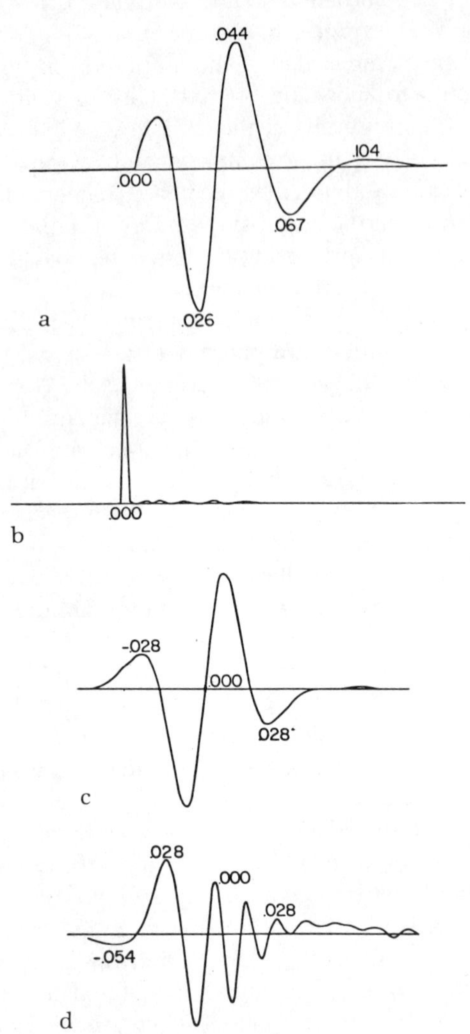

FIGURE 7.22. *Comparison of Wiener processor in minimum-phase and nonminimum-phase cases. (a) Minimum-phase wavelet with 16–32 Hz bandwidth. (b) Convolution of Wiener inverse operator with minimum-phase wave-phase wavelet. (c) Wavelet with same amplitude spectrum as wavelet in (a), but with phase spectrum constant and equal to 90° for all frequencies. (d) Convolution of Wiener inverse operator with nonminimum-phase wavelet. (From Shugart 1973, courtesy of Geophysics.)*

Because b and b_m have the same autocorrelation function in the frequency domain, $B\bar{B} = B_m\bar{B}_m$, so that

$$X = \frac{\bar{B}}{B_m\bar{B}_m}. \tag{7.37}$$

Let x_m be the inverse of b_m; therefore,

$$X = \bar{B}X_m\bar{X}_m. \tag{7.38}$$

Define the Wapco operator $D = BX_m$. Then,

$$X = X_m\bar{D}. \tag{7.39}$$

Now the situation is clear. The reverberated seismic data $s = b * h * r + n$ should be convolved with x to perform signature deconvolution,

$$\begin{aligned} s * x &= b * x * h * r + n * x \\ &= h * r + n * x, \end{aligned} \tag{7.40}$$

followed by Wiener deconvolution to suppress the reverberations h. However, it should be noted that x_m is minimum-phase, so its effect on the data can be taken care of by the subsequent Wiener deconvolution. Therefore only $\bar{d}$, rather than x need be applied to s:

$$s * \bar{d} = b * \bar{d} * h * r + n * \bar{d}, \tag{7.41}$$

and Wiener deconvolution applied to $s * \bar{d}$.

Figure 7.23 shows the Wapco process applied to the signature b. The final result of convolving b with $\bar{d}$ and then by the Wiener operator w (obtained by autocorrelating $b * \bar{d}$) shows that $b * \bar{d} * w$ is essentially an impulse. This is what the Wapco process set out to do—to convert a nonminimum-phase signature into an impulse, using Wiener deconvolution.

Wavelet processing is a technique that uses signature deconvolution en route to modifying the source signal that is actually generating the recorded seismic data into a source signal that has more desirable and known characteristics, such as minimum-phase or zero-phase with specified bandwidth and rejection rates. To do this, the seismic pulse must be measured near the source or estimated from the recorded reflection data. The reflectiv-

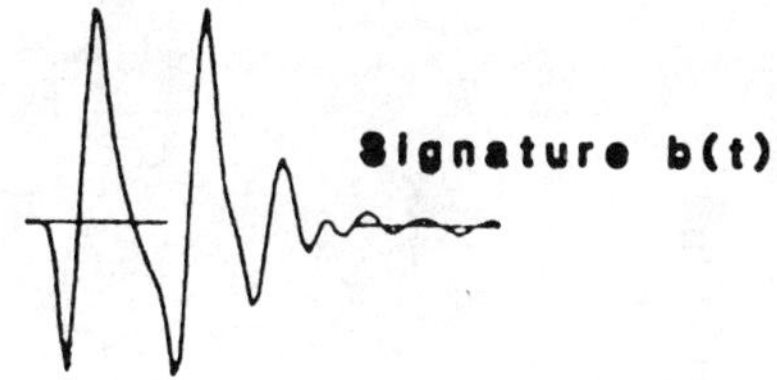

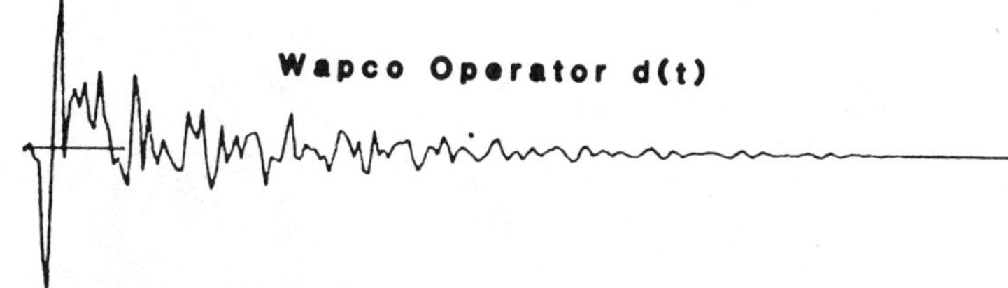

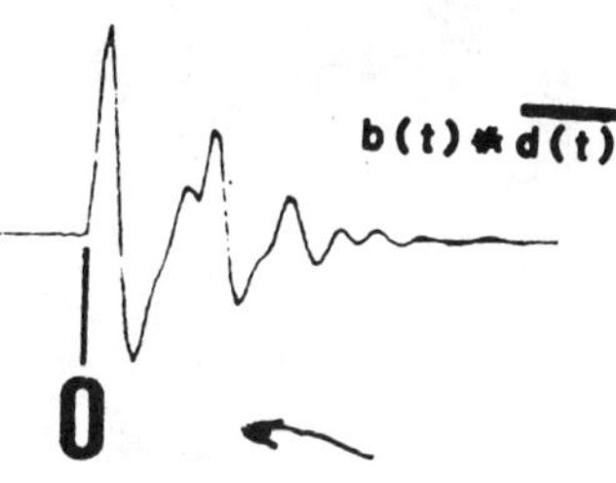

Wiener deconvolution operator w(t)

$b(t) * \overline{d(t)} * w(t)$

0

200ms

FIGURE 7.23. *Wapco processing of nonminimum-phase signature. (From Fourmann 1974, courtesy of CGG.)*

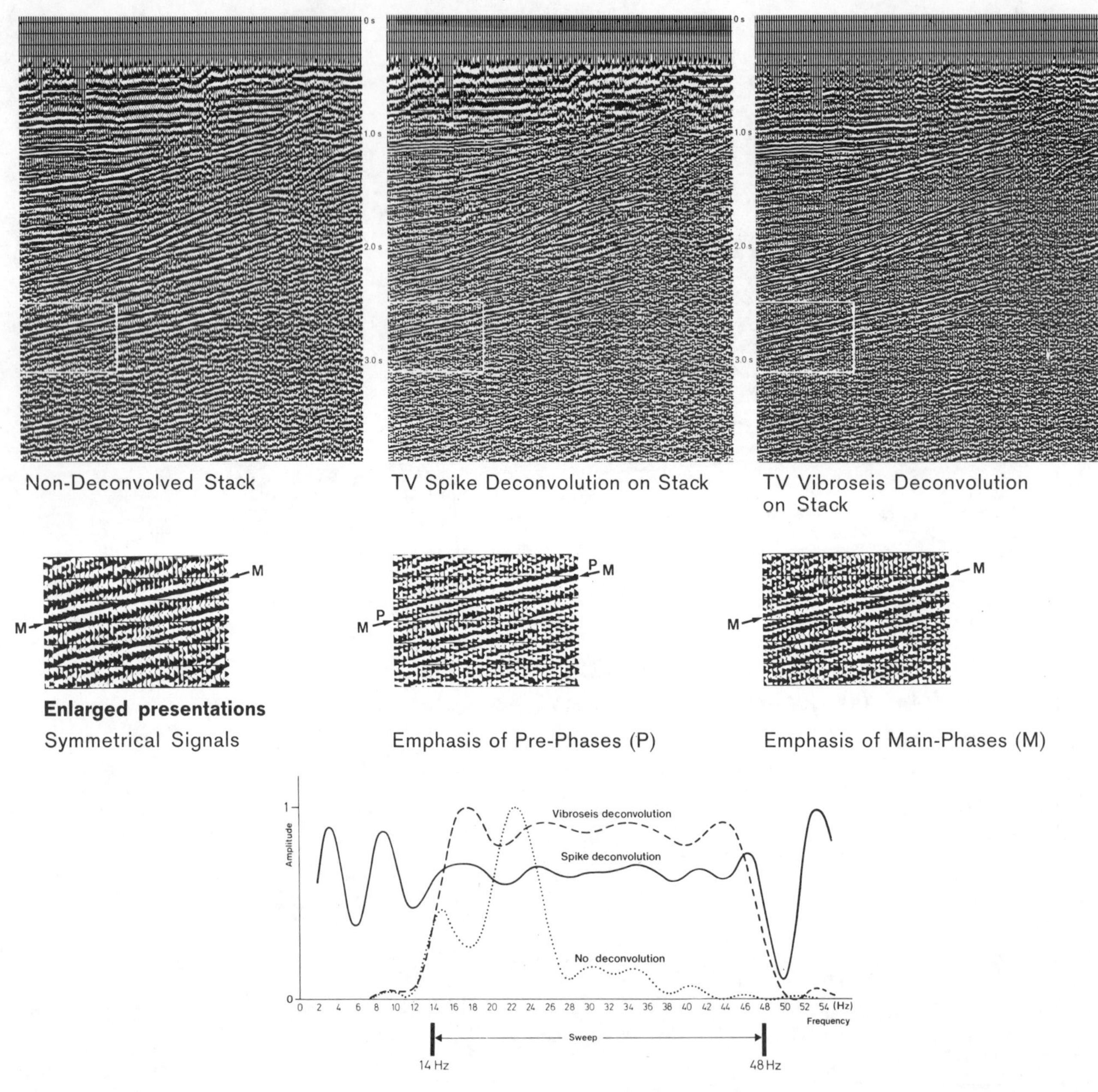

FIGURE 7.24. *Comparison of Wiener and zero-phase deconvolution of Vibroseis data. (Courtesy of Prakla-Seismos Report.)*

ity from an impedance (or velocity) log in a nearby well, if available, may be used in estimating the seismic pulse. If the reflectivity has the characteristics of white noise and the seismic pulse is time-invariant, then, in the noise-free case, cross correlation of the reflectivity with a seismic trace near the well will produce the seismic pulse that generated the trace, according to Lee (1960) (as discussed in Appendix D in the section on random inputs to linear systems). Inasmuch as the reflectivity is only approximately white, the seismic pulse is slowly time-variant, and the seismic trace is not noise-free, the cross correlation will give only an estimate of the seismic pulse that generated the trace, and the estimate will be only as good as the deviations from the ideal will allow.

OTHER TYPES OF DECONVOLUTION

Predictive deconvolution (Peacock and Trietel, 1969) is an application of Wiener processing that does not attempt to compress the distorted seismic pulse to an impulse, but instead preserves the waveform up to a certain time, called the prediction distance, and reduces the tail of the distorted pulse to zero. In effect, this combines inversion with bandlimitation and eliminates or at least reduces the need to apply another bandlimiting filter, such as the *d*-operator. Wiener deconvolution that attempts to compress the distorted pulse into an impulse is sometimes called spiking deconvolution. Spiking deconvolution is a special case of predictive deconvolution in which the prediction distance is one sample interval.

Vibroseis data present an enigma with respect to deconvolution, because the effective input pulse (the Klauder wavelet obtained by autocorrelation of the sweep signal) is zero-phase and all the rest of the seismic system, attenuation, distortion caused by ghosts and reverberations, detectors, instruments, and recording and aliasing filters are minimum-phase. Therefore, neither zero-phase deconvolution nor Wiener deconvolution is really appropriate. Experimental evidence by Prakla-Seismos (Report 3-74) shows that use of Wiener deconvolution with Vibroseis data produces precursors on reflections that do not exist on the stacked data, and that zero-phase deconvolution broadens the bandwidth without producing the precursors (Figure 7.24).

The restriction to minimum phase is removed with homomorphic deconvolution (Oppenheim and Schafer, 1975). This technique has been used in speech and photographic processing. A homomorphic system is a generalized linear system. Recall that a linear system is defined by the properties of superposition and homogeneity:

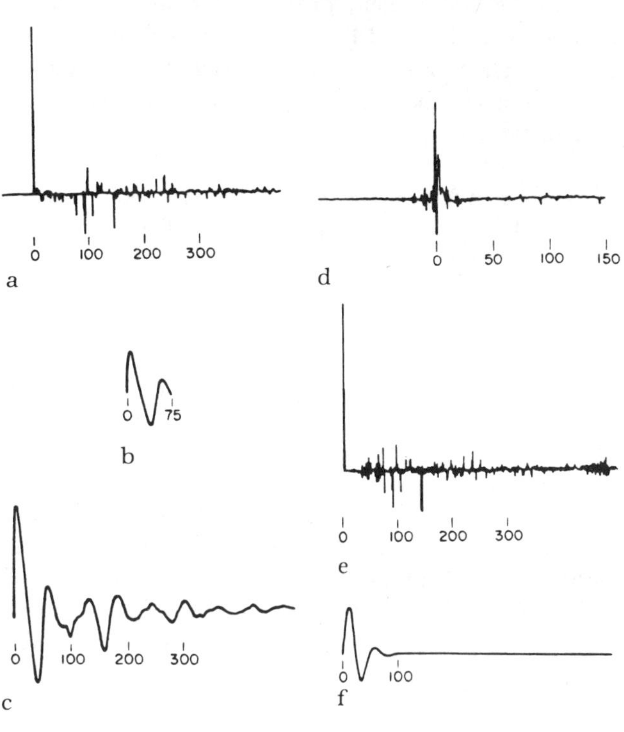

FIGURE 7.25. *Homomorphic deconvolution of synthetic data. (a) Reflectivity near Leduc, Alberta (after O. Jensen). (b) Seismic source pulse. (c) Synthetic seismogram. (d) Complex cepstrum of synthetic seismogram, exponentially weighted with $\alpha = 0.985$. (e) High-pass output. (f) Low-pass output. (From Ulrych 1971, courtesy of Geophysics.)*

$$\text{Superposition:} \quad \mathscr{L}[g + h] = \mathscr{L}[g] + \mathscr{L}[h], \tag{7.42}$$

$$\text{Homogeneity:} \quad \mathscr{L}[cg] = c\mathscr{L}[g], \tag{7.43}$$

where g and h are functions and c is a scalar.

In a homomorphic system, the addition operations in the superposition property may be replaced by any mathematical operations that satisfy the algebraic postulates of vector addition, and the multiplication operations in the homogeneity property may be replaced by any operations that satisfy the postulates of scalar multiplication. Then the homomorphic system transformation $\mathcal{H}$ that replaces the linear transformation $\mathcal{L}$ is an algebraic linear transformation from an input vector space to an output vector space.

In the seismic case, the complex cepstrum of the data, which is the inverse Fourier transform of the logarithm of the Fourier transform of the data, is operated on by a linear operator to separate the seismic pulse from the reflectivity. Ulrych (1971) shows the effectiveness of the method on noise-free synthetic data (Figure 7.25). He also used homomorphic methods to estimate the seismic wavelet from a teleseis event originating in Venezuela and measured in Alberta, Canada.

The results of processing seismic data with homomorphic deconvolution have not been very useful to date, and considerably more work must be done to establish its applicability. The two problems of noise and time-variance that plague Wiener deconvolution probably have their counterparts with homomorphic deconvolution.

SUMMARY

The linear filter model of the seismic reflection process provides a basis for interpreting field seismograms in terms of fine subsurface detail (McDonal and Sengbush, 1966). The synthetic seismogram, prepared within the framework of this model, provides the link between the seismogram and the interpretation of geological features such as pinchouts, facies change, sand lenses, and reefs. This chapter extends the concept of the linear filter model to include Wiener deconvolution to suppress distortions associated with reverberations and ghosts and compensate for changes in the source pulse waveform that may occur from record to record.

Chapter 8 Velocity Estimation

INTRODUCTION

Good velocity estimates of velocity as a function of depth are of paramount importance in the interpretation of seismic data. The time-honored technique of shooting a well for velocity gives good estimates of the gross velocity layering. Fine detail is provided by continuous velocity logs. These two methods provide good, accurate information in the wells, but what can be done where there are no wells? Experience has shown that velocity functions such as a linear function with depth, $V = V_0 + az$, where V_0 is the velocity at datum and a is constant, have some merit in velocity estimation (Slotnik, 1959). An intriguing idea, however, is using surface measurements to estimate the velocity from the normal moveout of seismic events.

Velocity estimation from surface seismic measurements using Δt analysis based on normal moveout has been done for many years (Green, 1938; Steele, 1941; Gardner, 1947; Brustad, 1953; Pfleuger, 1954; Slotnik, 1959). The weaknesses in these techniques arise because (1) small errors in Δt result in large errors in V and consequently in the calculated depth, and (2) straight-line paths in media with uniform velocity are often assumed, rather than using least-time paths in layered media that obey Snell's Law.

Into this arena, where the importance of velocity was well recognized but the techniques for estimation were inaccurate, rode Professor C. Hewitt Dix, with his expanding spread technique for accurate estimation of interval velocities from surface seismic measurements.

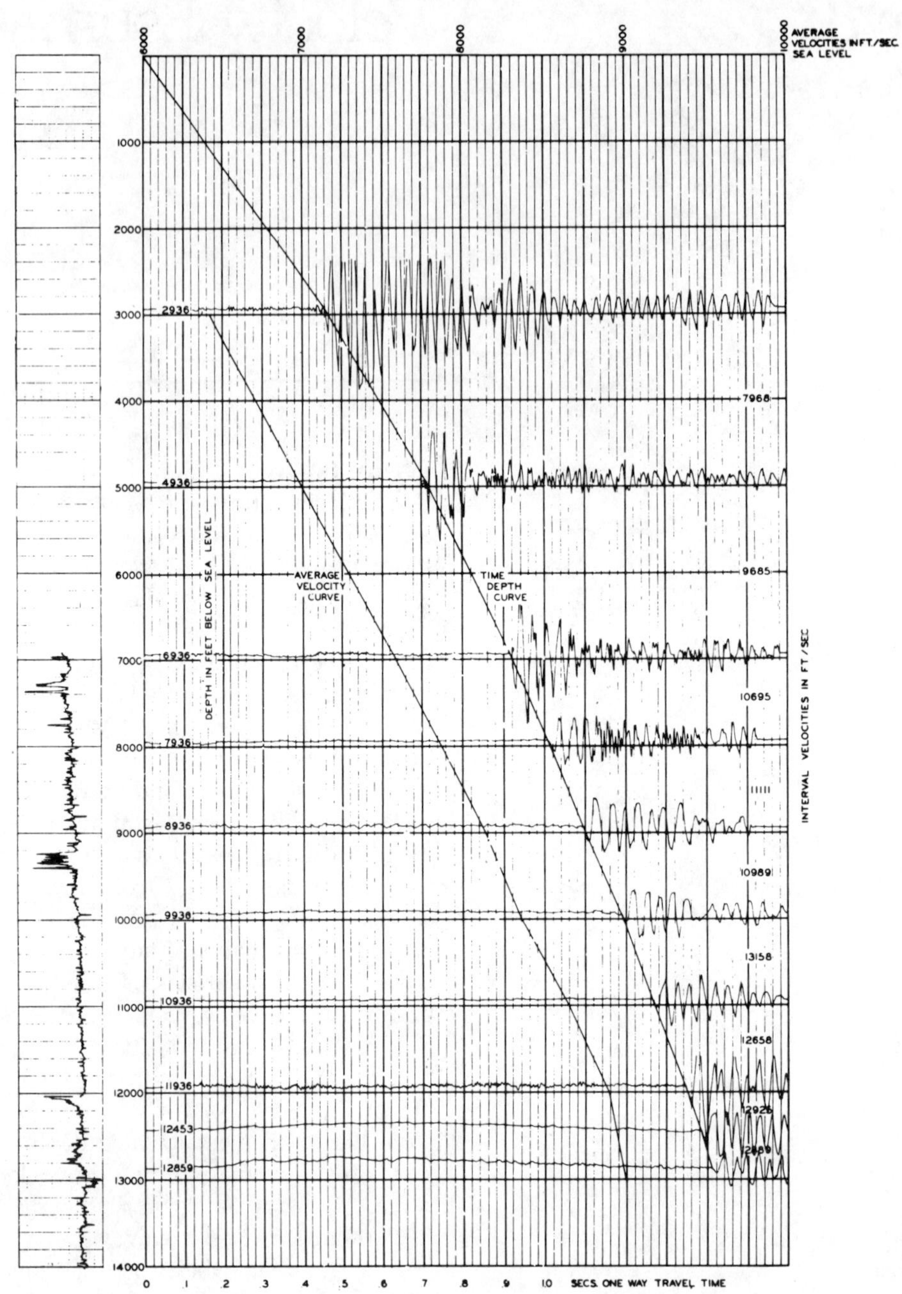

FIGURE 8.1. *Time-depth measurements from seismic velocity survey.*

Dix's technique had a limited life because of the subsequent development of horizontal stacking, but his theoretical development is the cornerstone to an understanding of velocity estimation from surface measurements. The widespread use of horizontal stacking has led to an abundance of velocity data that have proved useful for lithologic studies, in addition to their primary purpose of enhancing the stack to increase the primary-to-multiple ratio and to improve the signal-to-noise ratio.

SHOOTING A WELL

Estimates of velocity in boreholes by shooting a well involve measurements of traveltime of seismic waves from sources at or near the surface to a detector located at various depths in the well. To obtain vertical traveltimes, the measurements must be corrected for angularity and refraction effects, because the source must necessarily be offset from the well bore. Well shooting gives good estimates of the gross velocity layering.

Figure 8.1 shows the results of shooting a well. The detector depths below sea level at which measurements were made are listed on the left. These depths are usually called the checkpoints. The received signals at the checkpoints are plotted on the graph. The time-depth pairs (t_i, z_i) form the time-depth curve, and the ratios z_i/t_i give the average velocity versus depth. For example, the average velocity to the base of the second layer is

$$V_{\text{ave},2} = \frac{z_2}{t_2} = \frac{4936}{0.711} = 6942 \text{ ft/sec.} \tag{8.1}$$

The interval velocity between adjacent checkpoints is listed on the right side of the figure. The interval velocity in the ith layer is given by

$$V_i = \frac{z_i - z_{i-1}}{t_i - t_{i-1}}. \tag{8.2}$$

For example, in the second layer between $z_1 = 2936$ ft and $z_2 = 4936$ ft, the one-way times are $t_1 = 0.460$ sec and $t_2 = 0.711$ sec. Therefore, the interval velocity in the second layer is

$$V_2 = \frac{z_2 - z_1}{t_2 - t_1} = 7968 \text{ ft/ms.} \tag{8.3}$$

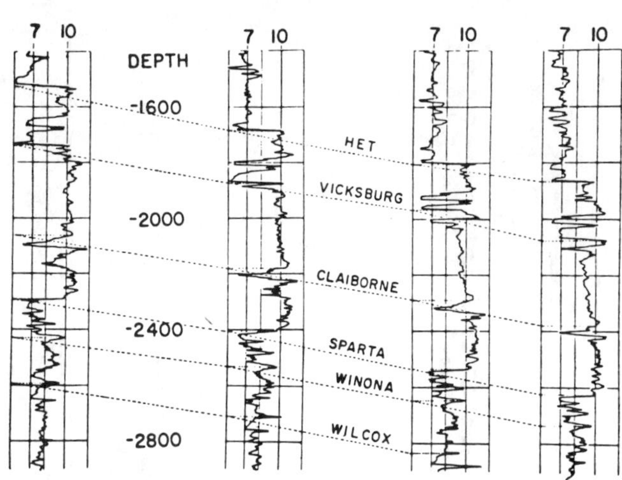

FIGURE 8.2. *Velocity log correlation, Pearl River County, Mississippi. (From Summers and Broding 1952, courtesy of Geophysics.)*

The relationship between the interval velocities V_i and the average velocity to the base of the nth layer is given by

$$V_{\text{ave},n} = \sum_{i-1}^{n} T_i V_i \Big/ \sum_{i-1}^{n} T_i. \tag{8.4}$$

CONTINUOUS VELOCITY LOGS

The invention of the continuous velocity log (CVL) by Summers and Broding (1952) provided, for the first time, the fine detail of the velocity layering. This type of log also is a very good correlation tool (Figure 8.2) and is useful in computing formation parameters, such as porosity.

The CVL tool consists of an acoustic transmitter and either one or two pressure detectors, with acoustic insulators separating the elements by known fixed distances (Figure 8.3). The source usually is an electromechanical device that produces acoustic pulses in the 10–20 kHz frequency range. By using a pulse repetition rate of about 20 pulses/sec and moving the tool slowly up the hole, the set of traveltime measurements produces essentially a continuous measure of transit time, either between transmitter and receiver or between the two receivers, depending on the tool configuration.

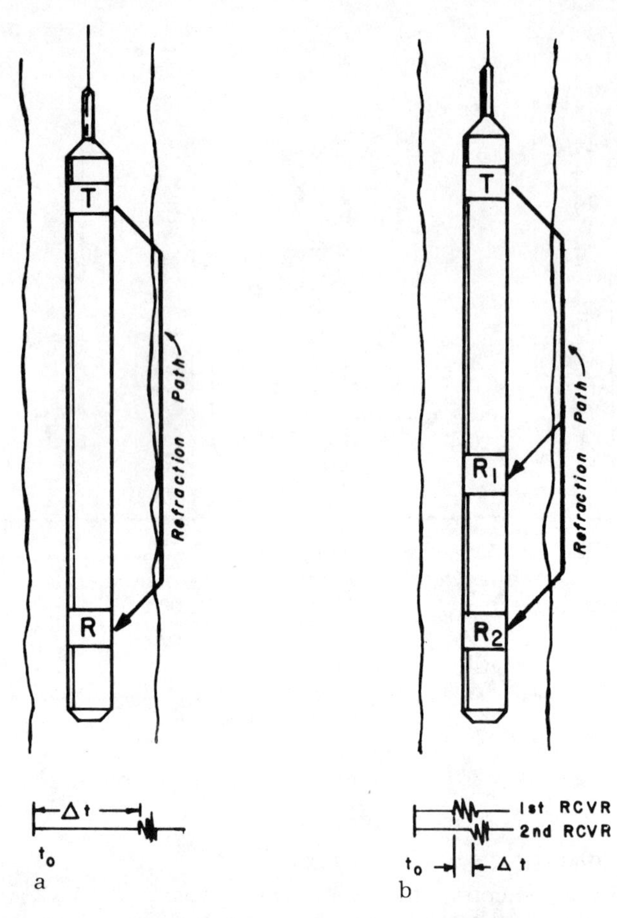

FIGURE 8.3. *CVL boreholed instruments. (a) One receiver. (b) Two receivers.*

The velocity obtained from the CVL measurements is the velocity averaged over the separation distance between elements. To obtain good velocity estimates, the refraction effects must be compensated for. This is especially important with the one-receiver tool, whereby the pulse refracts into the formation at the source and out of the formation at the receiver. With the two-receiver tool, the refraction effects are minimized, because only differences in the refraction paths out of the formation at the two receivers must be considered. This has led to almost universal usage of the two-receiver system.

By integrating the transit-time curve, a time-depth function can be constructed. Errors in the transit-time measurements caused by formation damage during drilling, borehole cavities and other irregularities, low signal amplitude because of absorption, calibration inaccuracies, and so on, will introduce errors in the time-depth computation. For this reason, the integrated CVL curve usually is tied to a well survey in order to produce a time-depth function that is accurate throughout the entire section of open hole that is being logged. The CVL cannot measure formation velocities that are less than the borehole fluid velocity, such as may happen in gas sands.

DIX'S EXPANDING SPREAD TECHNIQUE

A very significant breakthrough in precision of estimating interval velocities from surface measurements was Dix's (1955) expanding spread technique (Figure 8.4). The field procedure, described by Hansen (1947), had been used by Western Geophysical before 1941 and by Romberg in 1938. By expanding the distance between source and detectors in such a way that large horizontal separation between source and detectors is achieved while illuminating only the subsurface coverage of a symmetrical split-spread, the slope of the reflection arrivals in (T_x^2, x^2) space can be determined with considerable accuracy; this slope, which is related to the velocity distribution, applies in the vicinity of the midpoint of the split spread.

Under the assumption of horizontal layering (no dip), Dix showed that the least-time path through layered media for small offsets leads to the concept of the rms velocity, V_{rms}. The rms velocity in the n-layer case is given by

$$V_{\text{rms},n} = \left(\sum_{i=1}^{n} T_i V_i^2 \Big/ \sum_{i=1}^{n} T_i \right)^{1/2}, \tag{8.5}$$

where T_i is the two-way traveltime and V_i is the interval velocity in the ith layer. The importance of the rms velocity, which is a mathematical quantity devoid of physical meaning, is that the interval velocity layering is measurable from the rms velocity distribution. The interval velocity in the nth layer is obtained by subtracting the equation for the rms velocity in the $(n-1)$th-layer case

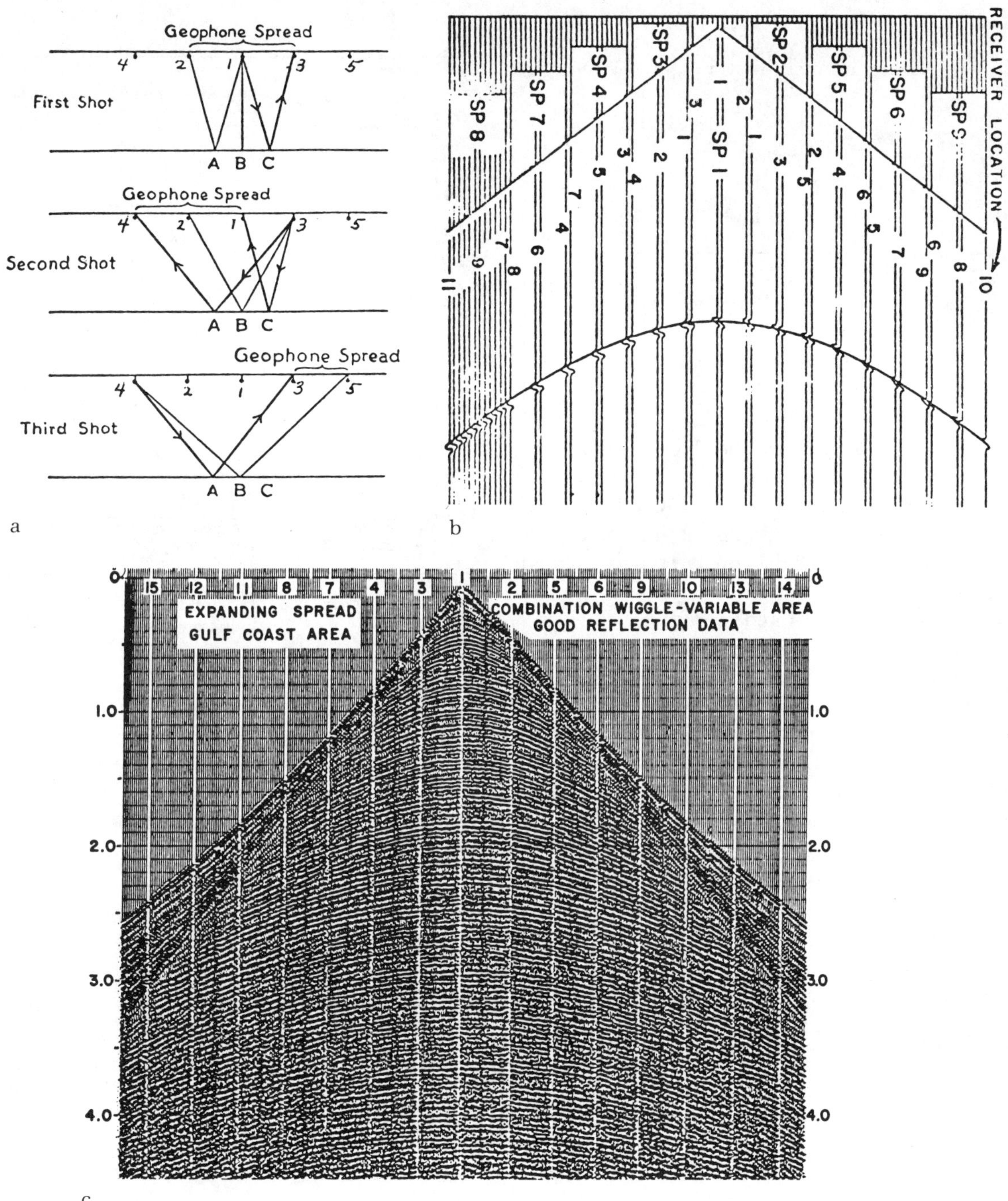

FIGURE 8.4. *Expanding spread technique. (a) Geometry of expanding spread. (From Dix 1955, courtesy of Geophysics.) (b) Record section display of expanding spread. (From Musgrave 1962, courtesy of Geophysics.) (c) Expanding spread cross section in lower Gulf Coast. (From Musgrave 1962, courtesy of Geophysics.)*

from the rms velocity in the nth-layer case. The result is

$$V_{\mathrm{rms},n}^2 \sum_{i=1}^{n} T_i - V_{\mathrm{rms},n-1}^2 \sum_{i=1}^{n-1} T_i = T_n V_n^2. \tag{8.6}$$

The interval velocity in the nth layer, V_n, is then

$$V_n = \sqrt{\left(V_{\mathrm{rms},n}^2 \sum_{i=1}^{n} T_i - V_{\mathrm{rms},n-1}^2 \sum_{i=1}^{n-1} T_i\right) \Big/ T_n}. \tag{8.7}$$

The expanding spread data are analyzed and the internal velocities calculated using equation (8.7). All quantities on the right-hand side are measurable from the data; the rms velocities are the slopes of reflection arrival times in (T_x^2, x^2) space, and $\Sigma_i T_i$ are the arrival times of the events, duly compensated for filter delay. From the interval velocities and interval times, the average velocity to the base of the nth layer is calculated using equation (8.4).

In Figure 8.5, the interval and average velocities and vertical time versus depth curves obtained from an expanding spread are compared with those quantities obtained from a conventional well survey (Musgrave, 1962). There are differences in the interval velocities from the two methods that create small differences in the average velocities and almost imperceptible differences in the time versus depth curve.

The rms velocity is always greater than or equal to the average velocity, as shown most clearly by Al-Chalabi (1974), who relates the two velocities through a quantity he calls the heterogeneity factor, H. The relationship is

$$V_{\mathrm{rms}} = (1 + H)^{1/2} V_{\mathrm{ave}}, \tag{8.8}$$

where the heterogeneity factor to the nth interface, H_n, is given by

$$H_n = 1/z_n^2 \sum_{i=1}^{n-1} z_i \sum_{j=i+1}^{n} z_j (V_i - V_j)^2 / V_i V_j. \tag{8.9}$$

The heterogeneity factor is nonnegative and equals zero if and only if $V_i = V_j$ for all (i,j), that is, when the earth is homogeneous.

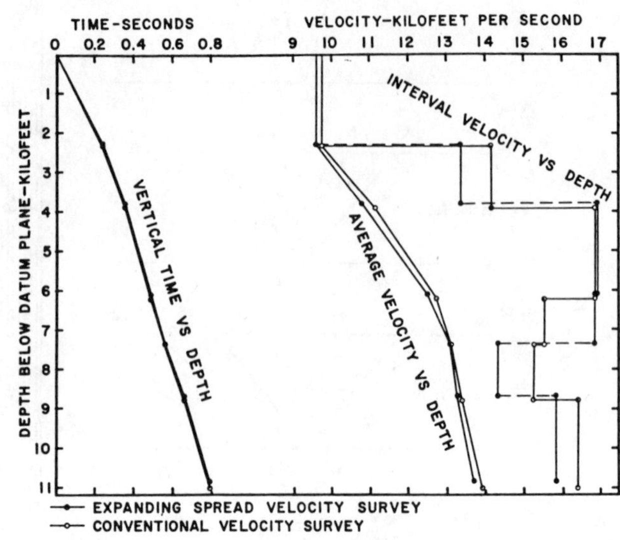

FIGURE 8.5. *Comparison of expanding spread with conventional well survey. (From Musgrave 1962, courtesy of Geophysics.)*

Dix's rms velocity is derived on the assumption of nondipping beds. Shah (1973) extended the concept of rms velocity to dipping beds, with arbitrary and nonuniform dips. He derives what he calls the NMO velocity, defined by

$$V_{\mathrm{nmo},n}^2 = 1/(T_0 \cos^2 \beta_0) \sum_{i=1}^{n} V_i^2 T_i \prod_{j=1}^{i-1} (\cos^2 \alpha_j / \cos^2 \beta_j), \tag{8.10}$$

where α_j is the angle of incidence and β_j is the angle of transmittance at the jth interface, and β_0 is the emergence angle of the raypath that is normal to the nth interface. For nondipping beds, the NMO velocity reduces to Dix's rms velocity.

In the case of a single dipping layer, V_{nmo} reduces to a result previously given by Levin (1971):

$$V_{\mathrm{nmo}} = V_1 / \cos \beta_0. \tag{8.11}$$

In the dipping case with all layers having the same dip,

$$V_{\text{nmo}}^2 = 1/(T_0 \cos^2 \beta_0) \sum_{i=1}^{n} V_i^2 T_i \quad \text{and}$$

$$V_{\text{nmo}} = V_{\text{rms}}/\cos \beta_0. \tag{8.12}$$

In the case where horizontal layers overlie the dipping nth layer, $\alpha_1 = \beta_0$, $\alpha_2 = \beta_1, \ldots,$. Therefore,

$$\prod_{j=1}^{i-1} (\cos^2 \alpha_j/\cos^2 \beta_j) = \cos^2 \alpha_1/\cos^2 \beta_{i-1})$$

and

$$V_{\text{nmo}}^2 = 1/T_0 \sum_{i=1}^{n} V_i^2 T_i/\cos \beta_{i-1}. \tag{8.13}$$

Interval velocities cannot be obtained by subtracting the NMO velocity to the $(n-1)$th layer from the NMO velocity to the nth layer in analogy with the Dix rms velocity technique, because the angles α_k and β_k to the nth layer differ from the angles to the $(n-1)$th layer.

Expanding spreads proved to be very useful for accurately determining interval velocities from which gross lithology could be deduced and for identifying multiples (Musgrave, 1962). However, some modifications in field procedures were required, which meant added costs to the routine seismic survey. The introduction of horizontal stacking procedures (Mayne, 1962) gave rise to techniques for velocity estimation directly from data that are being collected routinely. This allows a dense collection of velocity estimation points without added field costs and has completely replaced the expanding spread technique. However, Dix's theory remains a cornerstone of velocity estimation.

VELOCITY SPECTRA OBTAINED IN CONJUNCTION WITH HORIZONTAL STACKING

Horizontal stacking of data collected at ever-increasing separation of source and receiver about a common surface point has proved to be an extremely powerful processing procedure in improving the signal-to-noise ratio and is especially useful in suppressing coherent multiples. In the horizontal layering case, the common surface point coincides with the projection of the common depth point to the surface. In the dipping case, the stacked data do not have a common depth point; rather, they have reflections from a subsurface segment whose length increases with dip. In order to enhance primaries and suppress multiples, it is essential to stack the data along the NMO curves appropriate for the primary reflection. In the normal case of increasing velocity with depth, the moveout at any fixed record time T_0 will be less for the primary reflection than for any coherent multiple arriving at that time. The larger the differential NMO between primary and multiple reflections, the easier it is to suppress the multiples differentially. In the case of velocity inversion with depth, it may be difficult to identify primaries and multiples, and it may be impossible to separate them by horizontal stacking.

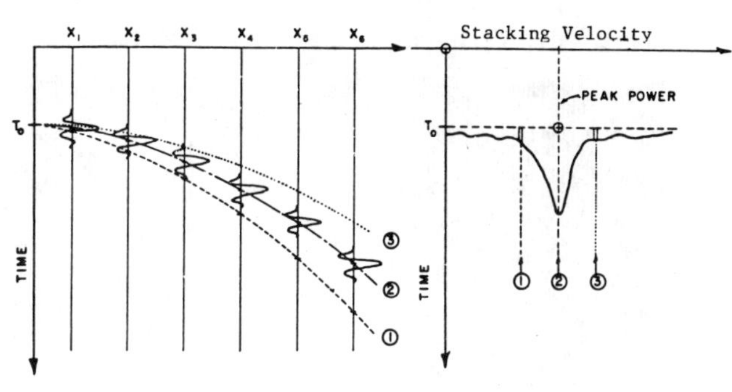

FIGURE 8.6. *Schematic of hyperbolic sweep of common surface point gathers to determine stacking velocity at reflection time* T_0.

A vast literature has grown up around velocity estimation from common surface point data, including, among the early papers, Garotta and Michon (1967), Taner and Koehler (1969), and Guinzy and Ruehle (patent filed 1969). The paper by Taner and Koehler is most instructive in the procedures and defines what they call the velocity spectrum.

All traces with common surface point are gathered together and subjected to a velocity search procedure (Figure 8.6). At a fixed zero-offset time T_0, the set of n traces with different offset distances x_i, $i = 1, 2, \ldots, n$, is swept with a collection of hyperbolas satisfying the equa-

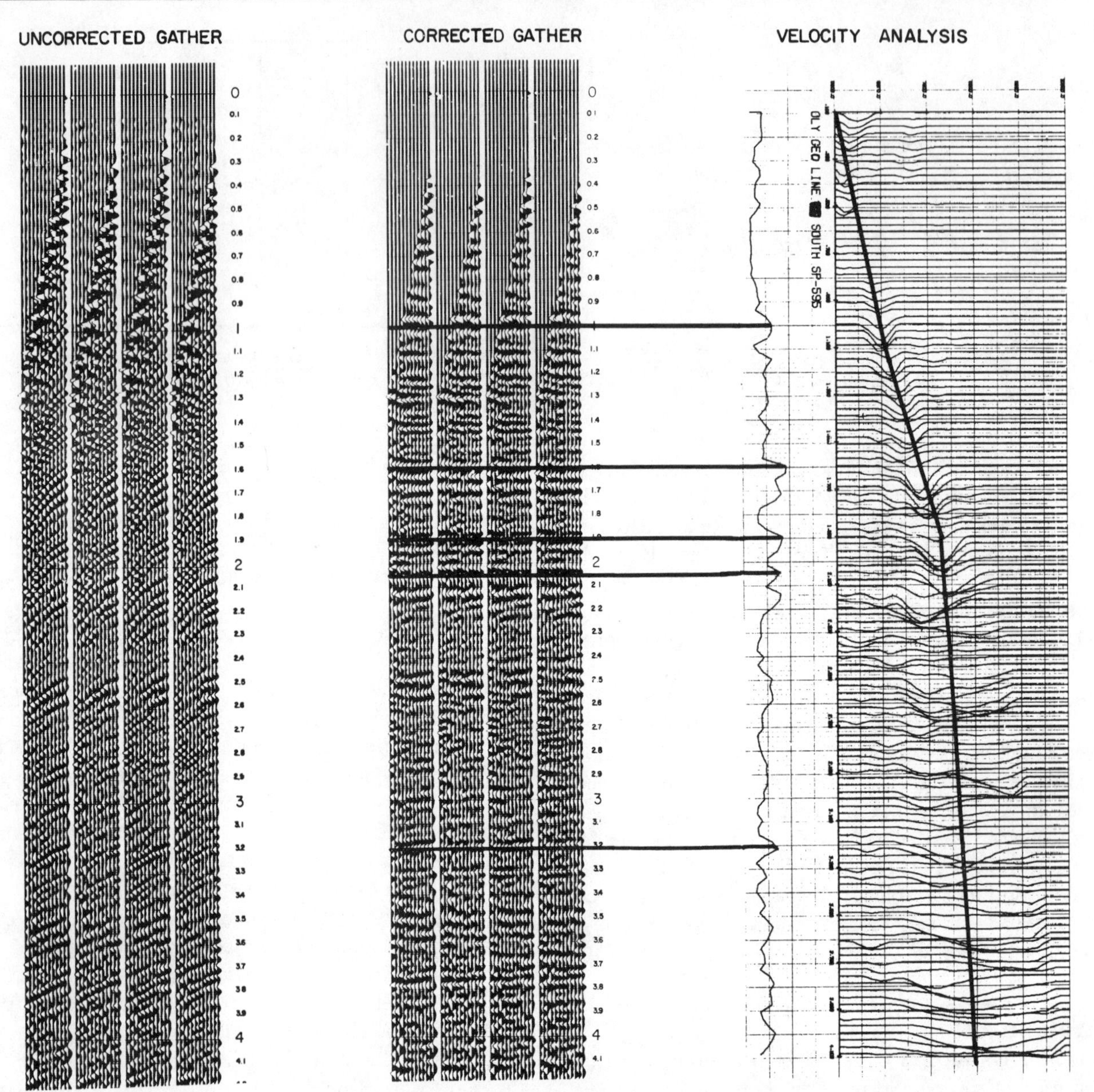

FIGURE 8.7. *Stacking velocity applied to correct gathers. (From Cosgrove, Edwards, and Grigsby 1969, courtesy of Grant Geophysical Company.)*

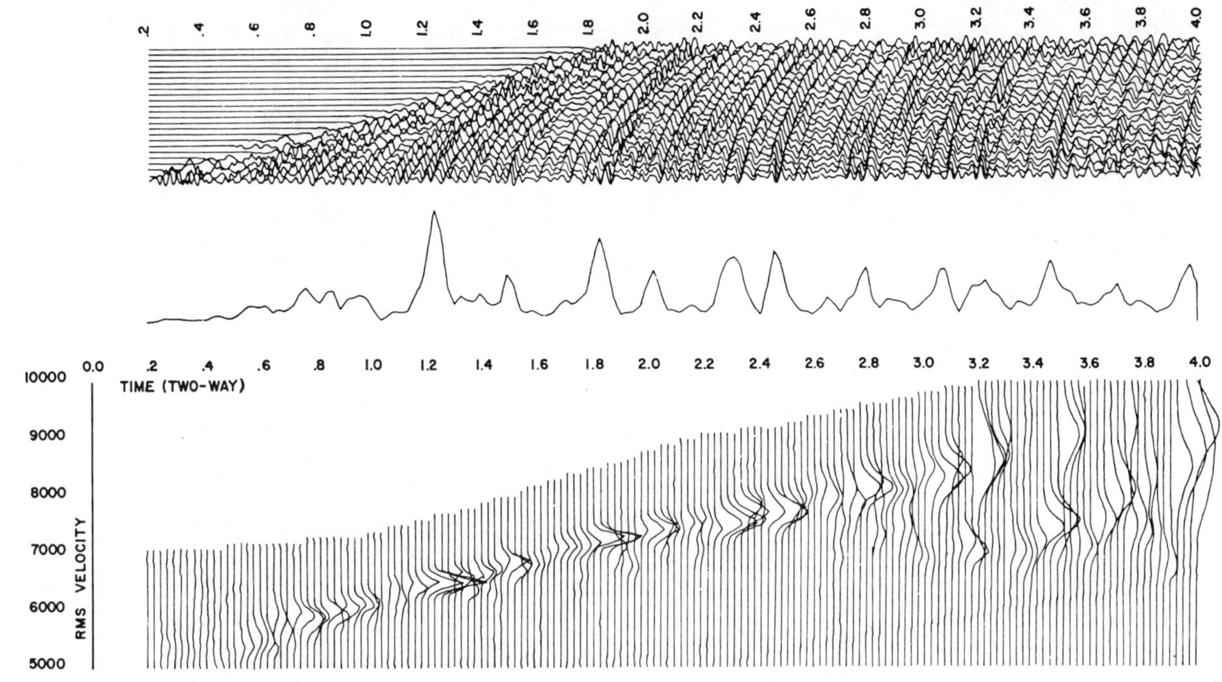

FIGURE 8.8. *Velocity spectrum. (From Taner and Koehler 1969, courtesy of Geophysics.)*

tion $T_x^2 = T_0^2 + (x/V)^2$, where the collection is indexed on V. Summing the trace amplitudes along each hyperbola and squaring the result will give an amplitude-squared (energy) function of V. The velocity that produces the maximum energy will produce the best stack at time T_0; this velocity is called the stacking velocity V_s at T_0. This procedure is repeated for a collection of zero-offset times, generally equally spaced in time at increments of Δt. The result is called the velocity spectrum at a common surface point. It consists of a three-dimensional display of stacked energy as a contour in (V_s,T_O) space, or some equivalent display. A trace of the stacked energy maxima as a function of zero-offset time T_0 is useful in identifying the coherent events.

Figure 8.7 (from Cosgrove, Edwards, and Grigsby, 1969) shows the uncorrected gathers that are swept to produce the three-dimensional velocity spectrum. Then the best stacking velocity function is drawn through the energy maxima and the uncorrected gathers are moveout-corrected, using this stacking velocity function. The resulting corrected gathers show that the primaries are properly corrected and, when the corrected gathers are stacked (all traces in each corrected gather summed to produce one trace of output data), the random noise and multiples are suppressed.

The collection of gathered traces corrected with the stacking velocity function obtained by connecting the stacked energy maxima associated with primary reflections is a useful display to determine the correctness of the stacking velocity function. Often, a collection of gathers stacked with a set of constant velocities is used as the prime data from which the stacking velocity function is deduced.

Several schemes have been used to measure the stacked energy as a function of stacking velocity, including use of normalized and unnormalized correlation functions, semblance, and squaring the sum. These coherency measures are discussed by Neidell and Taner (1971). For the velocity spectrum shown in Figure 8.8, Taner and Koehler (1969) use semblance, which they

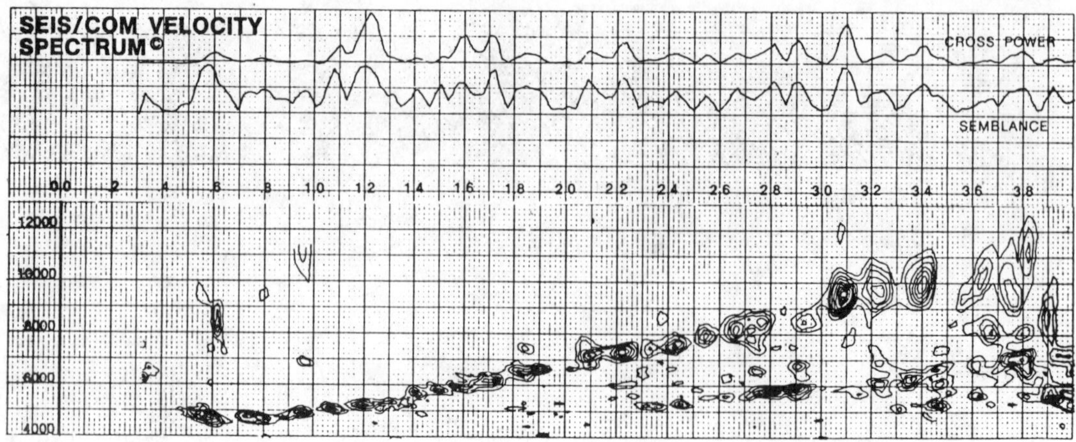

FIGURE 8.9. *Contoured velocity spectrum. (From Taner and Koehler 1969, courtesy of Geophysics.)*

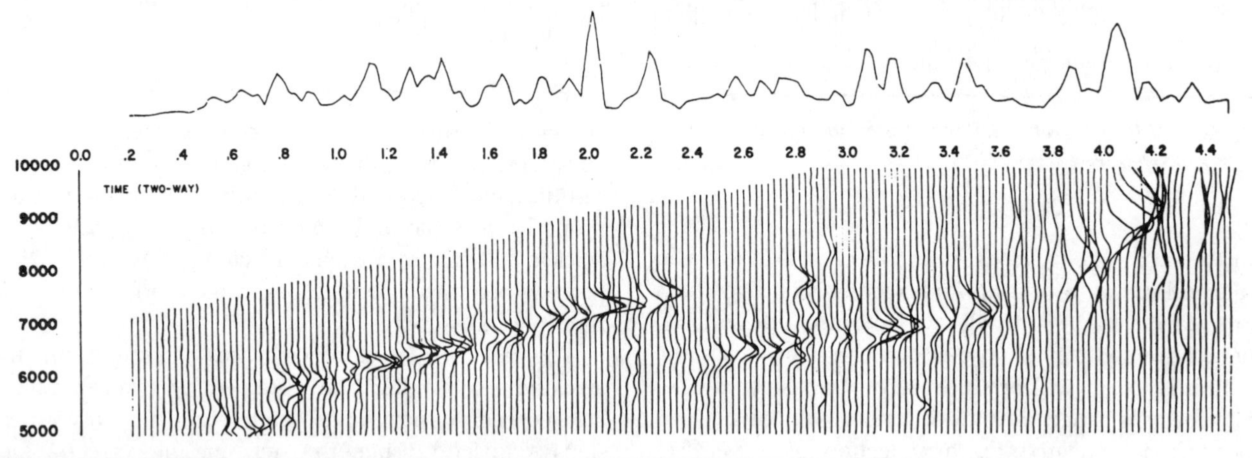

FIGURE 8.10. *Velocity spectrum in region that has large shale mass from 2.3–4.0 sec. (From Taner and Koehler 1969, courtesy of Geophysics.)*

describe as a normalized output/input energy ratio. Most investigators do not believe that there is a significant difference between measures, and squaring the sum is often used because it uses the least computer time.

The velocity spectra in Figures 8.7 and 8.8 plot energy as a downward deflection on the traces in (V_s,T_0) space. An alternative and superior display is a three-dimensional display such as the one shown in Figure 8.9, where the energy surface in (V_s,T_0) space is contoured.

The major applications of velocity spectra are the following:

(1) To determine velocity function for optimal stacking;
(2) To identify multiple reflections;
(3) To estimate interval velocities;
(4) To predict lithology, such as predictions of shale and limestone masses, predictions of sand/shale ratios, and predictions of anomalous overpressured zones.

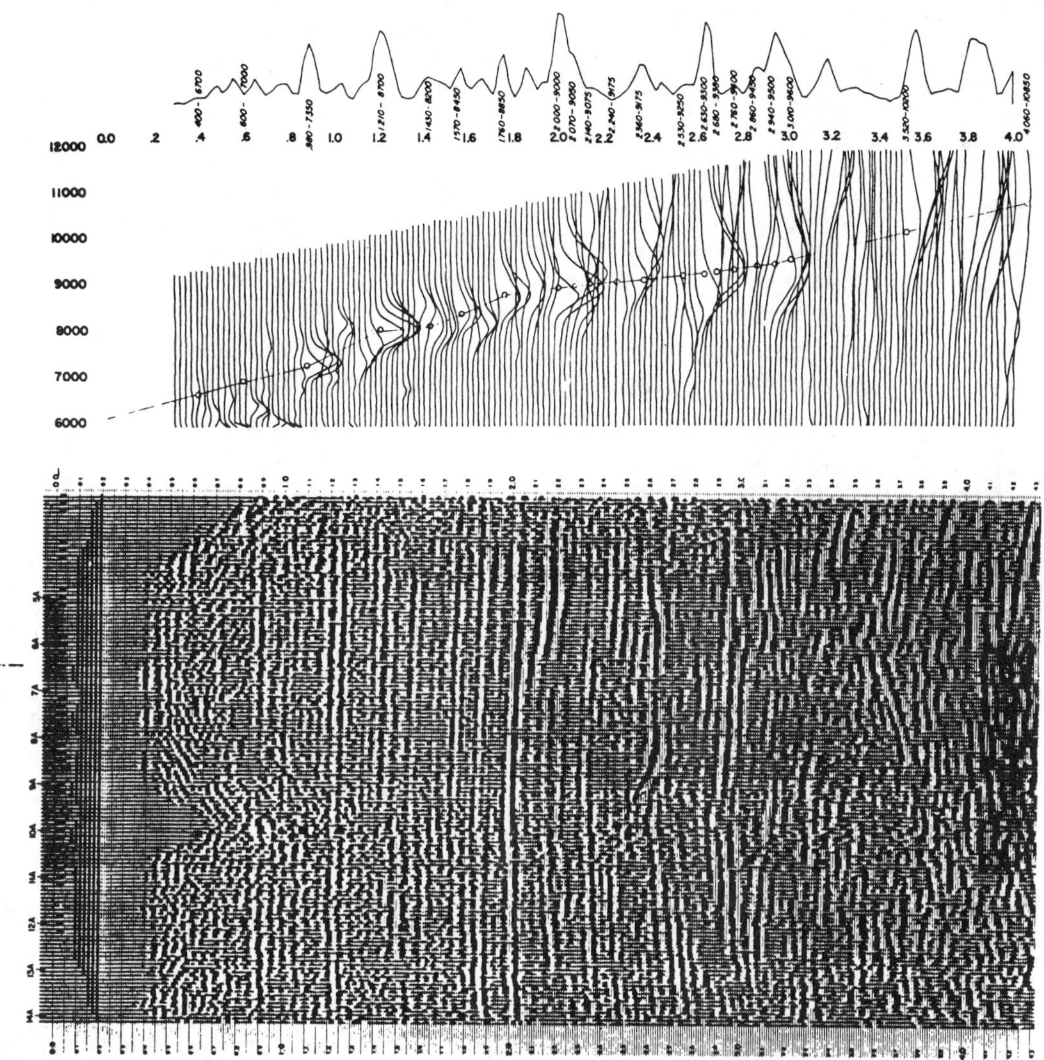

FIGURE 8.11. *Velocity spectrum in South Texas onshore. (From Taner and Koehler 1969, courtesy of Geophysics.)*

A series of examples from Taner and Koehler show some uses of these displays. In the first example, a thick shale mass from 2.24–4.0 sec has no primary reflections in that interval, and the velocity spectrum (Figure 8.10) shows lower stacking velocity, indicative of multiples. The fact that the stacking velocity within that zone at a given time—say, 2.6 secs—is the same as the stacking velocity at half that time—1.3 sec—indicates that the multiple at 2.6 sec is probably the *W*-multiple of the primary at 1.3 sec. A velocity spectrum for land data in South Texas (Figure 8.11) shows that the events between 2.0 and 3.0 sec are probably primaries and not multiples. Drilling confirmed that these reflections were primaries from sand layering. Velocity spectra in areas where there

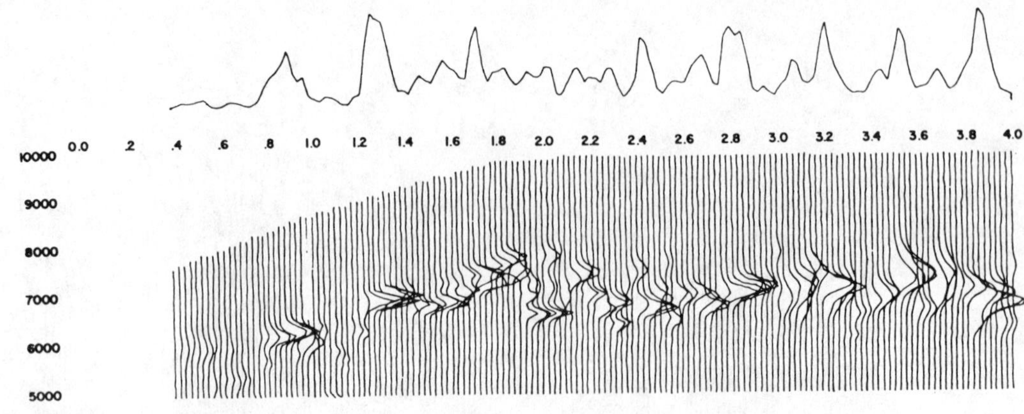

FIGURE 8.12. *Velocity spectrum in area of velocity inversion. (From Taner and Koehler 1969, courtesy of Geophysics.)*

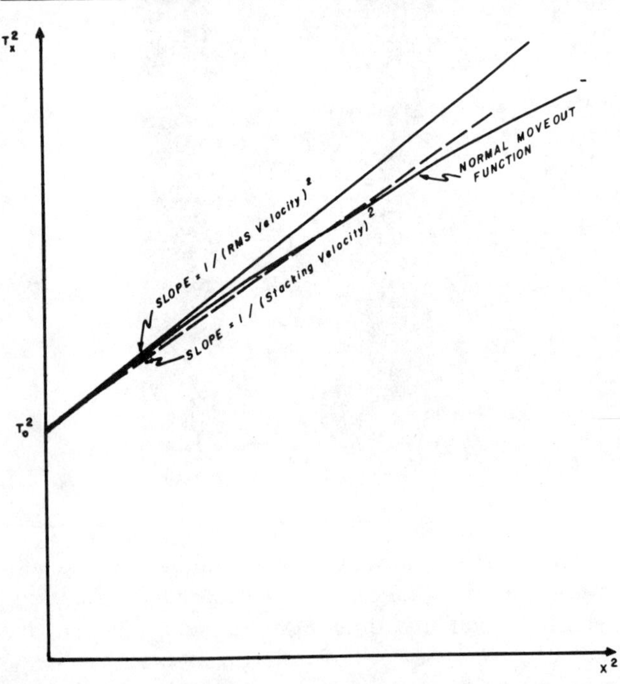

FIGURE 8.13. *Comparison of rms and stacking velocities in* $x^2 - T_x^2$ *space.*

are velocity inversions are difficult to interpret, because primary events may be mistakenly identified as multiples, as shown in Figure 8.12.

Much confusion has resulted from the use of stacking velocity to estimate interval velocities. Although the stacking velocity is related to Dix's rms velocity, they are not the same entity. The rms velocity is valid for small dip and short offset distances. The stacking velocity, which increases as the offset distance increases, is always greater than or equal to the rms velocity, even when there is no dip. As the dip increases, the difference between the stacking and rms velocities increases.

The difference between stacking and rms velocities is demonstrated in Figure 8.13. Because of Fermat's principle of minimal time paths and anisotropy, the NMO function bends downward in $(T_x{}^2, x^2)$ space as x increases. The procedures used in generating velocity spectra always result in a straight-line approximation to the NMO function, and its slope is the reciprocal of stacking velocity squared. The slope of the NMO function at $x^2 = 0$ is, by definition, the reciprocal of the rms velocity squared, and its slope is always greater than the aforementioned straight-line approximation. Hence, V_{rms} is always less than V_s.

Because of these differences, the insertion of stacking velocities into the Dix equation for determining interval velocities always results in interval velocities that are too large. Al-Chalabi (1973) has most clearly related stacking and rms velocities by determining the bias B in the

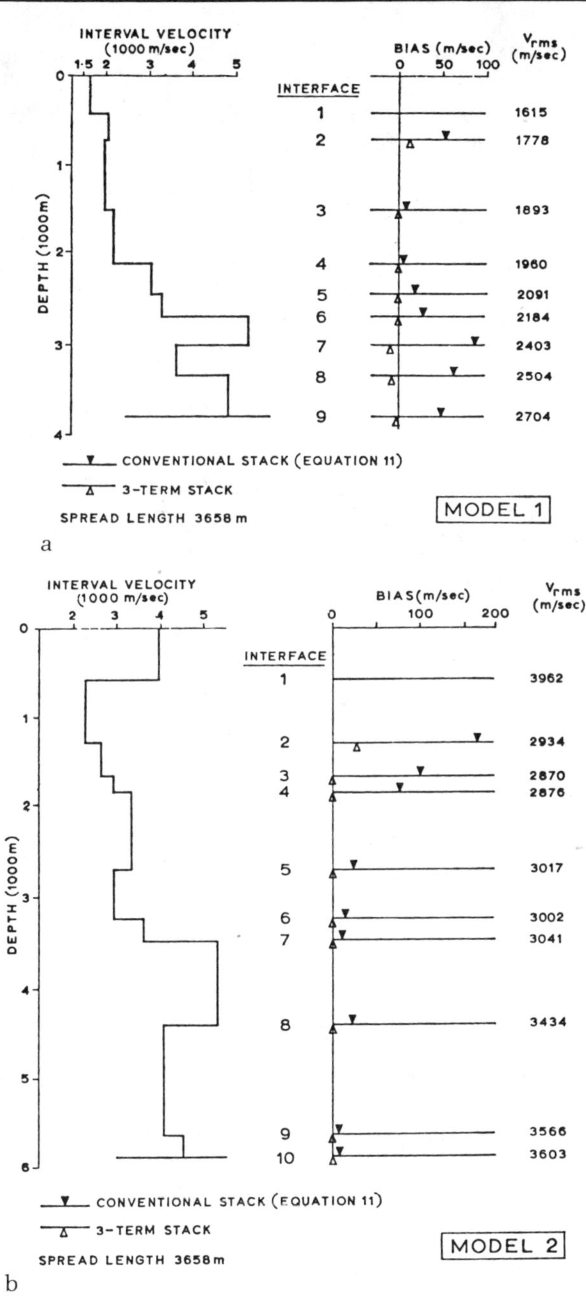

FIGURE 8.14. *Bias between stacking and rms velocities. (a) Model 1: North Sea. (b) Model 2: Alaska. (From Al-Chalabi 1974, courtesy of Geophysical Prospecting; Blackwell Scientific Publications Limited.)*

estimate of rms velocity from the stacking velocity, $B = V_s - V_{\text{rms}}$ (Figure 8.14). He shows that the bias B is a nonnegative quantity that increases as the quantity $\Sigma_{k=1}^{n} F_k p^{2k}$ increases. F_k is a complicated nonnegative function of the velocities and thicknesses of the layers, which increases as the heterogeneity factor H increases, and p is the ray parameter. Normally, the bias decreases with depth because the ray parameter decreases with depth and its contribution swamps out any possible increase due to F_k. However, if there is an increase in heterogeneity with depth, the increase in F_k may overcome the decrease due to the ray parameter. An example from the North Sea shows an increase in bias with depth because of greatly increasing heterogeneity with depth. Another example, from Alaska, shows large bias near the surface because of the permafrost layer.

In the case of arbitrary nonuniform dips, it is necessary to go to ray tracing using Snell's Law, building the model layer by layer and iterating to minimize the difference between observed and computed arrival times. Taner, Cook, and Neidell (1970) show the results of iterative modeling to compute dip and interval velocity (Figure 8.15). An example of the misuse of stacking velocities in Dix's equation in a model line across a synclinal subsurface shows calculated interval velocities at three locations of 8069, 5550, and 9041 ft/sec when the actual interval velocity is 8000 ft/sec. Examples such as these abound in the literature. Everett (1974) also discusses an iterative procedure that, in the presence of arbitrary dips, converges on the correct interval velocities, dips, and depths from the stacking velocities, reflection times, and time slopes.

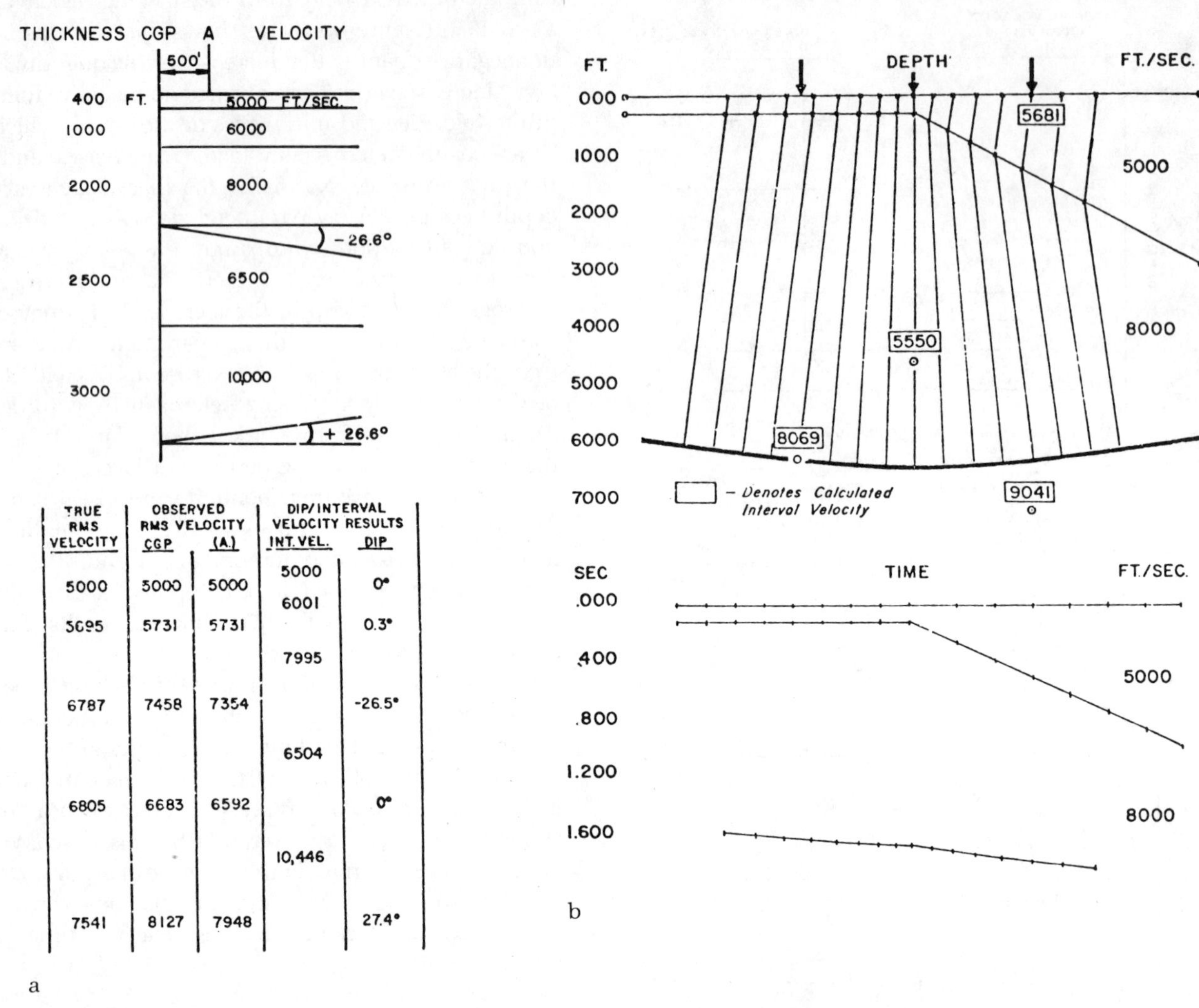

TRUE RMS VELOCITY	OBSERVED RMS VELOCITY CGP	(A.)	DIP/INTERVAL VELOCITY RESULTS INT. VEL.	DIP
			5000	
5000	5000	5000		0°
			6001	
5695	5731	5731		0.3°
			7995	
6787	7458	7354		-26.5°
			6504	
6805	6683	6592		0°
			10,446	
7541	8127	7948		27.4°

FIGURE 8.15. *Interval velocity calculations. (a) Iterative procedure to determine dip and interval velocity. (b) Erroneous application of Dix equation to stacking velocities in dipping case. (From Taner, Cook, and Neidell 1970, courtesy of Geophysics.)*

APPLICATIONS OF VELOCITY SPECTRA IN PREDICTING LITHOLOGY

PREDICTION OF SHALE AND LIMESTONE MASSES

Shale masses have been predicted on the basis of abnormally low stacking velocity through the shale interval, and by the lack of primaries and the presence of prominent multiples within the shale interval, as shown previously.

Limestone masses in West Texas have been separated from limestone stringers emanating from such masses, using interval velocities derived from expanding spreads. Davis (1972) used velocity spectra in studying velocity variations around Leduc reefs in Alberta, Canada. The known increase in interval velocity in the Blairmore-Ireton interval above the reefs and the increase in rms velocity to the Ireton, known from studies of continuous velocity logs, was substantiated by velocities derived from velocity spectra. An increase in interval velocity of about 7 percent was obtained by each method.

PREDICTION OF SAND/SHALE RATIOS

The prediction of sand/shale ratios is based on the known linear increase in velocity as a function of sand percentage in a sand-shale sequence at a given depth and an experimentally derived increase in velocity with depth for a given sand/shale ratio as discussed by Tegland (1972). Assuming that the interval velocity varies as an exponential function of depth, $V_i = \alpha z^{\beta}$, where α is a lithologic index representing the combined effects of lithology, age of formation, pressure, and possibly other factors, and β is related to compaction properties, Tegland derived empirical curves that range from $V_i = 1121z^{1/4}$ for 100 percent sand to $V_i = 931z^{1/4}$ for 100 percent shale, and velocity curves of interval velocity as a function of sand percentage, S, given by $V_i = 2100S + 9100$ at a depth of 9580 ft, for typical Gulf Coast formations (Figure 8.16).

Deriving interval velocities from stacking velocity is hazardous, and the resulting inaccuracies often negate any possible prediction of sand/shale ratio. However, massive stacking velocity data smoothed in space, time, and velocity have proved useful in lithologic studies, as discussed by Saugy and Engels (1975) in their application of continuous velocity analysis applied to a growth anticline in the Niger delta onshore. The results of this study showed the following positive results: (1) the horizontal gradient of the velocity field brought out an unexpected closure on the depth map of a prominent reflection that corresponds to hydrocarbon-bearing levels known to exist in the area; (2) the normalized interval velocities show lateral variation in the lithology of the Agbada series, probably due to variation in sand/shale ratio; and (3) the top of the overpressured Akata shale was located with confidence.

Severe problems are introduced by the presence of calcareous sands or siliceous shales into a section that was thought to be a normal sand-shale sequence. The abnormal velocities so introduced would probably make sand/shale ratio studies meaningless.

PREDICTION OF OVERPRESSURED SHALES

The presence of overpressured shale zones is a severe problem in drilling. These zones have abnormally low interval velocity because they are uncompacted. Pennebaker (1968) presents a thorough discussion of the effect of pore pressure, lithology, geologic age, and depth on interval velocity or its reciprocal interval traveltime, as measured with sonic logs. The interval traveltime T is exponentially related to depth by the empirically derived relation, $T = \alpha z^{-\beta}$, where α depends on pressure lithology and geologic age, and $\beta = 1/4$ in normal sand-shale sequences. This equation establishes a baseline for normal sequences of sand and shale. Sudden departures from this baseline with depth herald changes in pressure or lithology. Geologic age will not be anomalous but will shift the baseline.

Examples from Pennebaker show the effects of pressure, lithology, and geologic age on interval traveltime versus depth function. In Figure 8.17(a), a well in Kleberg County, Texas shows abnormal pressure below 8500 ft. Examples from Gonzales and Kendall Counties, Texas, in Figure 8.17(b) show anomalies due to limestone, dolomite, and thick shale zones. Thick shale zones shift the normal compaction line to the right, to longer interval traveltimes or lower interval velocities. Limestone shifts the normal line to the left, and dolomite shifts it still farther to the left. In many areas of South Texas and Louisiana, a section of calcareous sand and shale overlies deep overpressured shale zones, as illustrated by two such examples in Figure 8.17(c). The effect of geologic age on the normal compaction line is shown

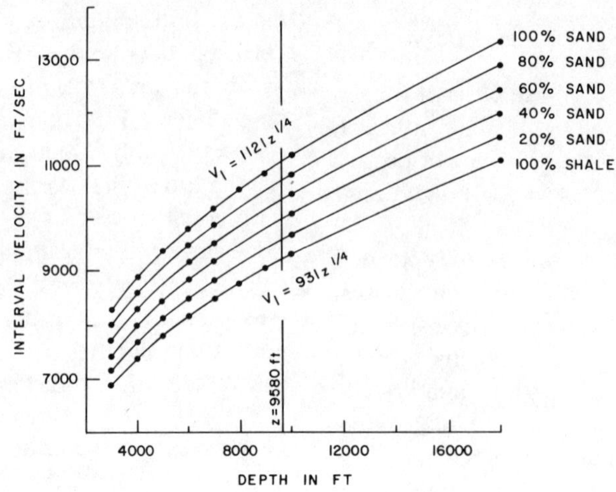

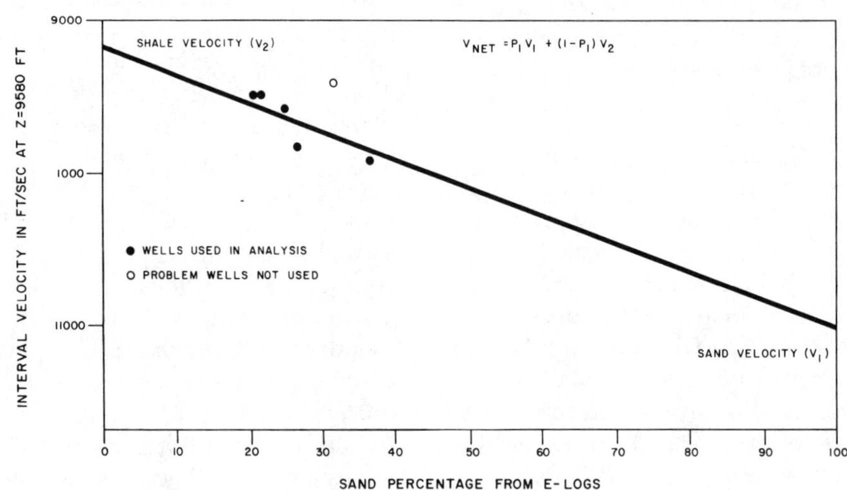

FIGURE 8.16. *Velocity functions used in sand/shale ratio studies in Gulf of Mexico sediments. (From Tegland 1972, courtesy of SEG.)*

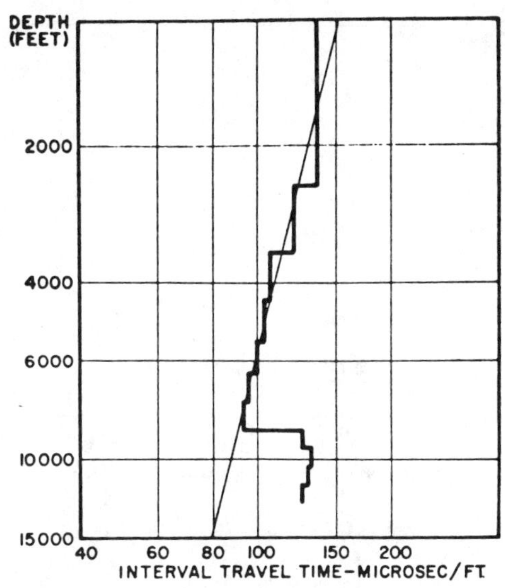

a

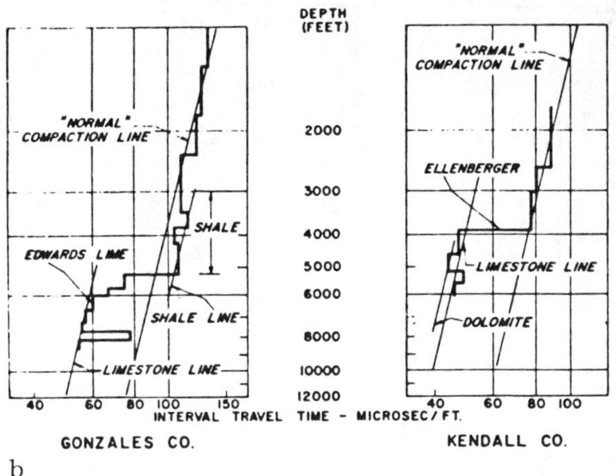

b

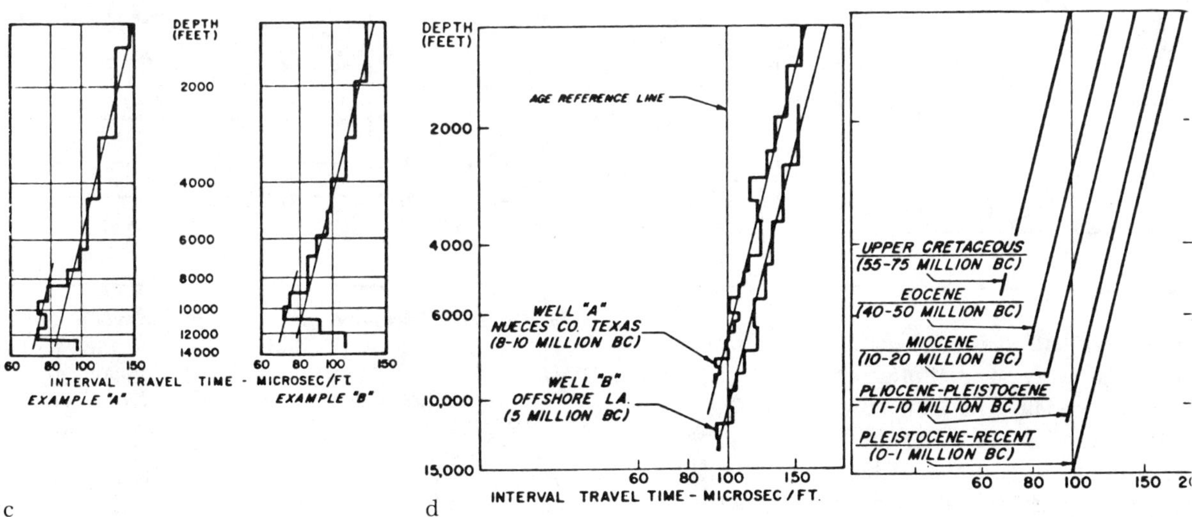

c

d

FIGURE 8.17. *Effect of overpressure, lithology, and geologic age on compaction baseline on interval traveltime versus depth functions. (a) Kleberg County, Texas. (b) Gonzales and Kendall Counties, Texas. (c) Calcareous sand and shale above overpressured zone, South Texas and Louisiana. (d) Nueces County, Texas, and Louisiana. (From SPE paper #2165 presented at the 43rd Annual Technical Conference and Exhibition of the Society of Petroleum Engineers of AIME, Houston, Sept. 29–Oct. 2, 1968 by Pennebaker; courtesy of SPE-AIME, copyright SPE-AIME.)*

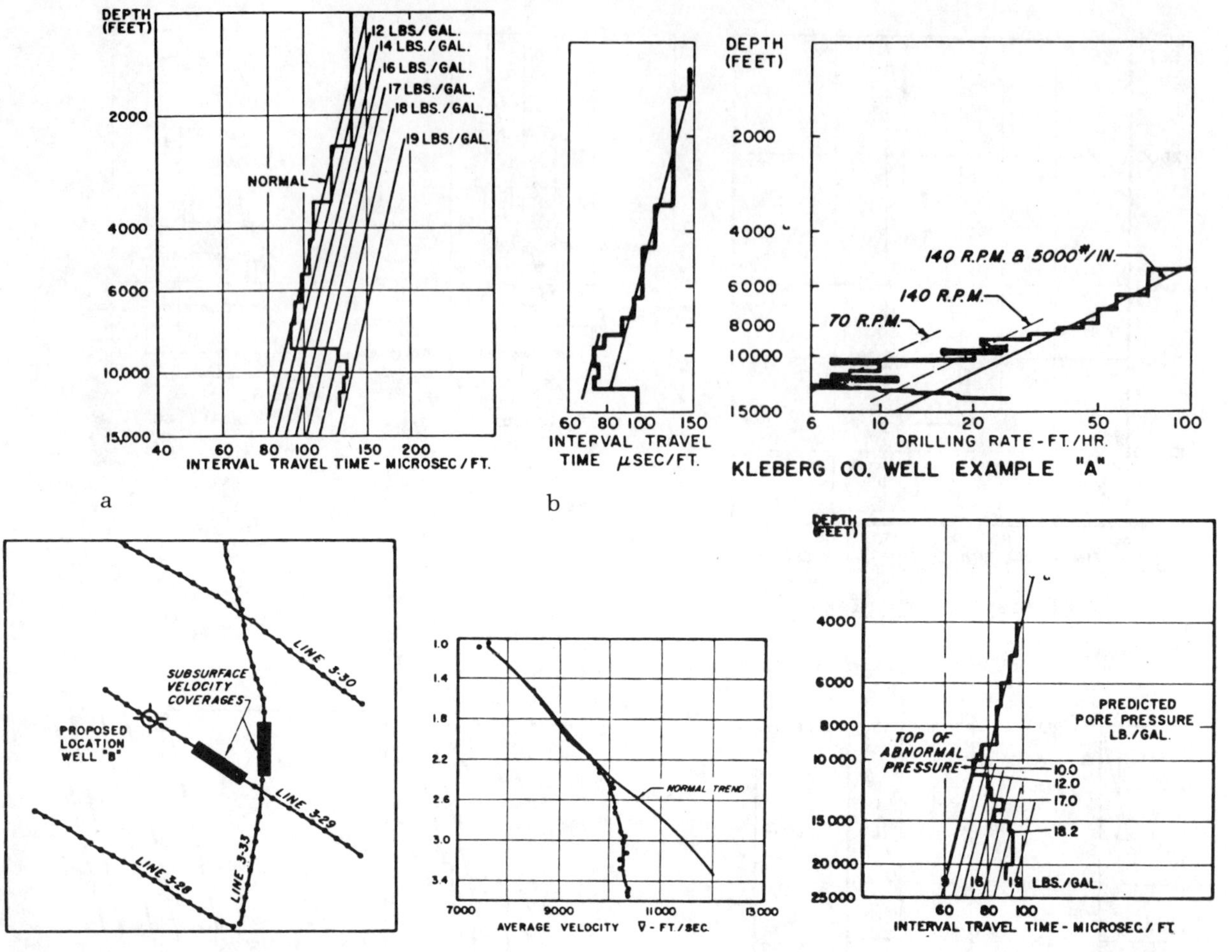

FIGURE 8.18. *Design of drilling programs based on interval velocities. (a) Overlay of abnormal pressure on interval traveltime function. (b) Drilling rate as a function of compaction. (c) Drilling program design based on seismic velocities. (From SPE paper #2165 presented at the 43rd Annual Technical Conference and Exhibition of the Society of Petroleum Engineers of AIME, Houston, Sept. 29–Oct. 2, 1968 by Pennebaker; courtesy of SPE-AIME, copyright SPE-AIME.)*

in Figure 8.17(d) by comparison of data from wells in Nueces County, Texas, and offshore Louisiana. The normal compaction lines have the same slope, but the line in the Nueces well is shifted to the left to higher interval velocity with respect to the offshore well, corresponding to a shift to higher interval velocity, because it penetrated much older rocks.

The degree of departure from the normal compaction line in overpressured zones is directly proportional to the abnormal pressure. An overlay of lines of constant pore-pressure gradient expressed in equivalent mud weight directly shows the mud weight required as a function of depth. In the Kleberg County well, the normal mud weight of 9 lb/gal must be increased to 18 lb/gal at the top of the overpressured zone, as seen in Figure 8.18(a). Pore-pressure estimates must be within 1 lb/gal to be useful in designing a drilling program. The most critical

TABLE 8.1. *Comparison of predicted and actual drilling programs*

	Estimated	Actual
Surface casing set	2,500 ft	2,505 ft
Top of abnormal pressure	10,500 ft	10,200 ft
12.0 lb/gal pore pressure	11,100 ft	11,000 ft
Protective casing setting	11,100 ft	10,875 ft
Liner setting	—	11,640 ft
Pore pressure gradients		
10,525 ft	—	10.6 lb/gal
11,100 ft	12.0 lb/gal	—
11,306 ft	—	14.8 lb/gal
12,311 ft	—	16.6 lb/gal
13,000 ft	17.0 lb/gal	—
13,965 ft	—	17.3 lb/gal
Maximum mud weight	18.1 lb/gal	18.2 lb/gal
Drilling rate at 4000 ft	70 ft/hr	76 ft/hr
Drilling rate at 9000 ft	14 ft/hr	19.5 ft/hr

phase in drilling into overpressured zones is the setting of protective casing at a point that will permit the use of subsequent higher mud weight without fracturing the formation and losing circulation.

In Figure 8.18(c), Pennebaker (1968) shows an example of drilling program design based on velocity information obtained from surface seismic measurements. Computed interval velocities indicated that the overpressured zone would be encountered at about 10,500 ft, and pore pressures of 12.0 lb/gal would be found at 11,000 ft, 17.0 lb/gal at 13,000 ft, and 18.2 lb/gal at 16,000 ft. Estimates of drillability from the velocity data indicated that the drilling rate would decrease from 70 ft/hr at 4000 ft depth to 14 ft/hr at the top of a calcareous sand/shale interval beginning at 9000 ft depth. Average drilling rate through the 2000 ft thick calcareous zone would decrease to less than 8 ft/hr. A marked increase in drilling rate is expected in the overpressured zone. Estimates of fracture gradients from the velocity data indicate that protective casing should be set at 11,100 ft in 12 lb/gal pore pressure, and that the fracture gradient below the casing would be 18.7 lb/gal. Thus, the maximum mud weight of 18.2 lb/gal would contain the overpressured zone without fracturing and losing circulation. The comparison of the predicted and actual drilling programs is summarized in Table 8.1.

A statistical evaluation by Reynolds (1974) of Conoco, comparing the actual tops of overpressured zones in 35 wildcat wells drilled worldwide in a variety of basins (mostly offshore Tertiary basins), with the tops predicted from seismic methods, showed only two total failures. One was slightly overpressured without being predicted, and the other was predicted to be overpressured at 2700 ft but was not. The failure of the latter well was due to adverse effects of calcareous shale on the normal compaction baseline. Of the remaining 33 wells, 10 predictions of no overpressured zones were substantiated by drilling. Of the 23 wells correctly predicted to be overpressured, the top of the overpressured zone was predicted within ±500 ft on 60 percent of the wells, and the pressure magnitude was predicted within ±1.5 lb/gal on 75 percent of the wells.

SUMMARY

The dense collection of velocity data from surface measurements obtainable through widespread use of horizontal stacking and the development of the concept of rms velocity and its relationship to interval velocities have constituted a major advance in seismic interpretation. They provide a source of additional information about the subsurface layering and an adjunct to the usual interpretation techniques. Contoured velocity sections that overlay seismic sections are valuable in gross lithologic studies. Their usefulness will increase as velocity estimation procedures are refined.

Chapter 9 Migration Techniques

INTRODUCTION

Geophysicists are well aware that a seismic record section is not a slice through the earth but is, instead, a transformation of the three-dimensional earth's reflectivity into a plane where each recorded event is located vertically beneath the source-receiver midpoint. Migration is the inverse transformation that carries the plane into the true three-dimensional reflectivity. Steep dip, curved surfaces, buried foci, faults, and other discontinuities each contribute their unique characteristics to the seismic record section and, in complexly faulted and folded areas, make interpretation of the geological layering from the seismic sections difficult unless the data are migrated.

Hagedoorn (1954) thoroughly described the process of migration in both two and three dimensions, using the concepts of surfaces of equal reflection time and surfaces of maximum convexity (commonly called diffraction surfaces). With the advent of digital computers, this work became the basis of a technique known as Kirchhoff summation, in which each point in the seismic section is replaced by the weighted sum of all points that lie on its surface of maximum convexity.

Recent studies led by Claerbout and his associates at Stanford University on solution of the wave equation using numerical techniques have led to a migration procedure called wave-equation migration (also called migration by finite differences). By this procedure, the observed wave field at the surface is continued downward by numerical solution of the wave equation. By knowing the wave field at the surface, the field is computed at Δz. Proceeding by recursive calculations, the field is computed step by step at $2\Delta z, 3\Delta z, \ldots, n\Delta z$. As the calculation proceeds, at any given step the wave field below the depth corresponding to that step is the same as would be obtained by recording at that depth, and all events from shallower depths will have been migrated into their correct position. This technique is superior to summation in low signal-to-noise cases, but it does not handle steep dips as well.

A third migration technique uses the two-dimensional Fourier transform of the wave equation rather than the finite-difference solution. The transform technique is called *f-k* migration, because frequency *f* and wavenumber *k* are the transforms of time and distance. Under low signal-to-noise conditions, *f-k* migration has good performance; it is superior to the finite-difference method in steep dip and is much faster computationally because the fast Fourier transform can be used.

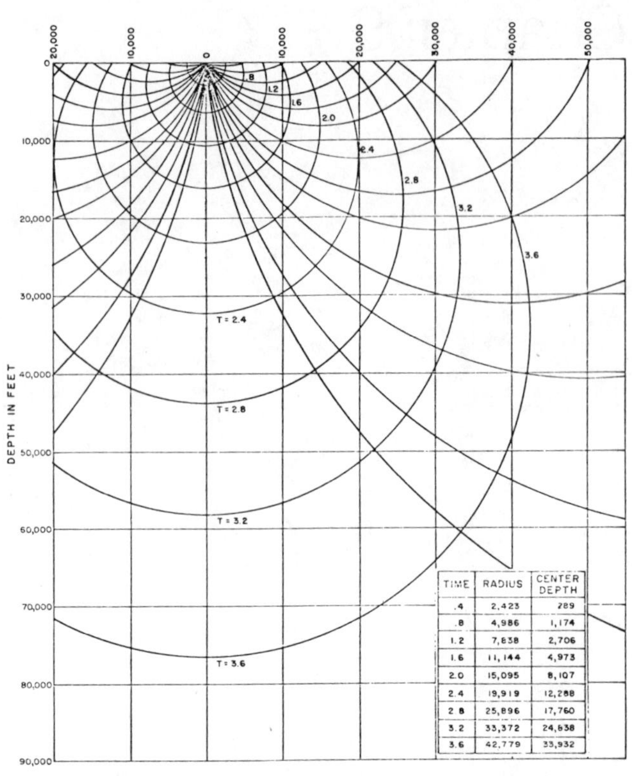

TIME	RADIUS	CENTER DEPTH
.4	2,423	289
.8	4,986	1,174
1.2	7,838	2,706
1.6	11,144	4,973
2.0	15,095	8,107
2.4	19,919	12,288
2.8	25,896	17,760
3.2	33,372	24,638
3.6	42,779	33,932

FIGURE 9.1. *Raypaths and wavefronts for linear increase in velocity with depth,* $V = V_0 + az$, *when* $V_0 = 6000$ *ft/sec and* $a = 0.6$. *(From Slotnik 1959, courtesy of SEG.)*

MIGRATION BY THE HAGEDOORN METHOD

Consider a half-space with source and receiver located on the surface. A reflected event received at a common source-receiver point at time *T* may have come from any point on a three-dimensional surface of equal time *T* within the half-space. A collection of surfaces of equal time is usually called a wavefront chart in two dimensions. The raypaths are perpendicular to the wavefronts. From the measurement at one source-receiver location, it is impossible to locate the subsurface reflection point. From a collection of such points, however, it is possible to deduce the subsurface structure.

The surface of equal reflection time depends on the velocity distribution. If the velocity *V* in the half-space is constant, the surface will be a half-sphere of radius *VT*. If the velocity increases linearly with depth, $V = V_0 + az$, where V_0 is the velocity at the surface and the wavefront surface at *T* is a sphere with center at $[x,z] = [0, (V_0/a)(\cosh aT - 1)]$ and radius $(V_0/a) \sinh aT$, as shown by Slotnik (1959).

The wavefront chart for $V_0 = 6000$ ft/sec and $a = 0.6$, which is the average characteristic of the Texas and Louisiana Gulf Coast, the San Joaquin Valley of California, and the Maturin Basin of Venezuela, shows the succession of circles expanding from points that become progressively deeper (Figure 9.1). Given the arrival time and the dip, it is possible to pinpoint the location of the reflection point, which is the point where the wavefront is tangential to the dipping bed.

Seismic record sections consist of seismic traces plotted vertically beneath the midpoint between source and receiver. The geometric effect of offset between source and receiver (normal moveout) and static variations are accounted for, and, in CDP presentation, the source-receiver points are common. A vertically located reflection point on the record section has no significance other than that it is a point on its surface of equal reflection time, which is a surface tangential to the actual reflector at some point in space.

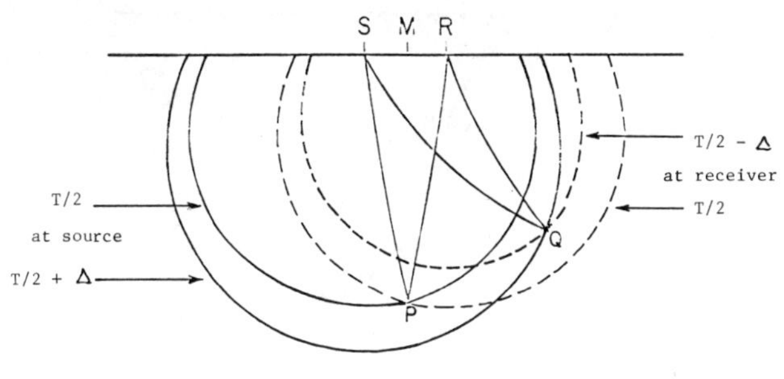

FIGURE 9.2. *Surfaces of equal reflection time. (From Hagedoorn 1954, courtesy of Geophysical Prospecting; Blackwell Scientific Publications Limited.)*

Consider the surfaces of equal reflection time in Figure 9.2. When the time from source to reflection point P and the time from reflection point P to receiver R are equal, P lies vertically below the midpoint M of source to receiver. Increasing the time from S to P by Δ and decreasing the time from P to R by Δ gives the same overall reflection time, with the reflection point now at Q.

Migration is the procedure for determining true reflection surfaces from surfaces observed on the record section. Figure 9.3 shows the relationship between dip α of the true surface in the earth and apparent dip β of the record surface. The relationship between α and β is developed as follows:

$$\tan \alpha = \frac{b}{d - a} = \frac{a}{b}$$

$$b^2 = ad - a^2$$

$$d = \frac{a^2 + b^2}{a} = \frac{c^2}{a}$$

$$\tan \beta = \frac{c}{d} = \frac{c}{c^2/a} = \frac{a}{c} = \sin \alpha.$$

Thus, the relationship between the angles is

$$\sin \alpha = \tan \beta. \tag{9.1}$$

Solving for the true dip α in terms of the apparent dip β gives the following result:

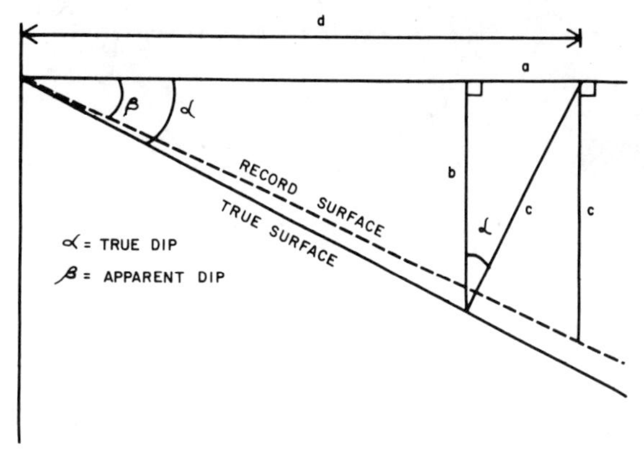

FIGURE 9.3. *Relationship between record surface and true surface.*

β		α	
$\beta =$	0°	$\alpha =$	0.00°
=	5	=	5.02
=	10	=	10.16
=	15	=	15.54
=	20	=	21.34
=	25	=	27.79
=	30	=	35.26
=	35	=	44.44
=	40	=	57.05
=	45	=	90.00

Hagedoorn (1954) thoroughly studied two- and three-dimensional migration of reflection data from record sections, using surfaces of maximum convexity derived from surfaces of equal reflection time (Figure 9.4). Consider the construction of the surface of maximum convexity through a reflection point P in the two-dimensional case. The event P from a source-receiver whose midpoint P' is directly above P is located in its correct position, at the nadir of its curve of equal reflection time T_p. The reflection from point P observed at an offset point Q' occurs at a time T_q, which is greater than T_p. Plotting its vertical position Q at the nadir of the T_q curve of equal reflection time specifies the point Q on the surface of maximum convexity through P. Moving Q' along the earth's surface traces out the surface of maximum convexity through P, with its apex at P. This curve is commonly called the diffraction curve for the diffraction point P.

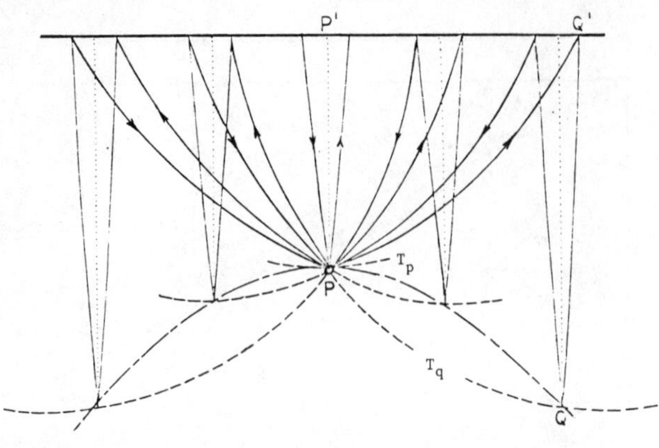

FIGURE 9.4. *Surfaces of maximum convexity. (From Hagedoorn 1954, courtesy of Geophysical Prospecting; Blackwell Scientific Publications Limited.)*

Hagedoorn's two-dimensional migration of a reflection plotted vertically on the record section at Q is based on finding the curve of maximum convexity that is tangential at Q to the horizon on the record section containing Q, as seen in Figure 9.5(a). Here, Q is migrated from the nadir of the curve of equal reflection time to the apex P of that particular curve of maximum convexity. The true dip of the horizon through Q on the record section is tangent to the curve of equal reflection time T_q at the point P.

Consider a reflector surface APB, with change in dip at P, as shown in Figure 9.5(b). The segments AP and PB transform into disjoint segments $A'P_1'$ and $P_2'B'$ on the record section that are joined together by the diffraction surface through the point P. All points at and between P_1' and P_2' migrate to P along the diffraction curve with apex at P; A' migrates to A along the diffraction curve with apex at A; and B' migrates to B along the diffraction curve with apex at B.

In principle, the extension from two- to three-dimensional migration is simple, and uses three-dimensional surfaces of equal reflection time and maximum convexity in place of the described two-dimensional curves.

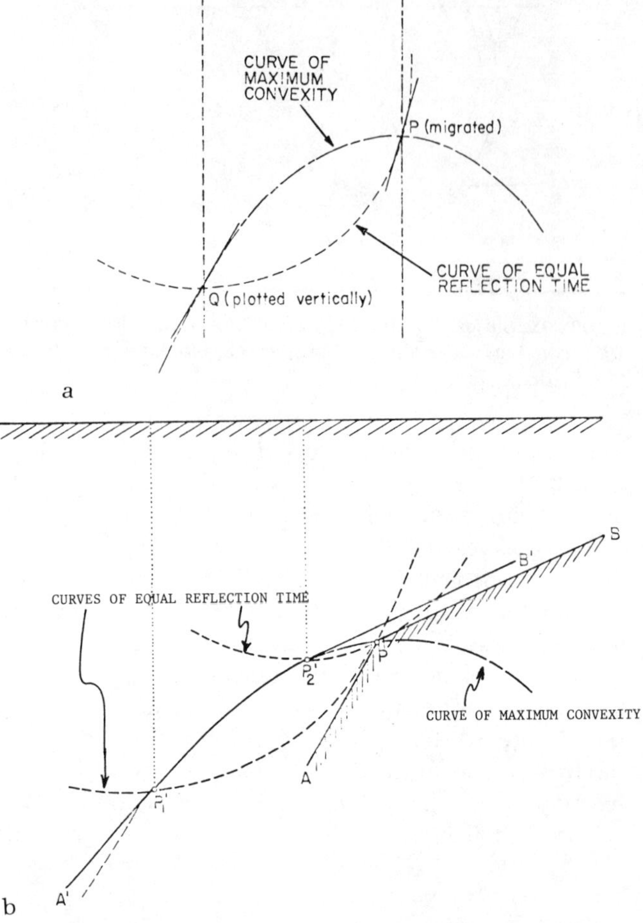

FIGURE 9.5. *Two-dimensional migration. (a) Two-dimensional migration of a reflection plotted vertically at Q. (b) Migration of plane surface that has a change in dip. (From Hagedoorn 1954, courtesy of Geophysical Prospecting; Blackwell Scientific Publications Limited.)*

MIGRATION BY SUMMATION METHODS

With the advent of digital computers, it became relatively simple to migrate events along diffraction curves to produce migrated sections (Figure 9.6). Each point in a record section is replaced by the weighted sum of contributions lying along the diffraction curve for that point. If a point is a diffraction point from a discontinuity, then

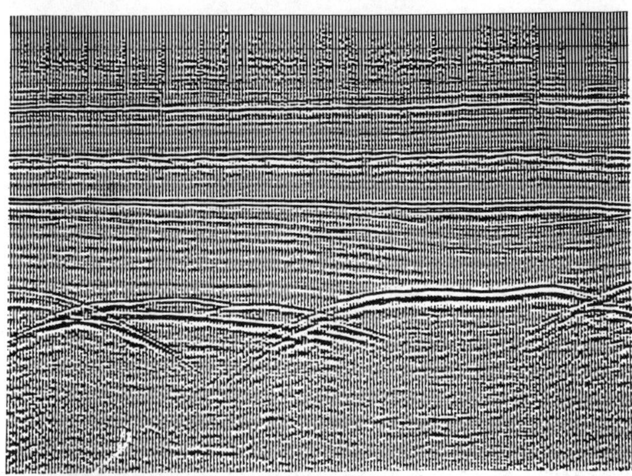

a

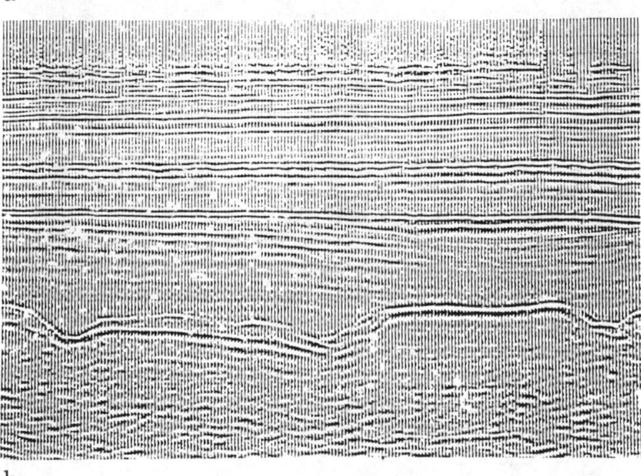

b

FIGURE 9.6. *Example of migration using summation. (a) Stacked section. (b) Migrated section using summation. (From Bortfeld 1974, courtesy of Prakla-Seismos Report.)*

the diffraction from that point follows its diffraction curve and summing collapses the diffraction back to its diffraction point. If a point is a reflection point from a locally continuous reflector, then the reflector contributes to the sum for a few traces on either side of the trace containing the point. For traces farther from this trace, the contributions to the sum consist of random amplitudes with zero mean, and the summation consists of the reflector plus low-amplitude diffraction noise. If a point is neither of the foregoing, the summation will consist solely of low-level diffraction noise.

The summation method depends critically on good estimates of the velocity distribution with depth in order to determine the correct surface of maximum convexity along which to sum. Provided that the velocity estimates are adequate and the geological layering is such that two-dimensional migration works, a number of parameters must be set for best results. The number of traces to be summed is most important. Too few traces results in a heavily mixed section. Noise bursts may appear to be reflection segments, faults may be blurred, and furthermore, diffractions are not collapsed. Too many traces may generate abnormal diffraction noise. The optimal number of traces to be summed increases with depth of the summing point because the curves of maximum convexity flatten with depth. Problems increase with depth due to decrease in signal-to-noise ratio and eventual loss of data when the record is terminated.

In Figure 9.7, a series of sections migrated using different apertures shows the effect of aperture size (from Prakla-Seismos Report 1-74). The sixfold CDP section shows that the Zechstein base reflection at 1.9 sec has several small irregularities. Above the Zechstein, in the range 1.3–1.9 sec, there is an indication of curved surface at the left of the section and a steeply dipping event near the middle. A small aperture (±20 traces) smears the irregularities in the Z-reflection and wipes out both the curved surface and the steep dip above the Zechstein. With ±40 traces, and somewhat better with ±80 traces, the irregularities on the Z-reflection show up as small faults and minor flexures, the curved surface develops its true synclinal structure, and the steep dip event is preserved. The largest aperture, ±160 traces, fails to improve the Z-reflection and may somewhat degrade the data below the Zechstein. The data above 1.2 sec appear to be best with ±20 trace aperture and progressively deteriorate with increasing aperture size because of diffraction noise.

In addition to controlling the aperture size and increasing the size with depth, efforts have been made to develop a variety of proprietary schemes to choose the summing weights. In Figure 9.8, for example, a weighted sum with weights chosen from coherency measurements shows somewhat better continuity of steeply dipping

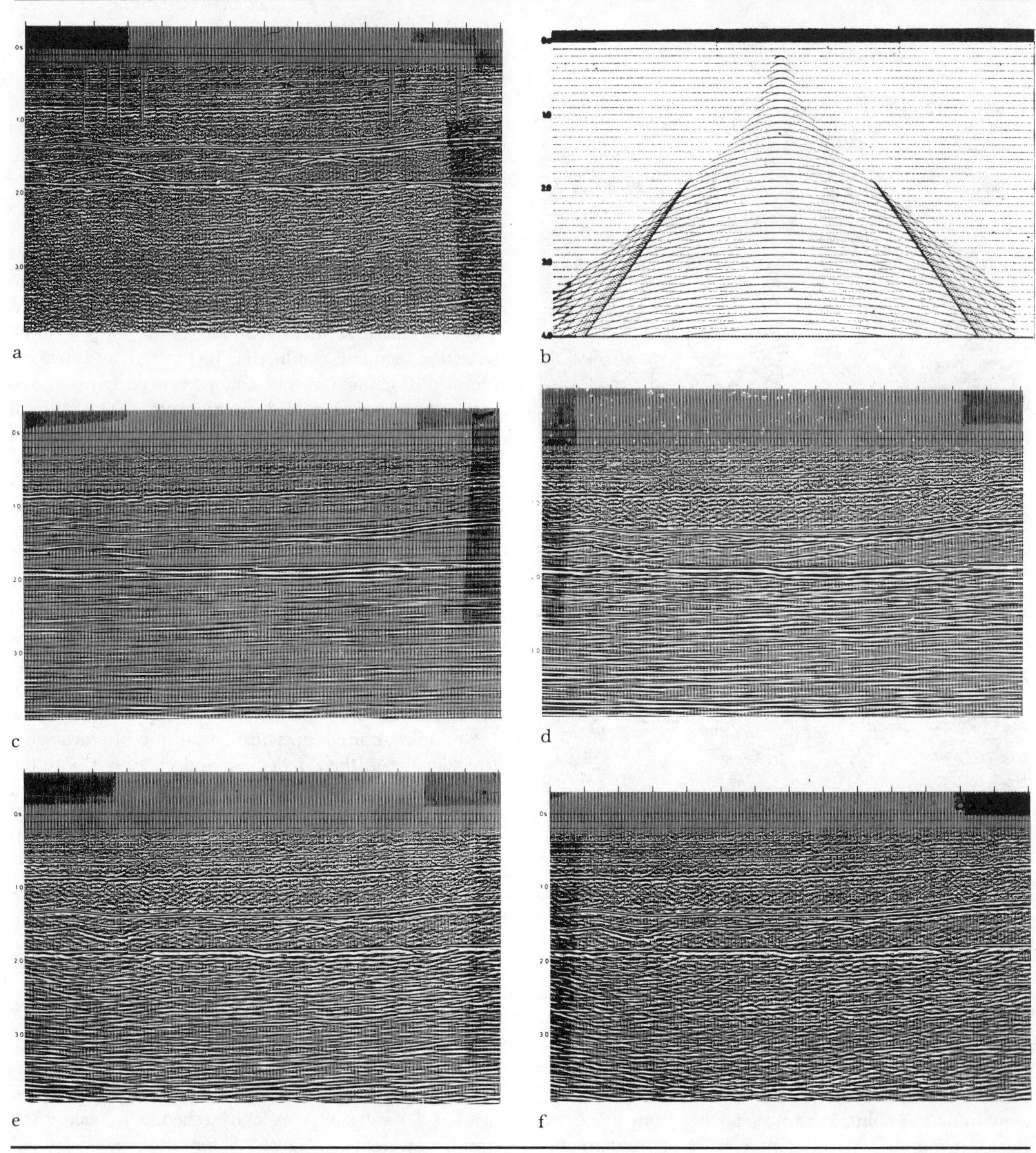

FIGURE 9.7. *Effect of aperture size on summation migration. (a) Stacked seismic section. (b) Diffraction curves. (c) ±20 trace aperture. (d) ±40 trace aperture. (e) ±80 trace aperture. (f) ±160 trace aperture. (Courtesy of Prakla-Seismos Report.)*

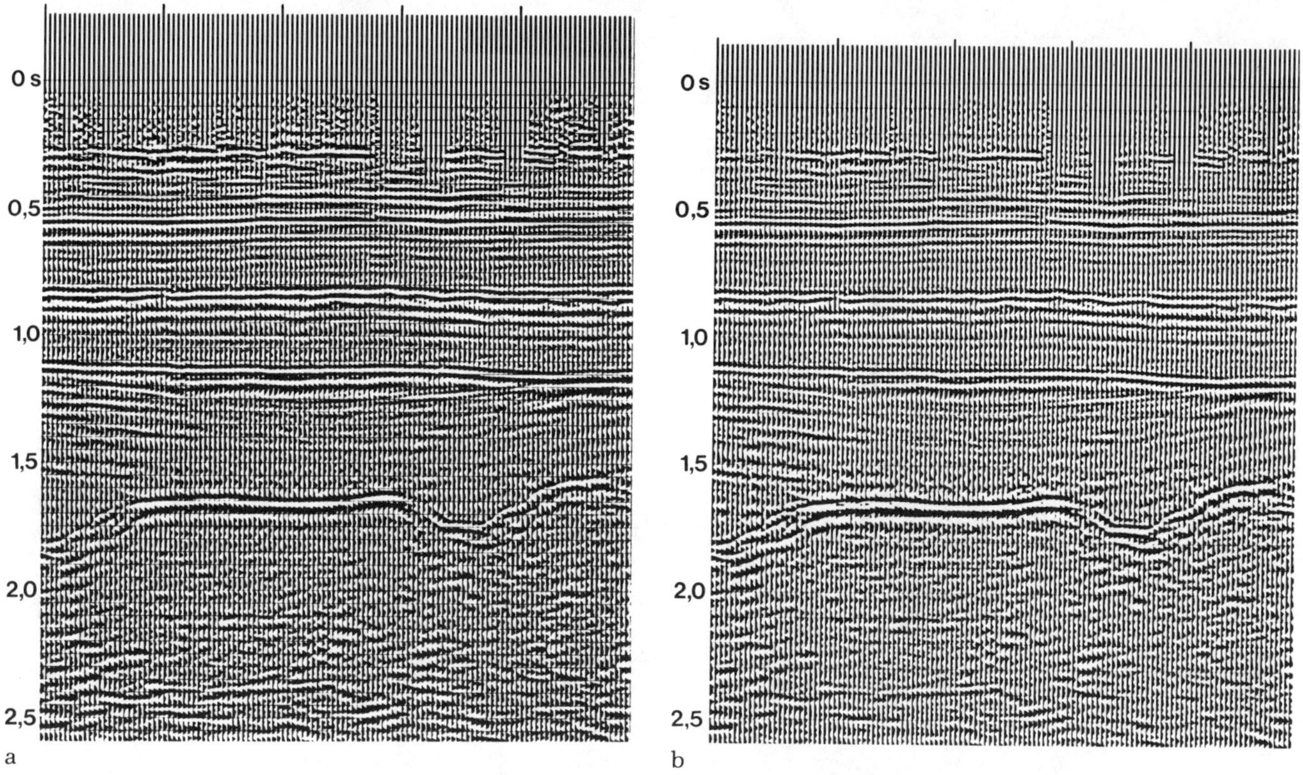

FIGURE 9.8. *Comparison of conventional summation with summation weights based on coherence, both using ±20 trace aperture. (a) Conventional summation. (b) Weights based on coherence. (Courtesy of Prakla-Seismos Report.)*

events than would be shown with conventional summation (from Prakla-Seismos Report 1-72).

One of the shortcomings of the summation method is the strong migration noise produced by noise bursts on the record section. Model results from Larner (1975) show the effect very well. The model in Figure 9.9(a) consists of a single curved horizon whose dip grades from zero to 45° and a single trace of bandlimited random noise. Kirchhoff summation in Figure 9.9(b) migrates the curved reflector back to its true position, but the single noise trace produces severe migration noise in a cone centered on the noise trace. The width of the cone increases with depth and indicates the aperture size used as a function of depth. Within the cone, the diffraction noise produces the curves of equal reflection time; that is, it produces a wavefront chart.

Migration noise on real sections produces the increasingly strong curves of equal reflection time as the depth increases, because sections are of finite length, and the summation must be stopped before the randomness in the sum can suppress the effect of deep events that have short lateral extent. These so-called smiles sometime give a false appearance to the migrated sections.

Despite these shortcomings, proper massaging of the data during summation gives very usable and often excellent migrated sections. However, the ad hoc selection of parameters may require some artistry as well as experience and intuition in order to optimize or at least prettify the migrated sections.

Since the development of wave-equation migration by finite differences, the summation method has been put on a more firm basis by relating it to the integral solution of the wave equation (Schneider, 1978).

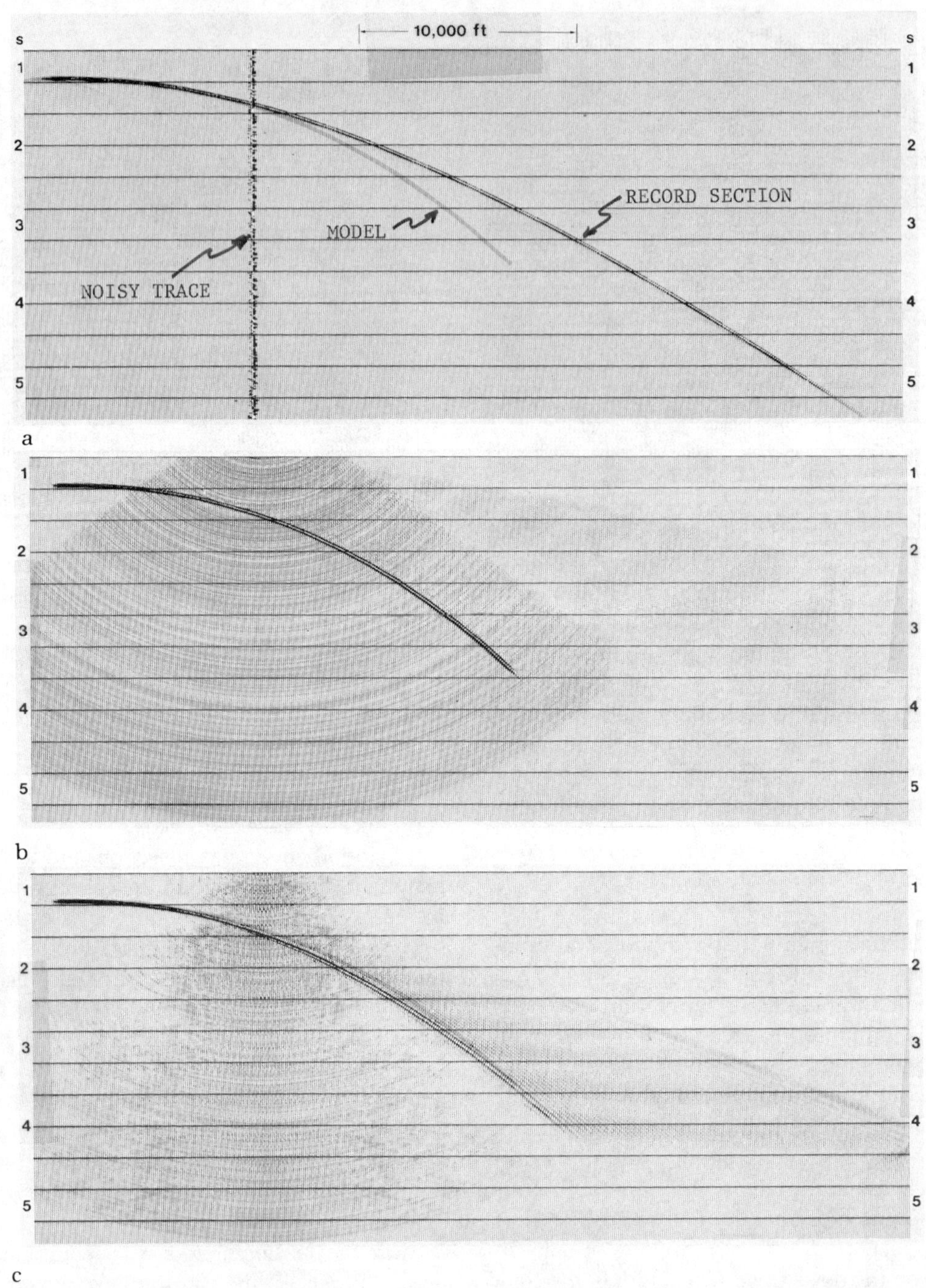

FIGURE 9.9. *Summation and finite-difference migration of curved reflector. (a) Model and seismic record section. (b) Kirchhoff summation migration. (c) Finite-difference migration. (From Larner and Hatton 1975, courtesy of Western Geophysical Company.)*

MIGRATION BY FINITE DIFFERENCES

The propagation of elastic waves is represented by solutions of the scalar wave equation,

$$\nabla^2\Phi = \frac{1}{V^2}\frac{\partial^2\Phi}{\partial t^2}, \tag{9.2}$$

where the ∇^2 operator stands for

$$\frac{\partial^2}{\partial x^2} + \frac{\partial^2}{\partial y^2} + \frac{\partial^2}{\partial z^2}. \tag{9.3}$$

Analytical solution of this formidable partial differential equation is possible only with simple layering. Consequently, the wave-equation model was shelved in exploration geophysics in favor of the far simpler geometric (raypath) model until digital methods came into widespread use.

The wave-equation model rested in limbo until Claerbout marshalled an assault upon this impregnable fortress using numerical methods. Work by his Stanford group, reported in several papers—including Claerbout and Johnson (1971), Claerbout (1971a,b), Claerbout and Doherty (1972), and Riley (1975)—is summarized in Claerbout's text, *Fundamentals of Geophysical Data Processing* (1976). These studies resulted in application of finite differences to migration. Loewenthal et al. (1974), Larner (1975), Reilly and Greene (1976), and Prakla-Seismos Report 1-76 describe this migration technique and show examples of its use.

Claerbout's method begins with the observable wave field $\Phi(x,z,t)$ at the surface, where Φ is the physical quantity measured; particle velocity, acceleration, displacement, or pressure. The wave field is continued downward by numerical solution of the wave equation. Knowing the field at the surface, $z = 0$, the field is computed at $z = \Delta z$. Then, by recursive methods, the field is progressively computed at $2\Delta z, 3\Delta z, \ldots, n\Delta z$. This is equivalent to moving the detector system successively deeper into the earth. Only when the detectors are at the level of a reflector is that reflector correctly mapped. Thus, when detectors are at a given level, all reflectors at and above this level are correctly mapped; that is, they are migrated to their true position. To overcome difficulties resulting from boundary reflections and to facilitate downward continuation, Claerbout uses a coordinate system that moves coincidentally with the wavefront.

Solution of the wave equation using finite differences can be accomplished using differing degrees of approximation. In the two-dimensional wave equation, the term $\partial^2\Phi/\partial z^2$ is small for waves traveling near vertical and can be ignored for dips less than 15°. Numerical solutions of the two-term wave equation

$$\frac{\partial^2\Phi}{\partial x^2} = \frac{1}{V^2}\frac{\partial^2\Phi}{\partial t^2}, \tag{9.4}$$

using the Crank-Nicholson numerical operator (Claerbout, 1976, p. 211) gives the so-called 15° wave-equation migration. For dips greater than 15°, the reflection wavelet is noticeably dispersed, because the phase error per iteration increases with dip and increases with frequency. The frequency content in the wavelet is moved different distances during migration. The lower frequencies are moved to the correct migrated position, but the movement is increasingly in error as the frequency increases. This disperses the wavelet during migration. Figure 9.9(c) shows the dispersion upon migration of the curved record surface in the model and also shows decreased diffraction noise compared to that generated by summation.

By including the $\partial^2\Phi/\partial z^2$ term, improved finite-difference operators have been derived that eliminate the problem of dispersion of the wavelet at steeper dips. One such operator, developed by Koehler, is shown by Reilly and Greene (1976) to have less phase error than the Crank-Nicholson operator and to be effective for dips up to 45°, as shown in the model study in Figure 9.10, which has horizons with dips varying from zero to 30°. The Koehler operator has no observable dispersion through 30°, while the Crank-Nicholson operator shows marked dispersion above 15°. The superior performance of the Koehler operator with respect to dispersion of steeply dipping events is confirmed using real data (Figure 9.11).

Finite-difference methods tend to be superior to summation methods for several reasons: (1) velocity inhomogeneities are more accurately and more easily handled; (2) diffractions are handled correctly; (3) diffraction noise is less objectionable and produces less ar-

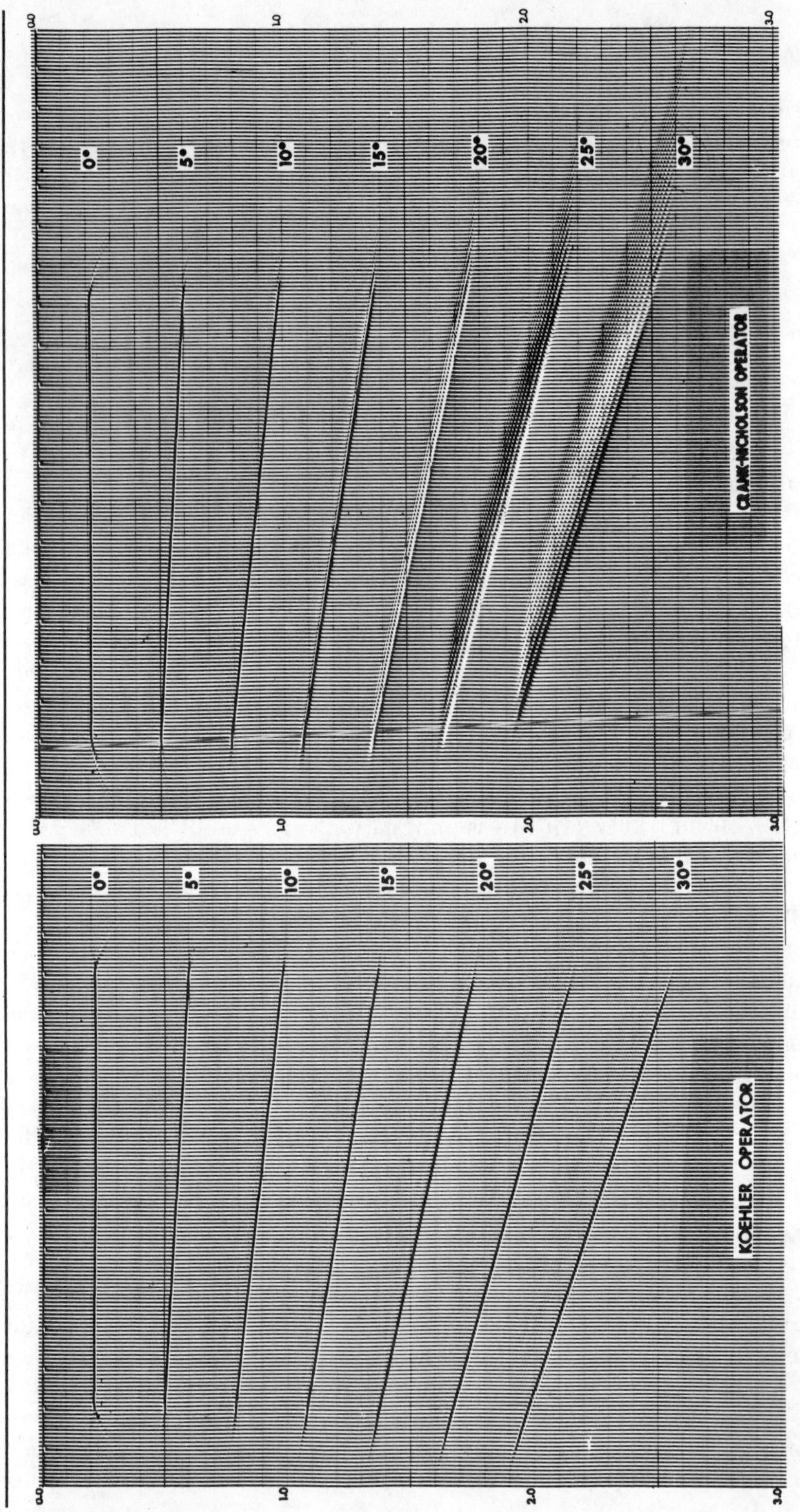

FIGURE 9.10. *Comparison of numerical operators in migrating dips varying from zero to 30°. (From Reilly and Greene 1976, courtesy of Seiscom-Delta Report.)*

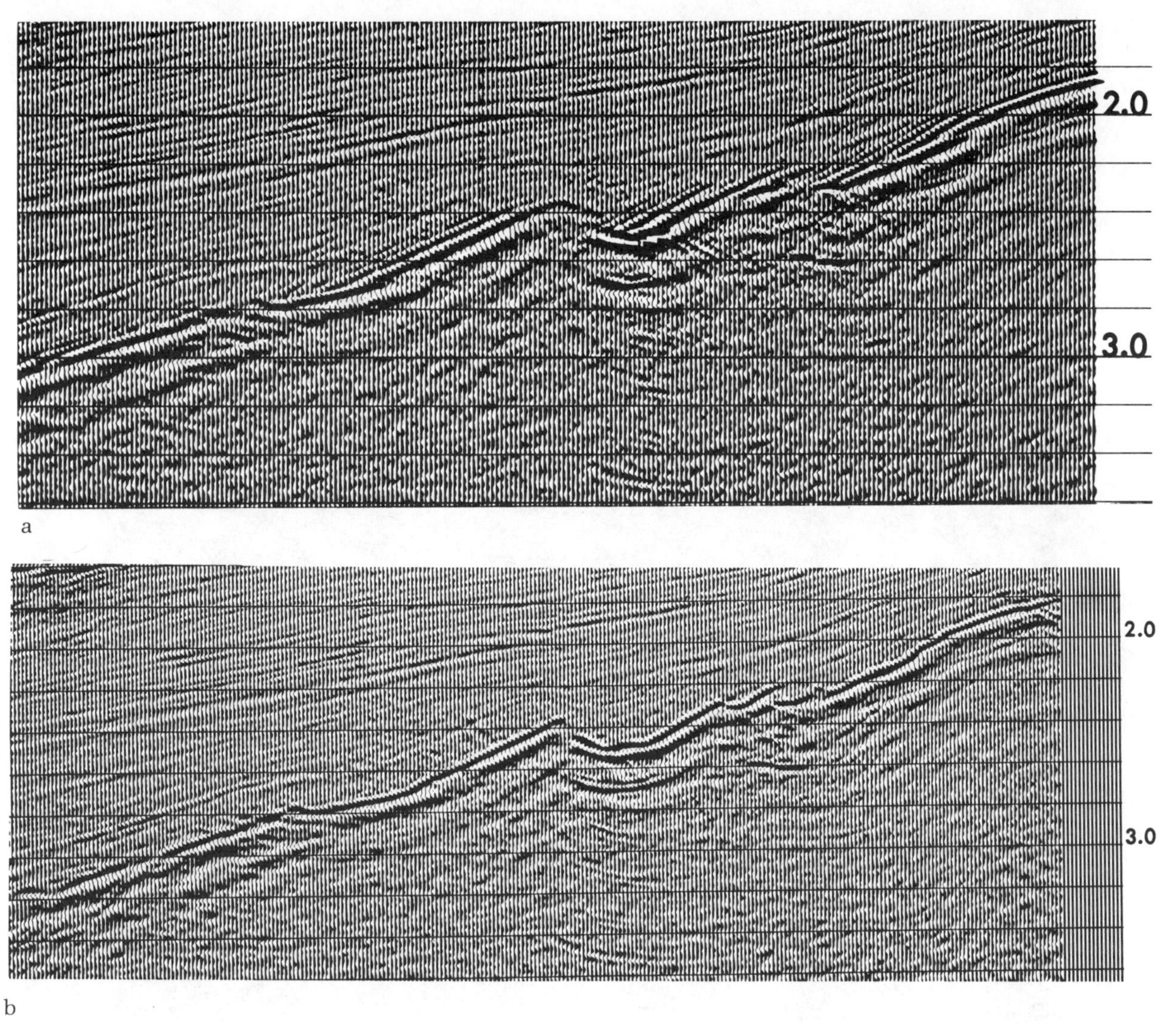

FIGURE 9.11. *Comparison of numerical operators in migrating real data. (a) Crank-Nicholson operator. (b) Koehler operator. (From Reilly and Greene 1976, courtesy of Seiscom-Delta Report.)*

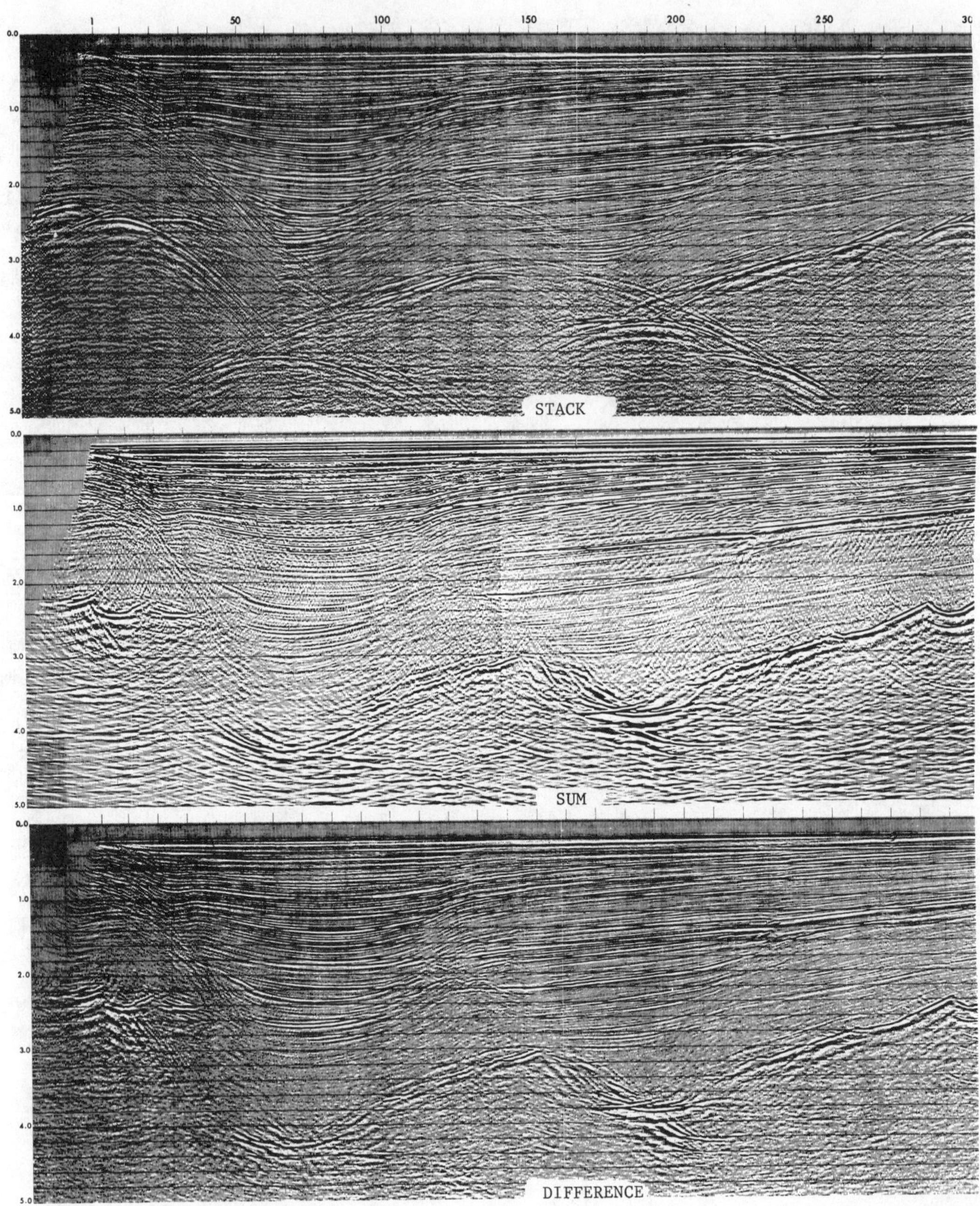

FIGURE 9.12. *Comparison of summation and finite-difference migration of seismic line from St. George Basin, Alaska. (From Reilly and Greene 1976, courtesy of Seiscom-Delta Report.)*

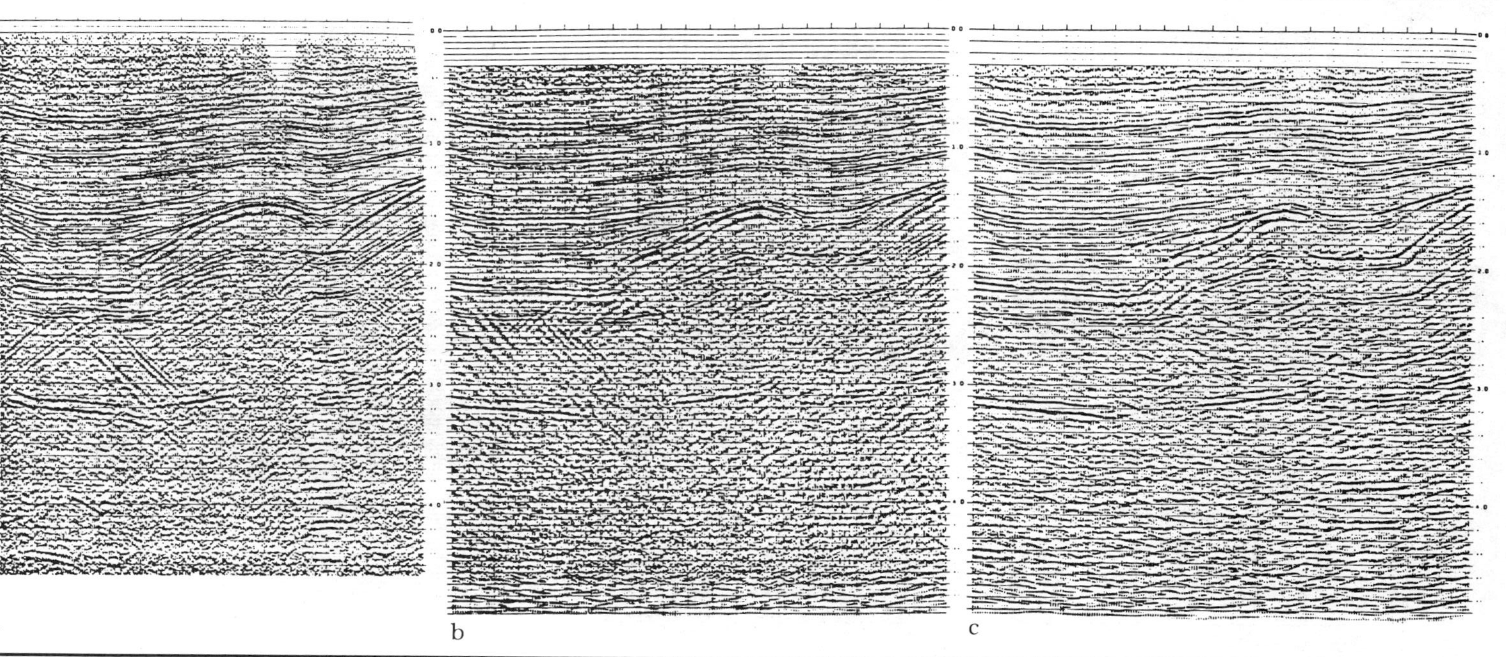

FIGURE 9.13. *Comparison of 15° wave-equation migration and* f-k *migration. (a) Tenfold CDP section. (b) 15° Claerbout migration. (c)* f-k *migration. (From Stolt 1978, courtesy of Geophysics.)*

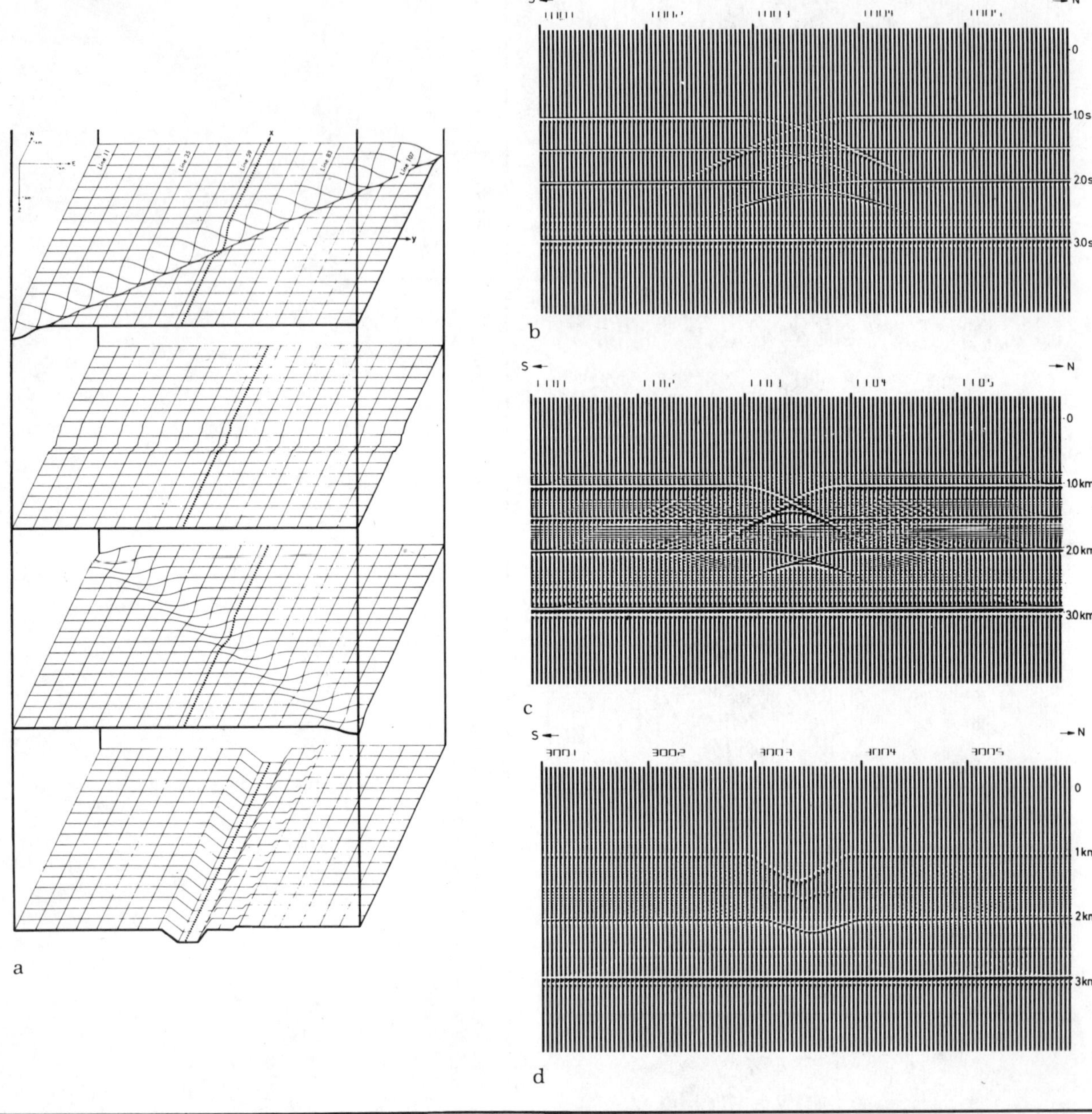

FIGURE 9.14. *Two- and three-dimensional migration using summation. (a) Seismic model. (b) N-S seismic profile. (c) Two-dimensional migration. (d) Three-dimensional migration. (From Bortfeld 1974, courtesy of Prakla-Seismos Report.)*

cuate effect, especially deep in the section; (4) reflection character is preserved; and (5) ad hoc design of parameters—aperture size, weights, and so on—is not necessary. An example from Seiscom-Delta, comparing summation and finite-difference migration, is shown in Figure 9.12.

MIGRATION USING FOURIER TRANSFORMS

Wave-equation migration can be accomplished by transforming the wave equation into *f-k* space, using the two-dimensional Fourier transform. Subject to sampling limitation in x and t and the resulting Nyquist frequency $f_N = 1/2\Delta t$ and wavenumber $k_N = 1/2\Delta x$, dips of any angle can be migrated correctly and without dispersion. A further advantage over migration by finite differences is that the speed of computation is increased as much as 10:1. The only disadvantage appears to be in regions of rapidly varying velocity functions.

Figure 9.13 shows the comparison of 15° finite-difference migration with *f-k* migration, from Stolt (1978). Comparisons between 45° finite differences and *f-k* migration show virtually identical results, except in very steep dip areas, where *f-k* migration is superior. For that reason and because of increased computational speed, *f-k* migration has become the most widely used technique in the industry.

THREE-DIMENSIONAL MIGRATION

Effectiveness of two-dimensional migration depends on all diffraction events originating in the plane of the record section. Dip lines and lines at right angles to synclines and faults will be handled very well. However, diffraction events from out of the plane will not be collapsed with two-dimensional migration and can only be collapsed by using three-dimensional migration. Although there is no intrinsic problem with three-dimensional migration, either in summing along the three-dimensional surface of maximum convexity or solving the wave equation in three dimensions, three-dimensional data acquisition is still in its infancy.

An example from Bortfeld (1974) shows three-dimensional model data that are properly migrated with three-dimensional migration (Figure 9.14). The model has four horizons, each containing a syncline with orientation NE–SW, E–W, SE–NW, and N–S, respectively. The N–S seismic profile shows the typical multibranch signatures caused by buried foci for the upper three synclines. However, only the signature from the syncline at right angles to the line of profile (the second horizon) is collapsed into its true synclinal structure with two-dimensional migration. Three-dimensional migration correctly collapses all signatures into their true structure.

French (1975) has shown that, in the truly two-dimensional case, where the faults, synclines and anticlines are crossed at an angle θ with respect to the strike line, these data can be migrated correctly by using a velocity of $V/\cos\theta$ in the migration procedures.

SUMMARY

Of the available migration methods, it is becoming increasingly evident that wave-equation migration is superior, is based on principles that are more sound, and is amenable to improvements through scientific studies. Also, when Fourier transform methods are used, it is much faster computationally. Summation methods tend to be improved by ad hoc procedures to improve the appearance of migrated sections, but what works in one instance may not work in another.

Wave-equation methods are in their infancy, and they promise to lead to solutions of many problems other than migration. This area of research is the latest in an impressive series of developments since the digital revolution inexorably changed the complexion of exploration geophysics.

Chapter 10

Direct Detection of Hydrocarbons on Seismic Data

INTRODUCTION

The realization that hydrocarbons can be detected directly on seismic data has caused a profound revolution in seismic acquisition, processing, and interpretation throughout the oil industry. A key factor has been the use of information-preserving processing of the seismic data, which allows both the conventional balanced amplitude and the true amplitude to be displayed—the former for structural interpretation and the latter for more positive identification of the presence of hydrocarbons.

The Russian literature reported the use of direct methods long before they became widely used in the Western world. Ballakh, Kochkina, and Gruzkova (1970) reported that the first attempt to determine the presence of oil and gas in rocks by means of the seismic method was made in 1952–1953 on structures in the Rostov district by locating the water-oil and gas-liquid contacts on seismic time profiles. Similar profiles were constructed in the Mukhanoyo field in the Kuibyshev district. The results confirmed the feasibility of using seismic prospecting for direct exploration of oil and gas pools. Churlin and Sergeyev (1963) reported experimental work in one of the fields in the Kuban-Black Sea basin.

The most significant diagnostics for direct detection are those given by Churlin and Sergeyev: (1) strong reflections due to the presence of hydrocarbons, (2) interference of reflections at the edges of the reservoirs, (3) flat reflections associated with gas-oil, gas-water, and oil-water contacts, and (4) high absorption of seismic energy by gas reservoirs.

A paper by Diekmann and Wierczeyko of West Germany, presented at the 11th International Gas Conference in Moscow, 1970, described the use of seismic techniques to outline the spread of gas injected into an aquifer.

Before direct detection methods became widely used, the Western literature contained many papers describing the change in rock properties because of porosity and fluid content. In CVL studies of sand reservoirs, Hicks and Berry (1956) noted that the velocity in Miocene sands at 6000–8000 ft offshore Louisiana decreased by about 15 percent when the interstitial fluid changed from salt water to gas. Pan and DeBremaecker (1970) discussed the direct detection method. Marr (1971a) gave a case history of the Fairbanks-North Houston structure in Harris County, Texas, in which abnormally strong reflections were associated with the areas of gas production; this occurred on data taken in 1937 using a programmed-gain amplifier and minimum filtering—that is, information-preserving acquisition.

Despite the considerable evidence that direct detection of hydrocarbons on seismic data was possible, and that digital recording and processing had the capability to preserve amplitude and bandwidth, there was little outward evidence that true-amplitude techniques were being used before 1972. Contractors were not offering this type of processing, but it was an idea whose time had come. Bright spots quickly became the hottest topic in exploration, and with good reason, because they were effective in pinpointing gas reservoirs in the Gulf of Mexico and other areas with similar geologic characteristics. The origins of bright-spot technology remain shrouded in mystery, but pioneering work seems to have been done by Shell Oil Company.

DIRECT DETECTION TECHNIQUES

The seismic techniques for direct detection of hydrocarbons depend on the effect of porosity and fluid content on the velocity and density of the reservoir rocks, and on the associated dependence of the reflectivity on these properties. The changes in reflectivity become most visible on seismic sections when the reflection amplitudes are preserved during recording and processing (Figure 10.1).

The most prominent diagnostic in sand/shale sequences in Tertiary basins is the strong amplitude anomaly, the bright spot, corresponding to gas accumulation, as shown, for example, in a seismic section in the Pleistocene offshore Texas (Figure 10.2). The section across the High Island Block 330 field in Figure 10.3 shows the multitude of gas zones that produce there. Typical reflection coefficients for the strongest nonhydrocarbon-associated reflections in such basins are usually much less than 0.05 in magnitude, while the presence of gas may increase the reflectivity magnitude to 0.2 or more. The presence of oil may cause a significant change in reflectivity, but it will be much less than that observed because of the presence of gas.

The second significant diagnostic is the presence of interference and, possibly, phase reversal at the edges of the reservoir (Figure 10.4). Interference occurs because top and bottom reflections with opposite polarity merge as the reservoir thins; phase reversal occurs if the reflectivity changes sign in going off the reservoir. Phase reversals sometimes appear to be half-cycle faults that outline the reservoir.

The third important diagnostic is related to the fact that the gas/oil, oil/water, and gas/water contacts are practically flat, horizontal, acoustically smooth surfaces, and reflections from these contacts often exhibit dips at variance with the structural dips, as shown in a seismic section from the High Island area offshore Texas (Figure 10.5). The flat reflections tend to curl upward on the sections at the edges of reservoirs because of thinning of the hydrocarbon section. When the reservoir is thicker than the tuned thickness, both the bottom and top reflections are observable and the two-way time thickness is measurable. Often, the thickness is less than the tuned thickness, resulting in a composite reflection whose amplitude depends on the ratio of thickness (in two-way time) to reflection period. In this case, only the upper limit of the thickness (that is, the tuned thickness) is known. Sometimes, the upper interface does not produce a strong reflection; in that case, the flat event may be the most significant diagnostic. The reflection from the upper interface may be weak because the interface may be acoustically rough or the acoustic contrast may not be sharp because of diffusion of gas into the overlying formations.

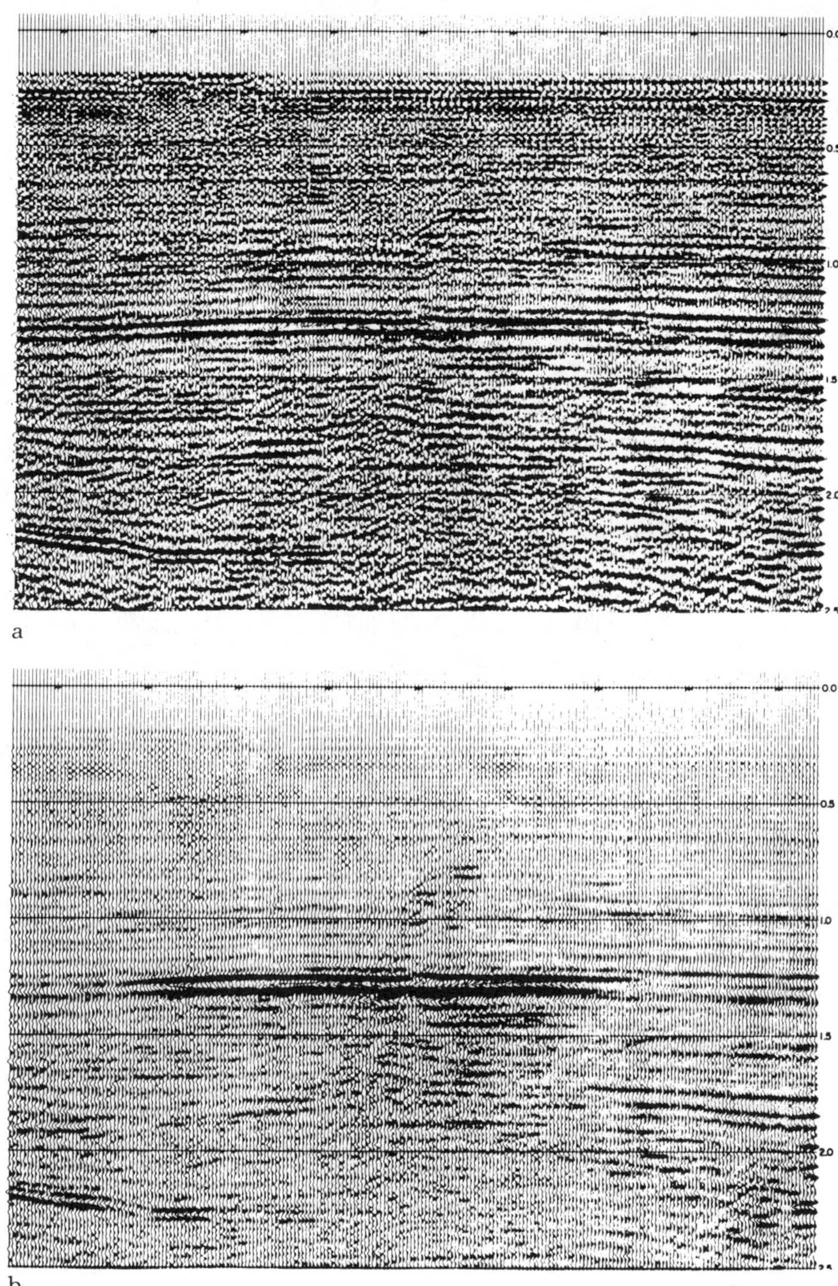

FIGURE 10.1. *Comparison of automatic gain control and true-amplitude processing, Gulf of Mexico. (a) Amplitude-destroying processing, using automatic gain control. (b) Amplitude-preserving processing. (From Barry and Shugart 1973, courtesy of Teledyne Exploration Company.)*

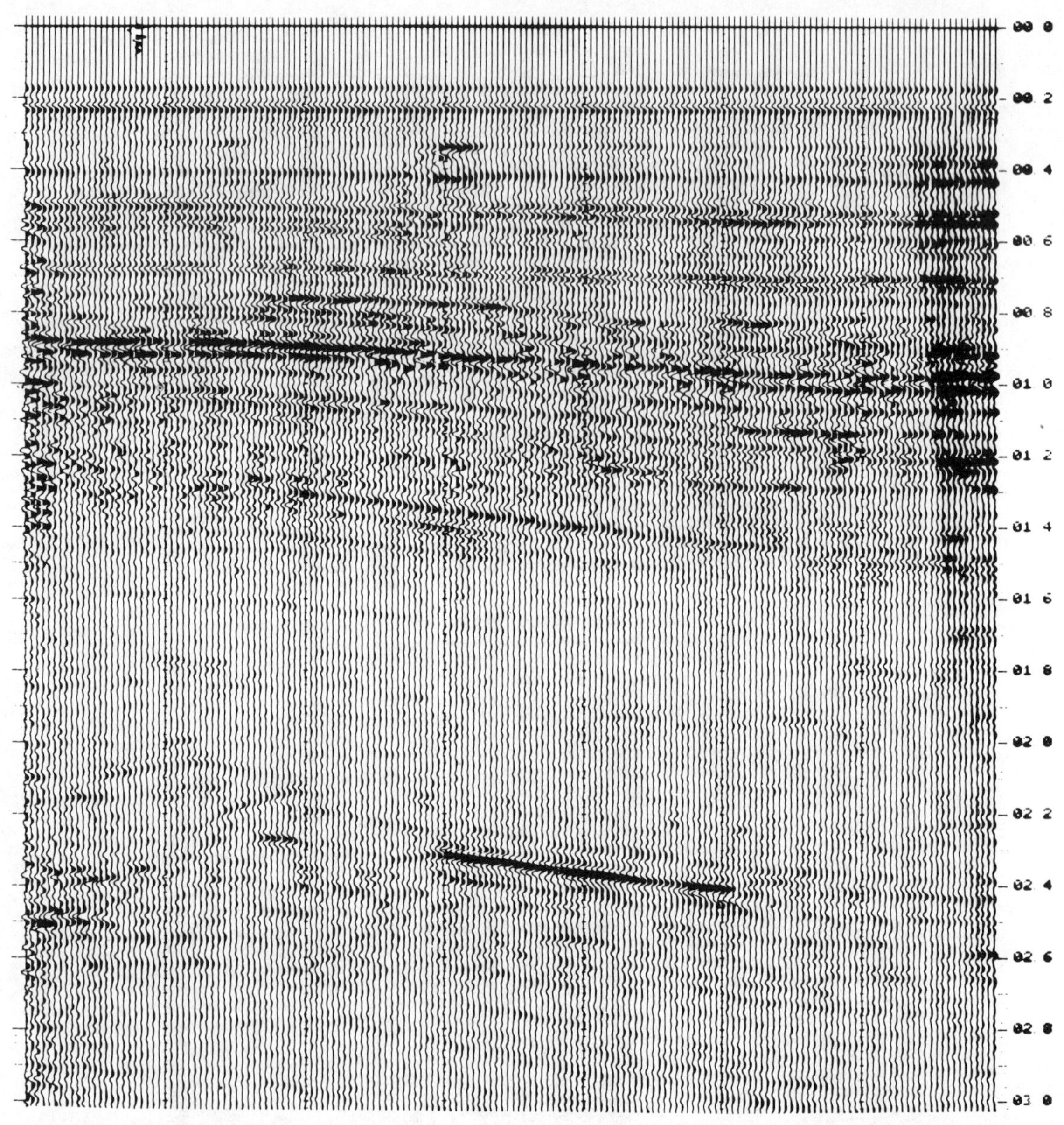

FIGURE 10.2. *True-amplitude seismic section in the Pleistocene, offshore Texas. (Courtesy of Pexcon Computing Company.)*

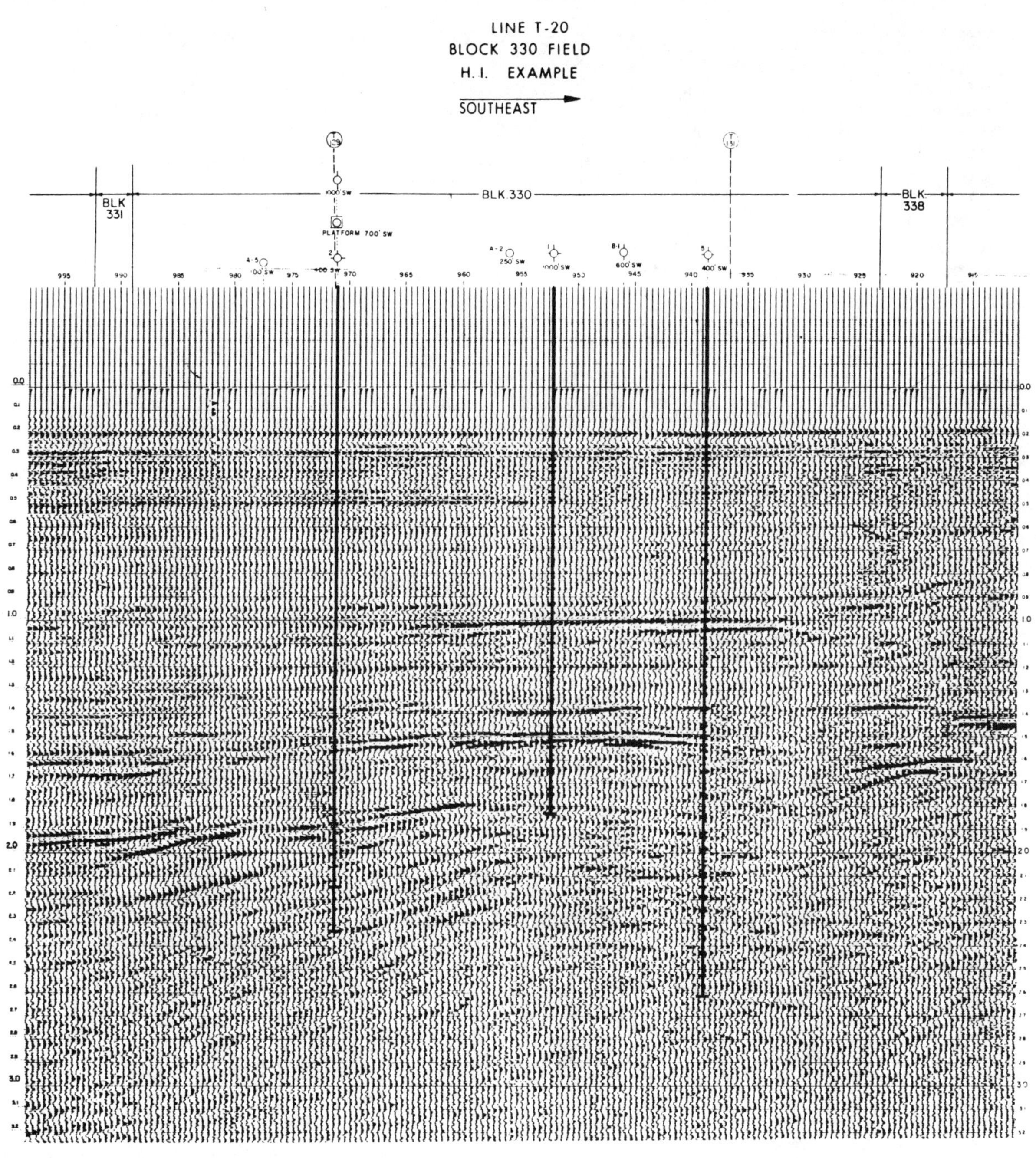

FIGURE 10.3. *True-amplitude seismic section across the High Island Block 330 field, offshore Texas.*

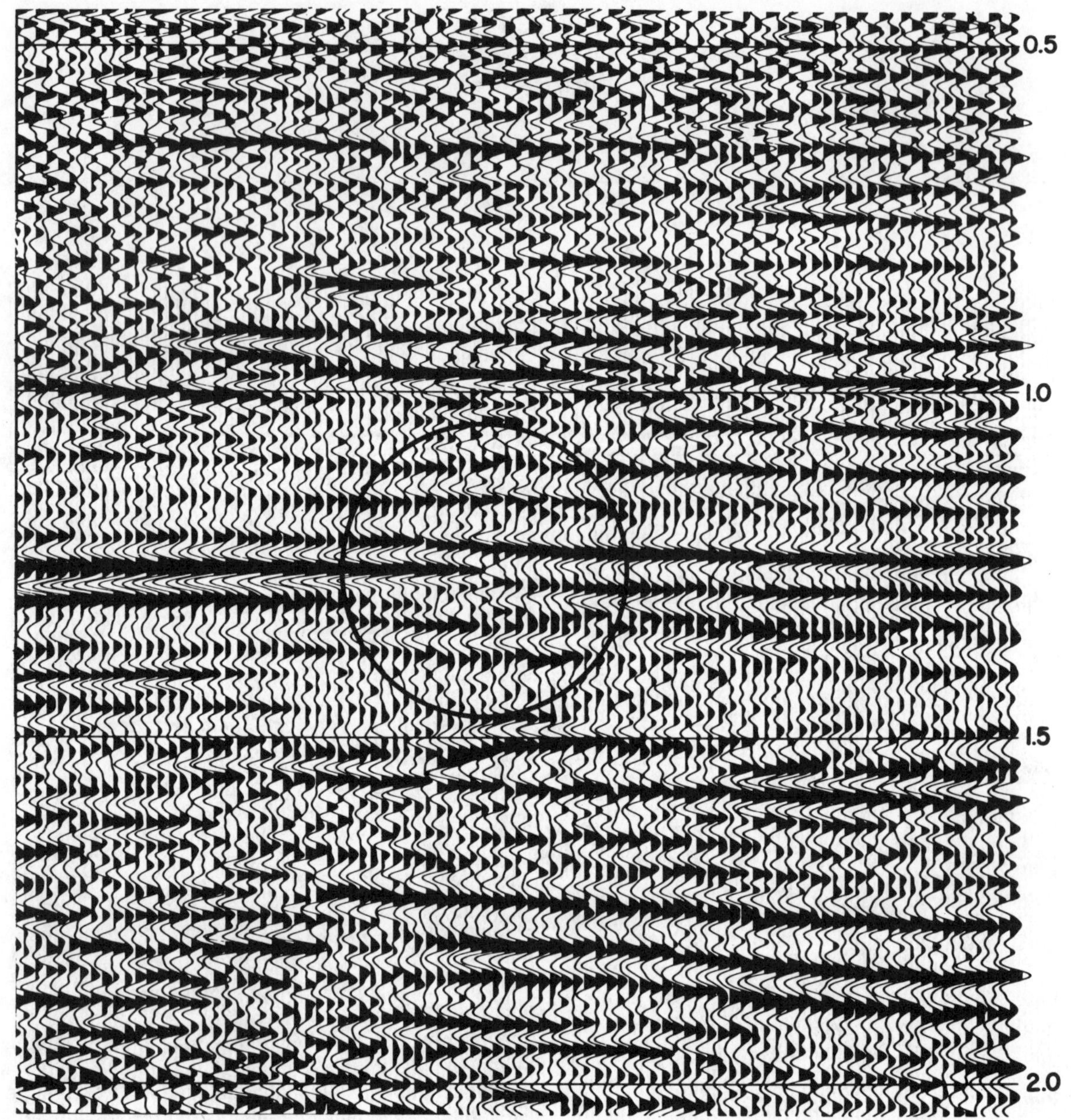

FIGURE 10.4. *Interference and phase reversal at the edge of a gas reservoir. (From Barry and Shugart 1973, courtesy of Teledyne Exploration Company.)*

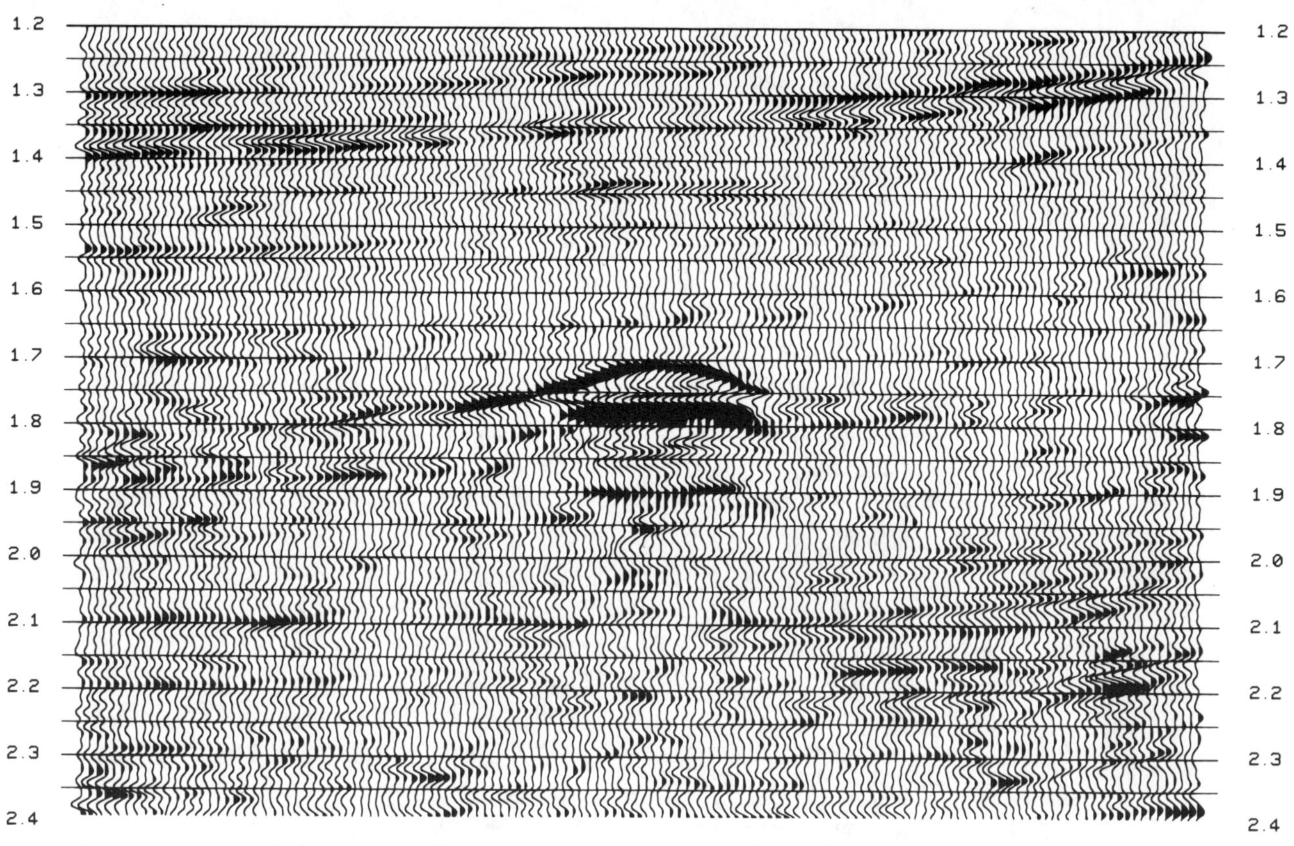

FIGURE 10.5. *True-amplitude seismic section showing classical bright spot at top of gas-filled anticline and flat spot at the gas/water contact. (Courtesy of Teledyne Exploration Company.)*

Frequency analysis is sometimes valuable in establishing some of the parameters associated with amplitude anomalies. A simple gas reservoir will have a reflectivity consisting essentially of a doublet, with the two reflections being of opposite polarity and spaced in time by the two-way traveltime through the reservoir. Such a reflectivity has notches in its frequency spectrum at intervals of $1/\Delta$, where Δ is the two-way time. Spectral analysis of a window containing the reflections from the reservoir may show these notches even when the thickness is less than the tuned thickness; if so, it will be a good measure of thickness.

The foregoing diagnostics of hydrocarbons are apparent on the bright spot in Figure 10.5. The top of the anticlinal gas reservoir is outlined by the strong peak (black) with apex at 1.710 sec. The flat trough (white) at 1.760 seconds is the gas/water contact at the base of the reservoir. The two-way time-thickness between top and bottom reflections of 50 ms is equivalent to a thickness of 175 ft, assuming gas-sand velocity of 7000 ft/sec. At the edge of the reservoir on the left, the basal reflection curls upward because of thinning of the reservoir, and there is interference between the top and bottom reflections at the edge, resulting in diminished amplitude of the reflection peak.

Another example, in Figure 10.6, also shows the diagnostics. The shallow reservoir at 1.1 sec is thin, as deduced from the high-frequency character of its seismic response and the lack of any evidence of tuning. It is

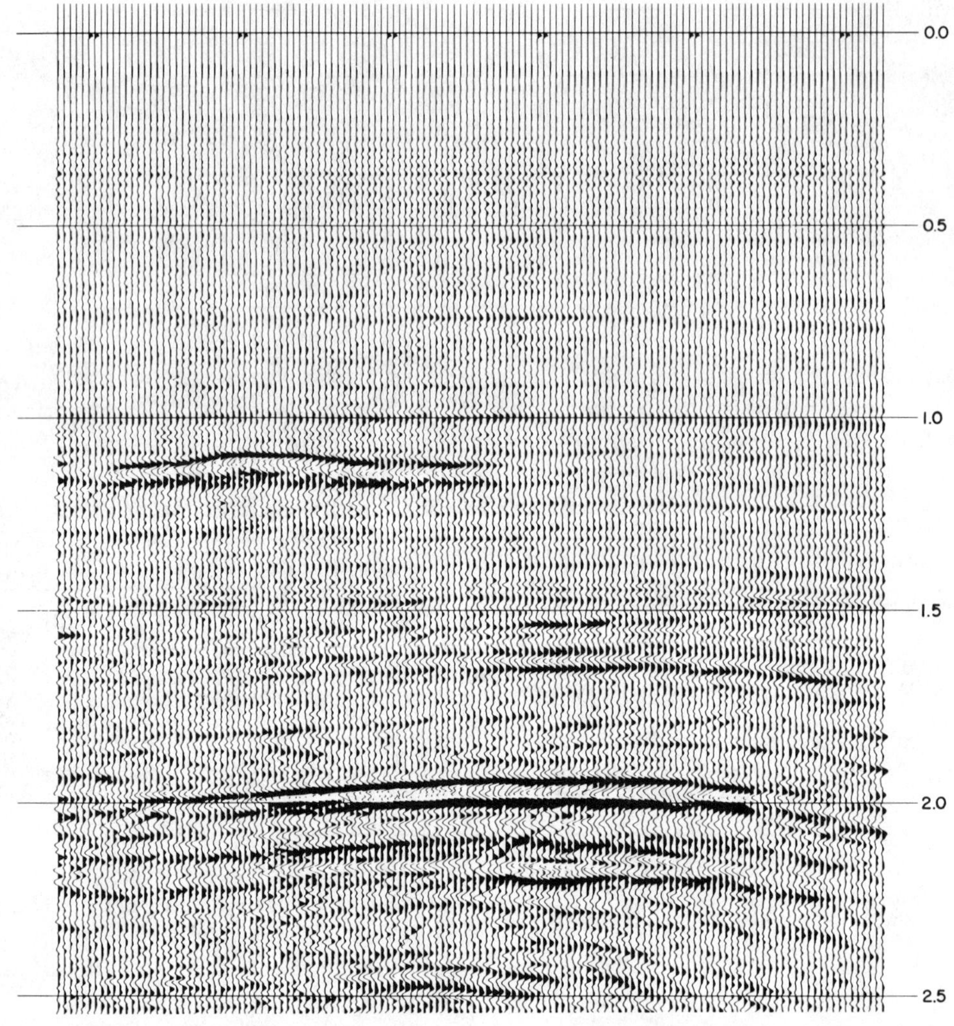

FIGURE 10.6. *Bright spots caused by gas accumulations in the Gulf of Mexico. (From Barry and Shugart 1973, courtesy of Teledyne Exploration Company.)*

thinner than the tuned thickness, $\Delta = T/2$. The period of the wavelet at this depth is about 24 ms, so the reservoir is thinner than about 12 ms two-way time, or thinner than about 42 ft, based on an expected velocity of 7000/ft sec at this depth. The deeper reservoir at 1.9 sec, is thick, with maximum thickness of about 50 ms two-way time, which corresponds to a thickness of about 185 ft, based on an expected velocity of 7500 ft/sec at this depth. By identifying the top of the reservoir as the strong peak at 1.9 sec, the bottom of the reservoir is then identified by the flat trough, at just less than 2.0 sec, that precedes the second strong peak. The reservoir is controlled by faulting to the right and by dip to the left. At the intersection of the projection of the flat trough and the top peak, the bright spot fades out, marking the edge of the reservoir.

A strong indirect indicator of significant reservoir

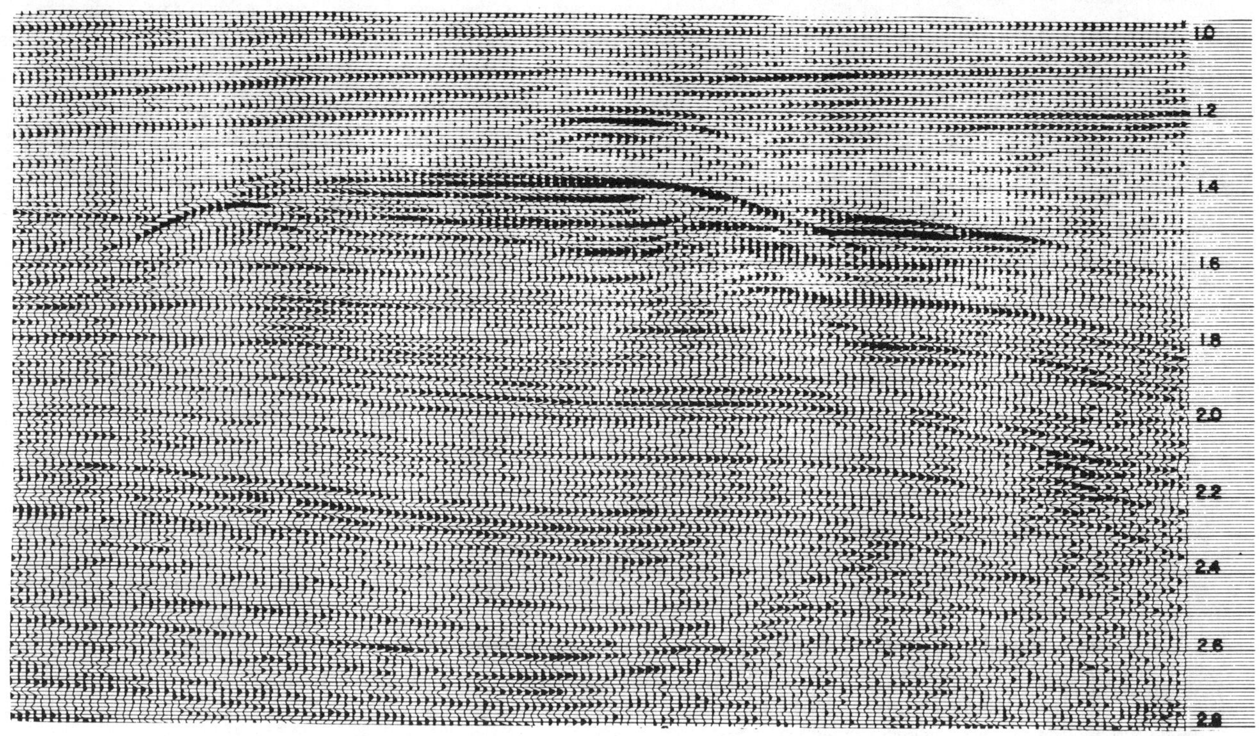

FIGURE 10.7. *Sag on reflections underlying structure, suggesting presence of thick hydrocarbon pay, offshore Nigeria. (True amplitude processing courtesy of Pexcon Computing Company.)*

thickness is the presence of time-sag on reflections from beneath the reservoir because of a decrease in velocity, resulting from hydrocarbons, as shown in Figure 10.7 on a section offshore Nigeria. Time-sag is most significant with gas reservoirs. The amount of time-sag Δt computed from the reservoir thickness Δz, the velocity of the reservoir rock containing hydrocarbons V_h, and its velocity when containing salt water, V_w, is

$$\Delta t = \frac{2\Delta z(V_w - V_h)}{V_w V_h}. \tag{10.1}$$

Another indirect indicator is the presence of a shadow zone of weak reflections beneath gas reservoirs because of increased absorption of the seismic energy. If shadow zones are evident, the ratio of multiple/primary reflection amplitude may increase in the shadow zone, because the multiples will not be affected by absorption, whereas the primaries will be.

A secondary diagnostic feature that is sometimes noticeable is the presence of diffractions from lateral changes in the acoustic impedance because of the termination of the reservoir. The diffractions may be faint and may be confused with diffractions from faults.

The study by Diekmann and Wierczeyko (1970) on the delineation of the extent of a gas storage reservoir shows all of the diagnostics given here (Figure 10.8). In the area where the aquifer contains gas, its reflection is very strong. The edge of the reservoir is clearly marked by an abrupt loss of amplitude. A diffraction can be seen emanating from the discontinuity at the edge of the reservoir. Sag on the deeper *P*-reflection is obvious, as is the shadow zone beneath the reservoir. The multiple from the gas reservoir is a very strong event.

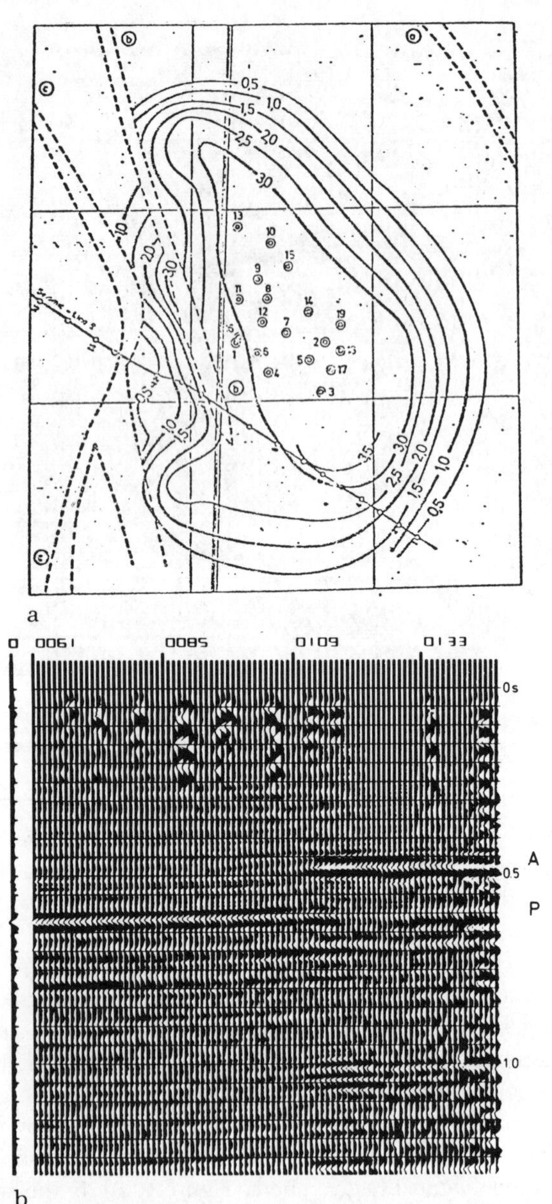

FIGURE 10.8. *Application of seismic method to determine extent of spread of gas injected into an aquifer. (From Diekmann and Wierczeydo 1970, courtesy of Prakla-Seismos Report.) (a) Contour map of A/P, reflection/amplitude ratio. (b) Absorption under the bright spot and diffractions at the edge. (From Bortfeld, 1974, courtesy of Prakla-Seismos Report.)*

Equally as important as the diagnostics is the requirement that, for an anomaly to be a hydrocarbon indicator, the interpretation must make geologic sense. Confidence in deducing hydrocarbon presence is greatly increased if the diagnostics fit the structural interpretation and occur on all seismic lines. Figure 10.9 shows a structural map in the High Island area offshore Texas. The seismic lines in Figures 10.10(a) and 10.10(b) that cross the anticlinal structure each show an interference pattern indicating the downdip limit of hydrocarbons at 1.975 sec. It is interesting to note that the interpretation of the extent of hydrocarbons here has been made on amplitude-balanced structural sections.

An example of reservoir delineation using true-amplitude processing is shown on the seismic sections in Figures 10.11(a) and 11(b), from offshore Trinidad. The upper reservoir clearly terminates downdip at 2 sec, and the lower reservoir terminates at about 2.1 sec.

As with any presumed panacea, caution must be exercised in equating bright spots with commercial hydrocarbon accumulations. Bright spots may be produced by low-saturated or thin gas sands, by calcareous and siliceous stringers, and by subsurface igneous and salt flows, and there are dry holes as proof of each of these bright-spot generators. Figure 10.12 shows a bright spot from a thin sand, and Figure 10.13 shows an amplitude anomaly associated with carbonates. The log in Figure 10.14, from a well in northeast Louisiana (Ramo and Bradley, 1977), shows the significant density contrast resulting from a rhyolite intrusion that caused a dry bright spot to be drilled. The bright spot showed termination and sag beneath to cause further expectation of hydrocarbons. The sag turned out to be structural and not due to the presence of low velocity above. Subsequent surface gravity and magnetic profiles across the feature corroborate the presence of an igneous intrusion here. The log in Figure 10.15, from southwest Arkansas, shows the significant velocity contrast resulting from a salt flow that caused a bright spot similar to what one might expect from a gas sand. Surface gravity and magnetic measurements indicated probable salt here.

Hydrocarbons are associated with dim spots in circumstances where the reflectivity is reduced by the presence of hydrocarbons, such as is found in porous limestone reservoirs (Figure 10.16). The reflectivity of a limestone overlain by shale decreases with porosity and decreases still further with increasing gas saturation.

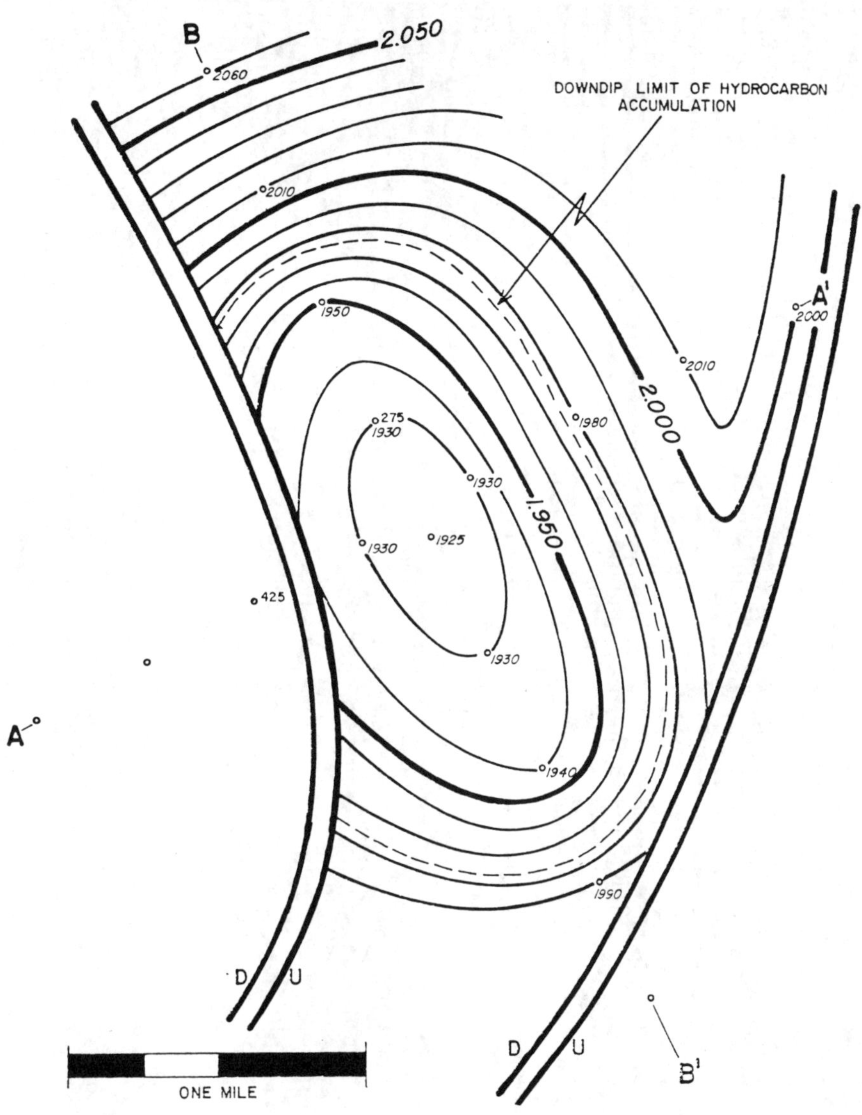

FIGURE 10.9. *Structure map offshore Texas showing that downdip limit of gas accumulation conforms with structure.*

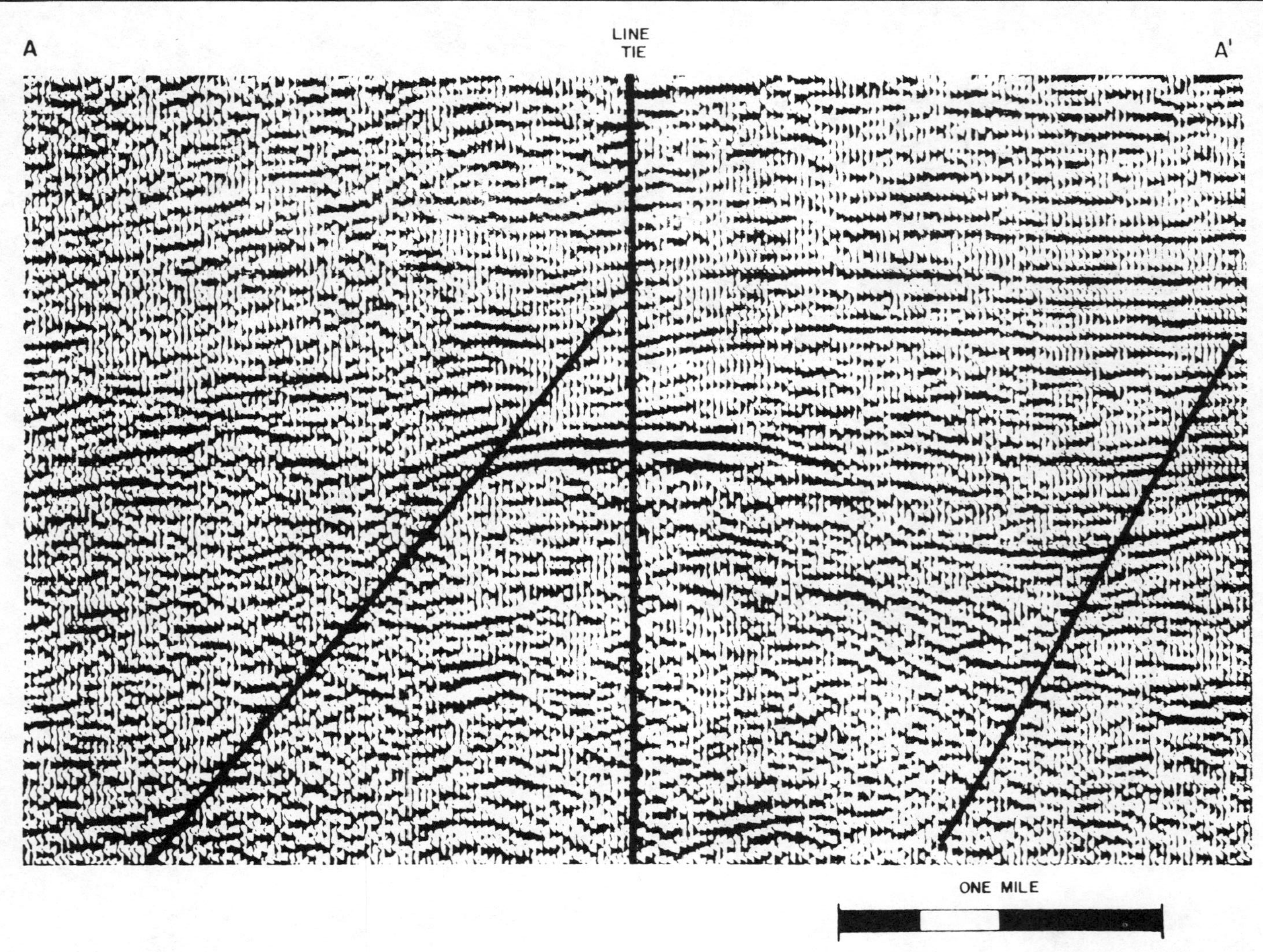

FIGURE 10.10(a). *Edge of reservoir mapped by change of amplitude and intersection with flat event on the right. (Courtesy of Grant Geophysical.)*

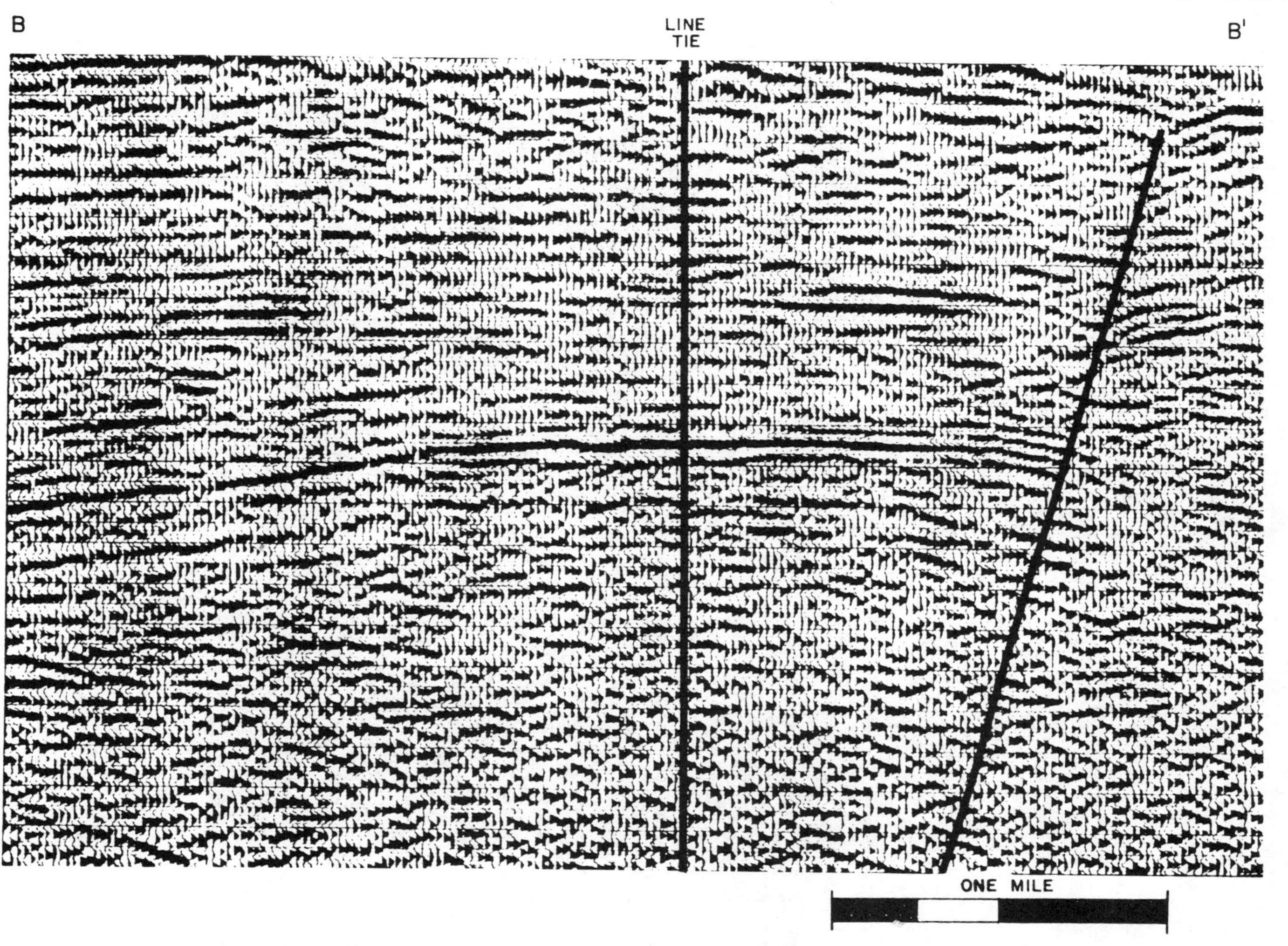

FIGURE 10.10(b). *Edge of structure mapped by flat event interfering with top reflector on left. (Courtesy of Grant Geophysical.)*

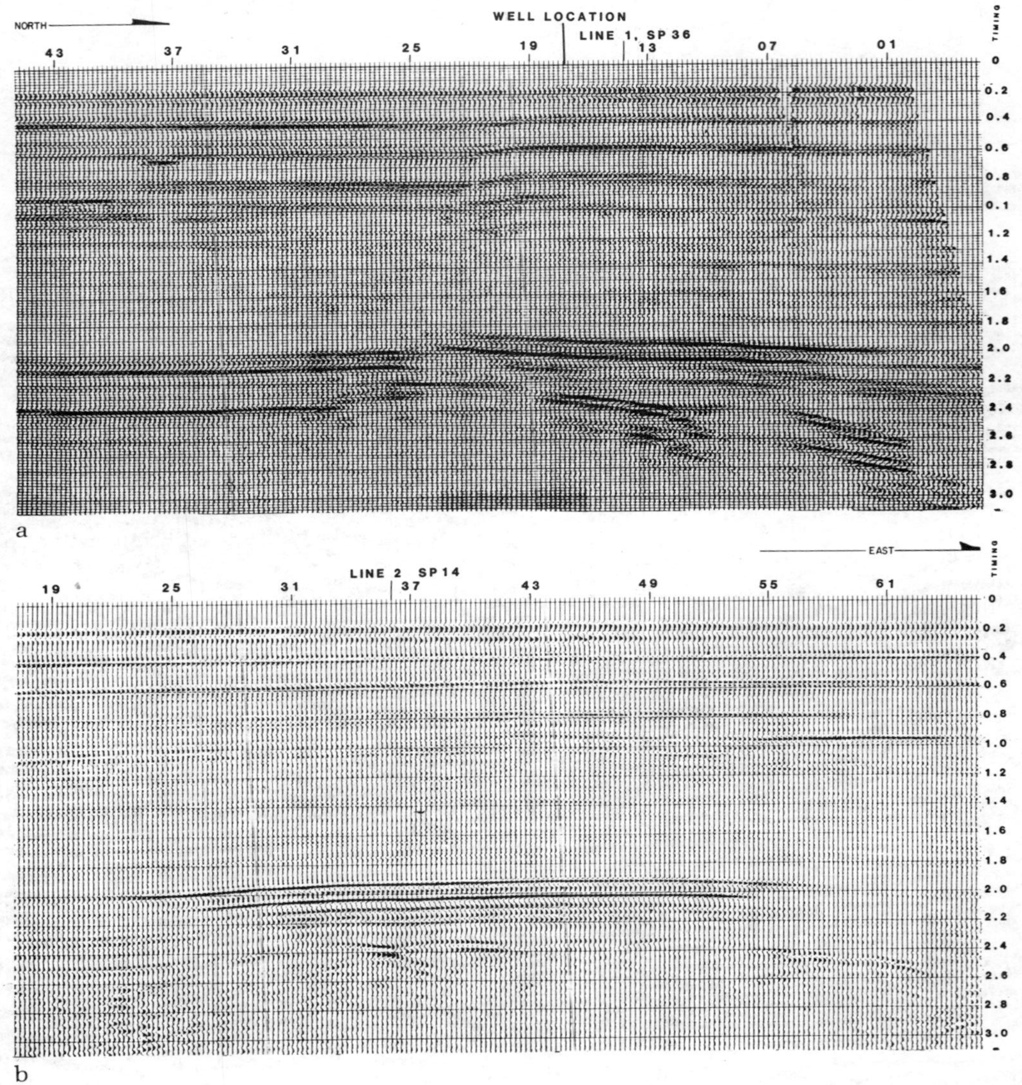

FIGURE 10.11. *(a) North-south cross section, offshore Trinidad. (b) East-west cross section, offshore Trinidad. (Courtesy of Pexcon Computing Company.)*

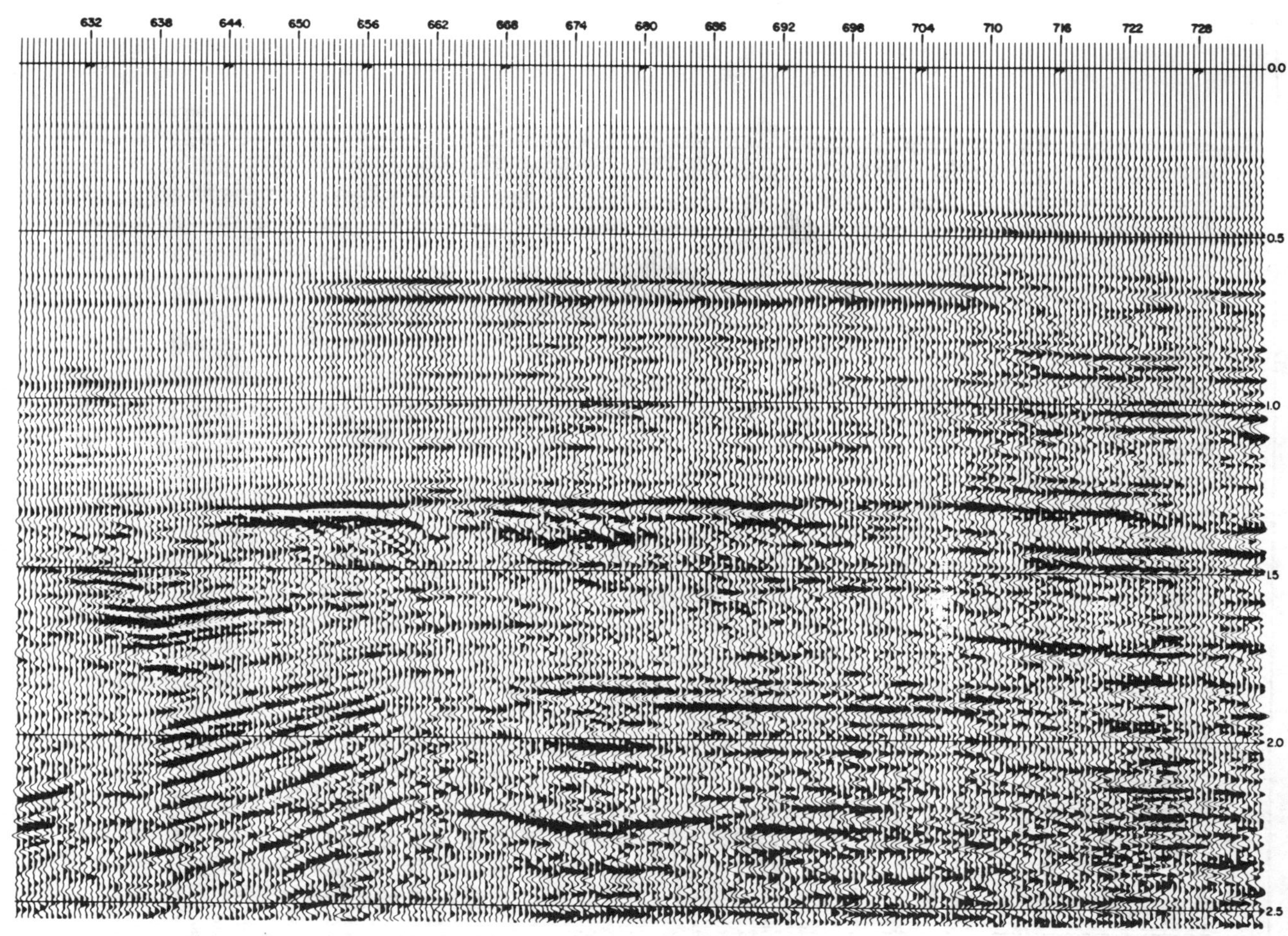

FIGURE 10.12. *Shallow bright spot at 0.7 sec from thin sand (15 ft), with 10 ft uncommercial gas pay. Prolific gas production exists in deeper zones from 1.3–2.1 sec between SP662-696, causing sag on deeper event at 2.3 sec. (From Barry and Shugart 1973, courtesy of Teledyne Exploration Company.)*

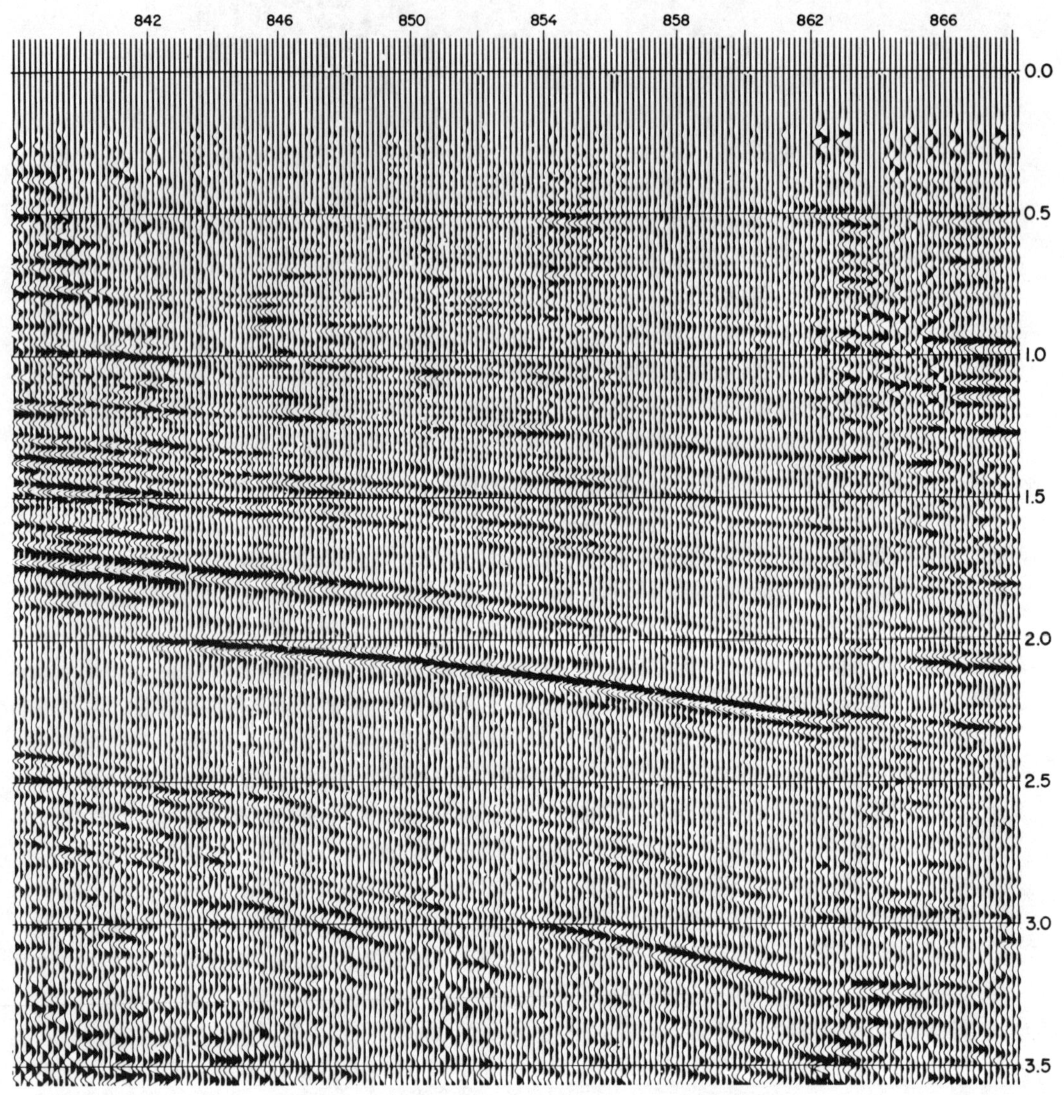

FIGURE 10.13. *Amplitude anomaly generated by the Heterostegina carbonate zone of the Anahuac formation within a comparatively low-velocity sand/shale sequence. This bright spot is not related to hydrocarbons. (From Barry and Shugart 1973, courtesy of Teledyne Exploration Company.)*

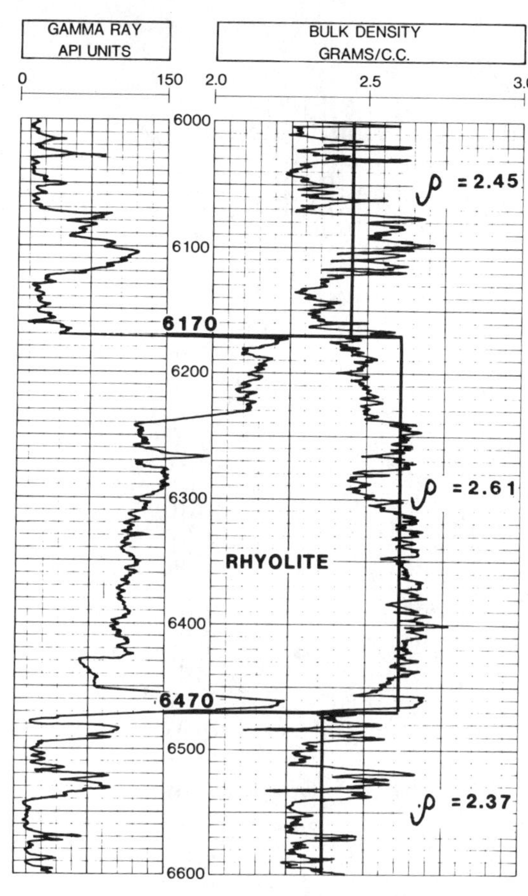

FIGURE 10.14. *Gamma ray and bulk density logs from a northeast Louisiana well, showing presence of rhyolite. (From Ramo and Bradley 1977, courtesy of Geophysics.)*

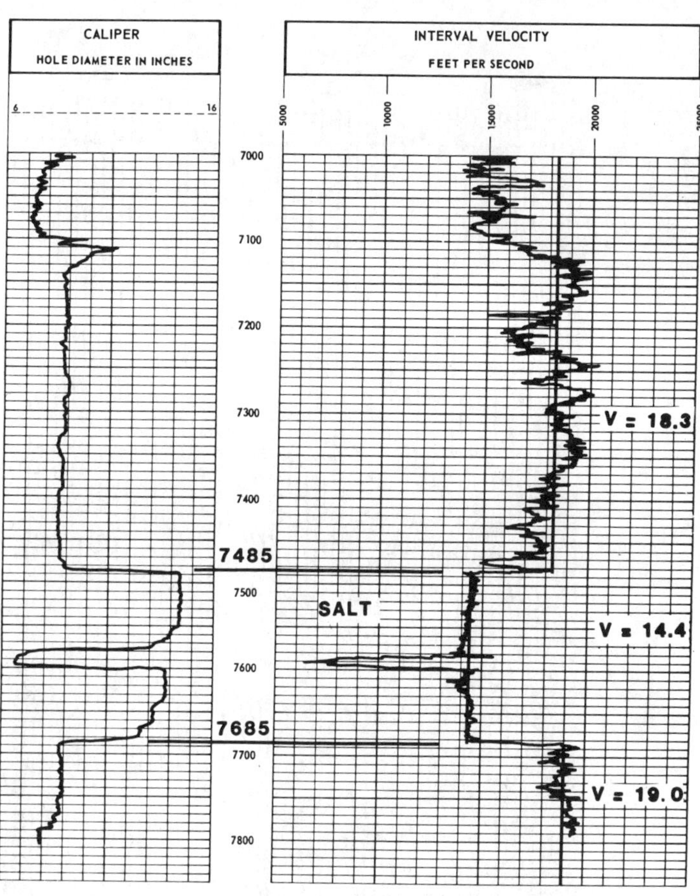

FIGURE 10.15. *Interval velocity and caliper logs from a southwest Arkansas well showing presence of salt stringer. (From Ramo and Bradley 1977, courtesy of Geophysics.)*

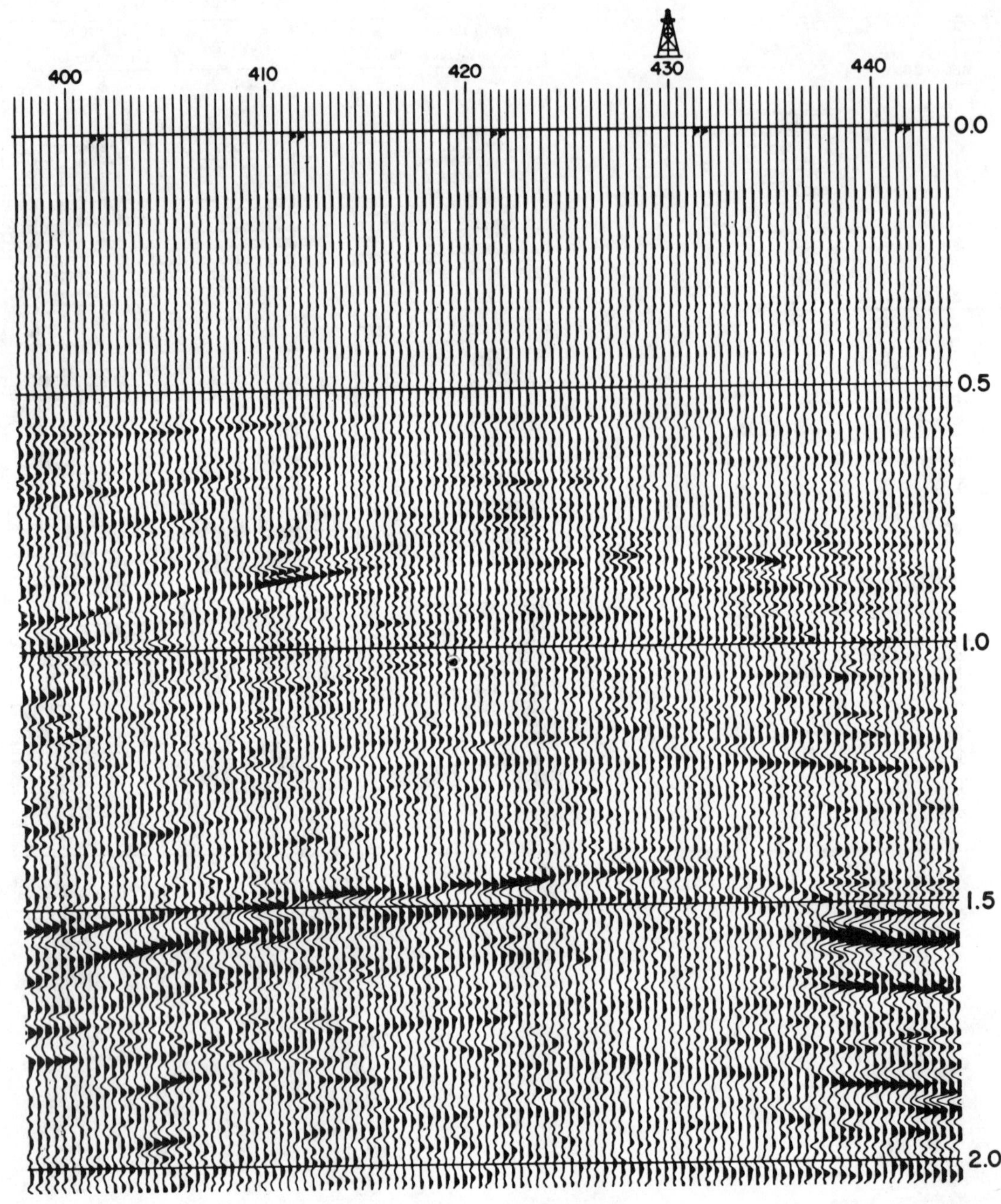

FIGURE 10.16. *Dim spot associated with gas-bearing porous carbonate overlain by interbedded sands and shales. (From Barry and Shugart 1973, courtesy of Teledyne Exploration Company.)*

THEORETICAL VELOCITY AND DENSITY CALCULATIONS

In 1956, Hicks and Berry presented a very important paper on the influence of fluid type and fluid saturation on the velocity in the Miocene section of the offshore Gulf Coast. This paper provided the basis for direct detection of hydrocarbons by seismic methods. The fact that 16 years elapsed until direct detection became routine is an indictment of the geophysical profession, whose mind was closed to the possibility of the seismic method being anything other than a structural tool and, therefore, only an indirect indicator of hydrocarbons.

Two of their examples are shown in Figure 10.17. In a well in the Eugene Island area, offshore Louisiana, a Miocene sand from 8098 to 8117 ft shows a low porosity sand (A) at the top of the section, followed by 22 ft of porous sand saturated with hydrocarbons (B), with a transition to complete water saturation in the lower part of the sand. The velocity in the water zone is about 9500 ft/sec and in the pay zone about 7900 ft/sec. The second example is from the West Cameron area, offshore Louisiana. The water sand velocity, zone E, is about 10,000 ft/sec. In contrast, zone A has velocity of only about 8500 ft/sec due to hydrocarbon saturation which is, in all probability, gas. Zone B is a thin, high-velocity stringer in the sand body and is interpreted as a zone of low porosity. Zone C is possibly 8 ft of additional pay sand. Zone D is an unusually high-velocity zone and is interpreted as a very tight, hard sandstone.

Theoretical solutions for the velocity of sound in porous media have been made under various assumptions about the matrix: Gassman (1951) considered hexagonal packing of spherical grains with the pore spaces occupied by a liquid; White and Sengbush (1953) considered cubical packing of spherical grains; and Brandt (1955) considered randomly packed, nonspherical particles with variable porosity. The three equations were used by Hicks and Berry (1956) to evaluate theoretically the effects of changes in various parameters. Of particular interest are the effects of changing compressibility of the saturating fluids. Water is less compressible than hydrocarbons. Hydrocarbon mixtures may have a wide range of compressibilities, depending on their composition, temperature, and pressure. In the cases of hexagonal and cubical packings of spheres, porosities are fixed

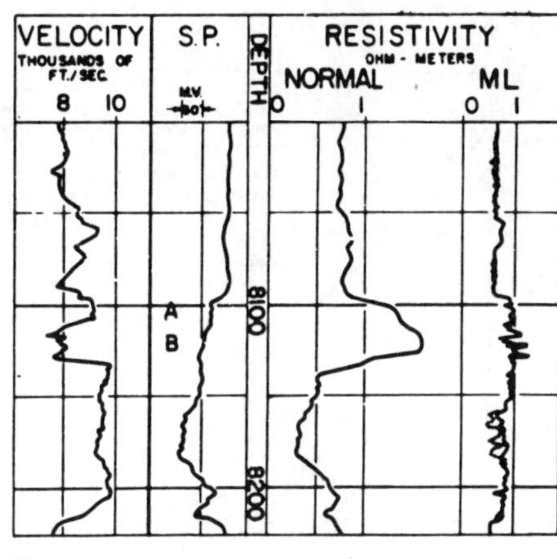

a

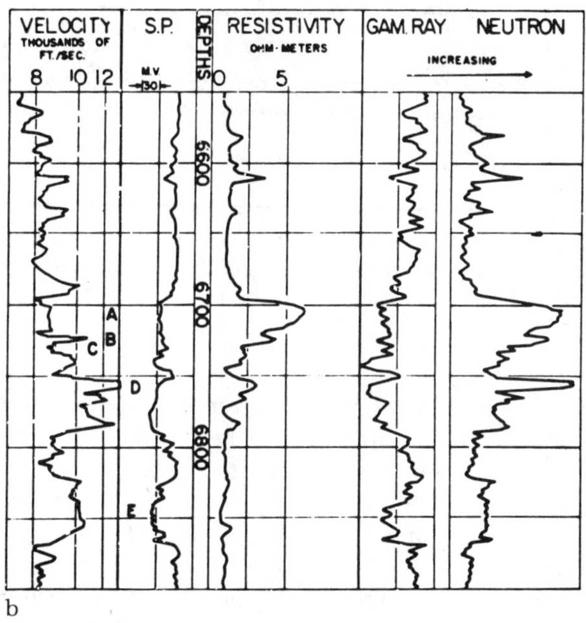

b

FIGURE 10.17. *Well logs through hydrocarbon-bearing zones in the middle Miocene, offshore Louisiana. (a) Eugene Island area. (b) West Cameron area. (From Hicks and Berry 1956, courtesy of Geophysics.)*

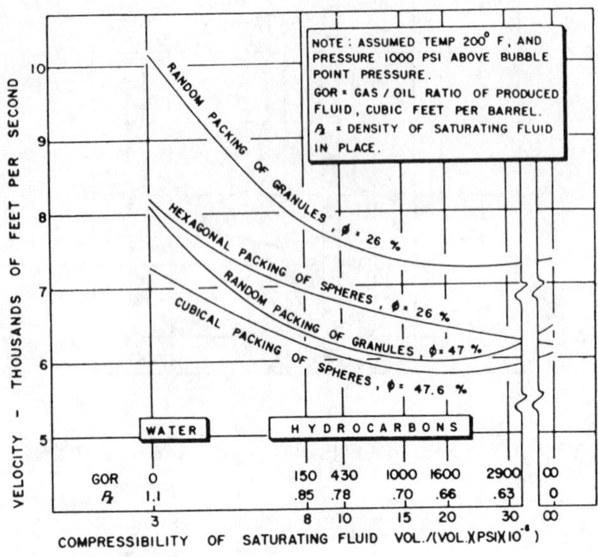

FIGURE 10.18. *Theoretical curves showing velocity through porous media as a function of compressibility of the saturating fluid. (From Hicks and Berry 1956, courtesy of Geophysics.)*

at 26 percent and 47 percent, respectively. In Brandt's equation, the porosity term may be varied, so those two values of porosity were chosen to permit a qualitative comparison of Brandt's results with those of Gassman and of White and Sengbush, as illustrated in Figure 10.18.

Figure 10.18 shows relatively high velocities when the saturating fluid is salt water with compressibility of 3×10^{-6} psi^{-1}. The velocities decrease appreciably as the compressibility of the saturating fluid increases. The more compressible fluids are hydrocarbons having the densities and gas-oil ratios shown. The curves are carried to the extreme case of infinite compressibility and zero density. The important feature of these curves is that the velocity is depressed about 15 to 20 percent as the saturating fluid changes from water to oil. All three equations have the same trend of decreasing velocity as the compressibility of the saturating fluid is increased. In making the computations, it is assumed that the pore spaces of the sands are completely saturated, either with water or with hydrocarbon fluids. In nature, however, there is nearly always some water present in the pores of reservoir rocks.

In estimating the effect of gas saturation on velocity, the time-average equation of Wyllie, Gregory, and Gardner (1956) was often used, although it is not very accurate, especially in unconsolidated rocks. The equation is developed by dividing the rock into its percentage of water volume, its percentage of hydrocarbon volume, and its percentage of rock matrix, and obtaining the total time through a unit thickness of the rock by summing the times through each constituent. It is a static equation, similar to the one used to calculate bulk density from the densities and volume percentages of the constituents. Expressed in terms of porosity ϕ, water saturation S_w, and the velocities of the components, the Wyllie time-average equation is

$$\frac{1}{V} = \frac{\phi S_w}{V_w} + \frac{\phi(1 - S_w)}{V_h} + \frac{(1 - \phi)}{V_m}, \tag{10.2}$$

where V_w is water velocity, V_h is hydrocarbon velocity, and V_m is matrix velocity.

A better estimate, attributed to Geertsma (1961) and based on the theoretical work of Biot (1956a,b), was used by Domenico (1974) to explain bright spots from noncommercial gas sands, a fact empirically derived by drilling some dry holes on bright spots. The Geertsma equation is

$$V^2 = \left[\frac{1}{C_b} + \frac{4}{3}\mu_b + \frac{(1 - \beta)^2}{(1 - \phi - \beta)C_m + \phi C_f}\right]\frac{1}{\rho_b}, \tag{10.3}$$

where C_m = compressibility of rock matrix material
C_b = compressibility of empty reservoir bulk material
C_f = compressibility of interstitial fluid
β = ratio C_m/C_b
μ_b = shear modulus of reservoir bulk material
ρ_b = density of reservoir bulk material.

When $\phi = 0$, equation (10.3) reduces to equation (2.2) for compressional velocity in elastic solids.

The bulk density is simply the weighted average of the densities of the constituents:

$$\rho_b = \phi S_w \rho_w + \phi(1 - S_w)\rho_h + (1 - \phi)\rho_m, \qquad (10.4)$$

where ρ_w, ρ_h, ρ_m, are water, hydrocarbon, and matrix densities, respectively.

Domenico (1974) calculated the reflectivity of shale/gas sand and shale/oil sand interfaces at depths of 2,000, 6,000, and 10,000 ft as a function of S_w, using Geertsma's equation for velocity and the bulk density equation. The results in Figure 10.19 show that, for gas reservoirs, the reflectivity decreases slowly as the water saturation increases from zero to 0.9 and then decreases rapidly as the water saturation increases from 0.9 to 1.0. (Table 10.1). In contrast, oil reservoirs exhibit a steadily decreasing reflectivity as the water saturation increases.

Because the bulk density is a linear function of water saturation, the foregoing results are due strictly to velocity changes as a function of water saturation until saturation reaches 0.9, then increasing rapidly as saturation approaches 1.0. In oil reservoirs, the velocity increases steadily as the water saturation increases from zero to 1.0.

Domenico made the following assumptions in his theoretical calculations:

TABLE 10.1. *Effect of water saturation on reflectivity of gas reservoirs*

	Reflectivity		
S_w	2000 ft	6000 ft	10,000 ft
0.0	0.275	0.177	0.136
0.9	0.216	0.130	0.098
1.0	0.012	0.026	0.031

TABLE 10.2. *Densities, velocities, and resulting reflectivities for 100% saturation*

	2000 ft	6000 ft	10,000 ft
Density (g/cm^3)			
100% gas	1.67	1.82	2.00
100% oil	1.91	2.02	2.16
100% water	2.04	2.13	2.23
Velocity in ft/ms			
100% gas	4.40	7.30	8.80
100% oil	5.40	7.50	8.80
100% water	6.25	8.40	9.80
Reflectivity			
100% gas/shale	0.275	0.177	0.136
100% oil/shale	0.105	0.125	0.100
100% wtr/shale	0.012	0.026	0.031

Water:	100,000 parts per million of dissolved salt		
Oil:	API gravity 35° containing gas of gravity 0.60		
Gas:	Methane		
Differential pressure:	0.53 psi/ft (difference between overburden pressure of 1.00 psi/ft and hydrostatic pressure of 0.47 psi/ft)		
Temperature gradient:	Linear with surface temperature 75°F, increasing at rate of 0.01°F/ft		
Density sand matrix:	2.650 at all depths		
Density water:	1.097 (2,000 ft)	1.089 (6,000 ft)	1.083 (10,000 ft)
Density oil:	0.755 g/cm^3	0.749 g/cm^3	0.742 g/cm^3
Density gas:	0.023	0.103	0.156
Density shale:	2.14	2.30	2.40
Velocity shale:	5.85 ft/ms	8.21 ft/ms	9.66 ft/ms
Porosity:	0.39	0.33	0.26

Based on these assumptions, the velocities and densities for 100 percent saturation by each of the fluids and the resulting reflectivities are as shown in Table 10.2.

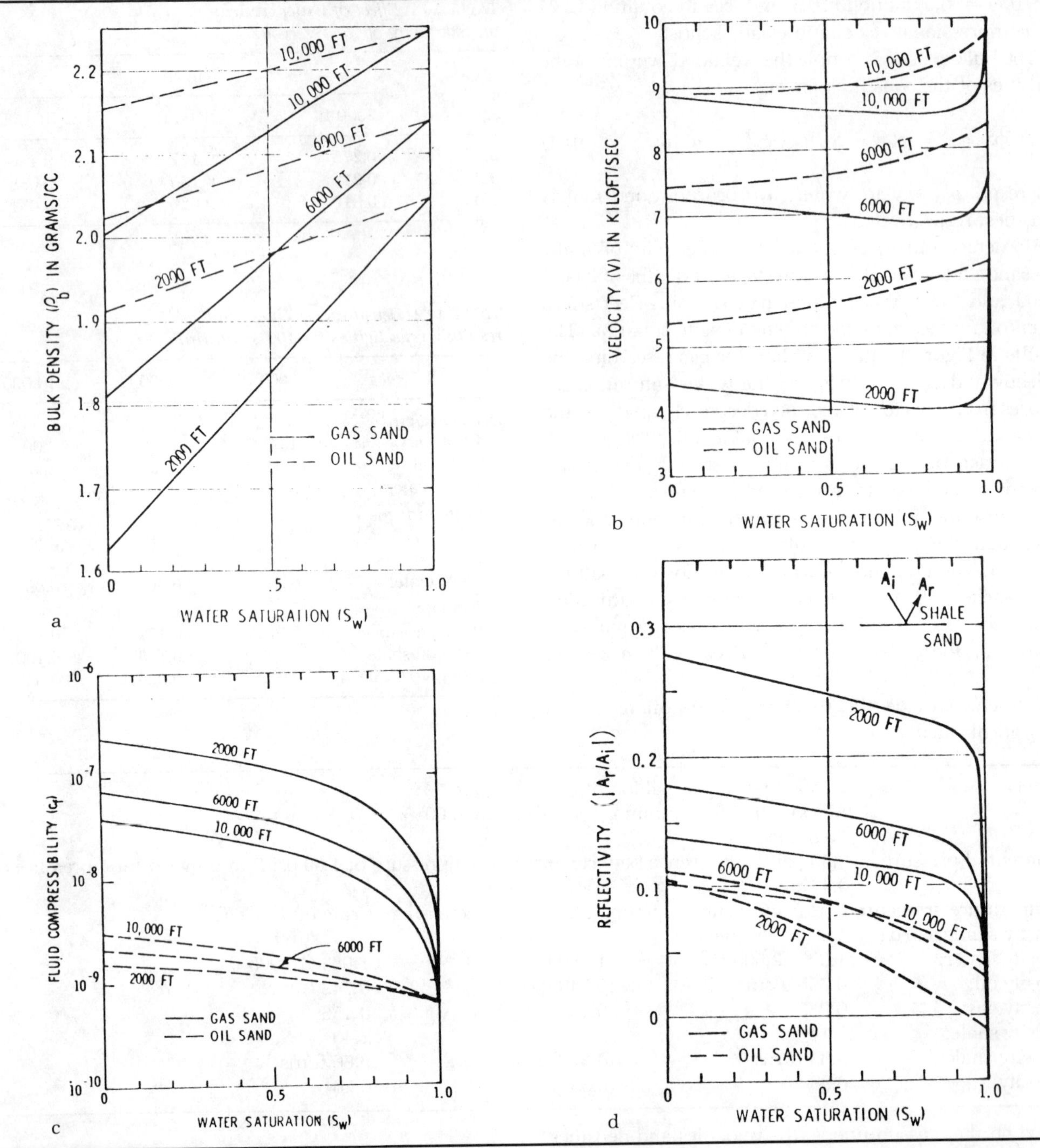

FIGURE 10.19. *Effect of water saturation on reflectivity of gas and oil reservoirs encased in shale. (a) Bulk density. (b) Velocity. (c) Fluid compressibility. (d) Reflectivity. (From Domenico 1974, courtesy of Geophysics.)*

TABLE 10.3. *Velocities, densities, and impedances of a gas/oil/water anticlinal model*

	Velocity (ft/ms)	Density (g/cm³)	Impedance
Shale	8.5	2.5	21.3
Gas sand	6.4	2.16	13.8
Oil sand	10.8	2.29	24.7
Water sand	12.1	2.33	28.2

MODEL STUDIES

Two-dimensional models of reservoirs and their seismic responses are valuable educational aids and may be useful in quantifying reservoir characteristics. They can be constructed by drawing the structure in depth and assigning an appropriate velocity and density to each layer. More advanced models have attenuation specified in each layer also. The reflectivity at each surface point consists of the set of reflection coefficients obtained from the acoustic impedance layering, properly spaced in time in accordance with the least-time path based on Snell's Law.

Consider an anticlinal model of a 400 ft sand with 20 percent porosity sandwiched between shale layers, with a reservoir consisting of a 250 ft gas cap over a 50 ft oil zone, as shown in Figure 10.20(a). The velocities and densities chosen are given in Table 10.3. From the acoustic impedance contrasts, the reflection coefficients are:

Shale/gas sand	−0.21
Gas/oil sand	+0.29
Oil/water sand	+0.06
Water sand/shale	−0.14
Shale/oil sand	+0.08
Shale/water sand	+0.14

The time layering in Figure 10.20(b) shows the upward curling of the tops of the gas and oil interfaces near the edge of the reservoir resulting from thinning of the low-velocity gas sand, and also shows the time-sag on the deep interface caused by presence of gas above.

The seismic response of the model in Figure 10.20(c) is the convolution of the reflectivity with an appropriate wavelet. In the middle of the structure, the reflections from the top and base of the gas sand are large, with

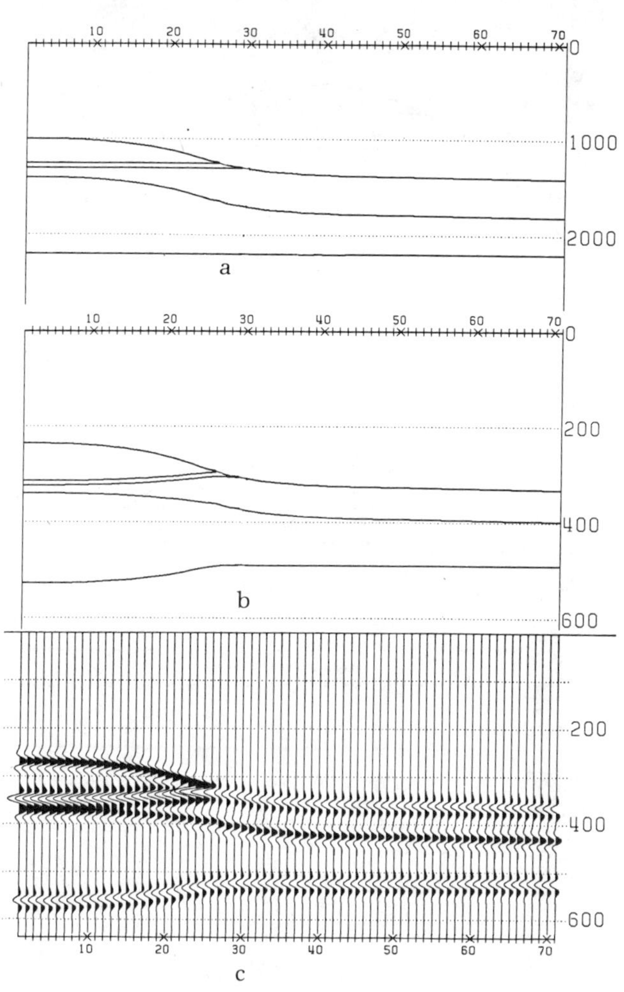

FIGURE 10.20. *A gas/oil/water anticlinal model and its seismic response. (a) Depth layering. (b) Time layering. (c) Seismic response.*

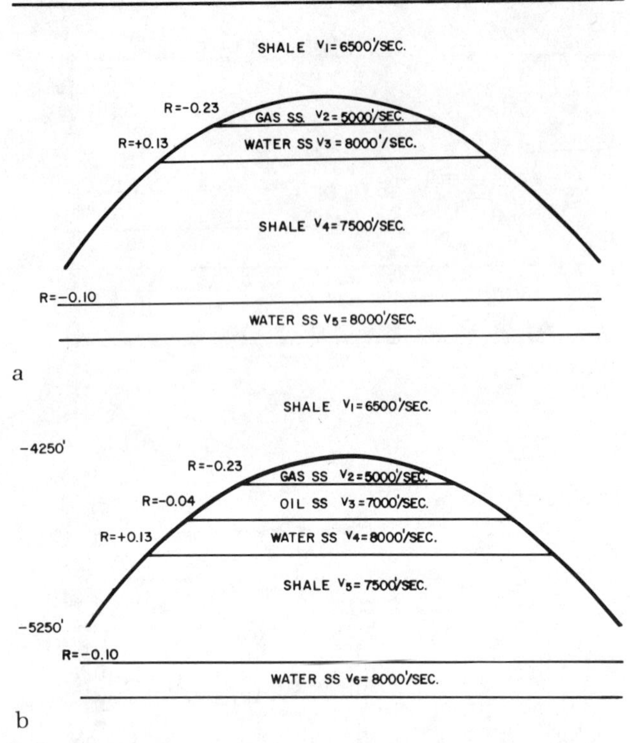

FIGURE 10.21. *Models of anticlinal reservoirs. (a) Gas reservoir. (b) Gas/oil reservoir.*

opposite polarity. The basal reflection, flat near the middle of the structure, curls up near the edge of the reservoir as the gas layer thins. As the gas layer thins, the two reflections interfere constructively at the tuned thickness, producing maximum amplitude in the composite reflection near the edge of the reservoir. Further thinning decreases the amplitude of the composite reflection until it disappears. The reflection from the top of the sand changes polarity as the fluid changes from gas to oil or water. The oil layer is not thick enough to allow observation of a distinct oil/water sand reflection, and it forms a composite reflection with the gas/oil sand reflection above. Both have the same polarity, so they will add constructively throughout the reservoir. A time-sag of 38 ms is observed on the reflection from a flat interface lying beneath the sand layer. Absorption and diffractions are not taken into account here.

Model seismic sections across two anticlinal structures (Figures 10.21–10.23), one gas-bearing and the other gas over oil, and a monocline with gas over oil (Figures 10.24, 10.25) show the diagnostic features that are most significant in the direct detection of hydrocarbons, bright spots, flat spots, and interference at edges of reservoirs. As in Figure 10.20, the models include neither diffractions nor absorption.

Another type of model that is very useful in evaluating real seismic data uses synthetic seismograms made from velocity or impedance logs that can be modified by stretching or compressing intervals and altering the velocity and density within intervals, using a digital processing procedure developed by Charles R. Saxon.

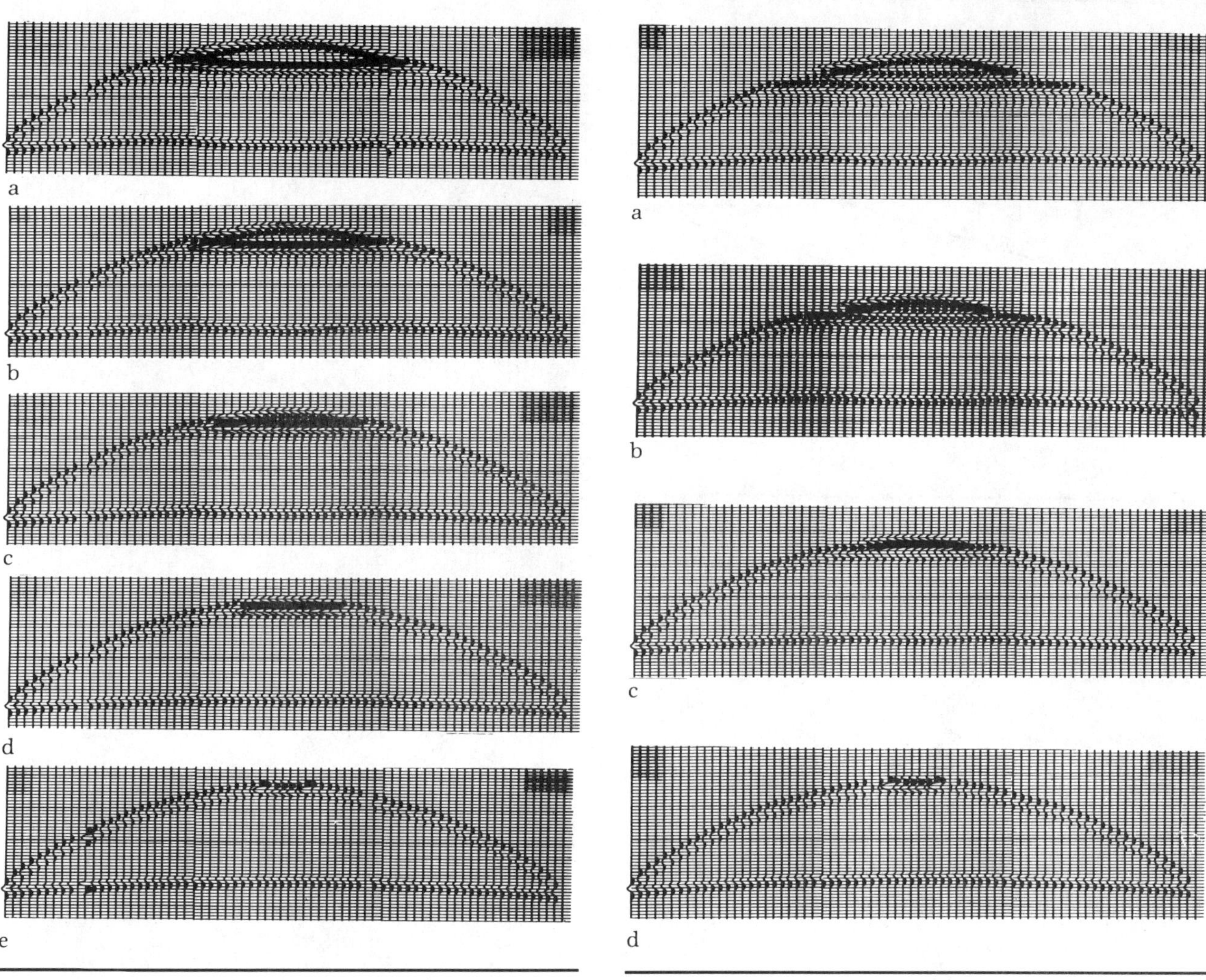

FIGURE 10.22. *Seismic sections across model anticlinal gas reservoirs. (a) 200 ft sand thickness. (b) 150 ft sand thickness. (c) 100 ft sand thickness. (d) 50 ft sand thickness. (e) 10 ft sand thickness.*

FIGURE 10.23. *Seismic sections across model anticlinal gas/oil reservoirs. (a) 150 ft sand thickness. (b) 100 ft sand thickness. (c) 50 ft sand thickness. (d) 10 ft sand thickness.*

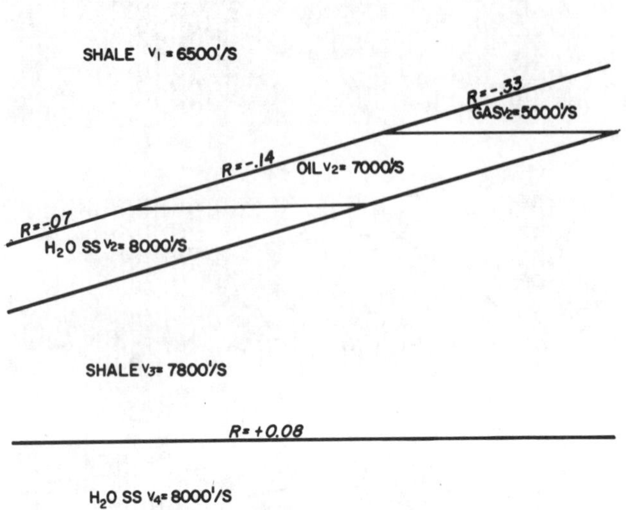

FIGURE 10.24. *Model of monoclinal gas/oil reservoir.*

An example from offshore Louisiana is shown in Figure 10.26. The unaltered impedance log has a 20 ft gas sand at the 7100 ft level (Trace 4), with its seismic response on Trace 8. The gas sand produces a strong reflection whose peak has been shifted earlier in time to line it up with the low-impedance sand layer. (Note: The synthetic traces are filtered with a realizable filter, and the shift has been made because of the resulting filter delay, as discussed in Chapter 5.) The 20 ft sand has been altered (1) by stretching the sand thickness to 43 ft and changing the fluid from gas to oil, (2) by stretching to 59 ft and changing the fluid from gas to water, and (3) by shaling out the gas sand. The velocity and density values for gas, oil, and water sands and for shale used here were measured on the logs. They are listed in Table 10.4, along with the reflectivities of sand/shales for the three fluid contents.

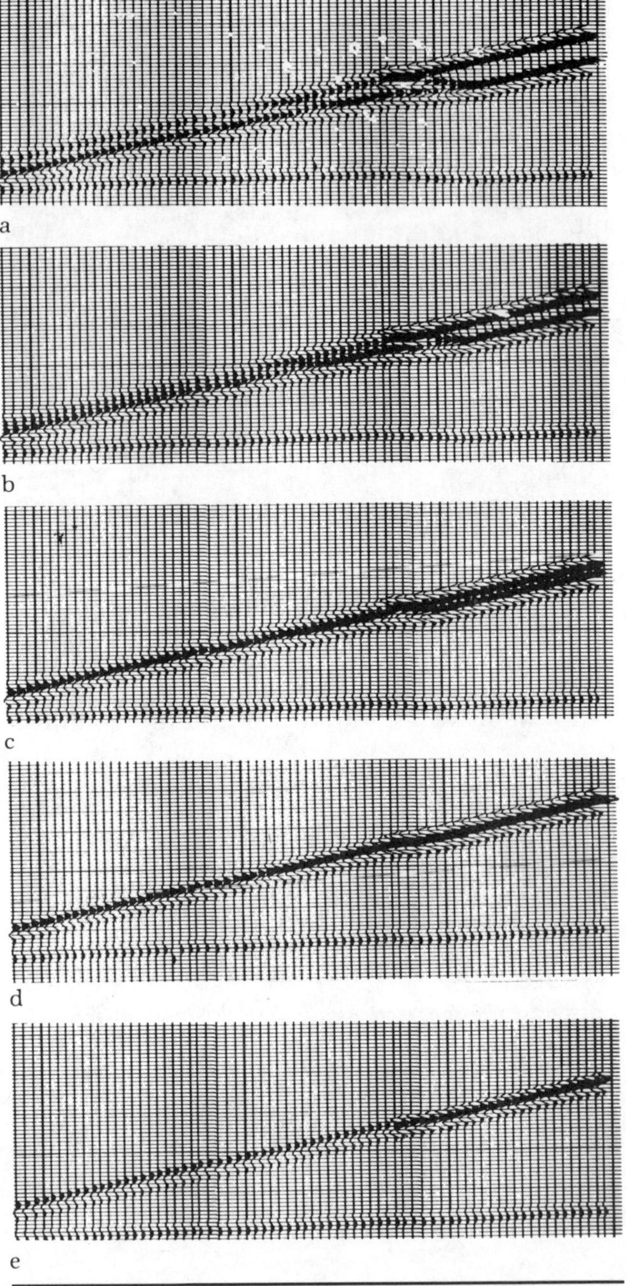

FIGURE 10.25. *Seismic sections across monoclinal gas/oil reservoirs. (a) 200 ft sand thickness. (b) 150 ft sand thickness. (c) 100 ft sand thickness. (d) 50 ft sand thickness. (e) 10 ft sand thickness.*

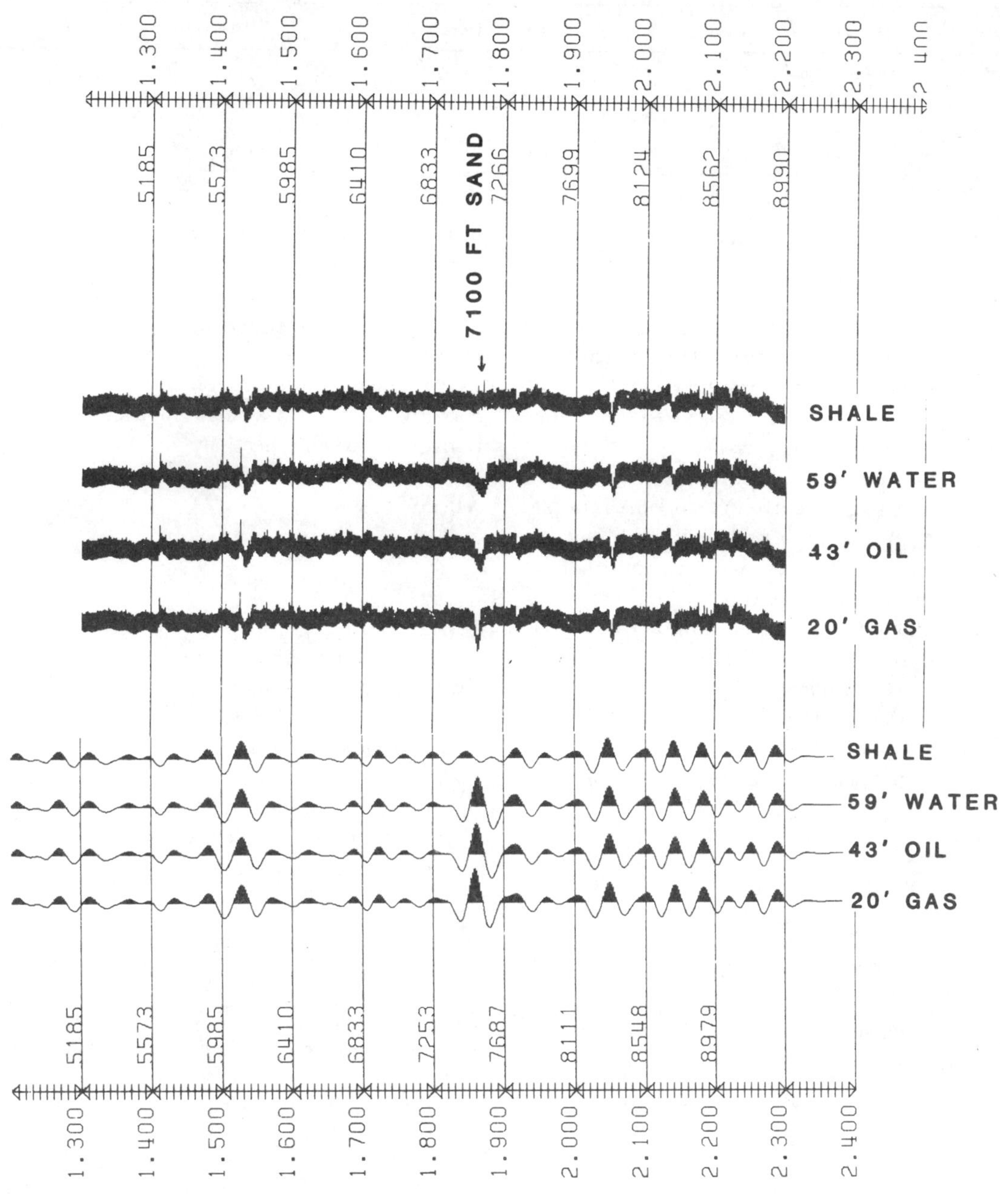

FIGURE 10.26. *Models showing equality of reflection amplitudes in cases of gas, oil, or water saturation in the Pleistocene, offshore Louisiana.*

TABLE 10.4. *Velocities, densities, impedances, and reflectivities of sand/shales, offshore Louisiana*

	Velocity (ft/ms)	Density (g/cm^3)	Impedance	Reflectivity
Gas sand	6.25	1.90	11.9	0.237
Oil sand	7.25	2.06	14.9	0.129
Water sand	7.52	2.12	15.9	0.097
Shale	8.33	2.32	19.3	—

The purpose of the foregoing model is to show the ambiguity that exists between thickness and fluid content when the layering is less than the tuned thickness, as can be seen by comparing the seismic responses on Traces 6–8 for the three different cases. The reflections in each case have virtually the same amplitude and character, as can be explained by use of the square-wave impedance layering discussed in Chapter 5.

To compare the reflection amplitude for the reflectivities in the three cases, the curve in Figure 5.2 is duplicated in Figure 10.27 for each reflectivity and is positioned vertically in accordance with its normalized reflectivity. Let the gas case be given by the normalized curve. Then the oil case will have amplitude values that are normalized to the ratio of R oil/shale to R gas/shale—0.129/0.237 = 0.544. Similarly, in the water case, the normalized value is the ratio of R water/shale to R gas/shale—0.097/0.237 = 0.409.

The two-way time thickness Δ in each case is given by

$$\begin{aligned} \Delta &= 2 * \Delta z/V, \\ \Delta_1 &= 2 * 20/6.25 = 6.4 \text{ ms (gas)}, \\ \Delta_2 &= 2 * 43/7.25 = 11.9 \text{ ms (oil)}, \\ \Delta_3 &= 2 * 59/7.52 = 15.7 \text{ ms (water)}. \end{aligned} \tag{10.5}$$

The period of the seismic pulse generating the synthetic traces in Figure 10.26 is 48 ms; therefore, the ratio Δ/T for each case is

$$\begin{aligned} \Delta_1/T &= 0.13 \text{ (gas)}, \\ \Delta_2/T &= 0.25 \text{ (oil)}, \\ \Delta_3/T &= 0.33 \text{ (water)}. \end{aligned} \tag{10.6}$$

Plotting the foregoing Δ/T values on the respective curves shows that each has a normalized amplitude of 0.8, so the models should generate the same reflection amplitude, as is shown experimentally on Traces 6–8 in Figure 10.26.

An analytical description of the reflection amplitude equality follows. For a single square-wave layer, the relationship between reflectivity and thickness that gives the same reflection amplitude in the case of two different reflectivities is

$$\Delta_2 = \frac{R_1}{R_2}\Delta_1, \tag{10.7}$$

where R_1 is the reflectivity and Δ_1 is the two-way time-thickness in Case 1 and R_2 and Δ_2 are similar quantities in Case 2, provided that both Δ_1/T and Δ_2/T are in the linear range of the curve—that is, $\Delta/T < 0.25$. Between 0.25 and the tuned thickness 0.5, the curve is not linear and the mathematical relationship is only approximate. In terms of depth thicknesses, velocities, and reflectivities, the relationship is

$$\Delta z_2 = \frac{V_2}{V_1} \cdot \frac{R_1}{R_2}\Delta z_1. \tag{10.8}$$

In the gas case, $\Delta z_1 = 20$ ft. Therefore, in the oil and water cases, the calculated thicknesses from equation (10.8) are

$$\Delta z_2 = \frac{7.25}{6.25} \cdot \frac{0.237}{0.129} \cdot 20 = 42.6 \text{ ft (oil)}, \tag{10.9}$$

$$\Delta z_3 = \frac{7.52}{6.25} \cdot \frac{0.237}{0.097} \cdot 20 = 58.8 \text{ ft (water)}. \tag{10.10}$$

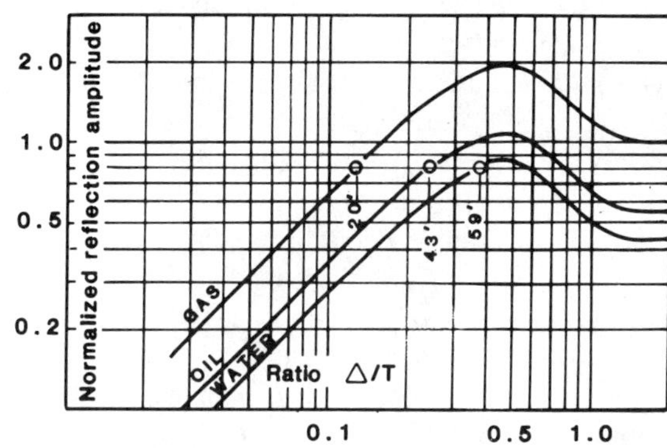

FIGURE 10.27. *Normalized reflection amplitude in cases of gas, oil, or water saturation.*

Because Δ/T for both gas and oil is in the linear range (less than 0.25), a 43 ft oil sand will produce the same reflection amplitude as the 20 ft gas sand. In the water case, Δ/T is just beyond the linear range, so a 59 ft water sand will produce approximately the same reflection amplitude as the 20 ft gas sand.

Model studies are important in illustrating the effects of porosity and fluid content on seismic data and are valuable in quantifying the parameters involved. The theoretical values of velocities and densities are good initial points in any iterative modeling procedure developed to zero in on the correct parameter values.

Another important aspect of modeling to quantify reservoir parameters is the estimation of the actual seismic pulse that is generating reflections on the real data.

CASE HISTORIES

EXAMPLE FROM CHURLIN AND SERGEYEV (1963)

This example from one of the fields in the Kuban-Black Sea basin is anticlinal, with a 25 m shale overlying a reservoir with an 80 m gas cap that overlies 22 m of oil sand. Porosity ranges from 13 to 35 percent. The overall

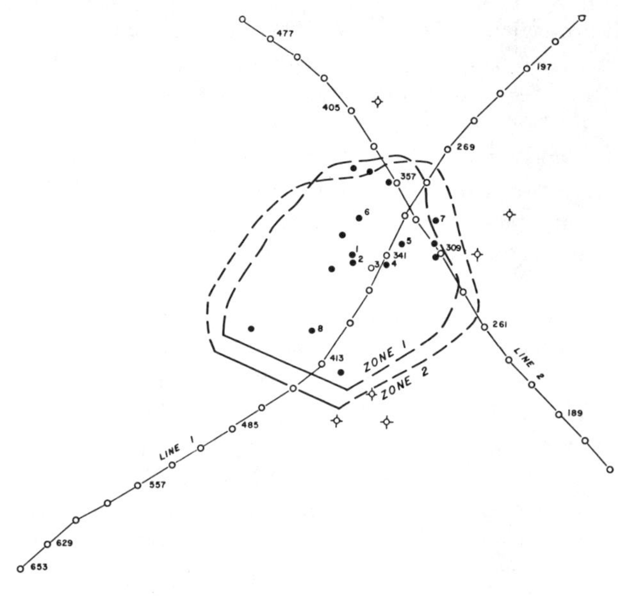

FIGURE 10.28. *Case history: areal extent of a Pleistocene gas field in the Gulf of Mexico, offshore Louisiana.*

lateral dimension of the reservoir is 2000 m, with the gas/oil contact at −1500 m. The reflection coefficients given by the authors are

Shale/gas sand	−0.24
Shale/oil sand	−0.01
Shale/water sand	+0.07
Gas/oil sand	+0.23
Oil/water sand	+0.08

The large coefficients from the shale/gas and gas/oil interfaces, with opposite polarity, were most significant.

This example is from land seismic data before reproducible recording, before digital processing and deconvolution, and before horizontal stacking. The flat events at the rim of reservoir were detected on the hand-plotted cross sections of reflection times taken from paper records. The evidence was sparse, but it was there, and the theoretical considerations of the Russian geophysicists were well in advance of the state of the art.

EXAMPLE FROM OFFSHORE LOUISIANA

This case history is from a Pleistocene area in the Gulf of Mexico, offshore Louisiana. From the seismic data presented before any well information, two gas zones based on bright-spot analysis were outlined, as shown in Figure 10.28. Two of the seismic lines are shown in Figures 10.29 and 10.30. These data have been processed by amplitude-preserving techniques. The full interpretation and final sections are not available for publication.

The bright spots are terminated by faults to the southwest on line 1 and to the northwest on line 2. As there is no evidence of pay thickness in either zone greater than the tuned thickness, the reflection amplitude decreases monotonically as the pay zone thins, in accordance with the relationship for thin beds developed in Chapter 5. Line 1 shows the upper reflection to be stronger than the lower reflection from the fault out to about SP362, indicating thicker pay in the upper zone than in the lower zone along this portion of the line. Near the projection of Well 8, there is noticeable sag and an apparent shadow zone on the lower reflection, indicating maximal thickness of the upper gas zone here. Beyond SP362 to the downtip terminations, the lower reflection holds its strength, while the upper reflection weakens, indicating that the upper pay zone is thinning, while the lower zone remains constant. Line 2 shows the upper reflection to be weak, indicating thinner pay in the upper sand along this line compared to line 1.

The logs in Well 6 (Figure 10.31) show the gas-producing zones. The upper zone produces gas from the top 40 ft of a 150 ft sand with top at 7300 ft measured depth. The porosity averages 30 percent and the permeability averages 750 mD. The lower zone produces from the top 70 ft of a 130 ft sand with top at 8120 ft measured depth. This zone has 31 percent porosity and 1000 mD permeability. Between the producing zones there is a 180 ft water sand with top at 7810 ft measured depth.

Well 7 shows thin sand development in the upper zone, with little or no indication of gas (Figure 10.31), as expected from the seismic response on line 2 to the west of the well, which shows that the reflection from the upper sand is much less bright—that is, has a much thinner producing zone—compared to either the upper reflection on line 1 or the lower sand reflection on either line.

Velocity and density logs in Well 8 through the upper reservoir (Figure 10.32) show the average velocity in the upper gas zone to be 7250 ft/sec and the average density to be 1.94 g/cm^3. The gas zone, 72 ft thick here, overlies 58 ft of water sand, with velocity 9090 ft/sec and density 2.28 g/cm^3. The shale above has velocity 7870 ft/sec and density 2.22 g/cm^3, and the shale below has velocity 8130 ft/sec and density 2.25 g/cm^3. The intermediate water sand is 186 ft thick, as shown in Figure 10.33. Its velocity and density averages 8850 ft/sec and 2.14 g/cm^3, respectively.

By using the average velocities, densities, and impedances measured in Well 8 and the depth thicknesses of the sands and shales in Well 6, a model of the reflectivity as a function of two-way time is constructed for Well 6, as given in Table 10.5. The two-way time-thickness Δt and the accumulative time t to the base of each stratigraphic unit are given in milliseconds. The reflection coefficient R is calculated at the base of each unit. The reflectivity $r(t)$ is then given by

$$r(t) = \sum_i R_i \, \delta(t - t_i) \tag{10.11}$$

where R_i and t_i are as given in Table 10.5 in the seventh and fourth columns, respectively.

The most striking aspect of the reflectivity is the very large reflection coefficients associated with gas sands, as compared to the small coefficients associated with water sands. This is a common characteristic in poorly consolidated sands in sand/shale sequences in the Gulf of Mexico, the Niger Delta, and areas with similar stratigraphy, where the shale and water sand impedances are very similar and the presence of gas reduces the impedance of the sand dramatically.

Figure 10.34 is a model synthetic showing how the seismic response changes as the thicknesses of the gas zones decrease and the thicknesses of the underlying water zones increase by a like amount, thus keeping the total sand thicknesses constant. The gas sands are progressively decreased by 10 percent on each successive trace; that is, the top sand decreases from 40, 36, 32, . . . , 4, 0 ft and the bottom sand decreases from 70, 63, 56, . . . , 7, 0 ft. The gas/water contact in each sand is flat in

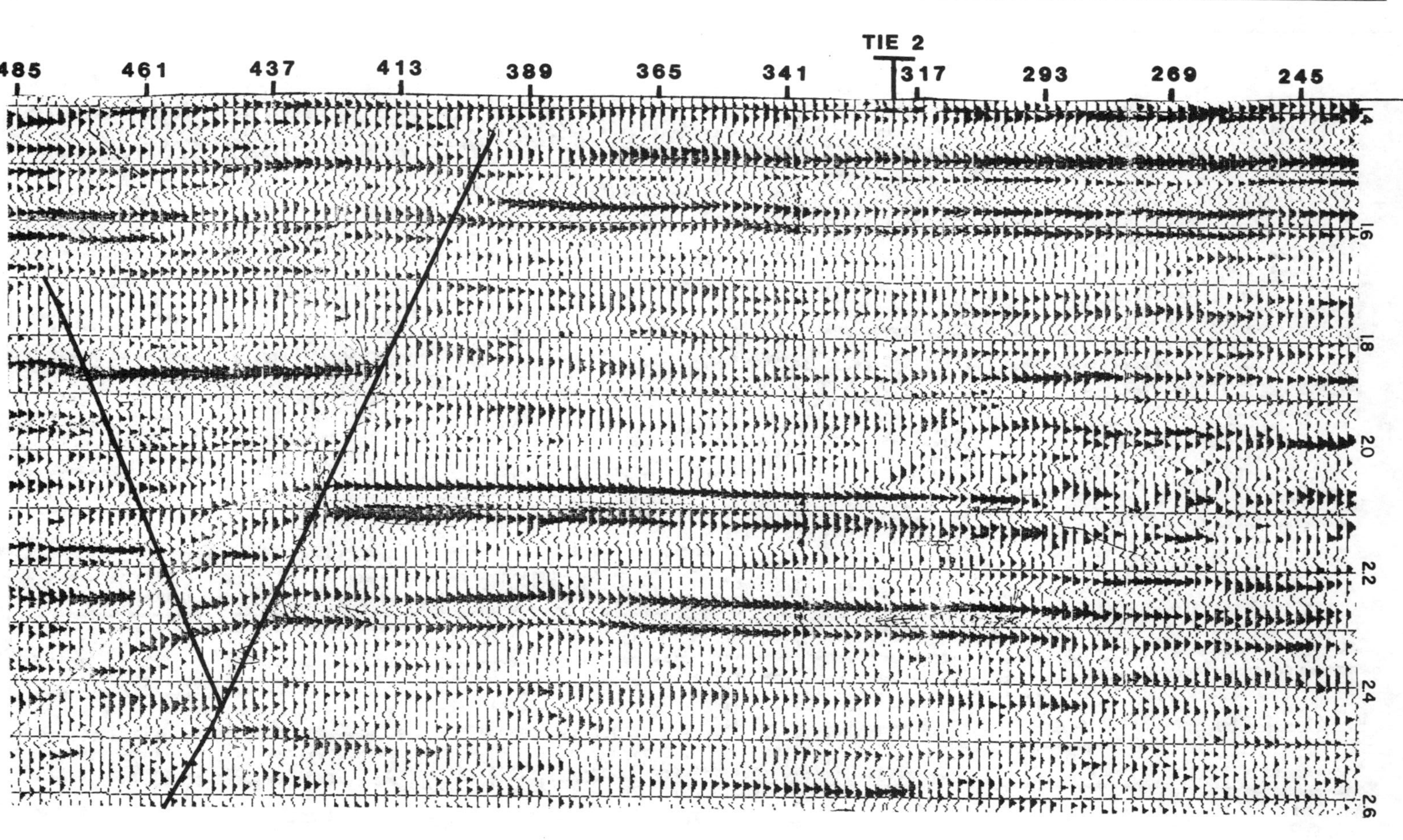

FIGURE 10.29. *Case history: seismic line 1.*

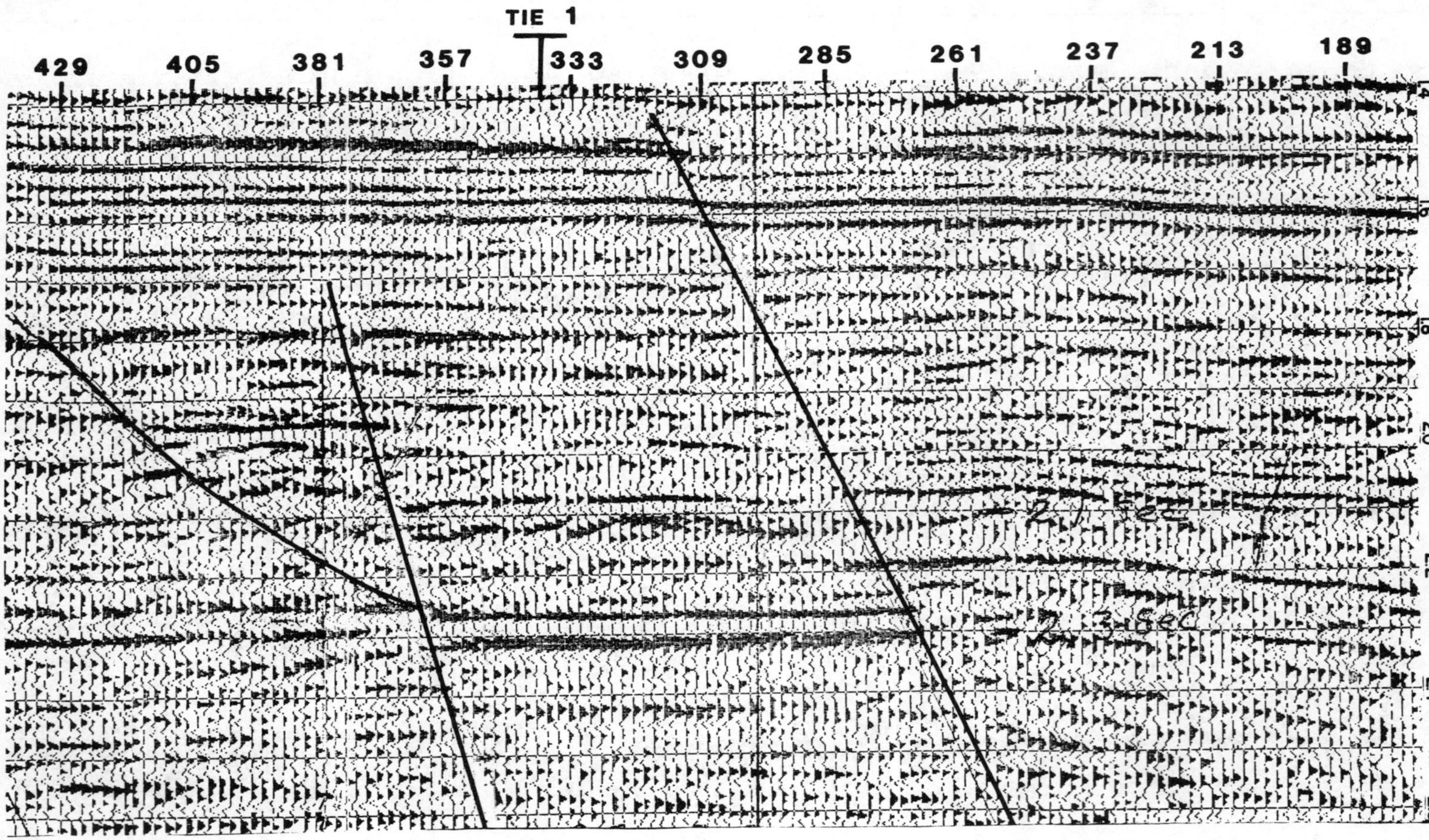

FIGURE 10.30. *Case history: seismic line 2.*

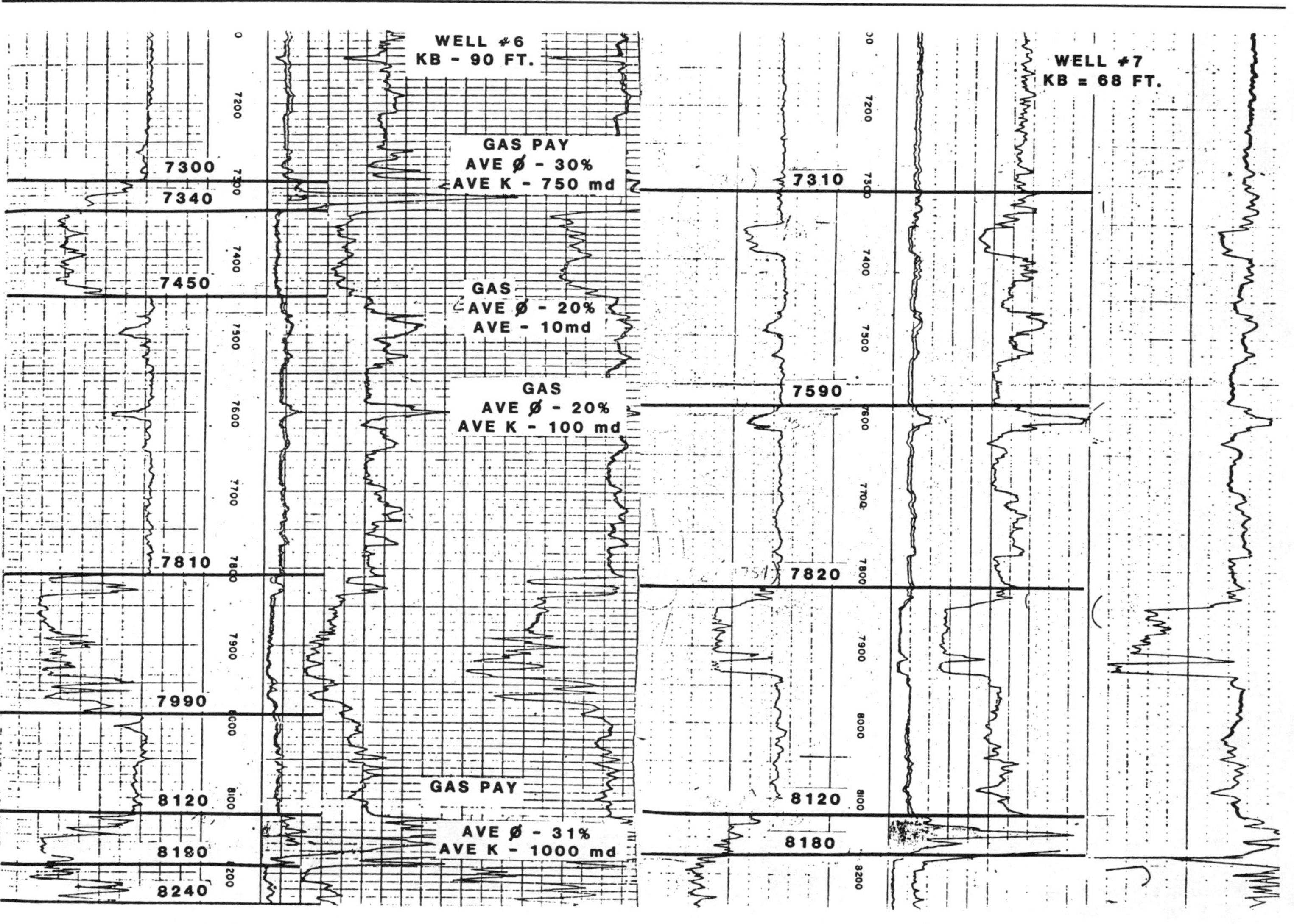

FIGURE 10.31. *Case history: electric logs in Wells 6 and 7.*

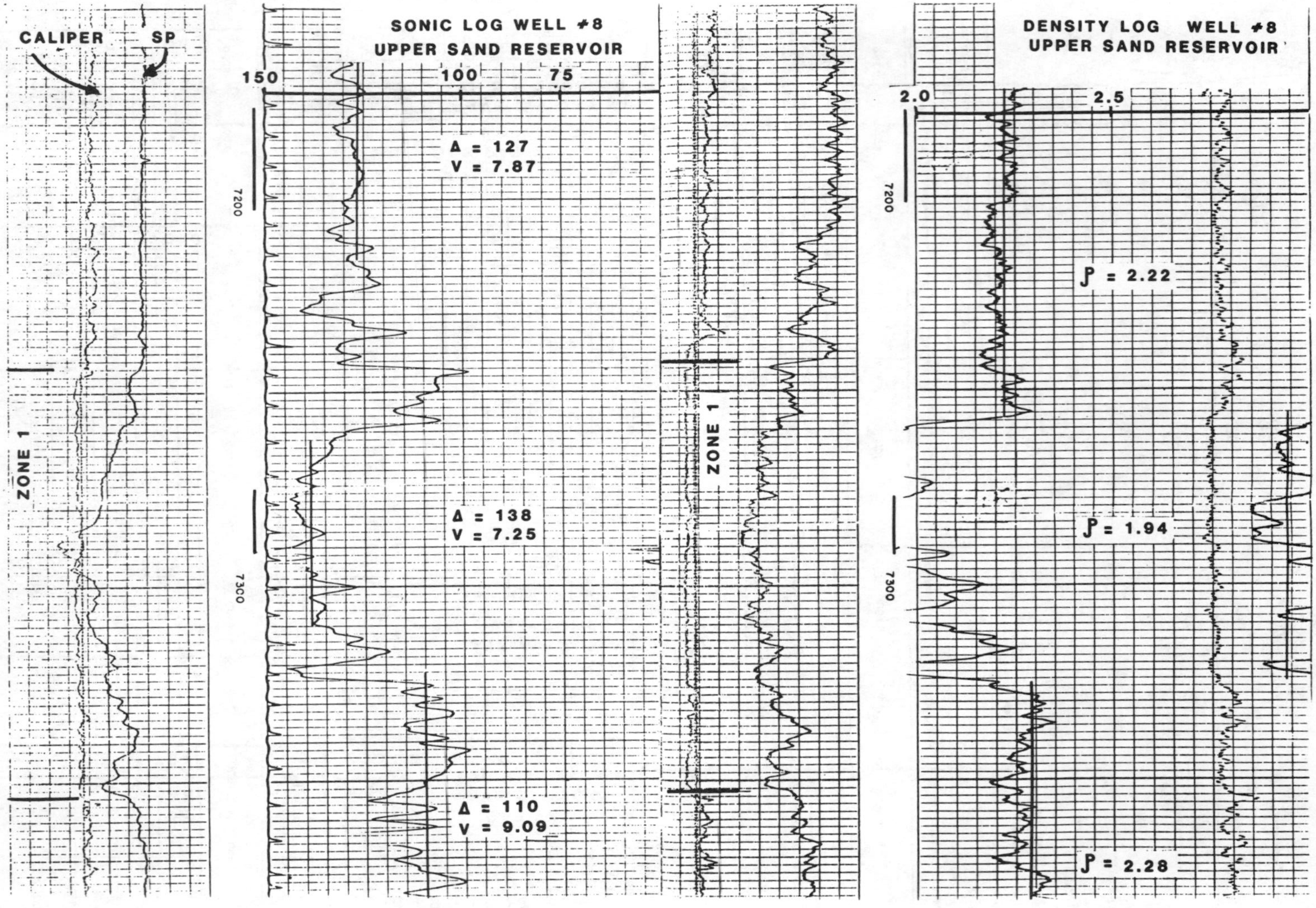

FIGURE 10.32. *Case history: sonic and density logs through upper gas sand.*

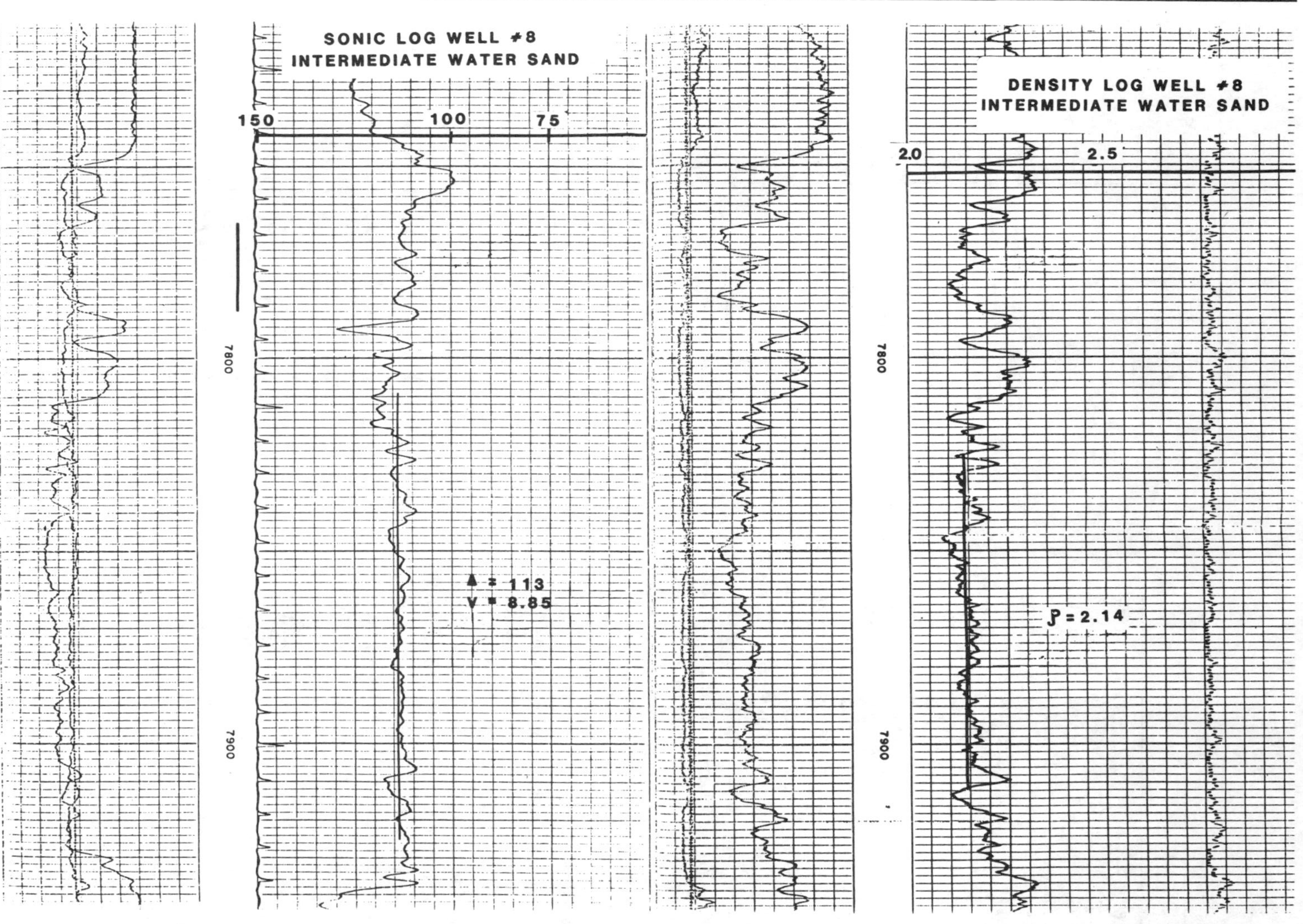

FIGURE 10.33. *Case history: sonic and density logs through water sand.*

TABLE 10.5. *Reflectivity of Well 6, offshore Louisiana*

	Δz	V	Δt	t	ρ	Z	R (base)
Shale	—	7.87	—	0	2.22	17.5	−0.11
Gas sand	40	7.25	11	11	1.94	14.1	+0.19
Water sand	110	9.09	24	35	2.28	20.7	−0.06
Shale	360	8.13	89	124	2.25	18.3	+0.02
Water sand	180	8.85	41	165	2.14	18.9	−0.02
Shale	130	8.13	32	197	2.25	18.3	−0.13
Gas sand	70	7.25	19	216	1.94	14.1	+0.15
Water sand	60	8.85	14	230	2.14	18.9	−0.02
Shale	—	8.13	—	—	2.25	18.3	—

depth. The reflection from the upper gas zone decreases as it thins because it is less than the tuned thickness at maximum thickness. The pulse period T is 33 ms; therefore, the ratio of thickness to period at maximum thickness $\Delta/T = 1/3$ compared to tuned thickness of ½. The lower gas zone is slightly greater than tuned thickness at maximum thickness—$19/33 = 0.58$—but is not thick enough to show evidence of top and bottom reflection. As this zone thins, its reflection decreases in amplitude from its tuned thickness, which occurs on Trace 2.

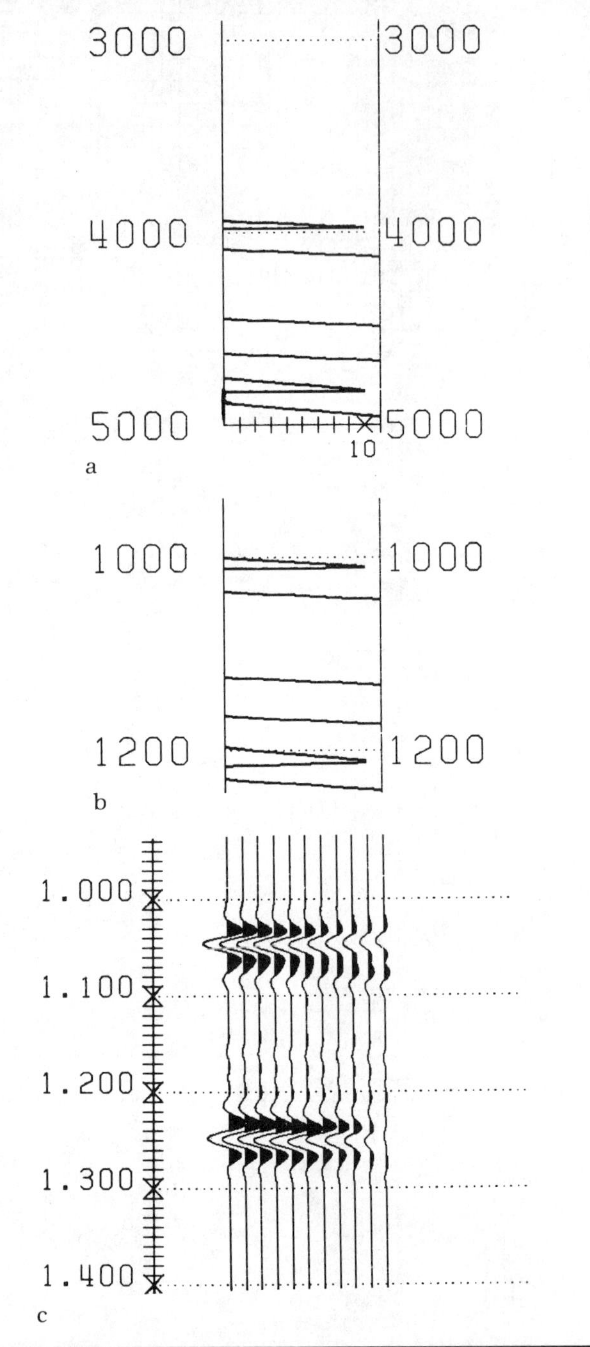

FIGURE 10.34. *Case history: seismic model of gas reservoirs. (a) Model in depth. (b) Model in time. (c) Seismic responses.*

EXAMPLE FROM BACKUS AND CHEN (1975)

The area is offshore in the Gulf of Mexico where the geologic section consists of young sand/shale sequences characterized by a series of growth faults. There are two gas reservoirs on the structure (Figure 10.35). The upper reservoir shows a structural nose with probable closure against a fault. The seismic section shows a flat spot corresponding to the gas/water contact in the upper reservoir. The lower reservoir is closed against the fault, and the seismic section shows a classic bright spot whose boundary conforms with the structural contour, indicating presence of hydrocarbons.

Self-potential and gamma ray logs in Figure 10.36 show that the upper reservoir sand is 140 ft thick and is encased in shale. The resistivity log shows that the upper 40 ft is full of gas. The gas/water contact occurs at 5010 ft, which corresponds to a seismic time of 1.469 sec. This agrees with the time of a flat spot on the seismic section.

On the seismic section in Figure 10.37, the upper arrow on the left points to a reflection peak corresponding to the shale/sand interface at the top of the upper sand. Updip, a somewhat stronger peak proceeds unconform-

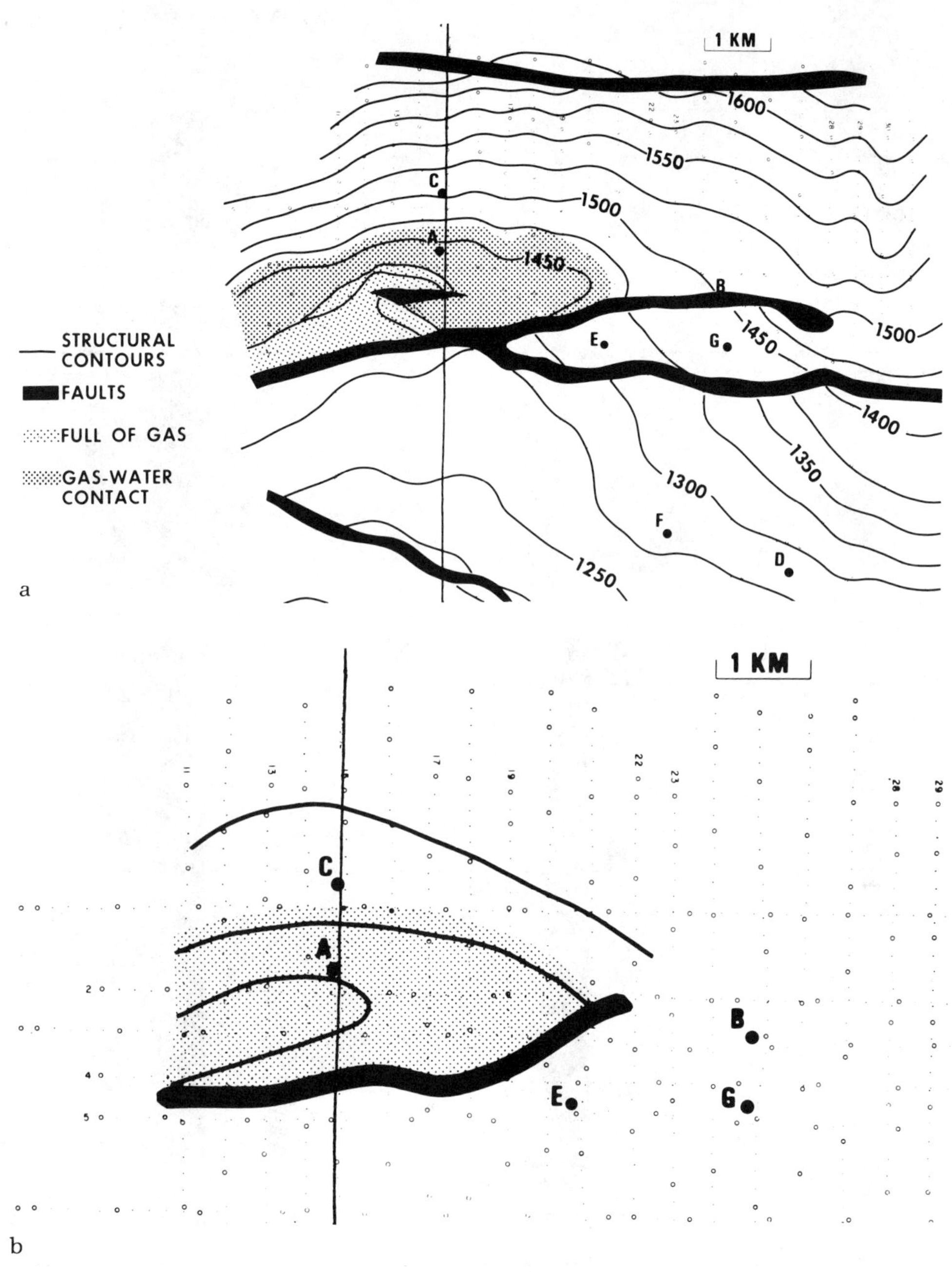

FIGURE 10.35. *Case history: structural contours on tops of reservoir sands and areal extent of reservoirs. (a) Upper reservoir. (b) Lower reservoir. (From Backus and Chen 1975, courtesy of Geophysical Prospecting; Blackwell Scientific Publications Limited.)*

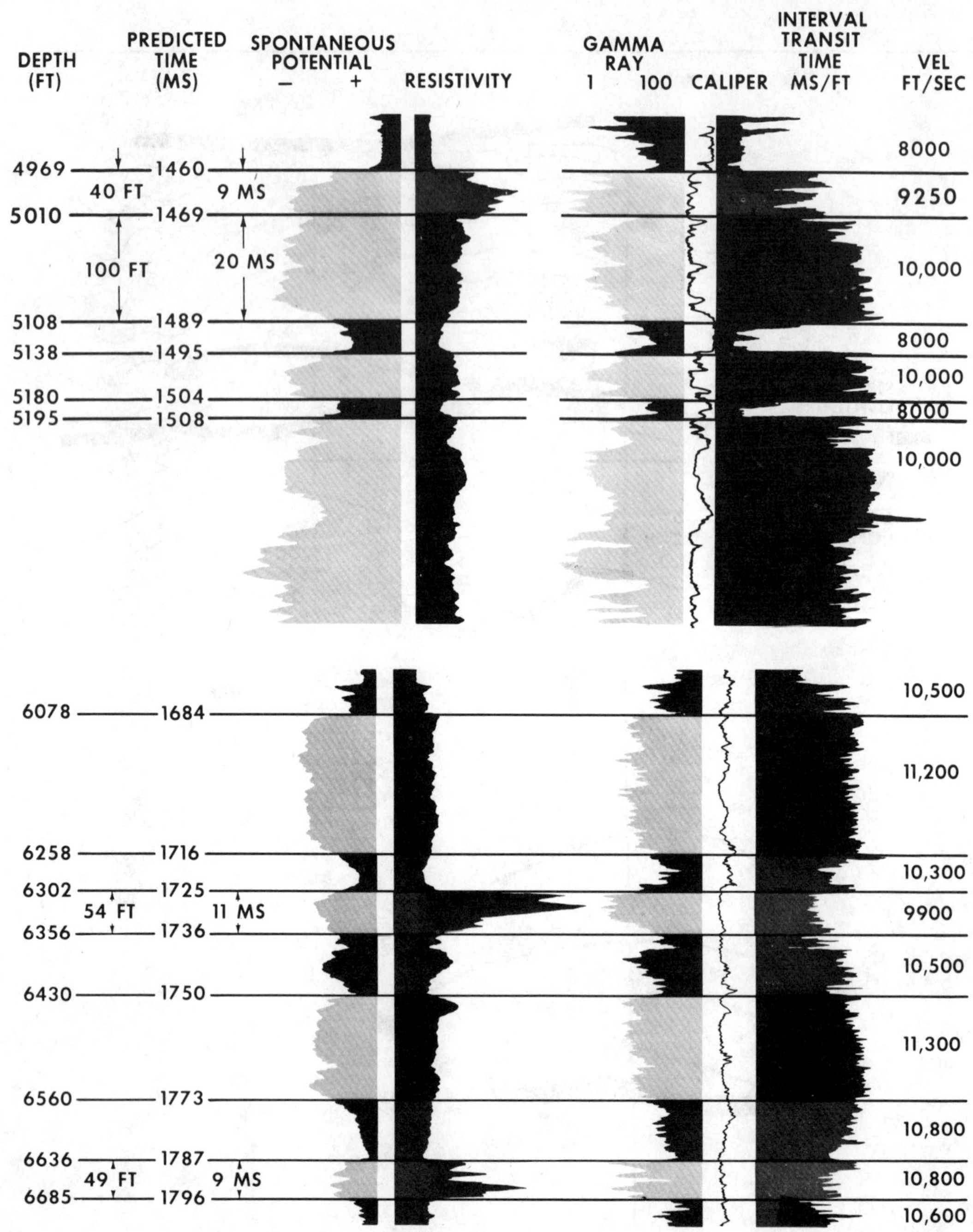

FIGURE 10.36. *Case history: well logs in Well A. (From Backus and Chen 1975, courtesy of Geophysical Prospecting; Blackwell Scientific Publications Limited.)*

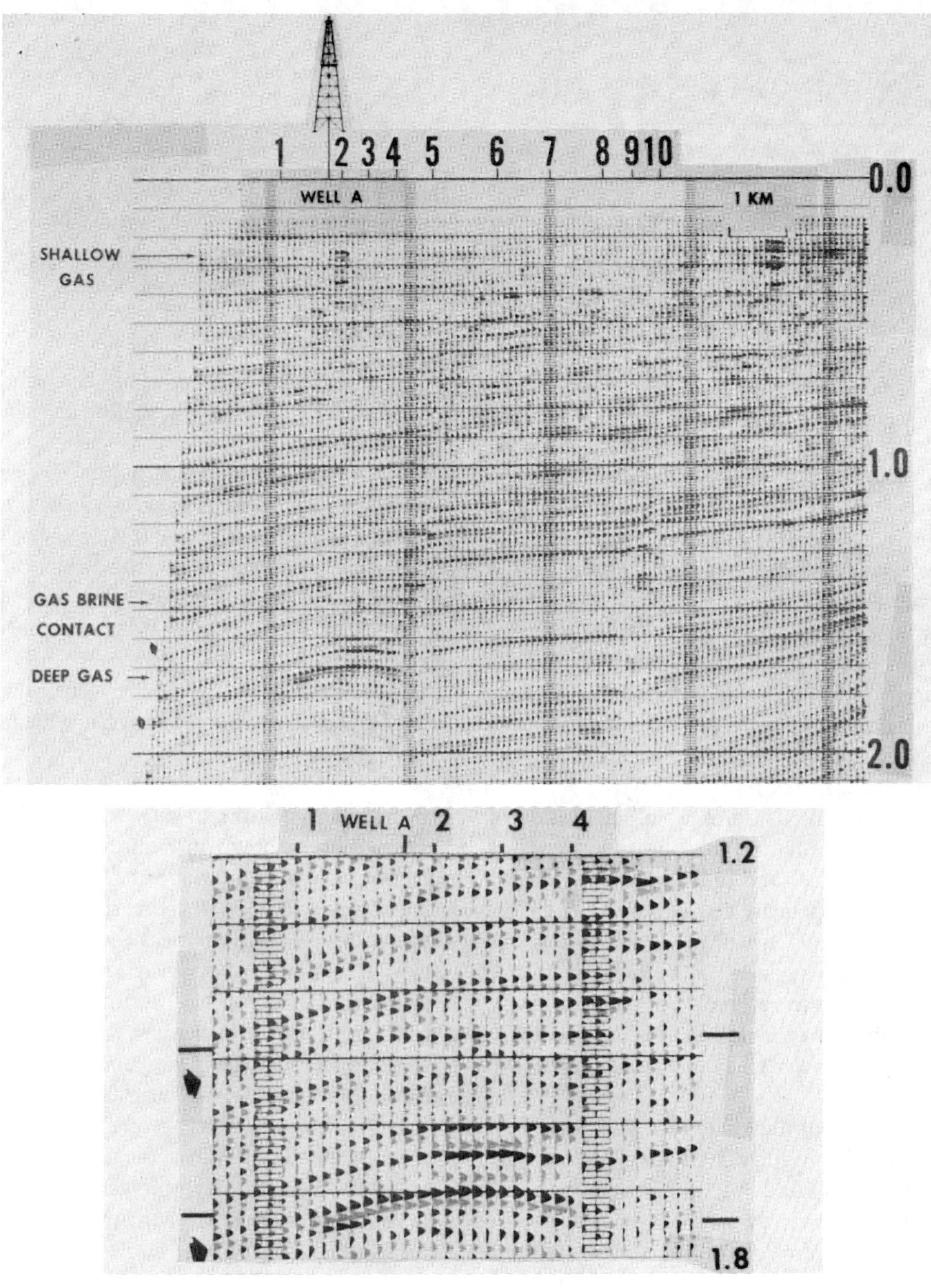

FIGURE 10.37. *Case history: seismic profile across structure, showing flat spots and bright spots. (From Backus and Chen 1975, courtesy of Geophysical Prospecting; Blackwell Scientific Publications Limited.)*

TABLE 10.6. *Case history: velocity and density values from well logs*

	V	ρ	Z
Upper reservoir			
Shale	8.00	2.45	19.6
Gas sand	9.25	2.01	18.6
Water sand	10.00	2.20	22.0
Lower reservoir			
Shale	10.00	2.45	24.5
Gas sand	9.80	2.00	19.6
Water sand	11.00	2.20	24.2

ably flat across the structure. This is the reflection from the gas/water interface in the reservoir. An interference zone exists at the left edge of the flat event, signifying the edge of the reservoir. The reflection from the top of the sand changes polarity as the fluid changes from water to gas in going updip across the reservoir. (Note that the seismic sections in Figure 10.37 are presented with full-wave rectification, with the peaks darker than the erstwhile troughs that are now peaks after rectification. On the original figures, colors were used to distinguish between the two types of peaks.)

The foregoing observations agree with calculation of reflectivity from the velocity and density measurements on well logs. The shale/gas sand reflectivity is small and negative (−0.025). The polarity changes and the reflectivity increases when the fluid changes to water. The shale/water sand reflectivity is +0.057. At the gas/water contact at the base of the reservoir, the reflectivity is +0.084. This explains why the flat event is the dominant reflection from this reservoir. There is no bright spot, but there is a clear flat spot.

The well logs show that the lower reservoir sand is 54 ft thick and is completely filled with gas. The seismic time to the middle of the sand is 1.73, which agrees with the time of a bright spot on the seismic section.

The lower arrow at the left of the seismic section points to the lower sand reflection. Following it updip shows that it becomes very strong suddenly at 1.74 sec and remains strong across the structure until it terminates against the fault.

The decrease in amplitude of the lower sand reflection off structure can be explained by the reflectivity calculations. On structure, the shale/gas sand reflectivity is −0.111. Off structure, the shale/water sand reflectivity is −0.06.

The two-way time-thickness of the sand is 11 ms on structure. This thickness corresponds to tuning at 45 Hz, which falls within the 20–60 Hz passband of the display filter. Large reflectivity at top and bottom of the gas reservoir, −0.111 and +0.106, respectively, and the tuned thickness combine to produce a very significant bright spot.

The reflectivities given here were calculated from velocity and density values taken from the logs, as shown in Table 10.6.

TABLE 10.7. *Case history: velocity and density values from well logs, and resulting reflectivities*

	V	ρ	Z	R
Upper shale	7.00	2.05	14.4	
				−0.12
Gas sand	5.50	2.05	11.3	
				+0.12
Water sand	6.90	2.05	14.1	
				+0.08
Lower shale	7.60	2.20	16.7	

EXAMPLE FROM BARRY AND SHUGART (1973)

The two lines shown in Figures 10.38 and 10.39 cross a gas reservoir at right angles to each other. The top of the sand occurs at 3800 ft, corresponding to about 1.3 sec record time. Sand thickness reaches a maximum of 140 ft, with 130 ft net pay with 32 percent porosity. Essentially no oil is present. The trapping mechanism results from gentle turnover against the fault on the left side of line B. Additional gas sands down to about 5500 feet are much thinner and less commercial.

The thick sand has strong amplitude standout because of the changes in acoustic impedance through the reservoir. Well logs give the values shown in Table 10.7 for velocity and density, from which the reflectivity is calculated.

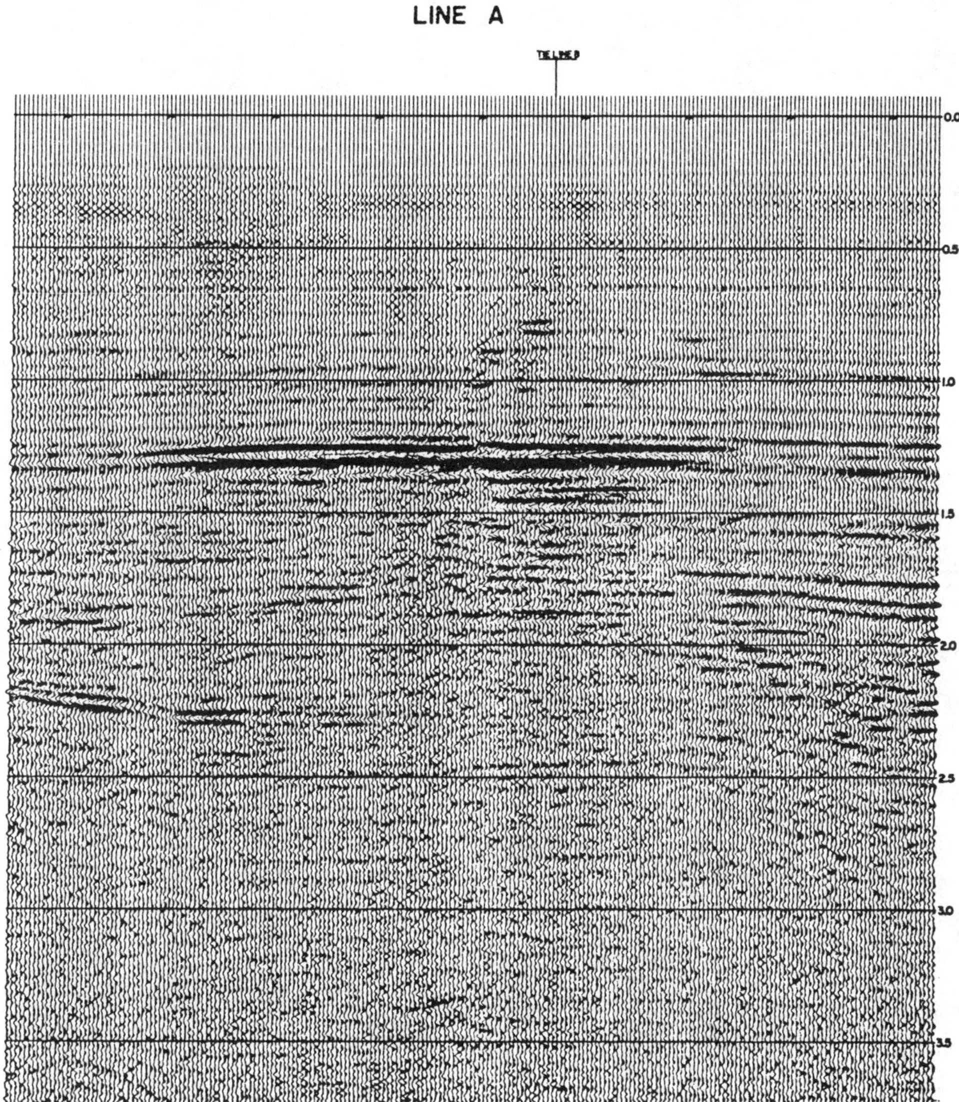

FIGURE 10.38. *Case history: line A. (From Barry and Shugart 1973, courtesy of Teledyne Exploration Company.)*

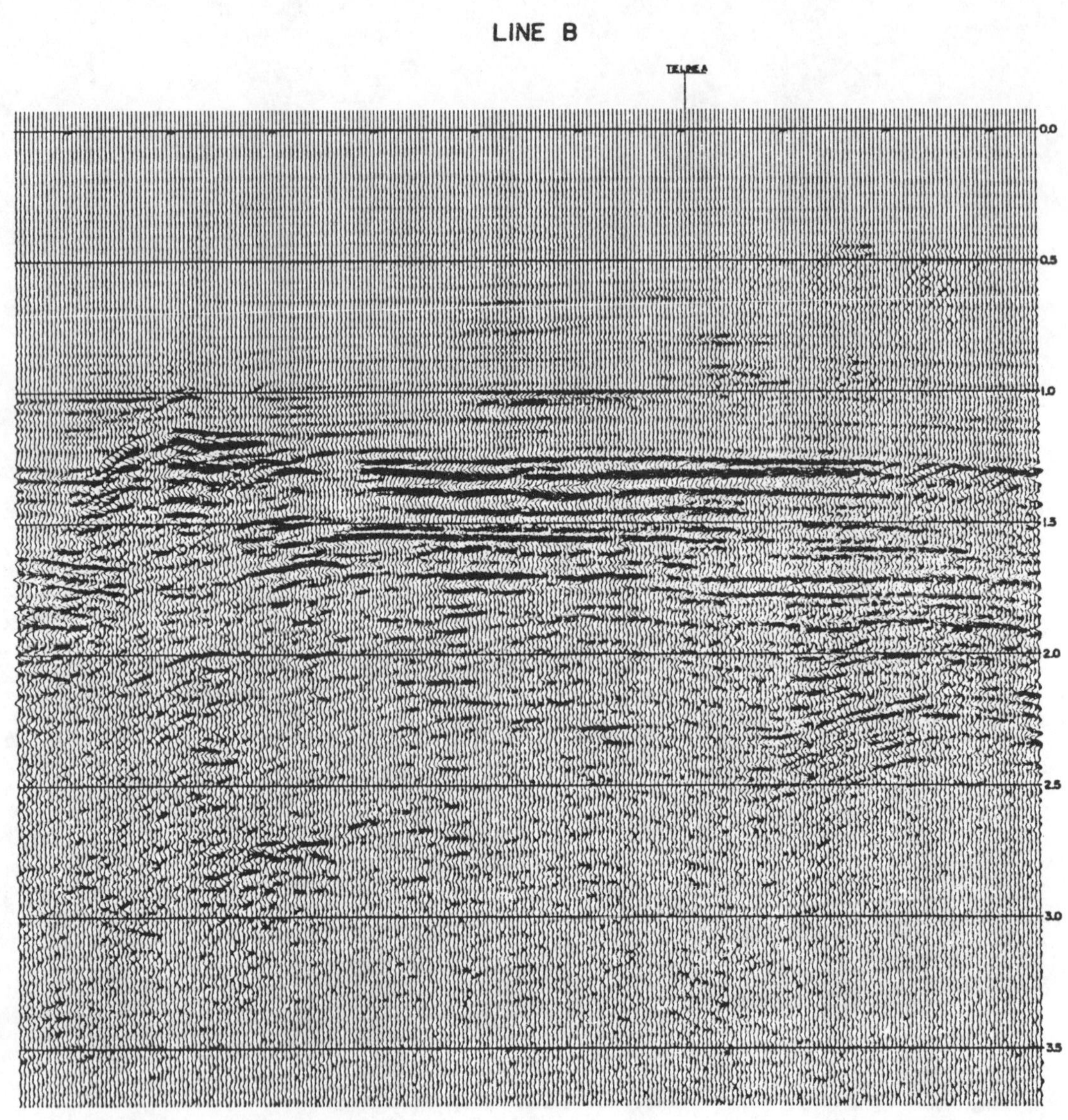

FIGURE 10.39. *Case history: line B. (From Barry and Shugart 1973, courtesy of Teledyne Exploration Company.)*

INFORMATION-PRESERVING PROCESSING

The objectives of information-preserving processing are (1) to correct the amplitude of the seismic data for spherical divergence, absorption, and oblique incidence, in an attempt to produce the plane wave response of a nonabsorbing earth; (2) to achieve the broadest possible bandwidth in the data consistent with objectives of the seismic survey; and (3) to positively establish the polarity convention being used on display.

In a medium with uniform velocity, the spherical divergence can be corrected by multiplying the trace amplitude by record time, because record time and depth are linearly related. In the real earth, the correction is accomplished by multiplying trace amplitude by a factor proportional to the product, $V_{\mathrm{rms}}^2 t$, where V_{rms} is the rms velocity and t is the record time.

After correcting for divergence, the amplitude decay due to absorption can be approximated by an exponential decay function, $\exp(-\alpha - \beta t)$. With consistent source/receiver conditions, such as in marine work, the ensemble average of the two parameters over all traces on a seismic line, or preferably on a grid of lines, will give the average absorption decay rate. Multiplying the trace amplitude by the factor, $\exp(\bar{\alpha} + \bar{\beta} t)$, where the bar indicates ensemble average of the parameters, will approximately correct for absorption. If information is available about the true reflection amplitude as a function of time, then another multiplicative function can be included to correct the trace amplitude to the true reflection amplitude.

The foregoing scheme for correcting for absorption can only be an approximation, because absorption is frequency-dependent. The resulting time variance on the field data can be approximately compensated by using time-variant deconvolution on the spherically corrected data.

The effect of oblique incidence can be compensated by making the ensemble average of the β parameter dependent on offset distance x and correcting the traces at each offset distance with the appropriate value $\bar{\beta}(x)$.

When the source/receiver conditions are not consistent, which is the usual case with land work, the overall level of the amplitude decay curve will shift (that is, the α parameter will vary because of source/receiver inconsistencies), but the decay rate (that is, the $\bar{\beta}$ parameter) will be relatively insensitive to such inconsistencies. It is appropriate in this case to correct each trace with its own α parameter and the ensemble average $\bar{\beta}$ parameter. Approximately the same result can be accomplished by correcting each trace to a constant rms energy over a long window and then computing the ensemble average of the absorption decay curve, as outlined previously in the consistent source/receiver case.

The second aspect of information-preserving processing is the preservation or enlargement of bandwidth, which is achieved by deconvolution. Usually, piecewise time-variant deconvolution is used in an attempt to counteract the time variance in the field data. The deconvolved data are then bandlimited before final display, usually with zero-phase filters to prevent phase distortion. Often, the bandlimiting filters are time-variant, with the spectral band shifting toward lower frequency with increasing record time. This technique of varying the filter to follow the time variance in the field data improves the signal-to-noise ratio in the final display and often produces better continuity for structural mapping, but it reduces the influence of the spectral content of the reflectivity in the final display; that is, it destroys information about the underlying reflectivity.

The third aspect of information-preserving processing is the establishment of the polarity used on the data display. In the old days of paper records, there was no confusion about the polarity convention being used. It was set by the seismic observer in the field and was readily observable on the seismic data. The first break, which is a compressional wave—that is, a positive pressure—causes the seismic trace to either break up or break down initially. With reproducible recording, either analog or digital, the seismic observer can no longer establish the polarity of the displayed data because of the many chances to reverse polarity in processing and display. Because it is often difficult to deduce the polarity convention being used from observations of the first breaks on the final seismic record sections, and because processing centers often do not consider polarity sufficiently important to state clearly on the record sections the polarity convention used to produce the display, the specification of polarity, which should be unambiguous, is often fuzzy and possibly unknown. This problem became so acute after bright spots came into prominence that the SEG formed a committee on polarity and established a polarity standard for the industry—that positive pressure breaks down initially (down equals white on variable-area record sections).

SUMMARY

Bright spots caused a profound revolution in geophysics. Amplitude and spectral content are meaningful in terms of bed thickness, lithology, and fluid content, and the importance of preserving these characteristics in the seismic data is now well accepted. No longer is the seismic method only a structural tool, although that is still its most important use; it is now a stratigraphic tool as well.

It is interesting to note that, long before bright spots and digital recording came into use, stratigraphic studies were being made using information-preserving acquisition and processing, as discussed by Sengbush (1962), for example. With the coming of digital processing, the subsequent development of deconvolution, and the widespread use of horizontal stacking, the seismic method has been perfected to the point at which it is no longer acceptable just to map structures; one also must look for fine detail in the rock layering (McDonal and Sengbush, 1966). Development of the seismic method is now at a plateau, being refined further and further, waiting for the next quantum jump. Bright spots were the last big jump. Will there be another? When?

Appendix A Transform Theory

INTRODUCTION

Transform theory as used in geophysics revolves around the Fourier transform that converts time functions into their frequency representation and space functions into their wavenumber representation. Although the functions of time and space are real functions of a real variable, their transforms are complex functions of a real variable (frequency and wavenumber). As with complex numbers, Fourier transforms have real and imaginary parts, or moduli and phase components. Any function that is absolutely integrable has a Fourier transform.

Periodic functions that are absolutely integrable over one period may be represented by their equivalent Fourier series, which are infinite sets of sine and cosine functions whose frequencies are integral multiples of their fundamental frequencies. The Fourier transform is often derived from the Fourier series in the limit as the periodicity becomes infinite; hence, the fundamental frequency and the harmonic spacing approach zero. Alternatively, the Fourier series coefficients of a periodic function can be derived from the Fourier transform of the corresponding transient function by use of the frequency-domain sampling theorem.

The Fourier transform is a special case of the more general Laplace transform, which transforms functions of a real variable into functions of a complex variable. The Laplace transform does not require absolute integrability of time and space functions. For any function that is Fourier transformable, its Fourier transform can be obtained directly from its Laplace transform by setting the real part of the complex variable to zero.

Time series that are sampled values at equal spacing have Fourier transforms that are periodic, with period equal to the reciprocal of the sampling interval. A convenient transform for time series is the Z-transform, which produces transforms that are polynomials in the transform variable. Hence, all the mathematical theory of polynomials is immediately applicable with Z-transforms.

The Hilbert transform is useful in finding the Fourier transform of a realizable time function from knowledge of either its real or its imaginary part. Minimum-phase functions have the property that the phase and the logarithm of the modulus of their Fourier transforms are Hilbert transforms of each other. Therefore, the modulus specifies the phase of a minimum-phase function.

It is the purpose of this appendix to explore some of the significant properties of Fourier, Laplace, Z, and Hilbert transforms without becoming enmeshed in and suffocated by the mathematical complications that are necessary for a thorough and complete understanding of the material.

FOURIER TRANSFORMS

The time and frequency domains, and the space and wavenumber domains, are related through Fourier transforms, as follows. The transforms are written in terms of time (t) and frequency (f). (Substituting space (x) for time and wavenumber (k) for frequency gives transforms in space and wavenumber domains.)

$$U(f) = \int_{-\infty}^{+\infty} u(t)e^{-j\omega t}\,dt, \tag{A.1}$$

$$u(t) = \int_{-\infty}^{+\infty} U(f)e^{j\omega t}df, \qquad \omega = 2\pi f. \tag{A.2}$$

$U(f)$ is called the Fourier transform of $u(t)$, and $u(t)$ is the inverse transform of $U(f)$. $U(f)$ is complex even if $u(t)$ is real, which is the usual case, as can be seen by substituting $\exp(-j\omega t) = \cos \omega t - j \sin \omega t$ into the defining equation for $U(f)$:

$$\begin{aligned} U(f) &= \int u(t)\,(\cos \omega t - j \sin \omega t)\,dt \\ &= \int u(t) \cos \omega t\,dt - j \int u(t) \sin \omega t\,dt. \end{aligned} \tag{A.3}$$

The Fourier transform $U(f)$ can be written in terms of its real part $\alpha(f)$ and its imaginary part $\beta(f)$, $U(f) = \alpha(f) + j\beta(f)$, where

$$\begin{aligned} \alpha(f) &= \int u(t) \cos \omega t\,dt, \\ \beta(f) &= -\int u(t) \sin \omega t\,dt. \end{aligned} \tag{A.4}$$

Because the Fourier transform is complex, it is often referred to as the complex spectrum of $u(t)$. The real part is called the cosine spectrum and the imaginary part is called the sine spectrum. A complex function can also be written in terms of its modulus $|U(f)|$ and its phase $\theta(f)$,

$$U(f) = |U(f)|\,e^{j\theta(f)}. \tag{A.5}$$

The modulus and phase are related to the real and imaginary parts by

$$\begin{aligned} |U(f)| &= \sqrt{\alpha^2(f) + \beta^2(f)}, \\ \theta(f) &= \tan^{-1} \frac{\beta(f)}{\alpha(f)}. \end{aligned} \tag{A.6}$$

The modulus is called the amplitude spectrum, and $\theta(f)$ is the phase spectrum.

In the case where $u(t)$ is real, the cosine spectrum must be even and the sine spectrum must be odd. This becomes apparent by substituting $\exp(j\omega t) = \cos \omega t + j \sin \omega t$ into the inverse transform and then requiring that the imaginary part of $u(t)$ vanish:

$$u(t) = \int (\alpha + j\,\beta)\,(\cos \omega t + j \sin \omega t)\; df$$

$$= \int (\alpha \cos \omega t - \beta \sin \omega t)\; df + j \qquad \text{(A.7)}$$

$$\int (\alpha \sin \omega t + \beta \cos \omega t)\; df.$$

This requires that both $\int_{-\infty}^{+\infty} \alpha(f) \sin \omega t\; df$ and $\int_{-\infty}^{+\infty} \beta(f) \cos \omega t\; dt$ must be zero, which, in turn, requires that the products $\alpha(f) \sin \omega t$ and $\beta(f) \cos \omega t$ must be odd functions because the integral of an odd function over symmetrical limits is zero. In order that $\alpha(f) \sin \omega t$ be odd, $\alpha(f)$ must be even, because $\sin \omega t$ is odd, and the product of an even function and an odd function is odd. Similarly, $\beta(f) \cos \omega t$ must be odd, hence $\beta(f)$ must be odd, because $\cos \omega t$ is even. This proves that, for a real function, $\alpha(f)$ is even and $\beta(f)$ is odd. Also note that, for real $u(t)$, the modulus is even because both $\alpha^2(f)$ and $\beta^2(f)$are even, the sum is even, and the square root of the sum is even. Also, the phase spectrum $\theta\ (f)$ is odd, because the quotient of an even and an odd function is odd and the arctan of an odd function is odd.

If, in addition to being real, $u(t)$ is also even, then $\beta(f) = 0$, because in equation (A.4) the product $u(t) \sin \omega t$ is odd because $\sin \omega(t)$ is odd. Therefore, a real, even function $u(t)$ has a real, even Fourier transform $U(f) = \alpha(f)$. Similarly, if $u(t)$ is real and odd, then by equation (A.4) $\alpha(f) = 0$, because the product $u(t) \cos \omega t$ is odd because $\cos \omega t$ is even. Therefore, a real, odd function $u(t)$ has an imaginary, odd transform $U(f) = j\beta(f)$. The phase spectrum for a real, even function is either zero or π radians, depending on whether $\alpha(f)$ is positive or negative, respectively, and the phase spectrum for a real, odd function is either $\pm\pi/2$, depending on the polarity of $\beta(f)$.

The value of $U(f)$ at the origin is found by substituting $f = 0$ into equation (A.1):

$$U(0) = \int_{-\infty}^{+\infty} u(t)\; dt. \qquad \text{(A.8)}$$

This shows that the spectral value at $f = 0$ is the net area under the time function $u(t)$. Similarly, the value of $u(t)$ at the origin is found by substituting $t = 0$ into equation (A.2):

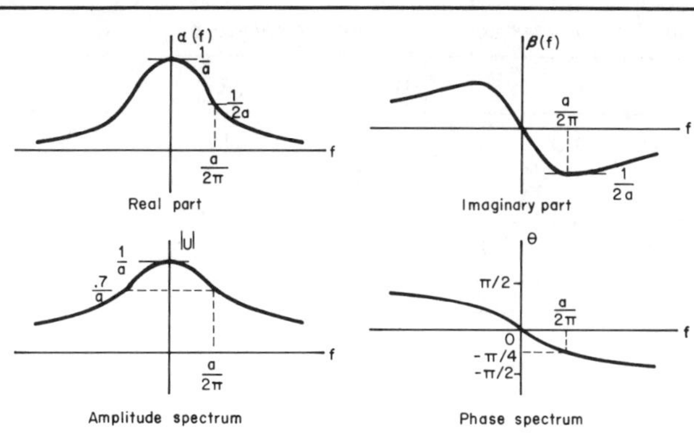

FIGURE A.1. *Complex Fourier spectrum of exponential function.*

$$u(0) = \int_{-\infty}^{+\infty} U(f)\; df. \qquad \text{(A.9)}$$

EXAMPLE

The exponential time function

$$u(t) = \begin{cases} e^{-at} & t \geqslant 0 \\ 0 & t < 0 \end{cases} \qquad \text{(A.10)}$$

has complex spectrum given by

$$U(f) = \frac{1}{a + j\omega}, \qquad \omega = 2\pi f. \qquad \text{(A.11)}$$

The real and imaginary parts and the amplitude and phase spectra in Figure A.1 are given by

$$\alpha(f) = \frac{a}{a^2 + \omega^2}, \qquad \beta(f) = \frac{-\omega}{a^2 + \omega^2}, \qquad \text{(A.12)}$$

$$|\,U(f)\,| = \frac{1}{\sqrt{a^2 + \omega^2}}, \quad \theta(f) = \tan^{-1}\left(\frac{-\omega}{a}\right). \qquad \text{(A.13)}$$

The Fourier spectral values at several frequencies are given in Table A.1. This example demonstrates that real-time functions have complex spectra whose real part is an even function of frequency and whose imaginary part is an odd function. Also, the amplitude spectrum is even and the phase odd. Thus, the negative frequencies con-

TABLE A.1. *Fourier spectral values at several frequencies*

ω	α	β	|U|	θ
0	$1/a$	0	$1/a$	0°
a	$1/2a$	$-1/2a$	$1/\sqrt{2}a$	−45°
$10a$	$1/101a$	$-10/101a$	$1/\sqrt{101}a$	−84.3°
$100a$[a]	$1/10000a$	$-1/100a$	$1/100a$	−90°

[a]For $\omega = 100a$, the values are approximate.

tain no new information beyond that contained by the positive frequencies in the case of real-time functions.

FOURIER TRANSFORM THEOREMS AND PROPERTIES

Only those properties of Fourier transforms necessary for an understanding of the material in this manual are included here. The theorems are stated without proofs as they are found in many textbooks. The most general conditions under which a function $u(t)$ has a Fourier transform are bypassed by considering only functions with finite energy. They always have Fourier transforms. The shorthand notation $\mathcal{F}u(t)$ will be used to mean the Fourier transform of $u(t)$.

LINEARITY THEOREM

(a) If $\mathcal{F}[u(t)] = U(f)$ and $\mathcal{F}[v(t)] = V(f)$, then $\mathcal{F}[u(t) + v(t)] = U(f) + V(f)$.

(b) If $\mathcal{F}[u(t)] = U(f)$, then $\mathcal{F}[Au(t)] = AU(f)$, where A is constant.

FOLDING THEOREM

If $\mathcal{F}[u(t)] = U(f)$, then $\mathcal{F}[u(-t] = U(-f)$.

This theorem states that, if a time function is folded, its transform is also folded. In the case of the exponential example, the time function and its spectrum are given by

$$u(-t) = \begin{cases} e^{-a(-t)} & t \leq 0 \\ 0 & t > 0 \end{cases}, \tag{A.14}$$

$$U(-f) = \frac{1}{a - j\omega}. \tag{A.15}$$

COROLLARY TO FOLDING THEOREM

If $u(t)$ is a real-time function and $\mathcal{F}[u(t)] = U(f)$, then $\mathcal{F}[u(-t)] = U^*(f)$, where $U^*(f)$ is the complex conjugate of $U(f)$.

With real-time functions, the amplitude spectrum is unaffected by folding the time function because $|U(f)| = |U^*(f)|$. Only the phase is affected, and it is merely changed in polarity.

SHIFTING THEOREM

If $\mathcal{F}[u(t)] = U(f)$, then $\mathcal{F}[u(t - T)] = e^{-j\omega T}U(f)$, where T is constant.

In the case of the exponential, the shifted time function and its spectrum are given by

$$u(t - T) = \begin{cases} e^{-a(t-T)} & t \geq T \\ 0 & t < T \end{cases}, \tag{A.16}$$

$$\mathcal{F}[u(t - T) = e^{-j\omega T}\frac{1}{a + j\omega}.$$

Writing $U(f)$ in its polar form gives

$$\mathcal{F}[u(t - T)] = |U(f)|\, e^{j\theta(f)}\, e^{-j\omega T}. \tag{A.17}$$

This shows that the amplitude spectrum is unaffected by the time shift, and the phase spectrum $\theta_s(f)$ of the shifted function is

$$\theta_s(f) = \theta(f) - \omega T. \tag{A.18}$$

Thus, a time shift of T introduces a phase shift of $-\omega T$, which is a phase shift that is linear with frequency. A linear phase shift does not distort the waveform; it merely translates it in time. The phase shift of the shifted exponential, shown in Figure A.2, is given by

$$\theta_s(f) = \tan^{-1}\left(\frac{-\omega}{a}\right) - \omega T. \tag{A.19}$$

DIFFERENTIATION THEOREM

If $\mathcal{F}[u(t)] = U(f)$, then $\mathcal{F}[du(t)/dt] = j\omega\, U(f)$.

INTEGRATION THEOREM

If $\mathcal{F}[u(t)] = U(f)$, then $\mathcal{F}[\int u(t)dt] = (j\omega)^{-1}\, U(f)$.

SCALE-CHANGE THEOREM

If $\mathcal{F}[u(t)] = U(f)$, then $\mathcal{F}[u(kt)] = 1/kU(f/k)$.

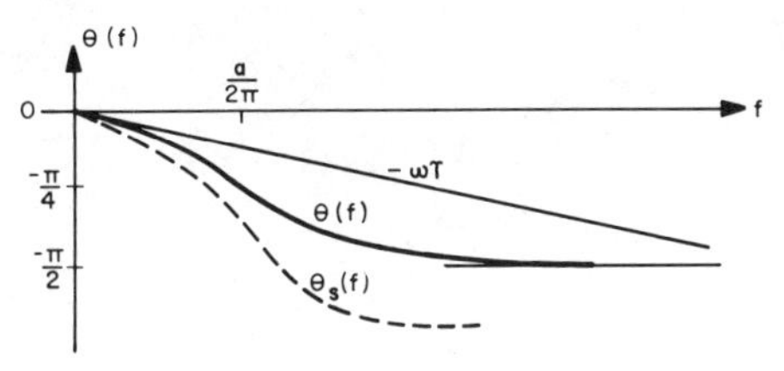

FIGURE A.2. *Phase spectra of shifted and unshifted exponentials.*

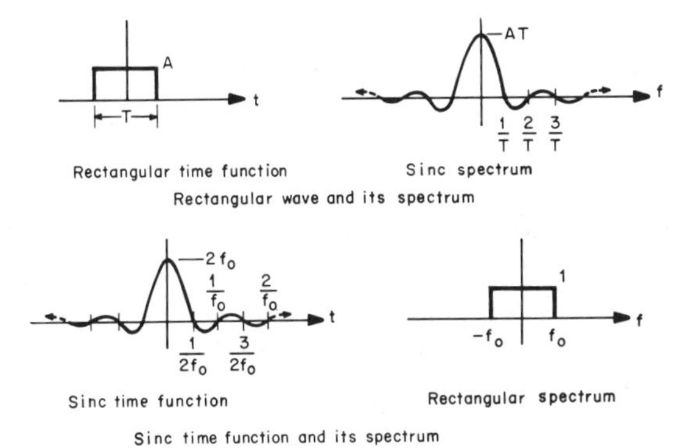

FIGURE A.3. *Rectangle/sinc transformations.*

CONVOLUTION THEOREM

If $\mathcal{F}[u(t)] = U(f)$ and $\mathcal{F}[v(t)] = V(f)$, then $\mathcal{F}[u(t) * v(t)] = U(f)V(f)$, where * means convolution.

PRODUCT THEOREM

If $\mathcal{F}[(t)] = U(f)$ and $\mathcal{F}[v(t)] = V(f)$, then $\mathcal{F}[u(t)v(t)] = U(f) * V(f)$.

SYMMETRICAL RECTANGULAR FUNCTIONS TRANSFORM INTO SINC FUNCTIONS

A symmetrical rectangular pulse of length T and height A (Figure A.3) has Fourier spectrum given by

$$U(f) = AT \operatorname{sinc} fT, \quad \text{where} \quad \operatorname{sinc} x \triangleq \frac{\sin \pi x}{\pi x}. \tag{A.20}$$

Its real and imaginary parts are

$$\alpha(f) = AT \operatorname{sinc} fT, \qquad \beta(f) = 0. \tag{A.21}$$

At the origin, $f = 0$, the spectrum $\alpha(f) = AT$, which is the net area of the time function. The spectrum $\alpha(f) = 0$ at integral multiples of $f = 1/T$ because of the sine term in the numerator of the sinc function.

Conversely, a sinc time function has a real rectangular spectrum. The time function whose $\alpha(f)$ spectrum is unity between $\pm f_0$ and whose $\beta(f)$ spectrum is identically zero is the following sinc function:

$$u(t) = 2f_0 \operatorname{sinc} 2f_0 t. \tag{A.22}$$

The value of the time function at the origin is the net area of the $\alpha(f)$ spectrum, and the time function is zero at integral multiples of $t = 1/2f_0$.

TRANSFORMS OF IMPULSES

A unit impulse $\delta(t)$ is defined by the following two properties:

(1) $\delta(t) = 0$ for $t \neq 0$.
(2) $\int_{-\infty}^{+\infty} \delta(t)\, dt = 1$.

Of course, no ordinary function can have these properties. Here we invoke the philosophy of Dirac (1958), who defended the use of this improper function by pointing out that, in itself, it does not result in lack of rigor, since all results can be obtained in a more cumbersome way without its aid.

The most important use of $\delta(t)$ is its sifting property,

$$\int_{-\infty}^{+\infty} h(t)\delta(t - T)dt = h(T), \tag{A.23}$$

provided that $h(t)$ is continuous at $t = T$.

Applying the sifting property in the defining equation for the Fourier transform of an impulse gives

$$\mathcal{F}[\delta(t)] = \int_{-\infty}^{+\infty} \delta(t)\, e^{-j\omega t} dt = 1. \tag{A.24}$$

The shifted impulse $\delta(t - T)$ has the transform

$$\mathcal{F}[\delta(t - T)] = \int_{-\infty}^{+\infty} \delta(t - T)\, e^{-j\omega t} dt = e^{-j\omega T}. \tag{A.25}$$

This result could have been obtained by applying the shifting theorem to $\mathcal{F}[\delta(t)]$.

FOURIER SERIES FOR PERIODIC FUNCTIONS

Periodic time functions with period T can be represented by a Fourier series whose coefficients can be obtained directly from the Fourier transform evaluated over one period. The Fourier series contains only harmonics of $1/T$. The complex coefficients $U(k/T)$ are given by

$$\begin{aligned} U(k/T) &= \int_{-T/2}^{T/2} u(t) \exp\left(-j\frac{2\pi kt}{T}\right)dt \\ &= \underbrace{\int u(t) \cos\frac{2\pi kt}{T}\, dt}_{} + j\underbrace{\int - u(t) \sin\frac{2\pi kt}{T}\, dt}_{} \\ &= \quad \alpha(k/T) \quad + j \quad \beta(k/T). \end{aligned} \tag{A.26}$$

The Fourier series representation of the periodic function is given by

$$u(t) = \frac{1}{T}\left[\sum_{k=-\infty}^{+\infty} \alpha(k/T) \cos\frac{2\pi kt}{T} + \sum_{k=-\infty}^{+\infty} \beta(k/T) \sin\frac{2\pi kt}{T}\right]. \tag{A.27}$$

In 1811, Fourier proved that, at points of continuity, the series converges to the function and, at ordinary discontinuities, the series converges to the arithmetic mean.

Periodic functions have spectra consisting of impulses whose weights are given by the products of the complex coefficients and the sampling interval, $(1/T)\ U\ (k/T)$:

$$U(f) = \frac{1}{T} \sum_{k=-\infty}^{\infty} U(k/T)\delta(f - kT). \tag{A.28}$$

CORRELATION FUNCTIONS AND ENERGY SPECTRA

The autocorrelation function of a function $g(t)$ with finite energy is defined by

$$\phi_{gg}(\tau) = \int_{-\infty}^{+\infty} g(t)g(t - \tau)\, dt. \tag{A.29}$$

The autocorrelation function has the following properties:

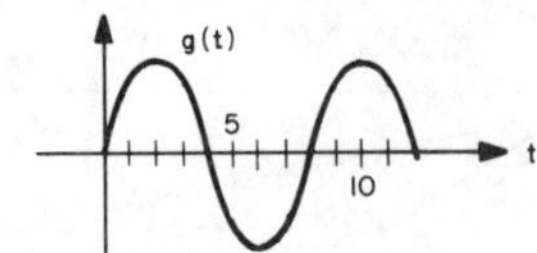

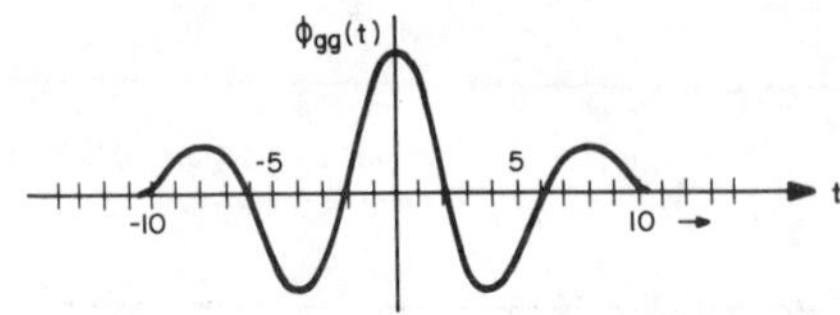

FIGURE A.4. *Autocorrelation of sine pulse.*

Symmetric: $\phi_{gg}(\tau) = \phi_{gg}(-\tau)$, (A.30)

Maximum at origin: $\phi_{gg}(0) \geqslant \phi_{gg}(\tau)$. (A.31)

Also, the value at the origin $\phi_{gg}(0)$ is equal to the total energy in the function $g(t)$, as shown by substitution of $\tau = 0$ into the autocorrelation equation:

$$\phi_{gg}(0) = \int_{-\infty}^{+\infty} g^2(t)dt. \tag{A.32}$$

An example of autocorrelating such a function is shown in Figure A.4.

The cross correlation of two finite energy functions $g(t)$ and $h(t)$ is defined by

$$\phi_{gh}(\tau) = \int_{-\infty}^{+\infty} g(t)h(t - \tau)dt. \tag{A.33}$$

The cross-correlation function has the following property:

$$\phi_{gh}(\tau) = \phi_{hg}(-\tau). \tag{A.34}$$

Cross correlation is equivalent to convolution without folding or, more precisely,

$$\phi_{gh}(\tau) = g(\tau) * h(-\tau). \tag{A.35}$$

Because both $g(t)$ and $h(t)$ have finite energy, they have Fourier transforms $G(f) = \mathcal{F}[g(t)]$ and $H(f) = \mathcal{F}[h(t)]$. Also, the cross-correlation function ϕ_{gh} has a

Fourier transform because it is the convolution of two finite energy functions, and the transform relation is

$$\mathcal{F}[\phi_{gh}(\tau)] = \mathcal{F}[g(\tau)] \cdot \mathcal{F}[h(-\tau)]. \tag{A.36}$$

By the folding theorem, if $\mathcal{F}[h(t)] = H(f)$, then $\mathcal{F}[h(-t)] = H^*(f)$, where the asterisk indicates the complex conjugate. Therefore, the Fourier transform of ϕ_{gh} is given by

$$\mathcal{F}[\phi_{gh}(\tau)] \triangleq \Phi_{gh}(f) = G(f) \cdot H^*(f). \tag{A.37}$$

In autocorrelation, $g(t) = h(t)$; therefore,

$$\mathcal{F}[\phi_{gg}(\tau)] \triangleq \Phi_{gg}(f) = G(f) \cdot G^*(f) = |G(f)|^2 \tag{A.38}$$

Because $|G(f)|$ is amplitude, its squared value is energy. Hence, $\Phi_{gg}(f)$ is called the energy spectrum of $g(t)$. It is a real, even, nonnegative function of frequency; that is,

$$\text{Real:} \quad \operatorname{Re} \Phi_{gg}(f) = \Phi_{gg}(f), \tag{A.39}$$

$$\text{Even:} \quad \Phi_{gg}(f) = \Phi_{gg}(-f), \tag{A.40}$$

$$\text{Nonnegative:} \quad \Phi_{gg}(f) \geq 0 \quad \text{for all } f. \tag{A.41}$$

Because the autocorrelation function is even, its Fourier transform $\Phi_{gg}(f)$ is given by the cosine transform,

$$\Phi_{gg}(f) = 2\int_0^\infty \Phi_{gg}(\tau) \cos 2\pi f\tau \, d\tau. \tag{A.42}$$

The inverse transform is also a cosine transform,

$$\theta_{gg}(\tau) = 2\int_0^\infty \Phi_{gg}(f) \cos 2\pi f\tau \, df. \tag{A.43}$$

LAPLACE TRANSFORMS

Laplace transforms are widely used in studies of transients in linear time-invariant systems. The Laplace transform $U(s)$ of a function $u(t)$ is defined by

$$U(s) \triangleq \mathcal{L}[u(t)] \triangleq \int_0^\infty u(t)\, e^{-st}\, dt, \tag{A.44}$$

where

$$s = \sigma + j\omega.$$

TABLE A.2. *Function-transform pairs*

One-sided time function	$\mathcal{L}$-transform
Unit impulse: $\delta(t)$	1
Unit step: 1	$1/s$
Exponential: e^{-at}	$1/(s + a)$
Sine: $\sin bt$	$b/(s^2 + b^2)$
Cosine: $\cos bt$	$s/(s^2 + b^2)$
Damped sine: $e^{-at} \sin bt$	$b/[(s + a)^2 + b^2]$
Damped cosine: $e^{-at} \sin bt$	$(s + a)/[(s + a)^2 + b^2]$
Powers of t: t^n	$n!/s^{n+1}$
Damped powers of t: $e^{-at}t^n$	$n!/(s + a)^{n+1}$

This is often called the one-sided transform because it is applicable only for one-sided time functions—that is, functions that are zero for $t < 0$. Such functions are said to be physically realizable. No significant problem exists in handling two-sided time functions, but the application of $\mathcal{L}$-transforms is largely in the realm of realizable systems and functions.

FUNCTION-TRANSFORM PAIRS

Functions are simplified through Laplace transformation. A unit step at the origin has transform $1/s$, an exponential has transform $1/(s + a)$, and a unit impulse has transform equal to unity. Table A.2 shows $\mathcal{L}$-transforms of some simple one-sided time functions.

The Laplace transforms in the table are polynomials in s and are characterized by the locations of their poles and zeros (poles are the zeros of the denominators) in the complex s-plane. The step function has a pole at the origin, the exponential has a pole on the real axis at $s = -a$, the sine and cosine functions have conjugate poles on the imaginary axis at $s = \pm jb$, and the damped sine and cosine functions have complex conjugate poles at $s = -a \pm jb$. All these poles are simple, or first-order. Functions that contain multiplication by nth powers of t have multiple poles of order $n + 1$.

The location of the poles of $U(s)$ determines the stability of the associated time function $u(t)$. A stable function converges to zero as t approaches infinity, whereas an unstable function diverges. A function that is bounded but does not converge to zero is called conditionally stable. Stability requires that all poles lie in the open left half-plane. One or more poles in the open right half-plane or multiple poles on the imaginary axis produce instability. Simple poles on the imaginary axis cause conditional stability.

The location of the zeros of $U(s)$ tells whether or not the inverse time function $u^{-1}(t)$ is stable. For $u(t)$ to have a stable inverse, it is also necessary that all zeros of $U(s)$ lie in the open left half-plane.

RELATIONSHIP TO FOURIER TRANSFORM

The Laplace transform is applicable to a wider class of functions than the Fourier transform because of the convergence factor σ built into the kernel of the Laplace transform. For a realizable function to be $\mathcal{F}$-transformable, all poles of its $\mathcal{L}$-transform must lie in the open left half-plane. Then the time function is stable—that is, has finite energy—and is $\mathcal{F}$-transformable.

Folding a realizable time function produces a function that is zero for $t > 0$. Folding the time function corresponds to folding its $\mathcal{L}$-transform about the imaginary axis. A stable function in positive time folds into a stable function in negative time, and all its poles that formerly were in the open left half-plane now lie in the open right half-plane. Such a function has a Fourier transform.

Two-sided time functions that are stable in both positive and negative time have Fourier transforms; the Fourier transforms of stable functions can be obtained from their Laplace transforms merely by substituting $j\omega$ for s.

Laplace transforms that are ratios of polynomials in s, $U(s) = A(s)/B(s)$, when evaluated along the imaginary axis, give the Fourier transform, as follows:

$$U(j\omega) = \frac{k\pi_i\ (s - z_i)}{\pi_k\ (s - p_k)}\bigg|\ s = j\omega, \tag{A.45}$$

where the zeros of $U(s)$ are at z_i and the poles at p_k. Each factor $(s - z_i)$ can be written in terms of its modulus and phase angle:

$$s - z_i = |s - z_i|\ e^{j\phi_i}, \tag{A.46}$$

where $s - z_i$ is the length of the vector from the zero z_i to the point where $U(s)$ is being evaluated, and ϕ_i is the angle between the vector and the positive horizontal half-line through z_i. Similarly, each factor $(s - p_k)$ can be written

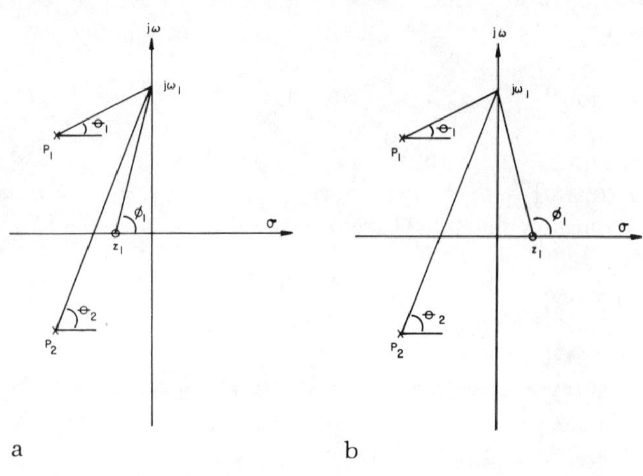

FIGURE A.5. *Vector evaluation of* U(s) *at the point* $j\omega_1$ *in the* s-*plane. (a) Minimum phase. (b) Nonminimum phase.*

$$s - p_k = |s - p_k|\ e^{j\phi_k}. \tag{A.47}$$

Thus, the modulus of $U(j\omega)$, which is the amplitude spectrum, and the phase spectrum are given by

$$|U(j\omega)| = \frac{k\pi_i\ |j\omega - z_i|}{\pi_k\ |j\omega - p_k|}, \qquad \Theta\ (j\omega) = \sum_i \phi_i - \sum_k \theta_k. \tag{A.48}$$

A realizable function must have all of its poles in the open left half-plane for stability, but its zeros need not be thus restricted. Transferring a zero from the point $z = a + jb$ in the left half-plane to its mirror point $z = +a + jb$ in the right half-plane will not affect the amplitude spectrum but will increase the phase lag, as can be seen by construction of the vector diagrams in Figure A.5 for a simple $U(s)$ function. Thus, location of all zeros in the left half-plane insures minimum phase lag for a given amplitude spectrum. This gives the definition of minimum-phase functions in terms of the Laplace transform: A stable realizable function is minimum-phase if and only if all the zeros of its Laplace transform lie in the left half-plane. A consequence of this definition is that, for a given amplitude spectrum, the minimum-phase function is the only one with a stable inverse.

Z-TRANSFORMS

Z-transforms play an important role in the analysis, representation, and synthesis of linear, time-invariant, discrete-time systems, similar to the role of the Laplace transform with continuous systems. Laplace transform properties and theorems have their counterpart in Z-transform theory.

A realizable time series $\{u\}$, which is a set of weighted impulses at equal spacing, defined by

$$\{u\} \triangleq \sum_k u_k \delta(t - k\Delta), \tag{A.49}$$

has Fourier transform given by

$$U(f) = \sum_k u_k \exp(-j2\pi f k\Delta). \tag{A.50}$$

Substituting the variable z $\exp(-j2\pi f\Delta)$ gives the polynomial $\Sigma_k u_k z^k$, which is defined as the Z-transform of the time series $\{u\}$ and is written $U(z)$.

An important class of Z-transforms is those represented by a ratio of polynomials in z. As with the Laplace transform, location of the poles and zeros gives valuable insight into the properties of systems or sequences generating Z-transforms. All poles of the Z-transform of a realizable sequence must lie outside the unit circle for stability. At least one simple pole on the unit circle gives conditional stability, and at least one within the unit circle produces instability. All zeros outside the unit circle are required for the realizable sequence to be minimum-phase (usually called minimum-delay).

Sequences in negative time require all their poles to be inside the unit circle for stability. Two-sided sequences that are stable in both positive and negative time have Fourier transforms as well as Z-transforms.

A minimum-phase sequence is a realizable sequence whose Z-transform has all of its poles and zeros outside the unit circle. Then the inverse Z-transform also has all of its poles and zeros outside the unit circle, and a stable realizable inverse sequence exists.

Another property of a minimum-phase sequence is that its accumulative energy buildup is faster than any other sequence that has the same amplitude spectrum. The accumulative energy is defined by

$$E_n = \sum_{k=0}^{n} |u_k|^2. \tag{A.51}$$

Any stable realizable sequence can be expressed in the form

$$H(z) = H_{\text{min}}(z) \cdot H_{\text{ap}}(z) \tag{A.52}$$

where $H_{\text{min}}(z)$ is the transform of the minimum-phase sequence that has the same amplitude spectrum as $H(z)$, and $H_{\text{ap}}(z)$ is an all-pass system that has unit amplitude response at all frequencies. All zeros of $H(z)$ inside the unit circle are reflected outside into the conjugate reciprocal location to produce $H_{\text{min}}(z)$. A sequence of length n has $(n-1)$ zeros, and 2^{n-1} sequences exist with the same amplitude spectrum.

Convolution of two time series $\{u\}$ and $\{v\}$ is simplified by taking Z-transforms. Convolution in time becomes multiplication of transforms, which in this case is the multiplication of polynomials in z:

$$Z\,[\{u\} * \{v\}] = U(z) \cdot V(z) = \sum_{k=0}^{\infty} u_k z^k \cdot \sum_{l=0}^{\infty} v_l z^l. \tag{A.53}$$

The resulting polynomial gives the Z-transform of the output, and the coefficients of the output time series are identical to the coefficients of the output Z-transform.

HILBERT TRANSFORMS

Hilbert transforms are useful in finding the Fourier transform of a realizable time function from knowledge of only the real part or the imaginary part of the Fourier transform. With minimum-phase functions, the phase and logarithm of the modulus of the Fourier transform are Hilbert transforms of each other.

Hilbert transforms are given by the following relationships:

$$\beta(\omega) = \frac{1}{\pi}\int_{-\infty}^{+\infty} \frac{\alpha(\nu)}{\nu - \omega}\, d\nu, \qquad \text{(A.54)}$$

$$\alpha(\omega) = \frac{1}{\pi}\int_{-\infty}^{+\infty} \frac{\beta(\nu)}{\nu - \omega}\, d\nu.$$

The discontinuity at $\nu = \omega$ requires that the integrals be split at that point and evaluated as principal values.

Signals applied to the quadrature filter produce Hilbert transforms of the signals. The quadrature filter has Fourier transform

$$H(f) = \begin{cases} +j, & f > 0 \\ -j, & f < 0 \end{cases}. \qquad \text{(A.55)}$$

Its amplitude spectrum is unity, and its phase is $\pi/2$ for positive f and $-\pi/2$ for negative f. Since it produces a $\pi/2$ phase shift, $\cos \omega_0 t$ as an input becomes $\sin \omega_0 t$ in the output—hence the name quadrature filter. Its impulse response is $h(t) = 1/\pi t$. Convolving $g(t)$ with $h(t)$ gives the Hilbert transform of $g(t)$:

$$g(t) * h(t) = \int g(\tau)h(t - \tau)d\tau, \qquad \text{(A.56)}$$

$$g(t) * \frac{1}{\pi t} = \int g(\tau)\frac{1}{\pi(t - \tau)}\, d\tau = \mathcal{H}\,[g(t)].$$

Because of this property, the quadrature filter is often called a Hilbert transformer. The foregoing transform relationships are useful in representing bandpass signals as complex signals.

Appendix B Linear System Theory

INTRODUCTION

Linear systems may be characterized in either the time domain or the frequency domain. In the time domain, transients such as impulses and step functions are used for inputs to such systems. These inputs result in characteristic output waveforms, called the impulse and step responses, respectively, which can be used with any input waveform to determine the resulting output waveform. In the frequency domain, the systems are characterized by their responses to sinusoids.

LINEAR SYSTEMS

A linear system is characterized in the time domain by two properties:

(1) Superposition: The output produced by the sum of a number of inputs is the sum of the outputs due to the individual inputs.
(2) Homogeneity: If the input is multiplied by a constant, the output will be multiplied by the same constant.

Linear systems that have the property that their response to a given input is independent of the time at which the input is applied are said to be *time-invariant*. The three properties of linear, time-invariant systems are written symbolically as follows. If $i_1(t) \rightarrow o_1(t)$ and $i_2(t) \rightarrow o_2(t)$, then

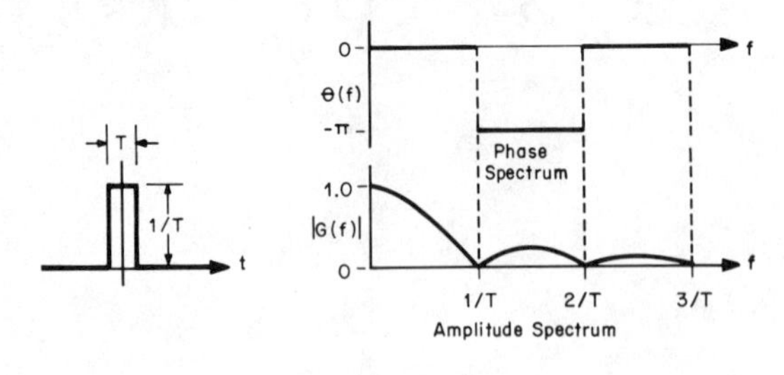

FIGURE B.1. *Fourier spectrum of rectangular pulse.*

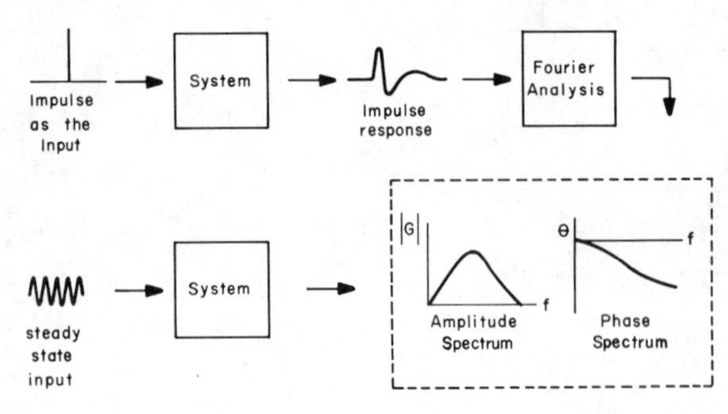

FIGURE B.2. *Comparison of impulse and steady-state responses.*

(1) $i_1(t) + i_2(t) \rightarrow o_1(t) + o_2(t)$.
(2) $Ai(t) \rightarrow Ao(t)$, A constant.
(3) $i(t - T) \rightarrow o(t - T)$, T constant.

A linear, time-invariant system is characterized in the frequency domain by its response to an ensemble of sinusoids. A pure sinusoid of a given frequency will pass through such a system without alteration of frequency or introduction of harmonics. Only the amplitude and phase may be modified. Using sinusoids with unit amplitudes and zero phase shifts as inputs and measuring the amplitudes and phase shifts of the outputs gives the amplitude and phase spectra of the system. These spectra constitute the frequency response, also called the steady-state response, of the system. The impulse response and the steady-state response are related through Fourier analysis, and knowledge of one gives implicit knowledge of the other.

IMPULSE RESPONSE

A unit impulse can be considered the limiting case of functions such as a rectangle with length T and height $1/T$. As $T \rightarrow 0$, this function approaches an impulse, and its area, being independent of T, remains finite and equal to unity (Figure B.1).

The frequency spectrum of the unit impulse is constant. Its amplitude spectrum is unity and its phase is zero for all frequencies. The spectrum of the aforementioned rectangle approaches the constant spectrum as the width $T \rightarrow 0$. The amplitude spectrum is unity at $f = 0$ and decreases to zero at $f = 1/T$. The phase is zero in this frequency range. As $T \rightarrow 0$, the frequency of the first zero in the spectrum, $f = 1/T$, approaches infinity. Thus, the amplitude spectrum of an impulse is constant and equal to unity, and the phase spectrum is zero for an arbitrarily large range of frequencies.

The great importance of the impulse response results from the fact that the frequency spectrum of a unit impulse is simply $I(f) = 1$. A unit impulse as the input into a system will produce an output whose spectrum $G(f)$ is identically the steady-state characteristic of the system (Figure B.2).

In the steady-state case, a series of constant-amplitude, zero-phase sinusoids is passed in succession through the system. The amplitude and relative phase shift of the output are measured as a function of frequency. The continuous amplitude and phase spectra are formed by a continuous curve through these points.

In the impulse case, the impulse applies all frequencies to the system simultaneously. All frequencies have unit amplitude and zero phase; hence, the impulse response, when analyzed by Fourier analysis, gives the same amplitude and phase spectra as those obtained by the steady-state method.

FILTERING IN THE TIME DOMAIN

A filter system can be characterized in either the frequency or the time domain. In the frequency domain, the amplitude and phase spectra of the system are determined using steady-state input signals. The impulse response is the equivalent representation in the time domain.

Likewise, the output of a system resulting from a given input waveform can be determined in either domain. In the frequency domain, the amplitude and phase spectra of the input must be calculated using Fourier analysis. Then the output amplitude spectrum is calculated by multiplying the input and the system amplitude spectra, and the output phase spectrum is the sum of the two corresponding phase spectra. To determine the output waveform from the amplitude and phase spectra of the output, it is necessary to combine the frequency components by Fourier synthesis.

This rather involved procedure of determining the output waveform in the frequency domain often becomes simpler and more meaningful in the time domain. The output waveform is found by combining the input waveform and the impulse response of the system, using convolution.

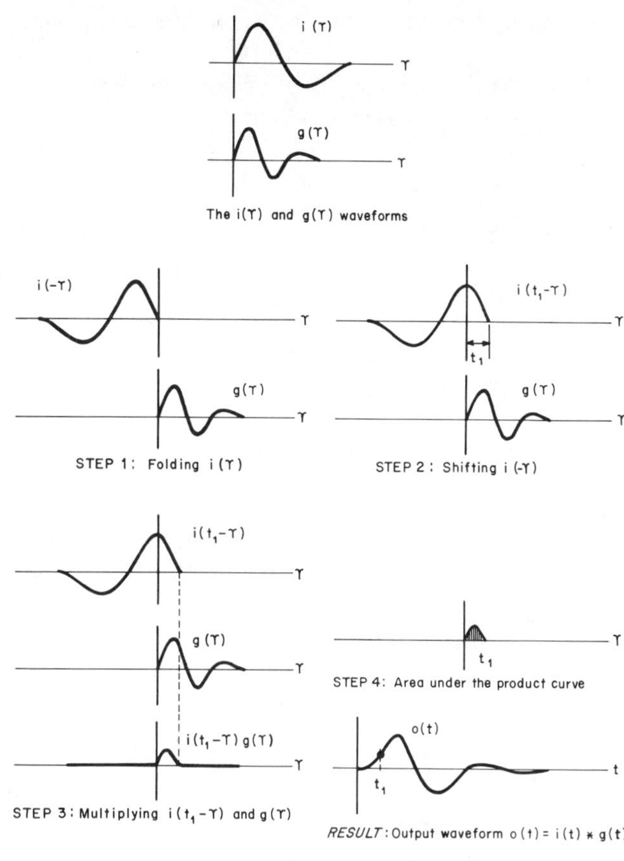

FIGURE B.3. *Solution of convolution integral.*

THE CONVOLUTION INTEGRAL

The mathematical description of filtering is the convolution integral,

$$o(t) \int_{-\infty}^{+\infty} g(\tau)i(t - \tau)d\tau. \qquad \text{(B.1)}$$

The input and the impulse response are functions of time, denoted $i(\tau)$ and $g(\tau)$, where τ serves as the dummy variable of integration. If both $g(\tau)$ and $i(\tau)$ are realizable, then the lower limit is zero and the upper limit is t.

The integral is solved by performing the following steps (Figure B.3):

Step 1, folding: The $i(\tau)$ waveform is folded about the origin by replacing τ with $-\tau$.

Step 2, shifting: The folded function is shifted to the right by t. At a specific time t_1, $(-\tau)$ is replaced with $(t_1 - \tau)$.

Step 3, multiplying: The folded, shifted input is multiplied by the impulse response.

Step 4, integrating: The area under the product curve is determined by integrating the product curve between the limits defined by the origin and t_1 in the realizable case. These limits result from the fact that the product curve is zero to the left of the origin, where $g(\tau)$ is zero, and to the right of t_1, where $i(t_1 - \tau)$ is zero.

The area under the product curve gives the output amplitude at the specific time t_1. At another time t_2, steps 2 through 4 must be repeated. Thus, the area under the product curve, which is the output amplitude, is determined as a function of t.

RELATION BETWEEN FILTERING IN TIME AND FREQUENCY DOMAINS

Filtering can be described in either the time or the frequency domain. In the time domain, the output waveform is obtained by the convolution of the input with the impulse response, which is written in shorthand notation, $o(t) = i(t) * g(t)$. In the frequency domain, the output spectrum is obtained by multiplying the input spectrum $I(f)$ with the steady-state filter response $G(f)$. The two equivalent statements of filtering are as follows:

$$\begin{aligned} i(t) * g(t) &= o(t), \\ I(f) \cdot G(f) &= O(f). \end{aligned} \tag{B.2}$$

The foregoing result, that convolution in the one domain (time in this case) is equivalent to multiplication in the other domain (frequency in this case), is a general property of Fourier analysis.

COMMUTATIVE PROPERTY OF CONVOLUTION

Convolution is commutative; that is,

$$i(t) * g(t) = g(t) * i(t). \tag{B.3}$$

The property shows that it makes no difference which waveform is folded. In terms of filters, this means that it does not matter which waveform is considered the input and which is considered the impulse response.

TYPES OF LINEAR SYSTEMS

PHYSICALLY REALIZABLE SYSTEMS

A physically realizable system has two necessary and sufficient conditions on its impulse response $g(t)$. Assuming that an impulse is applied at time zero, the conditions are:

(1) $g(t) = 0$ for all $t < 0$.

(2) $\int_{-\infty}^{+\infty} |g(\tau)|^2 d\tau < \infty$.

In other words, the system cannot respond before the impulse is applied, and the impulse response must have finite energy. The latter condition requires that the impulse response eventually die out; that is, $g(t) \to 0$ as $t \to \infty$. This property is called stability. An unstable system has impulse response that diverges as $t \to \infty$, and a conditionally stable system has a nonzero bound on its impulse response as $t \to \infty$.

MINIMUM-PHASE SYSTEMS

A subclass of physically realizable systems is the class of minimum-phase systems. Of all the physically realizable systems with a given amplitude spectrum, there exists one system whose phase lag is minimum at all frequencies. This unique system is called the minimum-phase system for the given amplitude spectrum. In phase space, the minimum-phase response forms the boundary between the realizable and nonrealizable systems that have a given amplitude response.

Minimum-phase systems have the following characteristics:

(1) The logarithm of the amplitude spectrum and the minimum-phase spectrum are Hilbert transforms of one another; knowing the amplitude response, the minimum-phase response can be calculated.

(2) Minimum-phase systems have the fastest accumulative energy buildup of their class, where the accumulative energy is defined by

$$E(t) = \int_0^t |g(\tau)|^2 d\tau. \tag{B.4}$$

(3) Minimum-phase systems are the only systems with stable inverses.

ZERO-PHASE SYSTEMS

Zero-phase systems have real, even Fourier transforms, which implies that their impulse responses are real, even time functions. Such systems cannot be realizable because their impulse responses are not zero for $t < 0$. This characteristic prevents using zero-phase filtering in real-time processing unless one is willing to allow the output to be delayed by at least half the length of the symmetrical impulse response. In processing captured data, realizability is of no consequence, because $t = 0$ is a parameter that can be chosen.

ANALOG AND DIGITAL FILTERS

Analog filters built from circuit elements—resistors, capacitors, and inductors—are minimum-phase, physically realizable systems. The amplitude and phase characteristics are interdependent; hence, given the amplitude spectrum, the minimum-phase spectrum is defined uniquely. Digital filters are not constrained by realizability or minimum-phase characteristics. Their amplitude and phase spectra can be chosen independently. Zero-phase filters are commonly used in digital processing of seismic data, because changes in the amplitude spectrum of such filters do not shift the peaks and troughs in the data nor produce character changes in the reflection wavelets.

Terms used with filter responses include the following:

(1) *Decibel:* A logarithmic measure of output-to-input amplitude ratio, defined by n dB $= 20 \log_{10} (A_{out}/A_{in})$.
(2) *Cutoff frequency:* Frequency at which the amplitude response drops to 3 dB below the plateau value (equivalent to 0.707 of the plateau value), also called the half-power point.
(3) *Octave:* A logarithmic measure of a ratio of frequencies, defined by m octaves $= \log_2(f_2/f_1)$.
(4) *Bandwidth:* With a bandpass filter, the bandwidth is measured in octaves, where f_2 and f_1 are the upper and lower cutoff frequencies, respectively.
(5) *Rejection rate:* Slope of the amplitude response outside the passband, often expressed in decibels per octave.
(6) *Time shift:* Related to phase shift by the relation $T = \theta/2\pi f$.

MINIMUM-PHASE FILTERS

The recording filters used in seismic exploration are usually minimum-phase filters of the type called Butterworth filters. These filters have the following characteristics:

(1) Of all minimum-phase filters, they have maximal flatness in the passband.
(2) The rejection rates are $6n$ dB per octave, where n is the order of the Butterworth filter.
(3) They have linear phase shift in the passband.

Butterworth filters have well-known analytical expressions for their impulse responses, and digital filters can be produced that have Butterworth characteristics. Figure B.4 shows the impulse response of a third-order bandpass Butterworth filter, with the bandwidth 25–50 Hz. A third-order filter has rejection rate of 18 dB/octave at both the high and low sides of the passband. The amplitude spectrum and the phase spectrum are shown in part (b) of the figure.

Impulse responses of a variety of bandpass digital Butterworth filters are shown in succeeding figures. Filters with the same bandwidth in octaves and the same rejection rates have impulse responses with the same shape but with different time scales, as shown in Figure B.5 for third-order bandpass filters whose bandwidths are 15–30 Hz, 20–40 Hz, and 30–60 Hz, respectively. The time scale is compressed as the passband shifts toward higher frequency in accordance with the Fourier scale change theorem. The impulse response becomes more leggy as the bandwidth narrows, as shown in Figure B.6 for three third-order Butterworth filters having bandwidths of two octaves, one octave, and one-half octave, respectively. Also, for a fixed bandwidth, the impulse response becomes more leggy as the rejection rates increase, as shown in Figure B.7 for 25–50 Hz Butterworth filters of third, sixth, and ninth order (that is, rejection rates of 18 dB/octave, 36 dB/octave, and 54 dB/octave, respectively).

Another minimum-phase filter, called the Chebyshev filter, is characterized by having equal ripple in the passband. In Figure B.8 (from Van Valkenburg, 1960), the amplitude response of a Chebyshev filter is compared with the response of a Butterworth filter of the same order under two conditions on a parameter that can be adjusted in the Chebyshev filter. In part (a), the parameter is set to give maximum ripple in the Chebyshev's passband. In this case, the Chebyshev response and the Butterworth response have the same value at the half-power point (the 0.707 amplitude point); in the rejection range, the Chebyshev response lies below the Butterworth response for all f greater than the frequency of the half-power point. In part (b) of the figure, the parameter is such that the ripple is less pronounced than in part (a); in this case, the Chebyshev filter has less rejection in the reject band than does the Butterworth. This shows that the parameter in a Chebyshev filter that controls the ripple in the passband also controls the response in the reject band. The rejection rate of a Chebyshev filter for any value of the parameter is $6n$ dB/octave, which is the same as the rejection rate of the Butterworth filter. In the foregoing examples, the filters are low-pass. Similar comments apply to bandpass filters.

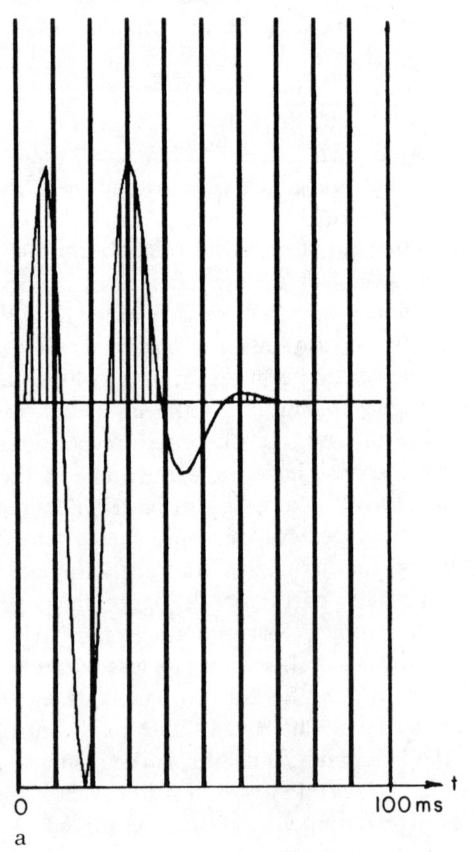

a

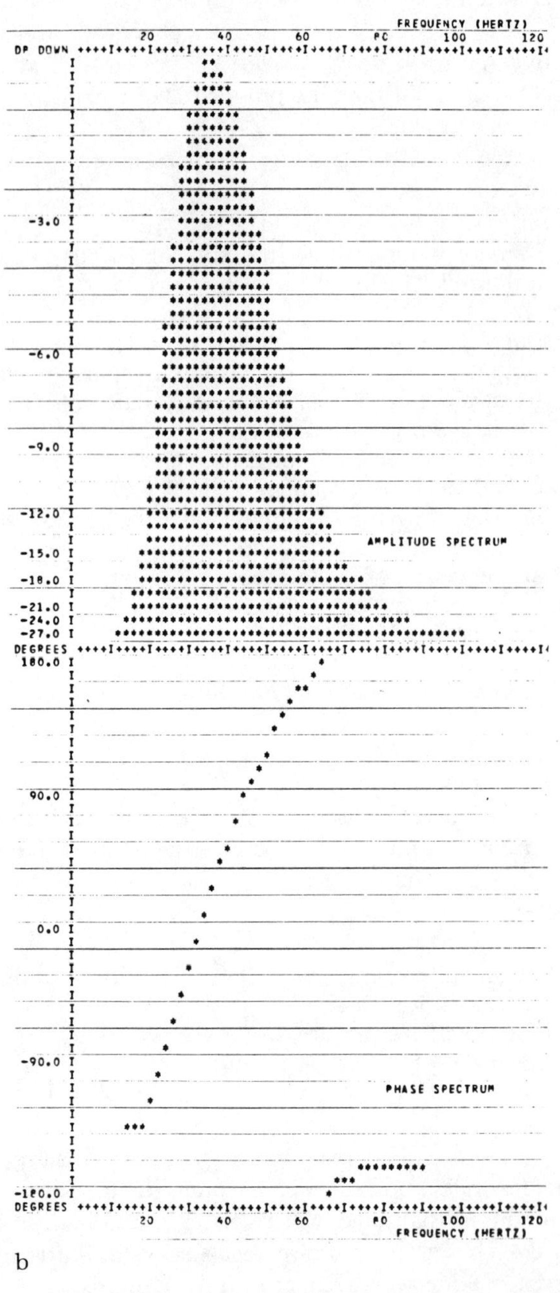

b

FIGURE B.4. *Characteristics of a third-order Butterworth filter whose bandwidth is 25–50 Hz. (a) Impulse response. (b) Fourier transform.*

15-30 Hz
20-40 Hz
30-60 Hz
t
0
100
200 ms

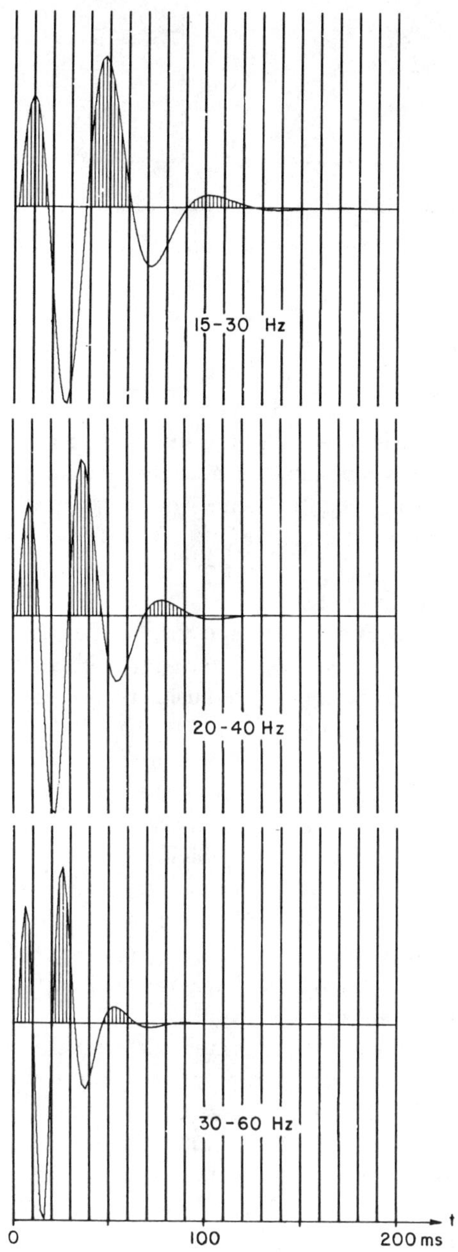

FIGURE B.5. *Impulse responses of third-order Butterworth filters that have one-octave bandwidths.*

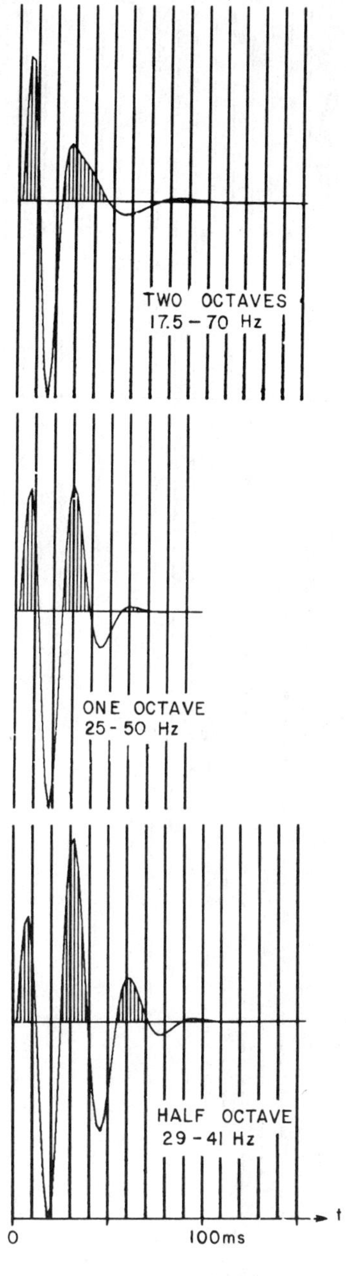

TWO OCTAVES
17.5 – 70 Hz
ONE OCTAVE
25 - 50 Hz
HALF OCTAVE
29 - 41 Hz
t
0
100ms

FIGURE B.6. *The effect of bandwidth on the impulse response of third-order Butterworth filters.*

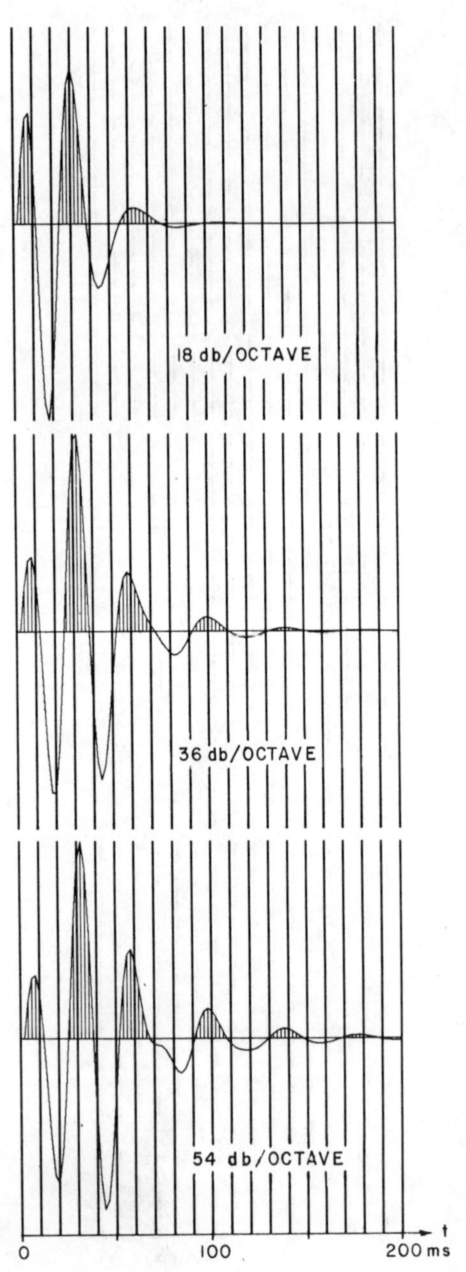

FIGURE B.7. *The effect of rejection rates on the impulse response of Butterworth filters.*

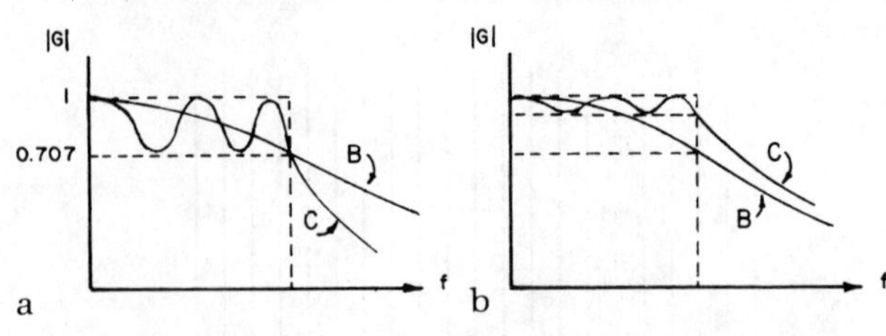

FIGURE B.8. *Comparison of the amplitude spectra of low-pass Butterworth and Chebyshev filters.*

ZERO-PHASE FILTERS

Zero-phase filters have real Fourier transforms; that is, their imaginary parts are zero. Consequently, their impulse responses are even time functions. Such filters are not physically realizable because their impulse responses are not zero in negative time. They respond to impulses before the impulses are applied. This presents no problem when processing captured data, because the time origin is a parameter.

Brickwall (Figure B.9) and trapezoidal (Figure B.10) filters belong to the class of zero-phase filters whose real transform can be constructed using rectangles, triangles, and constants (Sengbush, 1960). These filters are also described by Ormsby (1961) and are often called Ormsby filters. Their impulse responses are linear combinations of sinc functions, sinc-squared functions, and impulses. The impulse response of a high-pass filter can be obtained by subtracting the impulse response of a low-pass filter from an impulse at the origin. Similarly, the impulse response of a band-reject filter can be obtained by subtracting the impulse response of a bandpass filter from an impulse at the origin.

Brickwall filters have rectangular spectra and hence have infinite rejection rate at the cutoff frequencies. Trapezoidal filters have linear rejection rates. Because of the differences in rejection rates, the impulse responses of trapezoidal filters decay more rapidly with time than the responses from brickwall filters (Figure B.11). The impulse response becomes more leggy as the rejection rate increases and as the bandwidth decreases, which is characteristic of filters of any type.

Truncating the impulse response to produce a finite-length operator causes the spectrum to deviate from its

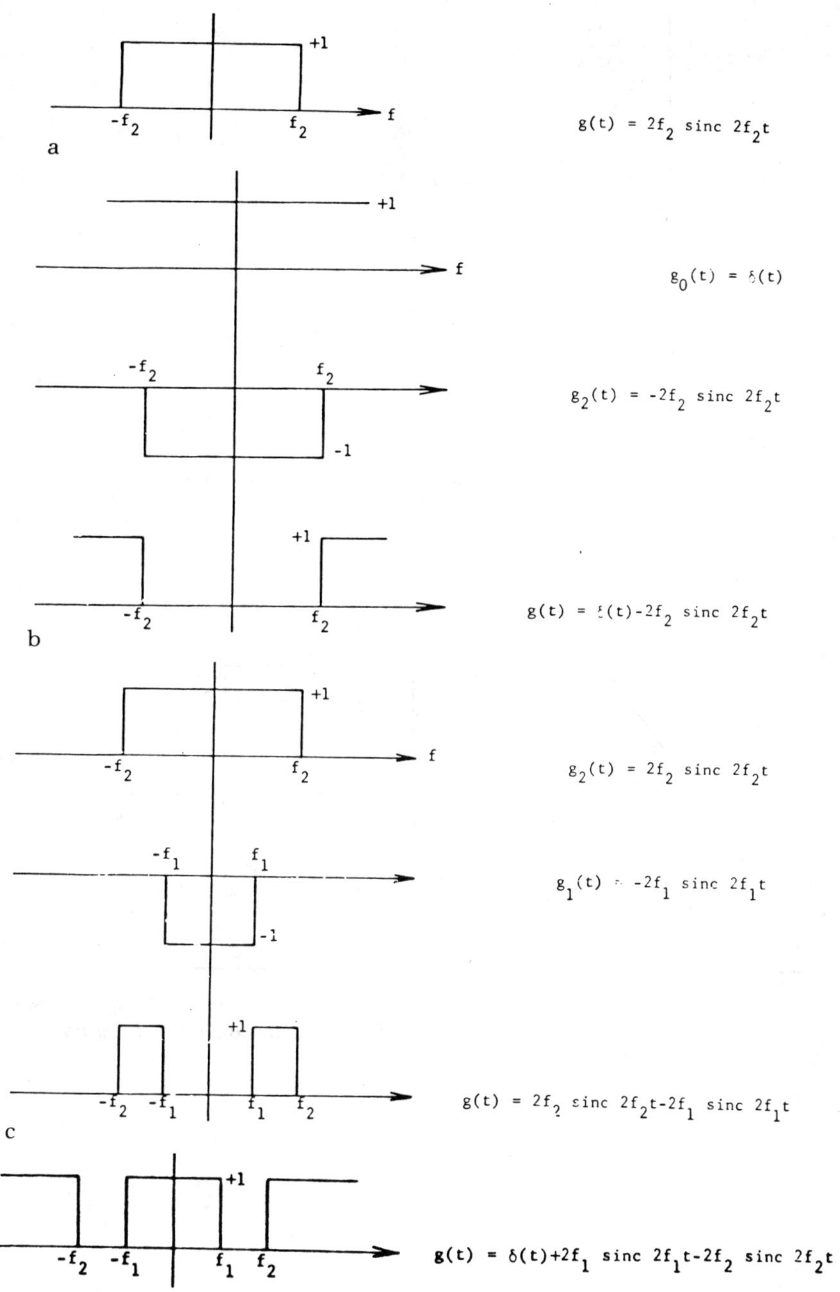

FIGURE B.9. *Spectra and impulse responses of ideal brickwall filters. (a) Low-pass. (b) High-pass. (c) Bandpass. (d) Band-reject. (From Sengbush 1960.)*

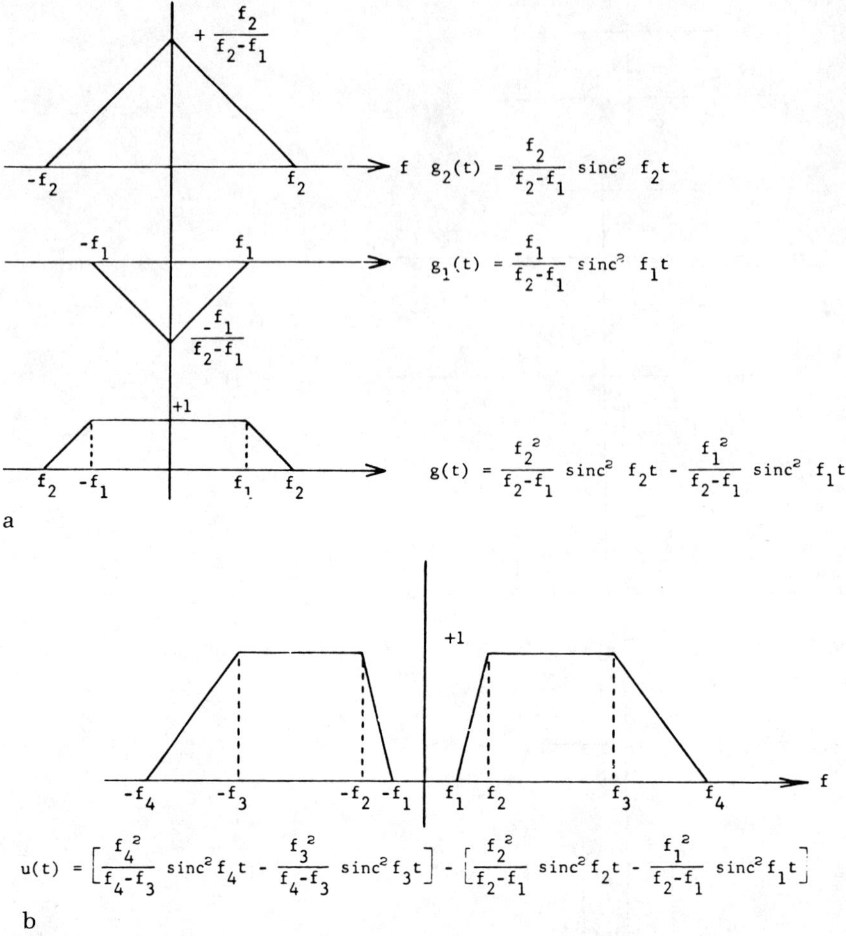

FIGURE B.10. *Spectra and impulse responses of ideal trapezoidal filters. (a) Low-pass. (b) Bandpass. (From Sengbush 1960.)*

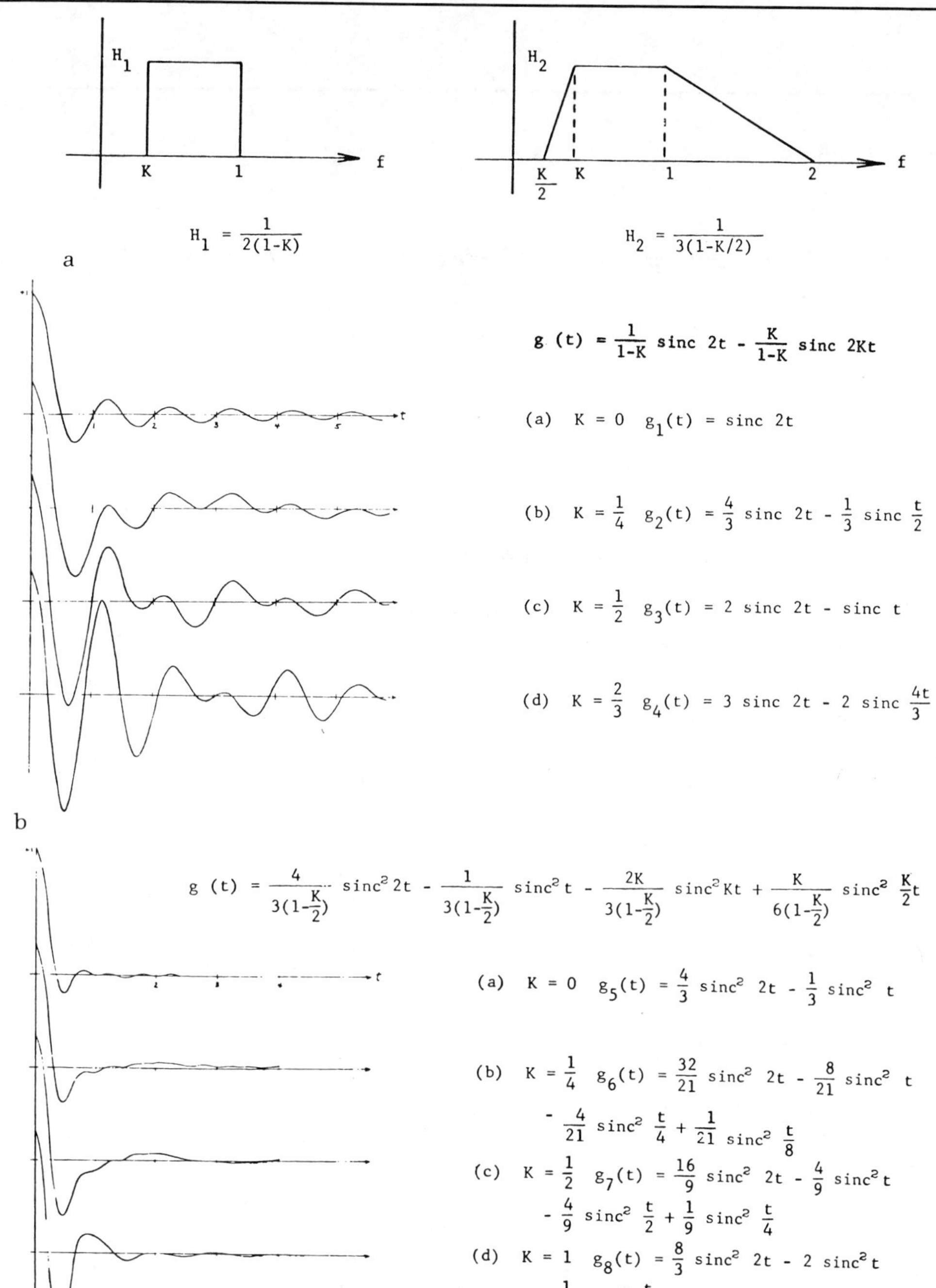

FIGURE B.11. *Impulse response waveforms from ideal filters. (a) Spectra of ideal filters. (b) Impulse responses of ideal brickwall filters. (c) Impulse responses of ideal trapezoidal filters. (From Sengbush 1960.)*

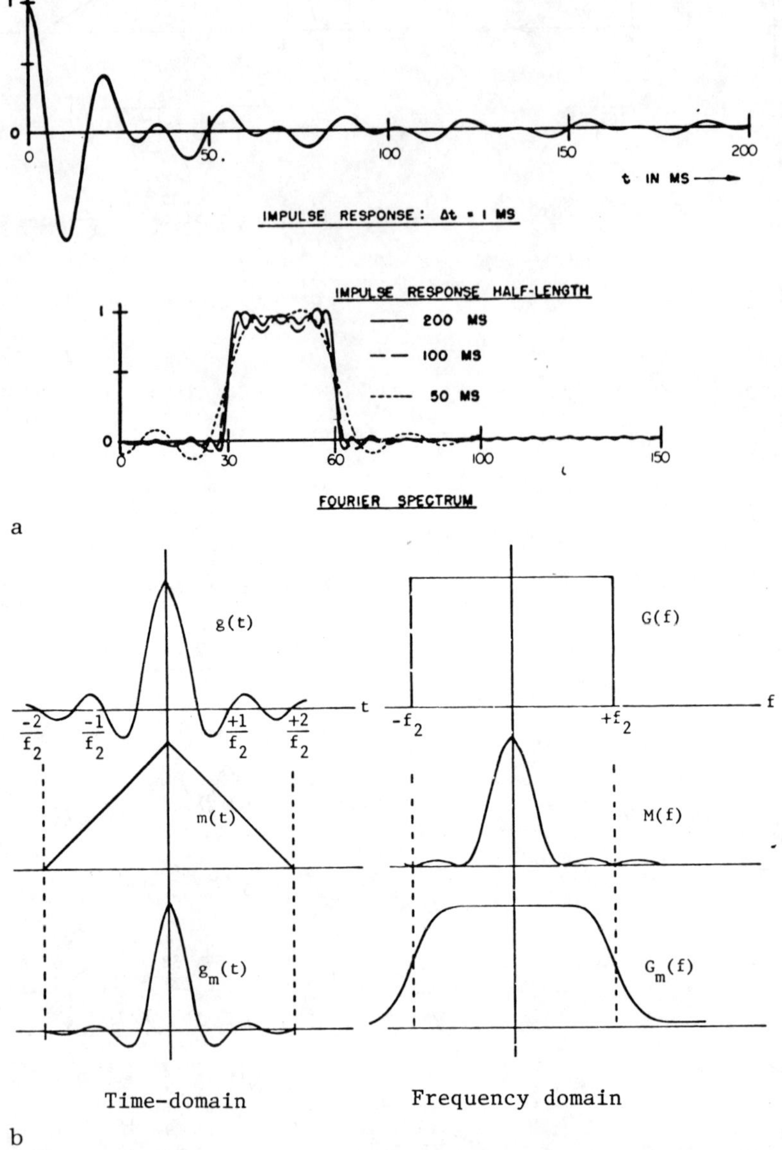

FIGURE B.12. *Effect of truncator length and truncator type on brickwall spectrum. (a) Effect of rectangular truncator length. (b) Effect of triangular truncator. (From Sengbush 1960.)*

ideal rectangular or triangular construction (Figure B.12). Increasing the truncator length improves the approximation to the ideal spectrum. Ripples in the spectrum caused by rectangular truncation of the impulse response can be suppressed by using other even functions, such as triangular, Gaussian, Hamming, or Hanning functions, as truncators. These functions, defined in the range $-T,T$, where T is the truncator length, and zero outside that range are as follows:

$$\begin{aligned} &\text{Triangular: } 1 - t/T \\ &\text{Gaussian: } \exp(-kt^2/T) \\ &\text{Hamming: } 0.54 + 0.46 \cos \pi t/T \\ &\text{Hanning: } 0.5 + 0.5 \cos \pi t/T \end{aligned} \tag{B.5}$$

Truncation by multiplying two time functions corresponds to convolving their spectra. Rectangular truncation causes the ideal spectrum to be convolved with a sinc spectrum. The other truncators have spectra smoother than a sinc function; hence, they produce smoother approximations to the ideal spectra and have lesser rejection rates.

Digital operators are obtained by sampling the truncated impulse responses. Brickwall filters have the following digital operators:

$$\text{Low-pass: } g_{LP}(k) = (f_2/f_N)\,\text{sinc}\,(kf_2/f_N); \tag{B.6}$$

$$\text{High-pass: } g_{HP}(k) = \begin{cases} 1 - (f_2/f_N), & k = 0; \\ -g_{LP}(k), & k \neq 0; \end{cases} \tag{B.7}$$

$$\text{Bandpass: } g_{BP}(k) = (f_2/f_N)\,\text{sinc}\,(kf_2/f_N) - (f_1/f_N)\,\text{sinc}\,(kf_1/f_N); \tag{B.8}$$

$$\text{Band-reject: } g_{BR}(k) = \begin{cases} 1 - (f_2 - f_1)/f_N, & k = 0; \\ -g_{BP}(k), & k \neq 0. \end{cases} \tag{B.9}$$

Brickwall band-reject filters have large sample values at the time origin compared to the other sample values (Figure B.13). The ratio is 6.35:1 with a 20–54 Hz filter and 249:1 with a 59–61 Hz notch filter. The impulse response of the notch filter is a very slowly decaying sinusoid with frequency 60 Hz. The suppression at 60 Hz is 51 dB.

Brickwall filters contain most of their energy in the passband, as compared to Butterworth filters, as shown in Table B.1.

TABLE B.1. *Comparison of brickwall and Butterworth filters*

Filter type	Low-pass, 0–60 Hz	Bandpass, 30–60 Hz
Brickwall		
T = 100 ms	—	96.1%
T = 200 ms	99.5%	98.6
T = 300 ms	—	99.0
Butterworth		
18 dB/octave	85.9	65.4
36 dB/octave	95.6	83.6
54 dB/octave	98.3	91.4

RECURSIVE FILTERS

Systems whose Z-transform transfer functions are rational functions of z have a negative feedback implementation that is computationally efficient, as compared to polynomial expansion of the denominator polynomial followed by convolution (Shanks, 1967). Suppose that the transfer function $G(z)$ is the ratio of polynomials in z,

$$G(z) = A(z)/B(z), \tag{B.10}$$

where

$$A(z) = \sum_{k=0}^{M} a_k z^k,$$

$$B(z) = \sum_{k=0}^{N} b_k z^k.$$

The first term b_0 in $B(z)$ can always be normalized to unity. Now let $X(z)$ and $Y(z)$ be the Z-transforms of the input and output time series; then,

$$Y(z) = G(z)X(z) \tag{B.11}$$

$$= [A(z)/B(z)]X(z),$$

$$B(z)Y(z) = A(z)X(z). \tag{B.12}$$

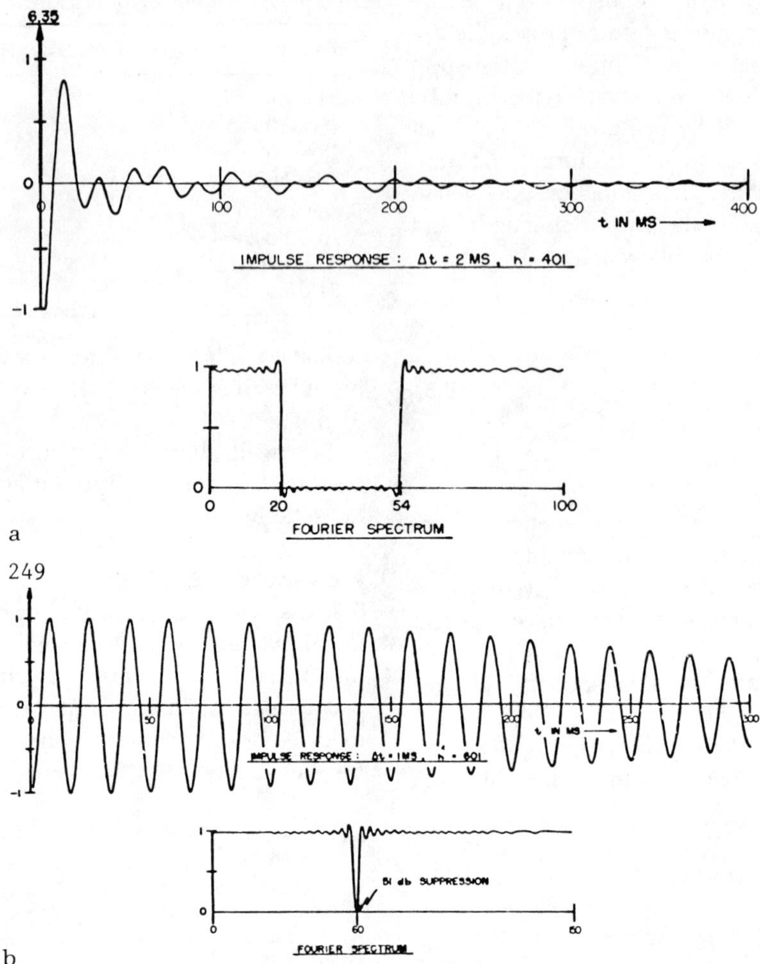

FIGURE B.13. *Digital band-reject filters. (a) Reject band 20–54 Hz. (b) Notch filter, reject band 59–61 Hz. (From Sengbush 1960.)*

Now, by writing $B(z)$ as a polynomial in z and rearranging terms, the function $B'(z)$ can be defined as follows:

$$
\begin{aligned}
B(z) &= 1 + b_1 z + b_2 z^2 + b_3 z^3 + \dots \\
&= 1 + z(b_1 + b_2 z + b_3 z^2 + \dots) \qquad \text{(B.13)} \\
&= 1 + zB'(z).
\end{aligned}
$$

Substituting for $B(z)$ in equation (B.12) gives

$$
\begin{aligned}
(1 + zB')Y &= AX, \\
Y &= AX - zB'Y. \qquad \text{(B.14)}
\end{aligned}
$$

Equation (B.14) is the negative feedback implementation of the rational transfer function $G(z)$ operating on the input $X(z)$ to produce the output $Y(z)$. The feedback block diagram is shown in Figure B.14. In the feedback loop, the operator z delays the output time series by one sampling interval, and the delayed output is convolved with the time series b', given by

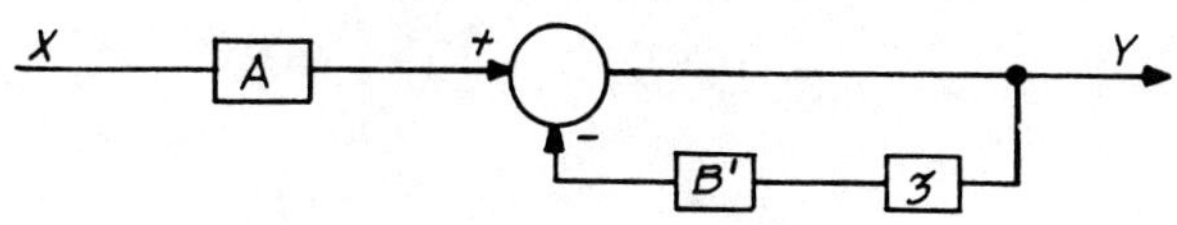

FIGURE B.14. *Block diagram for recursive filter.*

$$b' = \sum_{k=1}^{N} b_k \, \delta[t - (k - 1)\Delta]. \tag{B.15}$$

The recursive filtering relation is then given by

$$y_l = \sum_{k=0}^{M} a_k x_{l-k} - \sum_{k=1}^{N} b_k y_{l-k}. \tag{B.16}$$

Appendix C Sampling Theory

INTRODUCTION

Digital data recording and processing have revolutionized seismic exploration over the past decade. The analog techniques that preceded digital methods have several significant shortcomings:

(1) Amplitude is imprecisely known at various stages in the processing because of imprecise knowledge of amplitude characteristics of such devices as programmed gain control. Also, the response of an analog system changes because of aging components, replacement circuit elements, and the like. (The impulse response of an analog device is slowly time-variant in an unknown way.)
(2) Frequency content of data is degraded in passing through the various analog devices. (The impulse response of an analog device is not an impulse.)
(3) It is difficult or impossible to build analog instruments that will perform many desired and useful mathematical operations.
(4) Analog data has limited dynamic range compared to that of digital data, whose dynamic range is limited only by word length.

Digital methods are capable of overcoming these shortcomings. By observing the cardinal rule of digitizing—that is, capture the data in digital form at the earliest possible stage with binary gain instruments—the continuous analog signals are replaced as early as possible with a discrete-time series of samples with precisely known amplitudes. From that point on, it is possible to keep track of the amplitudes and to preserve the frequency content. Important mathematical processes that are impossible or impractical with analog methods but are readily performed digitally include zero-phase filtering, correlation, deconvolution, velocity analysis, and migration.

The foundation for digital processing lies in the simple, elegant time-domain sampling theorem of Professor Claude Shannon (1948) of M.I.T., in which he shows that functions with limited-frequency bandwidth can be sampled at a rate inversely proportional to the upper bandlimit without loss of information.

It is interesting to reflect on the furor that surrounded the conversion of analog-recorded seismic data to discrete samples in the early days of the digital revolution. Some thought it couldn't be done without losing something of value in the data, never once considering that, in the space domain, the seismic data have always been sampled discretely. There are no continuous seismic detectors blanketing the earth.

This appendix discusses sampling in the time and frequency domains, the consequence of improper sampling, and some simple mathematical operations with time series, such as convolution, correlation, and Fourier analysis.

TIME-DOMAIN SAMPLING

In order to process continuous data in a digital computer, it is first necessary to sample the data. The question arises whether or not the sampling can be done in such a way that the samples contain all the information contained in the continuous data. If the samples do contain all the information, then the continuous data can be reconstructed from the samples. In other words, the original continuous function can be selected out of the infinite set of continuous functions that pass through the sampling points.

Shannon's time-domain sampling theorem states the conditions that must be met in order to sample without loss of information: A continuous function $g(t)$ whose spectral content is zero at and above the frequency f_0 can be sampled at intervals of $\Delta \leq 1/2f_0$ without loss of information.

A function that has these spectral characteristics is said to be bandlimited to f_0.

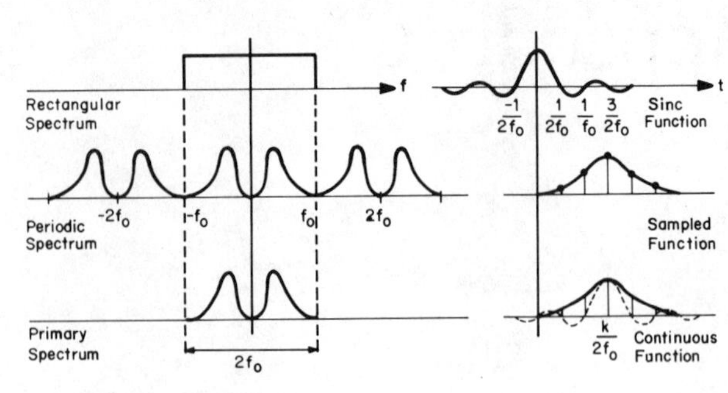

FIGURE C.1. *Time-domain sampling theorem.*

Now consider a continuous function $g(t)$ whose spectrum $G(f)$ is bandlimited to f_0. Sampling this function in accordance with the sampling theorem, $\Delta = 1/2f_0$, produces a set of weighted and delayed delta functions,

$$g_s(t) = \Delta \sum_k g(k\Delta)\delta(t - k\Delta). \tag{C.1}$$

The weight of the kth delta function is the product of the sampled value $g(k\Delta)$ and the sampling interval Δ.

The Fourier transform of $g_s(t)$ is given by

$$\mathcal{F}[g_s(t)] \triangleq G_s(f) = \Delta \sum_k g(k\Delta)e^{-j\omega k\Delta}. \tag{C.2}$$

Because the exponential $e^{-j\omega k\Delta}$ is periodic with period $f = 1/\Delta$, the spectrum of the sampled function is periodic with the same period. This is analogous to periodic time functions with period T that have spectral samples at multiples of $1/T$. These statements are in accordance with a basic property of Fourier transforms—that sampling in one domain corresponds to periodic functions in the other domain.

Because the sampling theorem is satisfied, no information about either $g(t)$ or its spectrum $G(f)$ is lost as a result of sampling. Multiplying $G_s(f)$ by a rectangular spectrum $H(f)$ that is unity within the frequency range $\pm f_0$ and zero outside produces the nonperiodic spectrum $G(f)$, as shown in Figure C.1.

Multiplication in the frequency domain corresponds to convolution in the time domain, giving the following pair of equations:

$$G(f) = G_s(f) \cdot H(f),$$
$$g(t) = g_s(t) * h(t). \qquad (C.3)$$

The time function $h(t)$ corresponding to $H(f)$ is the sinc function $h(t) = 2f_0 \operatorname{sinc} 2f_0t$. Substituting the sampled time function $g_s(t)$ and the sinc function $h(t)$ into the convolution gives

$$\begin{aligned} g(t) &= [\Delta \sum_k g(k\Delta)\delta(t - k\Delta)] * [2f_0 \operatorname{sinc} 2f_0 t] \\ &= \sum_k g(k\Delta)[\delta(t - k\Delta) * \operatorname{sinc} 2f_0 t]. \end{aligned} \qquad (C.4)$$

Convolving a weighted and delayed impulse with a sinc function produces a weighted and delayed sinc function:

$$\begin{aligned} g(k\Delta)\delta(t - k\Delta) * \operatorname{sinc} 2f_0t &= g(k\Delta) \operatorname{sinc} 2f_0(t - k\Delta) \\ &= g(k\Delta) \operatorname{sinc} (2f_0t - k). \end{aligned} \qquad (C.5)$$

Hence, the continuous function is given by

$$g(t) = \sum_k g(k\Delta) \operatorname{sinc} (2f_0t - k). \qquad (C.6)$$

Reconstruction of a continuous time function from its time samples is accomplished by replacing each sample with a sinc function of the same magnitude and then adding the set of sinc functions. The sinc function associated with the kth sample, $g(k\Delta) \operatorname{sinc} (2f_0t - k)$ has value $g(k\Delta)$ at $t = k\Delta$ and is zero at all other sampling points. Thus, it contributes nothing at the other sampling points. Conversely, all other sinc functions contribute nothing at the sample point $k\Delta$. The sum of all weighted and delayed sinc functions uniquely determines the function $g(t)$ between the sample points.

ALIASING: SAMPLING IS NOT FILTERING

Consider, now, the consequence of improper (too-coarse) sampling of a time function. The unique time function obtained from the samples when they are interpolated with sinc functions gives the correct time function only if the sampling interval is less than or equal to $1/2f_0$. Figure C.2 shows a function that is improperly sampled and the resulting function generated by sinc

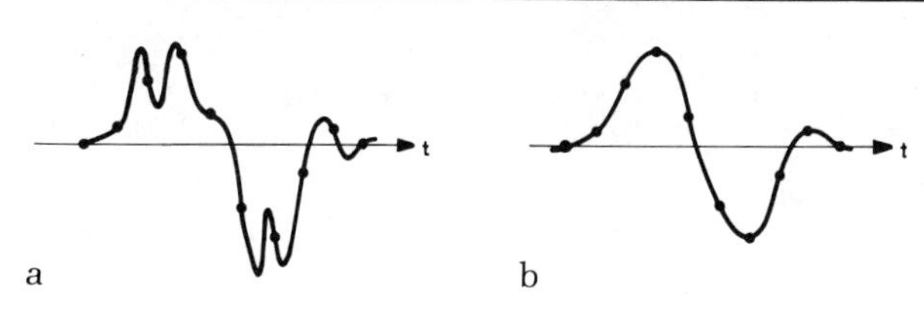

FIGURE C.2. *Effect of aliasing on time function. (a) Improper sampling of* g(t). *(b) Waveform obtained by sinc interpolation.*

interpolation. The latter is smoother than the original function, indicating loss of information about the high frequencies. It appears that the high frequencies are simply filtered out of the waveform, but this is not correct. The spectral content of the high frequencies is not lost, as it would be if sampling were filtering. Instead, the high-frequency spectral content is folded back and distorts the low-frequency spectral content.

Sampling in time at intervals of Δ results in a periodic spectrum with spectral period $1/\Delta$. The primary spectrum lying between $\pm f_N$, where $f_N = 1/2\Delta$, is repeated along the frequency axis. In effect, the frequency axis is folded like an accordion, with pleats at integer multiples of the folding (or Nyquist) frequency f_N.

The folding frequency depends solely on the sampling interval and is completely independent of the spectral content of the waveform being sampled. It is a general characteristic that samples at intervals of $\Delta = 1/2f_N$ cannot distinguish a frequency $f < f_N$ from frequency components $2kf_N \pm f$, for k any integer. For example, with $\Delta = 5$ ms, the samples cannot distinguish a frequency below $f_N = 100$ Hz, such as $f = 80$ Hz, from frequencies above f_N given by $200k \pm 80$ Hz for all k (Figure C.3).

To show folding of the spectral content into the primary passband below f_N, suppose that the continuous function $g(t)$ consists of the sum of sinusoids with frequencies $2kf_N \pm f$:

$$g(t) = A_0 \cos 2\pi ft + \sum_{k=1}^{\infty} A_k \cos 2\pi(2kf_N - f)t + \sum_{k=1}^{\infty} B_k \cos 2\pi(2kf_N + f)t. \qquad (C.7)$$

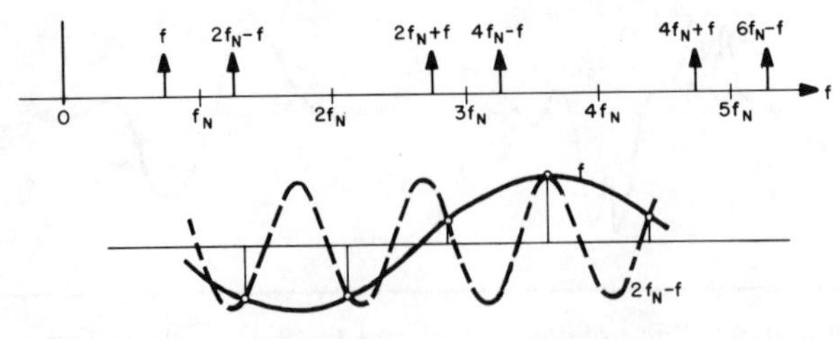

FIGURE C.3. *Frequency components indistinguishable from samples.*

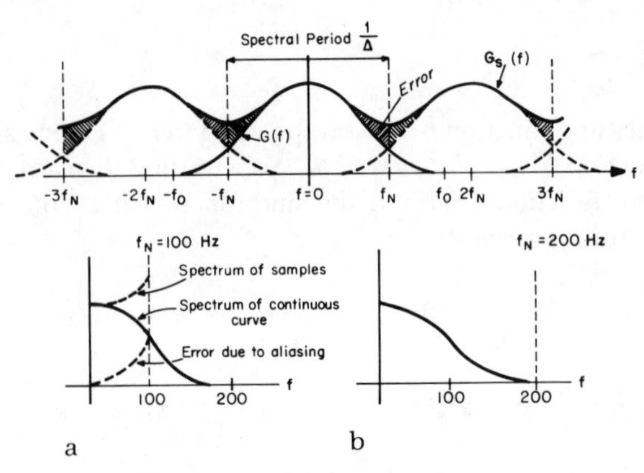

FIGURE C.4. *Aliasing in frequency domain caused by too-coarse sampling. (a) Samples at Δ = 5 ms. (b) Samples at Δ = 2.5 ms.*

Sampling $g(t)$ at $t = l\Delta$ gives the sampled function g_s whose lth sample is

$$g_s(l\Delta) = A_0 \cos 2\pi f l\Delta + \sum_k A_k \cos 2\pi(2kf_N - f)l\Delta + \sum_k B_k \cos 2\pi(2kf_N + f)l\Delta$$

$$= A_0 \cos \frac{\pi f l}{f_N} + \sum_k A_k \cos\left(2\pi lk - \frac{\pi f l}{f_N}\right) + \sum_k B_k \cos\left(2\pi lk + \frac{\pi f l}{f_N}\right). \quad \text{(C.8)}$$

Let $x = 2\pi lk$ and $y = \pi lf/f_N$, which, upon substituting into the trigonometric relation $\cos(x \pm y) = \cos x \cos y \pm \sin x \sin y$, gives $\cos x = 1$ and $\sin x = 0$ for all l and k. Therefore, the expression for $g_s(l\Delta)$ becomes

$$g_s(l\Delta) = \left[A_0 = \sum_k (A_k + B_k)\right] \cos \frac{\pi f l}{f_N}. \quad \text{(C.9)}$$

A possible conclusion from the samples is that $g(t)$ is a single sinusoid of frequency $f < f_N$ with amplitude $A_0 + \Sigma_k (A_k + B_k)$. From the samples, there is no way to unscramble this to determine how much is contributed by each sinusoidal component in $g(t)$. Therefore, because the samples do not uniquely determine the spectrum of the continuous function $g(t)$, the samples cannot specify $g(t)$. In other words, the correct continuous waveform cannot be obtained from the samples by sinc (or any other) interpolation because the samples do not contain enough information. This scrambling of the spectral content because of folding is called aliasing.

In sampling, the spectral content of frequencies above f_N is aliased (or folded back) into the primary spectral range between $\pm f_N$. Figure C.4 shows the distortion of the spectrum due to aliasing caused by the sampling interval being greater than $1/2f_0$. The spectral content between f_N and f_0 is folded back into the primary range between $\pm f_N$. The spectral error due to improper sampling is indicated by the cross-hatched area.

The only way to prevent aliasing is to make $f_N \geqslant f_0$ by sampling at intervals of $\Delta \leqslant 1/2f_0$. Then the spectral content is zero in the frequency range above f_N that is folded back into the primary spectral range. Thus, the samples contain the correct spectral information about the continuous function $g(t)$, and the continuous waveform can be obtained from the samples using sinc interpolation.

In digital processing, the data must be bandlimited before sampling in order to avoid aliasing. The usual practice involves passing the continuous data through an analog filter that suppresses the energy above f_N before sampling the now bandlimited data at $\Delta = 1/2f_N$. The bandlimiting filter is often called an anti-aliasing filter.

The minimum number of samples needed to characterize a bandlimited time function of length T is given by $T/\Delta = 2f_0T$. For example, seismic data bandlimited to 125 Hz can be sampled at intervals of 4 ms without loss of information. If the data are 6.0 sec long, then 1500 samples are required to characterize each trace.

FREQUENCY-DOMAIN SAMPLING

The discrete-line spectrum of a periodic function representing its Fourier series can be obtained by sampling the continuous spectrum of the corresponding transient at intervals of $\Delta f = 1/T$. The essence of the frequency-domain sampling theorem is that these spectral samples uniquely define the continuous spectrum from which they came. In other words, the continuous spectrum of a transient is defined uniquely by the line spectrum of its corresponding periodic function. Consequently, there is no additional information in the spectral values at frequencies intermediate to the sampling points.

The relationship between the Fourier series representation for a periodic function and the Fourier transform of its corresponding transient (one cycle of the periodic function) is shown by means of the frequency-domain sampling theorem:

If $G(f)$ is the Fourier spectrum of $g(t)$, where $g(t)$ is zero outside the finite time interval T, then $G(f)$ can be sampled at intervals of $\Delta f = 1/T$ without loss of information.

A function that is zero outside a finite time interval T is said to be time-limited to T. A periodic time function with period T is defined by

$$p(t) = \sum_{n=-\infty}^{+\infty} g(t - nT). \tag{C.10}$$

The corresponding transient time-limited to the interval T can be obtained by multiplying $p(t)$ by a unit rectangular function $h(t)$ that is zero outside the interval T. The periodic function and its corresponding transient, along with the real part of their spectra, are shown in Figure C.5.

The Fourier transform of a periodic function is given by

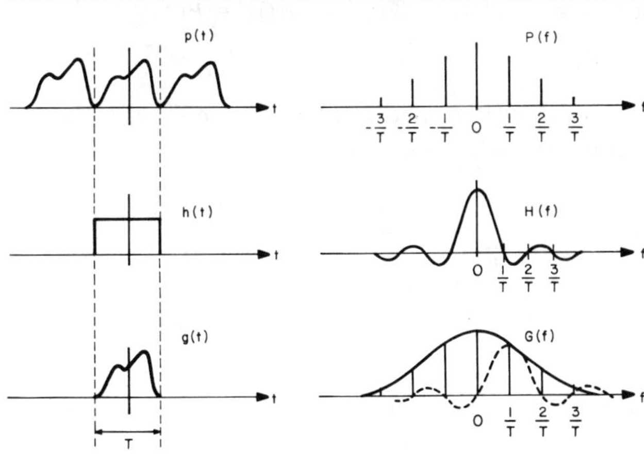

FIGURE C.5. *Frequency-domain sampling theorem.*

$$P(f) = \frac{1}{T} \sum_{n=-\infty}^{+\infty} \left[\alpha\left(\frac{n}{T}\right) + j\beta\left(\frac{n}{T}\right)\right] \delta\left(f - \frac{n}{T}\right), \tag{C.11}$$

where

$$\alpha\left(\frac{n}{T}\right) = \int_{-T/2}^{T/2} p(t) \cos \frac{2\pi nt}{T} \, dt,$$

$$\beta\left(\frac{n}{T}\right) = -\int_{-T/2}^{T/2} p(t) \sin \frac{2\pi nt}{T} \, dt.$$

The α and β coefficients are Fourier series coefficients that have the dimension of amplitude per fundamental frequency. Multiplying them by the fundamental frequency $1/T$ gives amplitude coefficients. The fundamental frequency is equal to the frequency-domain sampling interval; hence, the amplitude coefficients $(1/T)\alpha(n/T)$ and $(1/T)\beta(n/T)$ are weights of impulses. Thus, the line spectrum of a periodic time function is actually a set of complex impulses in the frequency domain.

Multiplication in the time domain becomes convolution in the frequency domain; therefore, the transform of the transient $g(t)$ is given by

$$G(f) = \mathscr{F}[g(t)] = \mathscr{F}[p(t) \cdot h(t)] = P(f) * H(f). \tag{C.12}$$

Substituting $P(f)$ and $H(f) = T \operatorname{sinc} fT$ into the frequency-domain convolution gives

$$\begin{aligned} G(f) &= \frac{1}{T} \sum_n \left[\alpha\left(\frac{n}{T}\right) + j\beta\left(\frac{n}{T}\right) \right] \delta\left(f - \frac{n}{T}\right) * T \operatorname{sinc} fT \\ &= \sum_n \left[\alpha\left(\frac{n}{T}\right) + j\beta\left(\frac{n}{T}\right) \right] \operatorname{sinc}\,(fT - n). \end{aligned} \tag{C.13}$$

The real and imaginary parts of the Fourier spectrum of the transient are given by

$$\begin{aligned} \operatorname{Re} G(f) &= \sum_n \alpha\left(\frac{n}{T}\right) \operatorname{sinc}\,(fT - n), \\ \operatorname{Im} G(f) &= \sum_n \beta\left(\frac{n}{T}\right) \operatorname{sinc}\,(fT - n). \end{aligned} \tag{C.14}$$

The spectrum of $p(t)$ consists of lines (samples) at intervals of $f = 1/T$, while the spectrum of the rectangular pulse is a sinc function that has zeros at all integral multiples of $1/T$. The spectrum of the transient is a continuous function of frequency formed by replacing each line with a sinc function of the same magnitude and then summing this set of sinc functions.

Consider the line $f_k = k/T$. The sinc function that replaces it has the same value as the line at f_k and has a value of zero at all other frequencies that are multiples of $1/T$. Hence, in the summation, this particular sinc function contributes nothing at the other sampling frequencies. Conversely, all other sinc functions contribute nothing at $f_k = k/T$. As a consequence, the spectrum at the sampling points uniquely determines the continuous spectrum by sinc interpolation.

The line spectrum specifies the periodic function throughout all time from $-\infty$ to $+\infty$, and also specifies the corresponding transient within T because the two functions are identical within that interval. Consequently, all the spectral values intermediate to the sampling points force the transient to zero outside the interval T and contribute nothing to the transient waveform because it lies wholly within T.

TIME-LIMITED AND BANDLIMITED FUNCTIONS

Time-limitedness and bandlimitedness are mutually exclusive properties in the strict mathematical sense. A bandlimited function must extend over an infinite time interval, and the spectrum of a time-limited function must extend over an infinite frequency range. However, it is possible to have a time-limited function whose energy content is almost completely contained within a given spectral band (or, conversely, a bandlimited function whose energy content is almost completely contained within a given time interval). Then $n = 2f_0T$ samples will specify the time function (or its spectrum) within an arbitrarily small error. Each such signal can then be represented as a point in n-dimensional space. This is a useful concept in information theory.

MATHEMATICAL OPERATIONS WITH TIME SERIES

CONVOLUTION

Consider the input $i(t)$ and impulse response $g(t)$ to be continuous functions that are sampled at intervals of Δ in accordance with the sampling theorem to give two time series, $\{i_0, i_1, i_2, \ldots\}$ and $\{g_0, g_1, g_2, \ldots\}$. The numbers represent the weights of impulses spaced in time by Δ. They are proportional to the amplitudes of the continuous functions at the sampling points.

The concise statement of convolution, using sampled data, is

$$O_j = \sum_{k=0}^{j} i_{j-k} g_k. \tag{C.15}$$

The index j represents output signal time. The index k is the summation index. For example, substituting $j = 3$ (corresponding to a time of 3Δ) into equation (C.15) gives

$$O_3 = \sum_{k=0}^{3} i_{3-k} g_k = i_3g_0 + i_2g_1 + i_1g_2 + i_0g_3. \tag{C.16}$$

Convolution of time series $\{i\}$ and $\{g\}$ corresponds to multiplication of their Z-transforms. A procedure for calculating the output Z-transform is shown in Table C.1. The input samples $\{i_0, i_1, i_2, \ldots\}$ and the impulse response samples $\{g_0, g_1, g_2, \ldots\}$ are listed on successive lines. Multiplying these as in regular multiplication, but

TABLE C.1. *Convolution of* $\{i\}$ and $\{g\}$

	$\frac{g}{i}$	$\frac{g}{i}$	$\frac{g}{i}$	$\frac{g}{i}$	$\frac{g}{i}$	....
	i_0g_0	i_0g_1	i_0g_2	i_0g_3	i_0g_4	....
	—	i_1g_0	i_1g_1	i_1g_2	i_1g_3	....
	—	—	i_2g_0	i_2g_1	i_2g_2	....
	—	—	—	i_3g_0	i_3g_1	....
	—	—	—	—	i_4g_0	....
Output	O_0	O_1	O_2	O_3	O_4	....

without carries, and summing the products in each column gives the coefficients of the output Z-transform.

CORRELATION

The cross correlation of time series $\{i\}$ and $\{g\}$ is given by the summation

$$\phi_k = \sum_{j=-\infty}^{+\infty} i_j g_{j-k}. \tag{C.17}$$

The index k represents correlation time, and the summation index j is the time index of the time series.

The autocorrelation of $\{g\}$ is given by

$$\phi_k = \sum_{j=-\infty}^{+\infty} g_j g_{j-k}. \tag{C.18}$$

The autocorrelation of $\{g\}$ does not specify $\{g\}$ uniquely. Consider the autocorrelation time series example from Robinson (1966):

$$\phi_0 = 17,\ \phi_1 = 0,\ \phi_2 = -4. \tag{C.19}$$

Each of the following three-point time series has the foregoing autocorrelation:

$$\begin{aligned} g_a &= \{4, 0, -1\}, \\ g_b &= \{2, 3, -2\}, \\ g_c &= \{-2, 3, 2\}, \\ g_d &= \{-1, 0, 4\}. \end{aligned} \tag{C.20}$$

From Fourier transform theory, the amplitude spectrum is invariant under folding and shifting. Therefore, it should be no surprise that g_a and g_d have the same autocorrelation function, and likewise for g_b and g_c. These time series differ in their cumulative energy buildup:

$$\begin{aligned} E_a &= \{16, 16, 17\}, \\ E_b &= \{\ 4, 13, 17\}, \\ E_c &= \{\ 4, 13, 17\}, \\ E_d &= \{\ 1,\ \ 1, 17\}. \end{aligned} \tag{C.21}$$

Series g_a has the quickest energy buildup in the set of time series that have the same autocorrelation function; hence, it is the minimum-delay series of the set. Series g_d has the slowest energy buildup; hence, it is the maximum-delay series of the set. All other series are said to have mixed delay.

A time series with n elements and $n - 1$ corresponding zeros has 2^{n-1} sequences with the same autocorrelation function, because the amplitude spectrum associated with a zero located at either z_0 or its conjugate reciprocal location $(1/z_0)^*$ is identical. The minimum-delay series from the set of all series that have a given autocorrelation is the one that has the quickest energy buildup. It is the only one of the set that has all of its zeros outside the unit circle; hence, it is the only one with a stable inverse.

To find the zeros of a time series, form its polynomial in z and set it equal to zero. In the case of g_a, the result is $G_a(z) = 4 - z^2 = 0$, so its zeros are located at $(2, -2)$. Since all are outside the unit circle, this is a minimum-delay series. The maximum-delay series has its zeros at the conjugate reciprocal locations of the zeros of the minimum-delay series, which, in the case of real zeros, are simply the reciprocal locations. Hence, the maximum-delay equivalent of g_a has its zeros at $(\frac{1}{2}, -\frac{1}{2})$. This places all its zeros inside the unit circle. The others have mixed delay, with the zeros of g_b at $(2, -\frac{1}{2})$ and those of g_c at $(\frac{1}{2}, -2)$.

The inverses of the foregoing time series, obtained by long division of the reciprocals of the polynomials in z, are

$$g_a{}^{-1} = \{1/4, 0, 1/16, 0, 1/64, \ldots\},$$
$$g_b{}^{-1} = \{1/2, -3/4, 13/8, \ldots\},$$
$$g_c{}^{-1} = \{-1/2, -3/4, -13/8, \ldots\}, \qquad \text{(C.22)}$$
$$g_d{}^{-1} = \{-1, 0, -4, 0, -16, \ldots\}.$$

Notice that only g_a has a stable inverse; hence, it is a minimum-delay time series.

When a time series has a finite number of points, its inverse is of infinite length and the product $G(z) \cdot G^{-1}(z) = 1$, which is exactly the desired output. Thus, $G^{-1}(z)$ is the exact inverse of $G(z)$. When the inverse is truncated, there is a residual error between the aforementioned product and the desired output. Consider the case of g_a with its inverse truncated to the same length as g_a. Then,

$$(4 - z^2)\left(\frac{1}{4} + \frac{1}{16} z^2\right) = 1 - \frac{1}{16} z^4. \qquad \text{(C.23)}$$

The rms error between the product and the desired output is 1/16. As the length of the inverse increases, the rms error decreases, provided that the inverse is stable—that is, provided that the original time series is minimum-delay.

Figure C.6 shows the waveform of a zero-phase Ricker wavelet (Ricker, 1953) and its minimum- and maximum-phase equivalents. The minimum-phase equivalent has its energy packed at the front end of the wavelet, and the maximum-phase equivalent, being its mirror image, has its energy packed at the tail end. As the wavelets have 99 points, these are but three of the $2^{98} = 3.169 * 10^{29}$ wavelets with the same autocorrelation function.

The concept of minimum delay in the discrete case (time series) is equivalent to the concept of minimum-phase functions in the continuous case. In 1945, Bode showed that, of the set of all realizable continuous functions that have a given autocorrelation function (hence, a given energy spectrum and, hence, a given amplitude spectrum), there is one function from the set whose phase lag is less at all frequencies than the phase lags of all other functions in the set.

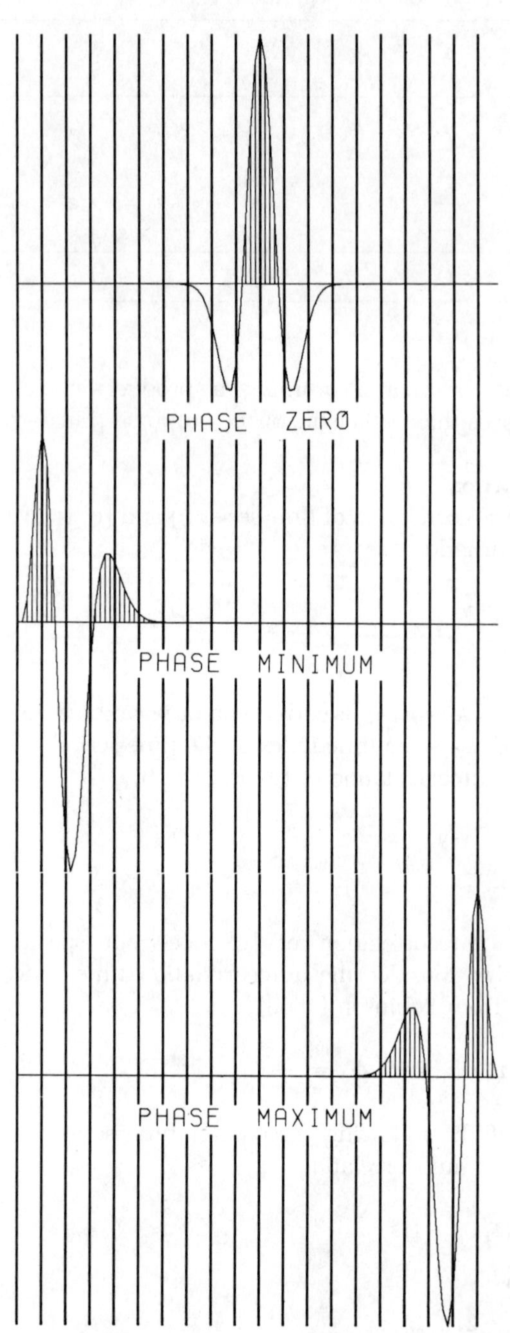

FIGURE C.6. *A zero-phase Ricker wavelet and its minimum- and maximum-phase equivalents.*

FOURIER ANALYSIS

The method of calculating the Fourier spectrum of a function given in Appendix A is useful when the waveform can be described analytically. However, with experimental data such as seismic traces, it is not possible to write down an analytical expression for the data. In this case, the procedure for calculating the spectrum is modified in light of the sampling theorems. The theorem in the frequency domain shows that the continuous spectrum of a transient of finite length T is completely specified by spectral samples taken at intervals of $\Delta f = 1/T$. If the transient is bandlimited to f_0, then it can be sampled at time intervals of $\frac{1}{2}f_0$ without loss of information. Therefore, the spectrum of an essentially bandlimited function of finite length can be obtained from its time samples by replacing the Fourier integrals for $\alpha(f)$ and $\beta\ (f)$ by the following summations:

$$\alpha_0 = \Delta t \sum_{k=1}^{n} u_k, \tag{C.24}$$

$$\alpha_m = \Delta t \sum_{k=1}^{n} u_k \cos \frac{2\pi mk}{n}, \tag{C.25}$$

$$\beta_m = \Delta t \sum_{k=1}^{n} u_k \sin \frac{2\pi mk}{n}, \tag{C.26}$$

where $m = 0, 1, 2, \ldots, (n - 1)/2$. This gives the same number of frequency samples as time samples.

Consider the wavelet of length $T = 100$ ms, with band limit $f_0 = 125$ Hz, in Figure C.7(a). The time-domain sampling theorem is satisfied with $\Delta T = 1/2f_0 = 4$ ms, and the frequency-domain sampling theorem is satisfied with $\Delta f = 1/T = 10$ Hz. Therefore, the Fourier spectrum is specified by the coefficients for all harmonics of 10 Hz out to the folding frequency of 125 Hz; that is, α_0, $\alpha_1, \ldots, \alpha_{12}, \beta_1, \ldots, \beta_{12}$. Note that 25 samples in either domain specifies the function.

Because the resulting amplitude spectrum is small as the folding frequency is approached, it is apparent that the time-domain sampling theorem has been satisfied, as it is in this example, shown in Figure C.7(b). The wavelet synthesized from the harmonics is periodic, with period $T = 100$ ms, because the fundamental frequency is $\Delta f = 10$ Hz.

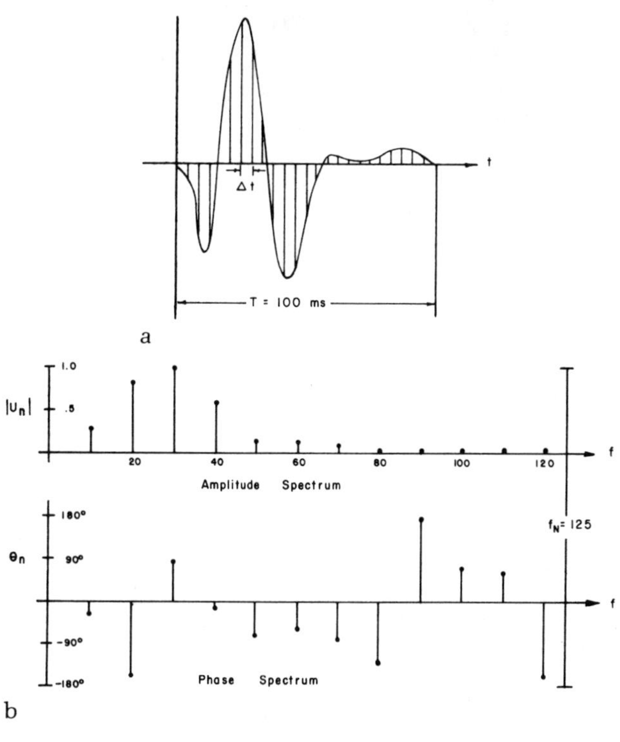

FIGURE C.7. *A seismic source pulse and its spectrum. (a) Source pulse. (b) Fourier spectrum of source pulse.*

In this spectral computation, the maximum possible frequency-sampling interval is used, and intermediate values can be obtained through use of sinc interpolation. However, it is often desirable in spectral analysis to use a finer frequency-domain sampling interval to avoid such interpolation. This can be accomplished by augmenting the time series by adding zeros at the end of the original time series, thereby increasing its length from T to T', thus decreasing the frequency-domain sampling interval Δf from $1/T$ to $1/T'$. Of course, with a digital computer, one merely specifies Δf to some small value, such as 1 Hz, and calls a subroutine to calculate the Fourier transform.

Appendix D Random Processes

INTRODUCTION

In the real world and in mathematical models of the real world, two classes of processes are clearly distinguishable: deterministic and random processes. Deterministic processes are controlled by well-known mathematical laws, and their behavior is known exactly for all time. With random processes, no simple laws are available to describe their behavior exactly, because of either the complexity or the incomplete understanding of the underlying causes of the processes, or both. Hence, future fluctuations of these processes cannot be precisely predicted.

A random process consists of an ensemble (collection) of random functions of a parameter, such as time or distance. Random functions extend forever into the past and are expected to continue forever into the future, and their behavior is known only through statistical properties. The two most important statistical properties are the arithmetic mean (average) and the correlation function. Correlation functions form a link between the process and its spectrum.

Examples of random processes include natural ground unrest in seismic prospecting and thermal noise in electronic instruments. The most significant contribution of random processes to seismic analysis, however, is the recent concept that the earth's reflectivity is a random process.

THEORY OF PROBABILITY

CLASSICAL DEFINITION

Consider a random experiment such as the throwing of a die. The outcome of each trial is independent of the results of other trials. The probability of a particular outcome—such as the occurrence of a six, written $p(6)$—is defined by

$$p(6) = \lim_{n\to\infty} \frac{m(6)}{n}, \tag{D.1}$$

where $m(6)$ is the number of times a six occurs and n is the total number of times the die is thrown. The ratio $m(6)/n$ is called the *frequency ratio* of the occurrence of the event six.

To be more general, if A is the event, the classical probability of A (due to Laplace) is given by

$$p(A) = \lim_{n\to\infty} \frac{m(A)}{n}. \tag{D.2}$$

This definition of probability has two major defects:

(1) Mathematical: This definition is not particularly useful in defining random variables, probability distributions, and the like. Also, the precise limit properties of sums or averages cannot be determined, because no mathematical basis exists for such calculations. Furthermore, it is not useful in defining a random process.
(2) Physical: No experiment can be performed indefinitely. In fact, some destructive experiments can be performed only once. Also, the conditions of the experiment may be time-variant. Finally, successive events are not always independent.

Even though the classical definition is not a suitable basis for extension of probability theory, however, a large number of independent trials will result in a value close to $p(A)$. In fact, Bernoulli's theorem proves that the probability that the frequency ratio (r) differs from the probability of the event (p) by less than a given positive member ϵ, however small, approaches unity as the number of trials approaches infinity:

$$\lim_{n\to\infty} P\{|r - p| < \epsilon\} = 1. \tag{D.3}$$

This theorem shows that the probability that r lies in the ϵ-neighborhood of p is dependent on the number of trials, and, as $n \to \infty$, r lies in the ϵ-neighborhood with probability one.

To overcome the shortcomings of the classical definition, Kolmogorov (1950 translation) formalized the axiomatic structure of probability theory based on the abstract integration theory of Lebesque. Each axiom has a counterpart in integration theory; in addition, each reflects properties that correspond to intuitive ideas about probability.

In the axiomatic structure, the classical definition appears as a result; it does not constitute the basis of probability theory. A meaningful definition of random variable, distribution function, random process, and other probability concepts can be given in terms of the axiomatic structure.

SET THEORY

The axiomatic structure is expressed in the algebra of sets; therefore, a brief review is given here.

(1) Set: a collection of points. The notation $x \in A$ means that point x belongs to set A.
(2) Union (of two sets): the collection of points belonging to either set or both, written $A \cup B$.
(3) Intersection (of two sets): the collection of points belonging to both sets, written $A \cap B$.
(4) Disjoint: having no points in common.
(5) Null set, ϕ: the set containing no points.
(6) Difference: the collection of points in A but not in B, written $A - B$.
(7) Subset: if all points in A are contained in B, written $A \subset B$; then A is a subset of B.
(8) Reference set, Ω: the union of all sets that appear in a given discussion.
(9) Complement, A^*: The collection of points in Ω but not in A.
(10) Class of sets: a collection of sets.
(11) Field, $\mathscr{F}$: A particular class of sets satisfying the following axioms:
   (a) $\Omega \in \mathscr{F}$
   (b) If $A \in \mathscr{F}$, then $A^* \in \mathscr{F}$
   (c) If $A_n \in \mathscr{F}$, where $n = 1, 2, \ldots$, then

$$\bigcup_{n=1}^{\infty} A_n \in \mathcal{F}$$

(12) Point function: to each point $x \in A$, a given real number $f(x)$ is a point function whose domain is A. As x ranges over A, the set of all values of $f(x)$ is called its range. In the usual notation of function, A is the set of real numbers.

(13) Set function: To each set $A_k \in \mathcal{S}$ ($\mathcal{S}$ is a class of sets), a given real number $F(A_k)$ is a set function whose domain is $\mathcal{S}$. For example, let $\mathcal{S}$ be the real line and let A_k be an interval on the real line. Then $F(A_k) = \| A_k \|$, where $\| A_k \|$ is the length of the interval, is a set function.

AXIOMATIC STRUCTURE OF THEORY OF PROBABILITY

The axiomatic structure is based on a probability space $(\Omega, \mathcal{F}, P)$.

(1) Ω is a reference measure space containing the collection of points ω.
(2) $\mathcal{F}$ is a field—a distinguished family of measurable subsets of Ω satisfying the axioms given in the preceding list.
(3) P is a set function, called the probability measure (or probability), satisfying the following axioms:
 (a) P is defined on the sets of $\mathcal{F}$.
 (b) P is a nonnegative function, $P(A) \geqslant 0$.
 (c) $P(\Omega) = 1$.
 (d) If A_k, $k = 1, 2, \ldots$, are disjoint, then

$$P\left(\bigcup_{k=1} A_k\right) = \sum_{k=1} P(A_k). \tag{D.4}$$

RANDOM VARIABLES

A random variable is a measurable point function $X(\omega)$ defined on Ω. It is measurable if and only if, for each real number x, the set of all ω that satisfy $\{X(\omega) < x\}$ constitutes a set A_x of $\mathcal{F}$. A_x is a measurable set.

PROBABILITY DISTRIBUTION AND DENSITY FUNCTIONS

The probability that $\{X(\omega) < x\}$ is called the *probability distribution function*, $q(x)$,

$$q(x) = P\{X(\omega) < x\}. \tag{D.5}$$

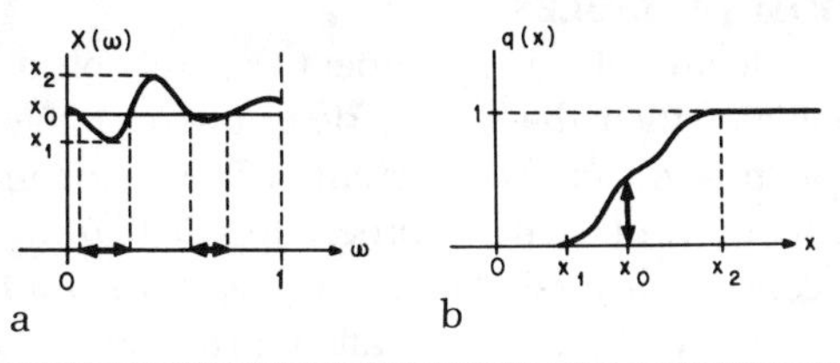

FIGURE D.1. *A random variable and its distribution function. (a) Random variable. (b) Distribution function.*

Let Ω be the real line from zero to one, inclusive. Then $q(x)$ is given by the sum of all intervals of Ω corresponding to $\{X(\omega) < x\}$. Figure D.1 shows a random variable and its distribution function.

It is intuitive that $q(-\infty) = 0$ and $q(\infty) = 1$, and that $q(x)$ is a nondecreasing monotonic function. If a random variable is continuous, $q(x)$ is continuous, and its derivative exists at all points (except in certain pathological cases). Its derivative is called the *probability density function*, $p(x)$,

$$p(x) = \frac{dq(x)}{dx}, \qquad q(x) = \int_{-\infty}^{x} p(x)dx. \tag{D.6}$$

If the random variable is discrete, $q(x)$ will consist of ascending stair-steps. By using impulses (and ignoring the pathological cases where the continuous part of $q(x)$ does not have a derivative), discrete random variables can be incorporated into the same formalism as continuous random variables:

$$\begin{aligned} p(x) &= \sum_k p(x_k)\delta(x - x_k), \\ q(x) &= \sum_k p(x_k)U_s(x - x_k). \end{aligned} \tag{D.7}$$

A mixture of continuous and discrete random variables will result in step-discontinuities in $q(x)$; that is, $q(x)$ will be piecewise continuous.

RANDOM VARIABLES

If the outcome of an experiment depends on a chance mechanism, then the proper description of the experiment is in terms of the different possible outcomes and the probability of each outcome. Suppose that the experiment consists of tossing a coin whose sides are marked -1 and 1. The outcome is called a random variable X, which may take on either of these values with probability of ½. (The probabilities are assigned by the experimenter, based on the state of his knowledge. Here the coin is considered to be fair.) The possible values of X form a discrete set of values, $x = -1$ and $x = 1$, and X is therefore called a discrete random variable. A discrete random variable is completely described by a listing of its possible values $\{x_1, x_2, \ldots\}$ and the associated probabilities $\{p(x_1), p(x_2), \ldots\}$. Each of the probabilities must lie in the range [0,1], and the sum of all probabilities must equal 1; that is,

$$0 \leqslant p(x_i) \leqslant 1 \quad \text{for} \quad i = 1, 2, \ldots, \qquad \text{(D.8)}$$
$$\sum p(x_i) = 1.$$

As a second example of a random variable, consider a roulette wheel that has a uniform scale from 0 to A marked on the circumference. The experiment consists of spinning the wheel and observing the value under the pointer when it comes to rest. The outcome is a random variable that may have any value between 0 and A with equal probability. In this case, X is called a continuous random variable. It is completely described by the probability density of the random variable, $p(x)$. In the roulette wheel example,

$$p(x) = \begin{cases} 1/A & 0 \leqslant x \leqslant A \\ 0 & \text{elsewhere} \end{cases} \qquad \text{(D.9)}$$

Probability density functions are nonnegative normalized functions,

$$p(x) \geqslant 0 \quad \text{for all } x, \qquad \text{(D.10)}$$
$$\int_{-\infty}^{+\infty} p(x)dx = 1.$$

The probability that X takes on a value in the interval $x \leqslant X \leqslant x + dx$ is defined by the incremental area $p(x)dx$.

Discrete random variables can be characterized by probability density functions consisting of impulses. In the coin-flipping experiment,

$$p(x) = \frac{\delta}{2}(x - 1) + \frac{\delta}{2}(x + 1). \qquad \text{(D.11)}$$

Consider now an ensemble of random functions. The amplitude of the ensemble at time t_1 is a random variable $X(t_1)$. The amplitude is specified by a probability density function. Assume, for example, that the amplitude may be any value between ± 1 with equal probability; then $X(t)$ is a continuous random variable described completely by the uniform probability density function,

$$p(x) = \begin{cases} ½ & |x| \leqslant 1 \\ 0 & |x| > 1 \end{cases} \qquad \text{(D.12)}$$

The probability density function that is most widely encountered in nature is the Gaussian, or normal, density function, defined by

$$p(x) = \frac{1}{\sqrt{2\pi\sigma^2}} \exp\left[\frac{-(x - m)^2}{2\sigma^2}\right]. \qquad \text{(D.13)}$$

The Gaussian function is a bell-shaped curve that is symmetrical about its mean, $x = m$; as its variance σ decreases, its width decreases and its peak value increases (Figure D.2). If a random process is Gaussian, the amplitude at t_1 is a Gaussian random variable $X(t_1)$. The amplitude may take on any value, but the probability of extreme values is very small. With $\sigma = 1$ and $m = 0$, for example, the probability that the amplitude lies in the range ± 3 is 0.9973.

A random process is a set of random variables $X(t_1)$, $X(t_2)$, $X(t_3)$, . . . that is completely described by a set of probability density functions. The simplest density function describes the ensemble at one point in time; for example, the random variable $X_1 = X(t_1)$ is described by $p(x_1)$. This is called a first-order density function. The next most complicated density function describes the statistical properties of pairs of random variables; for ex-

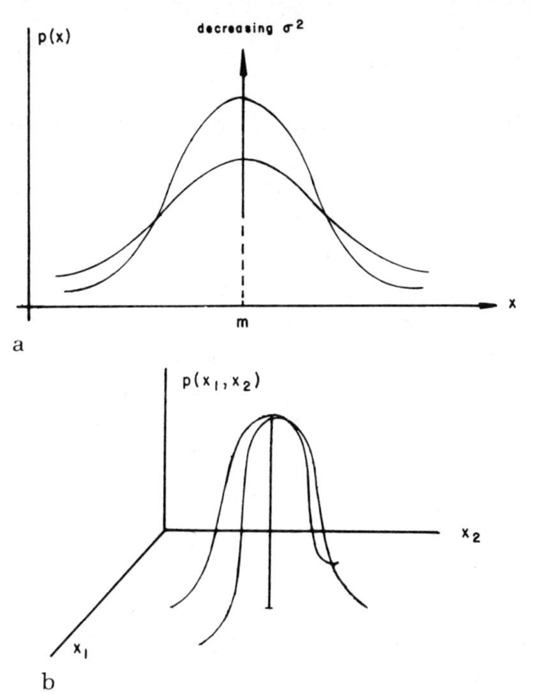

FIGURE D.2. *Gaussian probability density functions. (a) First order. (b) Second order.*

ample, the pair X_1, X_2 is described by the second-order probability density function $p(x_1, x_2)$. This function is a surface that lies above the (x_1, x_2) plane. The properties of $p(x_1, x_2)$ are that it is nonnegative and that its volume is equal to unity:

$$p(x_1, x_2) \geqslant 0 \quad \text{for all } x_1, x_2,$$
$$\int_{-\infty}^{+\infty} \int_{-\infty}^{+\infty} p(x_1, x_2)\, dx_1\, dx_2 = 1. \tag{D.14}$$

The joint probability that the amplitude X_1 lies in the interval $x_1 \leqslant X_1 \leqslant x_1 + dx_1$ and the amplitude X_2 lies in the interval $x_2 \leqslant X_2 \leqslant x_2 + dx_2$ is defined by the incremental volume $p(x_1, x_2)\, dx_1\, dx_2$.

Higher-order probability density functions describe the joint statistical properties of three, four, and so forth, random variables.

A stationary random process is defined as one whose statistical properties are independent of the time origin. Consequently, the first-order density function is independent of time; the second-order density function depends on time difference $t_1 - t_2$, rather than on the individual times t_1 and t_2; the third-order density function depends on two time differences, and so on. A process whereby the first- and second-order density functions are independent of the time origin is called a wide-sense stationary process.

ENSEMBLE AVERAGES

The simplest ensemble average of a random variable is its arithmetic mean, m_1. If X is a discrete random variable, then each of its possible values is weighted in accordance with its probability of occurrence, and the mean is the sum given by

$$m_1 = \sum_i x_i p(x_i). \tag{D.15}$$

In the coin-toss example, $x_1 = -1$, $x_2 = 1$, $p(x_1) = p(x_2) = ½$; therefore,

$$m_1 = (-1)½ + (1)½ = 0. \tag{D.16}$$

If X is a continuous random variable, then its mean is given by

$$m_1 = \int_{-\infty}^{+\infty} x p(x)\, dx. \tag{D.17}$$

In the roulette wheel example, where $p(x)$ was uniform between 0 and A, the mean value of X is

$$m_1 = \frac{1}{A} \int_0^A x\, dx = \frac{1}{A} \frac{x^2}{2} \Big|_0^A = \frac{A}{2}. \tag{D.18}$$

The average value of the square of a random variable X, m_2 (also called the mean square of X), is defined as follows:

$$\text{Discrete case: } m_2 = \sum_i x_i^2 p(x_i),$$
$$\text{Continuous case: } m_2 = \int_{-\infty}^{+\infty} x^2 p(x)\, dx. \tag{D.19}$$

In the two previous examples,

$$m_2 = \sum_{i=1}^{2} x_i^2 p(x_i) = (-1)^2(½) + (1)^2(½) = 1,$$

$$m_2 = \frac{1}{A}\int_0^A x^2\,dx = \frac{1}{A}\frac{x^3}{3}\bigg|_0^A = \frac{A^2}{3}. \qquad \text{(D.20)}$$

The variance of X, σ^2, is the average value of $(X - m_1)^2$. The variance measures the spread or dispersion of values of X about its mean. It is defined by

$$\text{Discrete case: } \sigma^2 = \sum_i (x_i - m_1)^2 p(x_i),$$

$$\text{Continuous case: } \sigma^2 = \int_{-\infty}^{+\infty} (x - m_1)^2 p(x)\,dx. \qquad \text{(D.21)}$$

If the mean is zero, as in the coin-toss example, then the variance equals the mean square. The variance of X in the roulette wheel example is given by

$$\sigma^2 = \int_0^A \left(x - \frac{A}{2}\right)^2 \frac{1}{A}\,dx = \frac{A^2}{12}. \qquad \text{(D.22)}$$

The two parameters m and σ^2 in the Gaussian density function are the mean and variance, respectively. As the variance decreases, the dispersion or spread of values about the mean decreases.

By expanding the integrand in the integral relationship defining σ^2, the following general relationship is found,

$$\sigma^2 = m_2 - m_1. \qquad \text{(D.23)}$$

The arithmetic mean of X is also called the expectation of X, written EX. The symbol E is an operator that operates on X. More generally, E operates on functions of a random variable $g(X)$, and the expectation of $g(X)$ is defined in the continuous case by

$$Eg(X) = \int_{-\infty}^{+\infty} g(x)p(x)\,dx. \qquad \text{(D.24)}$$

If $g(X) = X$, then $Eg(X)$ is the mean of X:

$$Eg(X) = EX = m_1. \qquad \text{(D.25)}$$

If $g(X) = X^2$, then $Eg(X)$ is the mean square of X:

$$Eg(X) = EX^2 = m_2. \qquad \text{(D.26)}$$

If $g(X) = (X - m_1)^2$, then $Eg(X)$ is the variance of X:

$$Eg(X) = E(X - m_1)^2 = \sigma^2. \qquad \text{(D.27)}$$

In the study of random processes, the two most important ensemble averages are the mean and the autocorrelation function. If the process is stationary, the mean value is independent of time and $EX(t) = m_1$ is a constant independent of t. The autocorrelation function as an ensemble average is defined as the expectation of the product $X(t)X(t - \tau)$. If the process is stationary, this average is dependent only on the time difference τ; that is,

$$EX(t)X(t - \tau) = \phi(\tau). \qquad \text{(D.28)}$$

If the random variables $X(t)$ and $X(t - \tau)$ are called X_1 and X_2, respectively, then

$$\phi(\tau) = EX_1X_2 = \int_{-\infty}^{+\infty}\int_{-\infty}^{+\infty} x_1x_2p(x_1, x_2)\,dx_1dx_2. \qquad \text{(D.29)}$$

If X_1 and X_2 have zero means, then EX_1X_2 is called a covariance function.

TIME AVERAGES

Although averages across the ensemble of random functions are fundamental to the theory of random functions, in practice, the statistical averages that are measured are time averages. If a stationary random process is ergodic, the averages with respect to time are equal to the ensemble averages, and any one of the ensemble can be chosen as representative of the process. Consider the function $x(t)$ to be a representative function. Its arithmetic mean, written $[x(t),]$, is defined by

$$[x(t)] = \lim_{T\to\infty} \frac{1}{T}\int_{T/2}^{T/2} x(t)\,dt. \qquad \text{(D.30)}$$

The mean value is the average value in the limit as the time duration approaches infinity. It is a constant.

The autocorrelation of $x(t)$ as a time average is the average value of the product $x(t)x(t - \tau)$ divided by the time duration as it approaches infinity, written $[x(t)x(t - \tau)]$ and defined by

$$[x(t)x(t-\tau)] = \lim_{T\to\infty}\frac{1}{T}\int_{-T/2}^{T/2} x(t)x(t-\tau)\,dt = \phi_{xx}(\tau). \tag{D.31}$$

The mean square value of $x(t)$ can be obtained by letting $\tau = 0$:

$$[x^2(t)] = \lim_{T\to\infty}\frac{1}{T}\int_{-T/2}^{T/2} x^2(t)\,dt = \phi_{xx}(0). \tag{D.32}$$

Autocorrelation functions have the following properties:

(1) $\phi_{xx}(0) \geqslant \phi_{xx}(\tau)$ for all τ.
(2) $\phi_{xx}(\tau) = \phi_{xx}(-\tau)$.
(3) $\phi_{xx}(\tau) \to 0$ as $\tau \to \infty$, if x does not contain a periodic component.

Similarly, the cross correlation of $x(t)$ and $y(t)$, written $[x(t)y(t - \tau)]$, is defined by

$$[x(t)y(t-\tau)] = \lim_{T\to\infty}\frac{1}{T}\int_{-T/2}^{T/2} x(t)y(t-\tau)\,dt = \phi_{xy}(\tau). \tag{D.33}$$

Cross-correlation functions have the following properties:

(1) $\phi_{xy}(\tau) = \phi_{yx}(-\tau)$.
(2) $\phi_{xy}(\tau) \to 0$ as $\tau \to \infty$, if x and y do not contain any common periodic components.

Correlations of finite-energy functions are defined similarly to the time averages of random functions, with the omission of the $1/T$ factor outside the integral, as discussed in Appendix A. Although these finite-energy correlation functions have all the properties of the time averages, they do not signify any statistical averaging.

SPECTRAL CONSIDERATIONS

Random functions of infinite duration have infinite energy; that is,

$$\lim_{T\to\infty}\int_{-T/2}^{+T/2} x^2(t)dt \to \infty.$$

Because Fourier transforms do not exist for functions with infinite energy, the spectral content of $x(t)$ cannot be determined by ordinary Fourier methods.

Wiener (1930) laid the mathematical foundation for harmonic analysis of random processes by showing that the spectral content of a random process can be derived from its autocorrelation function because it has a Fourier transform $\Phi_{xx}(f)$ given by

$$\Phi_{xx}(f) = 2\int_0^\infty \phi_{xx}(\tau)\cos 2\pi f\tau\,d\tau. \tag{D.34}$$

The cosine transform is appropriate because $\phi_{xx}(\tau)$ is an even function. The spectral function $\Phi_{xx}(f)$ is called the power spectrum of the process. It is a real, even, nonnegative function of frequency; that is,

(1) Real: Re $\Phi_{xx}(f) = \Phi_{xx}(f)$.
(2) Even: $\Phi_{xx}(f) = \Phi_{xx}(-f)$.
(3) Nonnegative: $\Phi_{xx}(f) \geqslant 0$ for all f.

The inverse transform is also a cosine transform

$$\phi_{xx}(\tau) = 2\int_0^\infty \Phi_{xx}(f)\cos 2\pi f\tau\,df. \tag{D.35}$$

The cross-correlation function $\phi_{xy}(\tau)$ has a Fourier transform $\Phi_{xy}(f)$, which is called the cross-power spectrum and is defined by

$$\Phi_{xy}(f) = \int_{-\infty}^{+\infty} \phi_{xy}(\tau)\exp(-j2\pi f\tau)\,d\tau. \tag{D.36}$$

The inverse transform is

$$\phi_{xy}(\tau) = \int_{-\infty}^{+\infty} \Phi_{xy}(f)\exp(j2\pi f\tau)\,df. \tag{D.37}$$

With the cross-correlation function, interchange of subscripts causes folding of the cross-correlation function:

$$\phi_{yx}(\tau) = \phi_{xy}(-\tau). \tag{D.38}$$

By the folding theorem for Fourier transforms,

$$\Phi_{yx}(f) = \Phi^*_{xy}(f), \tag{D.39}$$

where the asterisk indicates a complex conjugate.

EXAMPLE: WHITE RANDOM PROCESS

A white random process is defined as one whose power spectrum is constant for all frequencies. This requires its correlation function to be an impulse. Although no physical process can have this property, because it implies infinite average power in the process, it can be approached with an arbitrary degree of precision; that is, a bandlimited white process, where the spectrum is constant for frequencies within the range $\pm f_0$ and zero outside, has finite average power for all $f_0 < \infty$. In this case the autocorrelation function is a sinc function.

ESTIMATION OF POWER SPECTRUM

In any practical situation, one has only a finite length of data from which to estimate the power spectrum of the process. This topic is discussed in detail by Blackman and Tukey (1958), and the results are summarized here.

First, consider a technique for estimating the power spectrum that is intuitively satisfying but gives highly questionable estimates. A finite piece of data $x_T(t)$, defined by

$$x_T(t) = \begin{cases} x(t) & 0 \leq t \leq T \\ 0 & \text{elsewhere} \end{cases}, \tag{D.40}$$

has finite energy; hence, it has a Fourier transform:

$$X_T(f) = \int_0^T x_T(t) \exp(-j2\pi ft)\, dt. \tag{D.41}$$

The energy spectrum of $x_T(t)$ is equal to $|X_T(f)|^2$, and it seems reasonable that the power spectrum could be obtained from the relation

$$S(f) = \lim_{T\to\infty} \frac{|X_T(f)|^2}{T}. \tag{D.42}$$

This procedure, however, fails to give $S(f)$ for a large class of random processes, including real Gaussian random processes.

Blackman and Tukey describe an alternate technique that gives more accurate estimates. Simply stated, the technique obtains the spectral estimate from the transform of the truncated autocorrelation function. Theoretical considerations show that the autocorrelation of $x_T(t)$ is not an accurate estimate of the true autocorrelation of $x(t)$ for large values of the shift τ. Hence, the autocorrelation of $x_T(t)$ is truncated to a length much less than T by multiplying by an even function that is zero outside $\tau = \pm\alpha T$, where α is on the order of 0.1. Multiplication in the time domain becomes convolution in the frequency domain; hence, the energy spectrum $|X_T(f)|^2$ is smoothed by convolution with $M(f)$. The result is called the smoothed spectral estimate, $\hat{S}(f)$, of the process, given by

$$\hat{S}(f) = |X_T(f)|^2 * M(f). \tag{D.43}$$

The rectangular truncator of length $2\alpha T$, given by

$$m(\tau) = \begin{cases} 1 & |\tau| \leq \alpha T, \\ 0 & |\tau| > \alpha T, \end{cases} \tag{D.44}$$

transforms into a sinc smoothing operator in the frequency-domain:

$$M(f) = 2\alpha T \operatorname{sinc} 2\alpha Tf. \tag{D.45}$$

The width of this smoothing operator is $\Delta f = 1/\alpha T$. With $T = 2$ sec and $\alpha = 0.1$, $\Delta f = 5$ Hz. A variety of truncators have been suggested by various authors; some of them are given in equation (B.5). Each of these truncators can be modified in length so that their smoothing operation is essentially equivalent to that of a rectangle. For example, a triangular truncator that is twice the length of a rectangular truncator will have a sinc-squared smoothing operator that is the same width as the rectangular truncator's sinc smoothing operator.

This can be seen to be true by considering that a rectangular function of length L convolved with itself produces a triangular function of length $2L$. Therefore, by the convolution theorem of Fourier transforms, the transform of the triangular function is the square of the transform of the rectangular function. As the rectangular function transforms into a sinc function, the triangular function transforms into that sinc function squared.

RANDOM INPUTS TO LINEAR, TIME-INVARIANT SYSTEMS

INPUT-OUTPUT RELATIONS

If the input to a linear system is a random function $x(t)$, then the output $y(t)$ is also a random function given by the convolution of $x(t)$ and the impulse response $g(t)$:

$$y(t) = x(t) * g(t). \tag{D.46}$$

The autocorrelation of the output ϕ_{yy} is defined by the time average,

$$\phi_{yy}(\tau) = [y(t)y(t - \tau)]. \tag{D.47}$$

Carrying out the time average gives

$$\phi_{yy}(\tau) = \phi_{xx}(\tau) * \phi_{gg}(\tau). \tag{D.48}$$

The power spectrum of the output process Φ_{yy} is given by

$$\Phi_{yy}(f) = \Phi_{xx}(f) \cdot |G(f)|^2. \tag{D.49}$$

If the output $y(t)$ is correlated with the input $x(t)$, the cross-correlation function is defined as the time average,

$$\phi_{yx}(\tau) = [y(t)x(t - \tau)]. \tag{D.50}$$

After carrying out the indicated operation, the cross correlation is given by

$$\phi_{yx}(\tau) = \phi_{xx}(\tau) * g(\tau) \tag{D.51}$$

and the cross spectrum by

$$\Phi_{yx}(f) = \Phi_{xx}(f) \cdot G(f). \tag{D.52}$$

If the input is from a white random process, then $\phi_{xx}(\tau) = \delta(\tau)$, and

$$\phi_{yx}(\tau) = \delta(\tau) * g(\tau) = g(\tau). \tag{D.53}$$

This shows the remarkable result that a white random excitation can be used to obtain the impulse response of a linear, time-invariant system, as shown in Figure D.3, from Lee (1960).

MATCHED FILTERS

One of the practical applications of statistical communication theory is the optimum detection of a signal of known form that is imbedded in a background of random noise. If the noise process is stationary and white, then the optimum filter is called the matched filter (Figure D.4). It is optimum in the sense that, of all possible filters, it is the one that produces the maximum signal-to-noise ratio in the output.

Mathematical development shows that the maximum signal-to-noise ratio at time t_0 is produced by a filter whose spectrum $G(f)$ equals the product of the complex conjugate of the signal spectrum and a delay factor,

$$G(f) = S^*(f) \exp(-j2\pi f t_0). \tag{D.54}$$

In the time domain, the impulse response of the matched filter $g(t)$ is then given by

$$g(t) = s(t_0 - t). \tag{D.55}$$

This shows that the impulse response is the signal waveform folded and delayed by t_0. The signal component in the output, given by $g(t) * s(t)$, is the autocorrelation of $s(t)$ shifted by t_0:

$$g(t) * s(t) = \phi_{ss}(t - t_0). \tag{D.56}$$

The maximum amplitude of $\phi_{ss}(t - t_0)$ is located at $t = t_0$. This means that, if the signal arrives at some arbitrary time t, the maximum amplitude in the output is delayed by t_0 with respect to the signal arrival time. The delay time t_0 is a known constant.

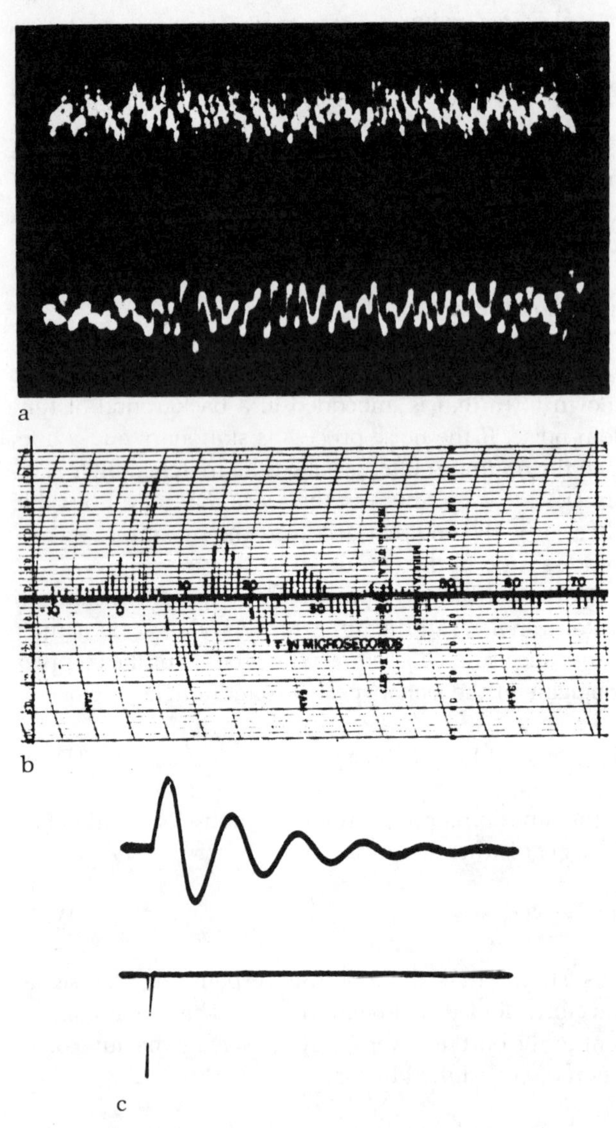

FIGURE D.3. *Comparing impulse response of a system using random input and impulse input. (a) White noise input and output. (b) Impulse response by cross correlation of white noise input and output. (c) System response to impulse. (From Lee 1960, courtesy of John Wiley & Sons.)*

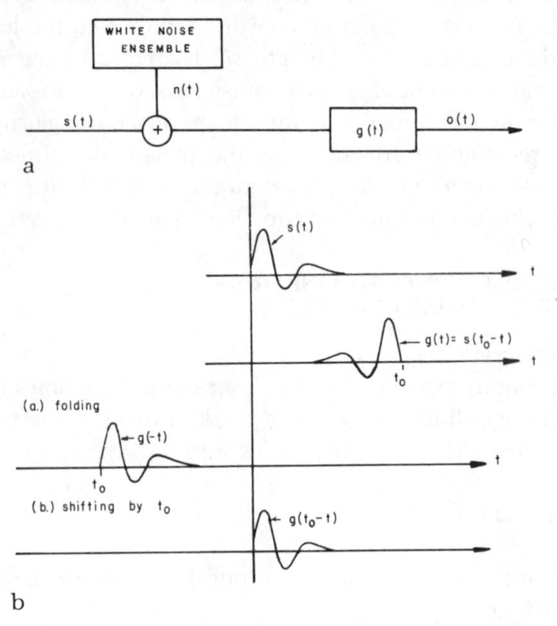

FIGURE D.4. *Block diagram of matched filter. (a) Block diagram. (b) Convolution of known signal with matched filter.*

Matched filtering is also called correlation detection because, in effect, the signal is cross-correlated with the data. When no signal is present in the data, there is little output, because the cross correlation between white noise and the signal is essentially zero. However, when a signal is present in the data, the cross correlation becomes maximum because of the signal autocorrelation component. Vibroseis processing is an example of matched filtering.

OPTIMUM WIENER FILTERS

The problem Wiener solved is the optimum separation of signal from noise when both the signal and noise are stationary random processes. In solving this problem, he made two lasting contributions to optimum filter design:

(1) Use of statistical description of signal and noise;
(2) Use of an objective measure of filter performance, known as the minimum mean square error criterion.

In the Wiener model (Figure D.5), the input data are

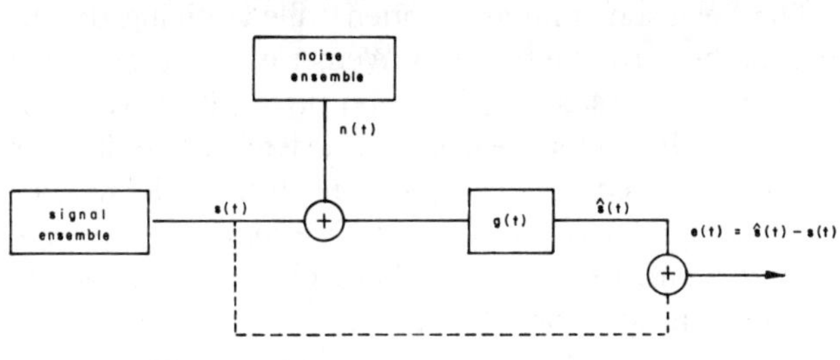

FIGURE D.5. *Block diagram of optimal Wiener filter.*

the additive mixture of signal $s(t)$ and noise $n(t)$. The input data are passed through filter $g(t)$ to produce a signal estimate $\hat{s}(t)$. The error $e(t)$ is the difference between the signal and its estimate. The objective is to find the filter that minimizes the mean square error. The time average of the square error (or error energy) is given by

$$[e^2(t)] = \lim_{T\to\infty} \frac{1}{T} \int_{-T/2}^{+T/2} e^2(t)dt. \qquad (D.57)$$

The time-averaged error energy is then averaged over the ensemble of all possible signals and noises, giving the mean square error. The mean square error is the expectation of $[e^2(t)]$. The optimum filter is the one that minimizes this expectation. The Wiener criterion, then, is

$$E[e^2(t)] = \text{minimum}. \qquad (D.58)$$

The optimum filter $g_0(t)$ is optimum for the ensembles of signal and noise and is not necessarily optimum for particular members of the ensembles.

Wiener's mathematical development shows that the only statistical information needed to solve this problem is contained in the correlation functions,

$$\begin{aligned} \phi_{ss}(\tau) &= Es(t)s(t-\tau), \\ \phi_{nn}(\tau) &= En(t)n(t-\tau), \\ \phi_{sn}(\tau) &= Es(t)n(t-\tau). \end{aligned} \qquad (D.59)$$

These functions must be known. They are either obtained from the assumed mathematical model or estimated from the data. Given the correlation functions, the minimization is carried out using the calculus of variations. The solution is known as the Wiener-Hopf equation, which has the engagingly simple form:

$$g_0(t) * \phi_{ii}(t) = \phi_{is}(t), \qquad t \geqslant 0, \qquad (D.60)$$

where ϕ_{ii} is the input autocorrelation function and ϕ_{is} is the input-signal cross-correlation function. These correlation functions can be written in terms of the noise and signal correlation functions:

$$\begin{aligned} \phi_{ii} &= \phi_{ss} + \phi_{nn} + \phi_{sn} + \phi_{ns}, \\ \phi_{is} &= \phi_{ss} + \phi_{ns}. \end{aligned} \qquad (D.61)$$

If the signal and noise are uncorrelated, which is the usual case, then $\phi_{ns} = 0$, and

$$\begin{aligned} \phi_{ii} &= \phi_{ss} + \phi_{nn}, \\ \phi_{is} &= \phi_{ss}. \end{aligned} \qquad (D.62)$$

The Wiener-Hopf equation then becomes

$$g_0(t) * [\phi_{ss}(t) + \phi_{nn}(t)] = \phi_{ss}(t), \qquad t \geqslant 0. \qquad (D.63)$$

If the realizability condition on $g_0(t)$ is waived, then the Wiener-Hopf equation can be solved in the frequency domain:

$$G_0(f) = \frac{\Phi_{is}(f)}{\Phi_{ii}(f)}. \qquad (D.64)$$

Assuming that the signal and noise are uncorrelated,

$$G_0(f) = \frac{\Phi_{ss}(f)}{\Phi_{ss}(f) + \Phi_{nn}(f)}. \qquad (D.65)$$

This result shows that, under the given assumptions, the optimum filter emphasizes the frequencies at which the signal-to-noise ratio is large and suppresses those at which it is low.

Levinson (1949) solved the Wiener-Hopf equation (D.60) in the discrete case. The matrix form of the solution is

$$\Phi_{ii} G = \Phi_{is}, \tag{D.66}$$

where Φ_{ii} is the following $n \times n$ autocorrelation matrix obtained by truncating the autocorrelation of the input data to n points,

$$\Phi_{ii} = \begin{bmatrix} \phi_0 & \phi_1 & \ldots & \phi_{n-1} \\ \phi_1 & \phi_0 & \ldots & \phi_{n-2} \\ & & \vdots & \\ \phi_{n-1} & \phi_{n-2} & \ldots & \phi_0 \end{bmatrix}; \tag{D.67}$$

G, the Wiener operator, is an $n \times 1$ vector,

$$G = \begin{bmatrix} g_0 \\ g_1 \\ \vdots \\ g_{n-1} \end{bmatrix}; \tag{D.68}$$

and Φ_{is} is the $n \times 1$ input-signal cross-correlation vector,

$$\Phi_{is} = \begin{bmatrix} 1 \\ 0 \\ \vdots \\ 0 \end{bmatrix}. \tag{D.69}$$

The solution of the matrix equation to define G merely requires inverting the correlation matrix and multiplying by Φ_{is}:

$$G = \Phi_{ii}^{-1} \Phi_{is}. \tag{D.70}$$

Because of the diagonal symmetry of the autocorrelation matrix (matrices with diagonal symmetry are called Toeplitz matrices), the inversion is much simpler than general matrix inversion. There is virtually no possibility that the inverse matrix will be unstable; however, a small value added to the diagonal element by changing ϕ_0 to $(1 + \epsilon)\phi_0$ will guarantee stability. Because the added term's only purpose is to produce stability and the Wiener operator is degraded as ϵ increases, it is advantageous to keep ϵ as small as possible, something on the order of 0.01. The ϵ-factor is usually called a white noise factor, because this is exactly the effect that would be produced on the input autocorrelation function if the input contained a white noise component.

The Levinson solution is often called spiking deconvolution because the resulting Weiner operator attempts to compress the signal in the input to a spike because of the chosen Φ_{is} vector. Peacock and Trietel (1969), in their procedure called predictive deconvolution, do not attempt compression to a spike and are content to preserve the first α samples of signal. They call $\alpha \Delta t$ the prediction distance. The autocorrelation matrix is the same as in the Levinson case, and the cross-correlation vector for predictive deconvolution is given by

$$\Phi_{is} = \begin{bmatrix} \phi_\alpha \\ \phi_{\alpha+1} \\ \vdots \\ \phi_{\alpha+n-1} \end{bmatrix}. \tag{D.71}$$

Predictive deconvolution is a generalization of Wiener-Levinson deconvolution; in the limit when the lag α becomes 1, predictive deconvolution becomes spiking deconvolution.

Appendix E Mathematics of Wave Propagation

STRESS AND STRAIN

In an x,y,z-coordinate system, where u, v, and w represent the displacement in the respective coordinate directions, the following definitions of terms apply.

Linear strain is the change in length per unit length, defined in the x-direction by

$$\epsilon_{xx} = \frac{(\Delta x + (\partial u/\partial x)\,\Delta x) - \Delta x}{\Delta x} = \frac{\partial u}{\partial x}. \tag{E.1}$$

The three components of linear strain are, by definition,

$$\epsilon_{xx} = \frac{\partial u}{\partial x}, \qquad \epsilon_{yy} = \frac{\partial v}{\partial y}, \qquad \epsilon_{zz} = \frac{\partial w}{\partial z}. \tag{E.2}$$

Volume strain, or *dilation*, is the change in volume per unit volume, defined by

$$\begin{aligned} \theta = [(\Delta x + (\partial u/\partial x)\,\Delta x)\,(\Delta y + (\partial v/\partial y)\,\Delta y)\,(\Delta z \\ + (\partial w/\partial z)\,\Delta z) - \Delta x\,\Delta y\,\Delta z]/\Delta x\,\Delta y\,\Delta z. \end{aligned} \tag{E.3}$$

Expanding and neglecting higher-order terms (which is equivalent to letting the volume $\Delta x\,\Delta y\,\Delta z$ approach zero) gives

$$\theta = \frac{\partial u}{\partial x} + \frac{\partial v}{\partial y} + \frac{\partial w}{\partial z} = \epsilon_{xx} + \epsilon_{yy} + \epsilon_{zz}. \tag{E.4}$$

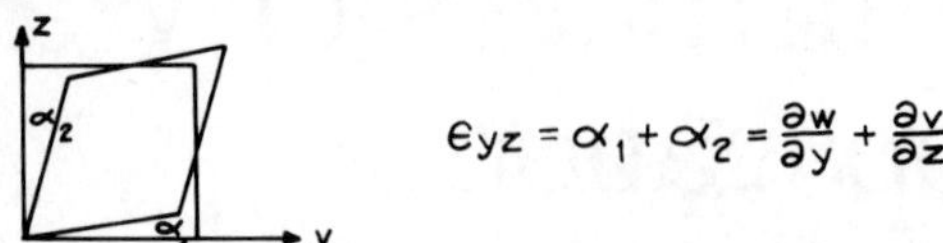

FIGURE E.1. *Shear strain.*

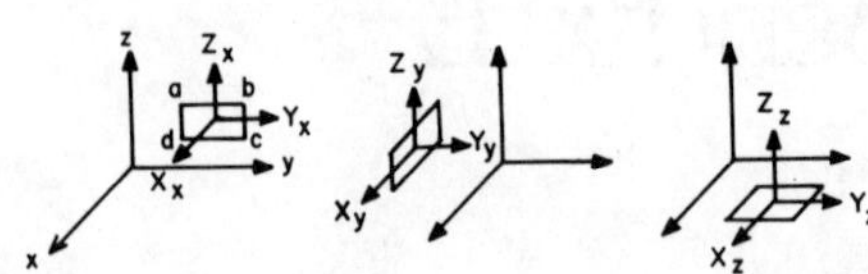

FIGURE E.2. *Components of stress.*

Shear strain is the shear displacement per unit length. From the diagram in Figure E.1 $\tan \alpha_1 = \partial w/\partial y$ and $\tan \alpha_2 = \partial v/\partial z$. For sufficiently small deformations, $\tan \alpha = \alpha$; therefore, the shear strain in the yz plane is defined by

$$\epsilon_{yz} = \alpha_1 + \alpha_2 = \frac{\partial w}{\partial y} + \frac{\partial v}{\partial z}. \tag{E.5}$$

By definition, the three components of shear strain are

$$\epsilon_{yz} = \frac{\partial w}{\partial y} + \frac{\partial v}{\partial z}, \quad \epsilon_{zx} = \frac{\partial u}{\partial z} + \frac{\partial w}{\partial x}, \quad \epsilon_{xy} = \frac{\partial v}{\partial x} + \frac{\partial u}{\partial y}. \tag{E.6}$$

Poisson's ratio is the ratio of lateral contraction to linear extension in a strained element. For extension in the x-direction,

$$\sigma = \begin{cases} \dfrac{\partial v/\partial y}{\partial u/\partial x} = \dfrac{\epsilon_{yy}}{\epsilon_{xx}} \\ \\ \dfrac{\partial w/\partial z}{\partial u/\partial x} = \dfrac{\epsilon_{zz}}{\epsilon_{xx}} \end{cases} \tag{E.7}$$

Thus, the linear strains in the three directions are related through Poisson's ratio,

$$\epsilon_{xx} = \frac{1}{\sigma}\epsilon_{yy} = \frac{1}{\sigma}\epsilon_{zz}. \tag{E.8}$$

Stress is defined as the ratio of force to area, in the limit, as the area approaches zero. Consider a rectangular surface *abcd*, which is perpendicular to the x-axis. The three components of stress acting on this surface are denoted X_x, Y_x, and Z_x. The subscript indicates that the surface is perpendicular to the x-axis.

The nine components of stress acting on a rectangular volume element are as follows (Figure E.2):

$$\begin{bmatrix} X_x X_y X_z \\ Y_x Y_y Y_z \\ Z_x Z_y Z_z \end{bmatrix}, \tag{E.9}$$

where the capital letter indicates the direction of stress and the subscript indicates the direction of the normal to the surface. For equilibrium, $X_y = Y_x$, $X_z = Z_x$, and $Y_z = Z_y$. These components constitute the shear stress, while X_x, Y_y, and Z_z are the compressional stress components. Thus, there are six independent stress components.

HOOKE'S LAW

Hooke's law states that stress is proportional to strain. In general, each of the six independent components of stress at a point are linear functions of the six independent components of strain. This results in 36 elastic constants, 21 of which are independent in the most general case, where no symmetry exists in the material. In a homogeneous and isotropic material, the number of independent elastic constants is reduced to the two Lamé constants, λ and μ, and the stress-strain relations become

$$\begin{aligned} X_x &= \lambda\theta + 2\,\mu\epsilon_{xx}, \\ Y_y &= \lambda\theta + 2\,\mu\epsilon_{yy}, \\ Z_z &= \lambda\theta + 2\,\mu\epsilon_{zz}, \\ Y_z &= \mu\epsilon_{yz}, \end{aligned} \tag{E.10}$$

$$Z_x = \mu\epsilon_{zx},$$

$$X_y = \mu\epsilon_{xy}.$$

From these equations, Young's modulus (E), Poisson's ratio (σ), and the bulk modulus (k) can be expressed in terms of λ and μ. The shear modulus is equal to μ, as noted in the latter three equations. Consider a purely compressional stress applied in the x-direction. In other words, X_x is applied and all other stress components are zero. After substituting $\theta = \epsilon_{xx} + \epsilon_{yy} + \epsilon_{zz}$, the first three equations become

$$X_x = (\lambda + 2\mu)\,\epsilon_{xx} + \lambda\,(\epsilon_{yy} + \epsilon_{zz}),$$
$$0 = (\lambda + 2\mu)\,\epsilon_{yy} + \lambda\,(\epsilon_{xx} + \epsilon_{zz}), \tag{E.11}$$
$$0 = (\lambda + 2\mu)\,\epsilon_{zz} + \lambda\,(\epsilon_{xx} + \epsilon_{yy}).$$

Solving for ϵ_{xx}, ϵ_{yy}, ϵ_{zz} gives

$$\epsilon_{xx} = \frac{\lambda + \mu}{\mu(3\lambda + 2\mu)} X_x,$$
$$\epsilon_{yy} = \epsilon_{zz} = \frac{\lambda}{2\mu\,(3\lambda + 2\mu)} X_x. \tag{E.12}$$

Young's modulus is defined as $E = X_x/\epsilon_{xx}$; thus,

$$E = \frac{\mu(3\lambda + 2\mu)}{\lambda + \mu}. \tag{E.13}$$

Poisson's ratio is defined as $\sigma = \epsilon_{yy}/\epsilon_{xx}$; thus,

$$\sigma = \frac{\lambda}{2(\lambda + \mu)}. \tag{E.14}$$

Subjecting the solid to a uniform hydrostatic pressure leads to

$$X_x = Y_y = Z_z = P,$$
$$X_y = Y_z = Z_x = 0. \tag{E.15}$$

The first three stress-strain equations give

$$\epsilon_{xx} = \epsilon_{yy} = \epsilon_{zz} = -\frac{P}{3\lambda + 2\mu}. \tag{E.16}$$

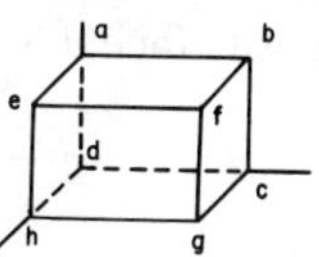

FIGURE E.3. *Elemental volume.*

The volume change $\theta = \epsilon_{xx} + \epsilon_{yy} + \epsilon_{zz}$; hence, the bulk modulus k is given by

$$k = \frac{P}{\theta} = \lambda + \frac{2}{3}\mu. \tag{E.17}$$

WAVE EQUATION

The wave equation is derived from Newton's equation of motion—force equals mass times acceleration—where the components of force are derived from the components of stress acting on an elemental volume (Figure E.3). The stress components in the x-direction on the face *abcd* is X_x, while on its opposite face, *efgh*, it is $X_x + (\partial X_x/\partial x)\Delta x$. The net stress in the x-direction acting on these two faces is the difference,

$$\left(X_x + \frac{\partial X_x}{\partial x}\Delta x\right) - X_x = \frac{\partial X_x}{\partial x}\Delta x. \tag{E.18}$$

The resulting force is the stress multiplied by the surface area on which it acts ($\Delta y\,\Delta z$). Hence, the force is

$$\left(\frac{\partial X_x}{\partial x}\Delta x\right)(\Delta y\,\Delta z). \tag{E.19}$$

The stress component in the x-direction on face *adeh* is X_y, while on its opposite face it is $X_y + (\partial X_y/\partial y)\,\Delta y$. The difference gives the net stress, which, when multiplied by the area of the face $\Delta x\,\Delta z$, gives a force of

$$\left(\frac{\partial X_y}{\partial y}\Delta y\right)(\Delta x\,\Delta z). \tag{E.20}$$

The stress component in the x-direction on the face $cdgh$ is X_z, while on its opposite face it is $X_z + (\partial X_z/\partial z)\,\Delta z$. The resulting force is

$$\left(\frac{\partial X_z}{\partial z}\,\Delta z\right)(\Delta x\,\Delta y). \tag{E.21}$$

Hence, the total force component in the x-direction is given by

$$F_x = \left(\frac{\partial X_x}{\partial x} + \frac{\partial X_y}{\partial y} + \frac{\partial X_z}{\partial z}\right)\Delta x\,\Delta y\,\Delta z. \tag{E.22}$$

Substituting F_x into Newton's law of motion, $F_x = ma_x$, where the mass is given by $\rho\Delta x\,\Delta y\,\Delta z$ and the acceleration in the x-direction is $a_x = \partial^2 u/\partial t^2$ gives

$$\left(\frac{\partial X_x}{\partial x} + \frac{\partial X_y}{\partial y} + \frac{\partial X_z}{\partial z}\right) = \rho\,\frac{\partial^2 u}{\partial t^2}. \tag{E.23}$$

Similar considerations of stresses in the y- and z-directions lead to the force components in the y- and z-directions and then to two additional equations of motion. The set of equations of motion for arbitrary stress-strain relations are

$$\begin{aligned}
F_x = ma_x &\rightarrow \frac{\partial X_x}{\partial x} + \frac{\partial X_y}{\partial y} + \frac{\partial X_z}{\partial z} = \rho\,\frac{\partial^2 u}{\partial t^2},\\
F_y = ma_y &\rightarrow \frac{\partial Y_x}{\partial x} + \frac{\partial Y_y}{\partial y} + \frac{\partial Y_z}{\partial z} = \rho\,\frac{\partial^2 v}{\partial t^2},\\
F_z = ma_z &\rightarrow \frac{\partial Z_x}{\partial x} + \frac{\partial Z_y}{\partial y} + \frac{\partial Z_z}{\partial z} = \rho\,\frac{\partial^2 w}{\partial t^2}.
\end{aligned} \tag{E.24}$$

These equations of motion hold for whatever stress-strain relations hold. To solve them requires substitution of the appropriate elastic relations. In a homogeneous, isotropic medium, its stress-strain relations are substituted into the equations of motion, resulting in the following set of equations of motion in homogeneous, isotropic media:

$$\begin{aligned}
(\lambda + \mu)\,\frac{\partial \theta}{\partial x} + \mu\left(\frac{\partial^2 u}{\partial x^2} + \frac{\partial^2 u}{\partial y^2} + \frac{\partial^2 u}{\partial z^2}\right) &= \rho\,\frac{\partial^2 u}{\partial t^2},\\
(\lambda + \mu)\,\frac{\partial \theta}{\partial y} + \mu\left(\frac{\partial^2 v}{\partial x^2} + \frac{\partial^2 v}{\partial y^2} + \frac{\partial^2 v}{\partial z^2}\right) &= \rho\,\frac{\partial^2 v}{\partial t^2},\\
(\lambda + \mu)\,\frac{\partial \theta}{\partial z} + \mu\left(\frac{\partial^2 w}{\partial x^2} + \frac{\partial^2 w}{\partial y^2} + \frac{\partial^2 w}{\partial z^2}\right) &= \rho\,\frac{\partial^2 w}{\partial t^2}.
\end{aligned} \tag{E.25}$$

For plane-wave propagating in the x-direction, the equations of motion simplify, because all partial derivatives with respect to y and z become zero. Hence, the equations of motion simplify to

$$\begin{aligned}
(\lambda + \mu)\,\frac{\partial \theta}{\partial x} + \mu\,\frac{\partial^2 u}{\partial x^2} &= \rho\,\frac{\partial^2 u}{\partial t^2},\\
\mu\,\frac{\partial^2 v}{\partial x^2} &= \rho\,\frac{\partial^2 v}{\partial t^2},\\
\mu\,\frac{\partial^2 w}{\partial x^2} &= \rho\,\frac{\partial^2 w}{\partial t^2}.
\end{aligned} \tag{E.26}$$

The term $\partial\theta/\partial x$ is further simplified to

$$\frac{\partial \theta}{\partial x} = \frac{\partial}{\partial x}\left(\frac{\partial u}{\partial x} + \cancel{\frac{\partial v}{\partial y}} + \cancel{\frac{\partial w}{\partial z}}\right) = \frac{\partial^2 u}{\partial x^2}. \tag{E.27}$$

This results in the following set of equations of motion for plane waves:

$$\begin{aligned}
\frac{\partial^2 u}{\partial t^2} &= \frac{\lambda + 2\mu}{\rho}\,\frac{\partial^2 u}{\partial x^2},\\
\frac{\partial^2 v}{\partial t^2} &= \frac{\mu}{\rho}\,\frac{\partial^2 v}{\partial x^2},\\
\frac{\partial^2 w}{\partial t^2} &= \frac{\mu}{\rho}\,\frac{\partial^2 w}{\partial x^2}.
\end{aligned} \tag{E.28}$$

The first of the three equations is the compressional wave equation, as it indicates particle motion in-line with the direction of propagation. The constant has the dimensions of velocity squared; hence,

$$V_p = \sqrt{\frac{\lambda + 2\mu}{\rho}}. \tag{E.29}$$

The general solution of this equation in terms of its particle displacement u is

$$u(t, x) = A_1 g_1 \left(t - \frac{X}{V_p}\right) + A_2 g_2 \left(t + \frac{X}{V_p}\right), \qquad \text{(E.30)}$$

where g_1 and g_2 are arbitrary functions traveling in the positive and negative directions, respectively.

The other two equations are shear wave equations, as they indicate particle motion transverse to the direction of propagation. The velocity is given by

$$V_s = \sqrt{\frac{\mu}{\rho}}. \qquad \text{(E.31)}$$

The general solution of v and w are likewise arbitrary functions that give the components of shear motion:

$$V = B_1 h_1 \left(t - \frac{x}{V_s}\right) + B_2 h_2 \left(t + \frac{x}{V_s}\right),$$

$$W = C_1 h_1 \left(t - \frac{x}{V_s}\right) + C_2 h_2 \left(t + \frac{x}{V_s}\right). \qquad \text{(E.32)}$$

The vector sum of v and w gives the amplitude and plane of polarization of the total shear motion.

BOUNDARY CONDITIONS: PLANE WAVES WITH NORMAL INCIDENCE

The general solution of the one-dimensional wave equation for plane compressional waves traveling in the x-direction is

$$u(t, x) = A_1 g_1 \left(t - \frac{x}{V}\right) + A_2 g_2 \left(t + \frac{x}{V}\right), \qquad \text{(E.33)}$$

where u is the particle displacement in the x-direction, g_1 and g_2 are arbitrary functions of (t, x) propagating with velocity V in the positive and negative x-directions, respectively, and A_1 and A_2 are their amplitudes. These waves propagate indefinitely in an infinite, isotropic, homogeneous medium.

When there is a plane interface separating two homogeneous, isotropic media that have different elastic properties, the incident wave will be partially reflected and partially transmitted. The amplitudes of the incident and transmitted waves can be calculated by satisfying two boundary conditions in the case of normal incidence: (1) normal displacement and (2) normal stress. Assuming that the two media are welded together, these two quantities must be continuous across the boundary; that is, they must be equal as the boundary is approached from either side. Let the elastic constants and displacements in the incident medium be indicated with subscript 1 and those in the second medium by subscript 2. The x-axis is positive downward and the interface is located at $x = 0$.

Consider the wave traveling in the positive x-direction to be the incident wave, designated $u_0\ (x,t)$, and let maximum displacement be A_0. In the special case of sinusoidal variation of displacement,

$$u_0 = A_0 \cos \omega \left(t - \frac{x}{V_1}\right). \qquad \text{(E.34)}$$

For mathematical simplicity, this equation is written as

$$u_0 = A_0 \exp \left[j\omega \left(t - \frac{x}{V_1}\right)\right], \qquad \text{(E.35)}$$

where it is assumed that u_0 is given by the real part of the complex exponential.

At the interface, the incident wave is partially reflected and partially transmitted. The reflected wave is designated $u_1(t,x)$, with maximum amplitude of A_1. It has the same sinusoidal variation as the incident wave, except that it is traveling in the opposite direction, which changes the sign of x, and has a 180° phase shift in displacement with respect to the incident wave, which is indicated by the negative sign prefixing A_1:

$$u_1 = -A_1 \exp \left[j\omega \left(t + \frac{x}{V_1}\right)\right]. \qquad \text{(E.36)}$$

The transmitted wave is designated $u_2(t,x)$, with maximum amplitude of A_2. It has the same sinusoidal variation as the incident wave, but it will travel with the compressional velocity of the second medium, V_2. It is traveling in the same direction as the incident wave; therefore, the sign of x is the same. The transmitted wave is

$$u_2 = A_2 \exp\left[j\omega\left(t - \frac{x}{V_2}\right)\right]. \tag{E.37}$$

The displacement in the incident medium at an arbitrary (x,t) is given by the sum of u_0 and u_1, and the displacement in the second medium is given by u_2.

The first boundary condition requires that, at $x = 0$, the normal displacements in the two media must be equal:

$$u_0\,(0,t) + u_1\,(0,t) = u_2\,(0,t). \tag{E.38}$$

Substituting the equations for u_0, u_1, and u_2 in equation (E.38) and letting $x = 0$ gives one equation in the two unknowns, A_1 and A_2 (A_0 is assumed known and is generally normalized by letting $A_0 = 1$):

$$A_0\, e^{j\omega t} - A_1 e^{j\omega t} = A_2 e^{j\omega t}. \tag{E.39}$$

This gives the normal displacement boundary condition:

$$A_0 - A_1 = A_2.$$

The second boundary condition requires that the normal stresses be equal on each side of the boundary. The normal stress X_x is given by

$$X_x = 2\mu \frac{\partial u}{\partial x} + \lambda\left(\frac{\partial u}{\partial x} + \frac{\partial v}{\partial y} + \frac{\partial w}{\partial z}\right). \tag{E.40}$$

In the case of plane waves with normal incidence, the displacements in the y- and z-directions are zero; therefore, their partial derivatives $\partial v/\partial y$ and $\partial w/\partial z$ are zero. The normal stress is then

$$X_x = (\lambda + 2\,\mu)\frac{\partial u}{\partial x}. \tag{E.41}$$

This equation can be written in terms of compressional velocity and density by making the substitution

$$\rho V^2 = \lambda + 2\mu,$$

$$X_x = \rho V^2 \frac{\partial u}{\partial x}. \tag{E.42}$$

The stress due to each wave—incident, reflected, and transmitted—is proportional to the partial derivative of the normal displacement with respect to x. Taking the partial derivative of the equations for u_0, u_1, and u_2 gives

$$\begin{aligned}
\frac{\partial u_0}{\partial x} &= \frac{-j\omega A_0}{V_1}\exp\left[j\omega\left(t - \frac{x}{V_1}\right)\right],\\
\frac{\partial u_1}{\partial x} &= \frac{j\omega A_1}{V_1}\exp\left[j\omega\left(t + \frac{x}{V_1}\right)\right],\\
\frac{\partial u_2}{\partial x} &= \frac{-j\omega A}{V_2}\exp\left[j\omega\left(t - \frac{x}{V_2}\right)\right].
\end{aligned} \tag{E.43}$$

Substituting into the equation for X_x gives the stress due to each wave. At $x = 0$, the boundary condition requires that the stresses be related as follows:

$$j\omega A_0 \rho_1 V_1 e^{j\omega t} + j\omega A_1 \rho_1 V_1 e^{j\omega t} = j\omega A_2 \rho_2 V_2 e^{j\omega t}. \tag{E.44}$$

This gives the normal stress boundary condition:

$$A_0 + A_1 = \left(\frac{\rho_2\, V_2}{\rho_1\, V_1}\right) A_2. \tag{E.45}$$

Combining the normal displacement and stress boundary conditions and solving for A_1 and A_2 gives the reflection and transmission coefficients:

$$\begin{aligned}
R &= \frac{A_1}{A_0} = \frac{\rho_2 V_2 - \rho_1 V_1}{\rho_2 V_2 + \rho_1 V_1},\\
T &= \frac{A_2}{A_0} = \frac{2\rho_1 V_1}{\rho_2 V_2 + \rho_1 V_1}.
\end{aligned} \tag{E.46}$$

ENERGY RELATIONS

A fundamental way of analyzing a physical problem is through the law of the conservation of energy. Applying this law to the elastic wave boundary problem requires that the energy flow away from a boundary be equal to the incident energy flow into that boundary. The energy flow away from the boundary must be partitioned into the reflected and transmitted P-waves and S-waves.

The intensity I, defined as the rate of flow of energy through a unit area, is given by the energy per unit volume times the velocity of propagation:

$$I = EV. \tag{E.47}$$

In the case of normal incidence of a P-wave, the energy flow into the boundary is given by

$$I_0 = E_0 V_1. \tag{E.48}$$

The energy flow in the reflected and transmitted wave is given by

$$I_1 + I_2 = E_1 V_1 + E_2 V_2, \tag{E.49}$$

where the subscript 2 indicates the transmitted wave and the subscript 1 indicates the reflected wave. Equating them gives the energy relations at the boundary:

$$E_0 = E_1 + \frac{V_2}{V_1} E_2. \tag{E.50}$$

Dividing through by E_0 gives the energy boundary conditions:

$$\frac{E_1}{E_0} + \frac{V_2}{V_1}\frac{E_2}{E_0} = 1. \tag{E.51}$$

The energy partitioned into the reflected P-wave is given by E_1/E_0, and that in the transmitted P wave is E_2/E_0.

In a sinusoidal wave with displacement given as $u = A \cos \omega\,[t - (x/V)]$, the particle velocity du/dt is given by

$$\frac{du}{dt} = -A\omega \sin \omega\left(t - \frac{X}{V}\right), \tag{E.52}$$

where $A\omega$ is the peak particle velocity. The kinetic energy is given by

$$E = \frac{1}{2} M (A\omega)^2, \tag{E.53}$$

where M is the mass. The mass per unit volume is the density; therefore, the energy per unit volume is given by

$$E = \frac{1}{2} \rho \omega^2 A^2. \tag{E.54}$$

Substituting into the boundary energy condition and canceling the $\omega^2/2$ factor common to all terms, one obtains

$$\rho_1 A_0^2 = \rho_1 A_1^2 + \frac{\rho_2 V_2}{V_1} (A_2)^2. \tag{E.55}$$

Dividing by $\rho_1 A_0^2$ gives

$$\left(\frac{A_1}{A_0}\right)^2 + \frac{\rho_2 V_2}{\rho_1 V_1}\left(\frac{A_2}{A_0}\right)^2 = 1. \tag{E.56}$$

The reflection coefficient is defined by A_1/A_0 and the transmission coefficient by A_2/A_0; therefore,

$$R^2 + \frac{\rho_2 V_2}{\rho_1 V_1} T^2 = 1. \tag{E.57}$$

Substituting equation (E.57) into the energy boundary condition gives the relation between energy and amplitude:

$$\frac{E_1}{E_0} = \left(\frac{A_1}{A_0}\right)^2, \qquad \frac{E_2}{E_0} = \frac{\rho_2}{\rho_1}\left(\frac{A_2}{A_1}\right)^2. \tag{E.58}$$

Whereas the reflected wave energy ratio is the square of the amplitude ratio, the transmitted wave energy ratio includes the density ratio as well.

References

Al-Chalabi, M., 1973, Series approximations in velocity and travel time computation: Geophys. Prosp., v. 21, p. 783.

——— 1974, An analysis of stacking, RMS, average, and interval velocities over a horizontally layered ground: Geophys. Prosp., v. 22, p. 458.

Angona, F. A., 1960, Two-dimensional modeling and its application to seismic problems: Geophysics, v. 25, p. 468.

Backus, M. M., 1959, Water reverberations, their nature and elimination: Geophysics, v. 24, p. 233.

Backus, M. M., and Chen, R. L., 1975, Flat spot exploration: Geophys. Prosp., v. 23, p. 533.

Ballakh, I. Y., Kochkina, M. V., and Gruzkova, G. L., 1970, The feasibility of direct exploration of screened oil and gas pools by seismic prospecting: Geolog. Nefti i Gaza, v. 14, p. 56.

Barry, K. M., and Shugart, T. R., 1973, Seismic hydrocarbon indicators and models: Publication of Teledyne Exploration Company.

Berryhill, J. R., 1977, Diffraction response for nonzero separation of source and receiver: Geophysics, v. 42, p. 1158.

Berryman, L. H., Goupillaud, P. L., and Waters, K. H., 1958, Reflections from multiple transition layers, Part I. Theoretical results: Geophysics, v. 23, p. 233.

Biot, M. A., 1956a, Theory of propagation of elastic waves in a fluid-saturated porous solid, I. Low-frequency range: J. Acoust. Soc. Am., v. 28, p. 168.

Biot, M. A., 1956b, Theory of propagation of elastic waves in a fluid-saturated porous solid, II. Higher frequency range: J. Acoust. Soc. Am., v. 28, p. 179.

Biot, M. A., 1962, Mechanics of deformation and acoustic propagation in porous media: J. Appl. Phys., v. 33, p. 1482.

Blackman, R. B., and Tukey, J. W., 1958, The measurement of power spectra: New York, Dover Publications, Inc.

Bode, H. W., 1945, Network analysis and feedback amplifier design: Princeton, NJ, Van Nostrand Co.

Bortfeld, R., 1974, Methods and trends in modern seismic exploration: Prakla-Seismos publication.

Brandsaeter, H., Farestveit, A., and Ursin, B., 1979, A new high resolution/deep penetration airgun array: Geophysics, v. 44, p. 868.

Brandt, H., 1955, A study of the speed of sound in porous granular media: Trans. ASME, v. 22, p. 479.

Brustad, J. T., 1953, Curved path Delta-T analysis: Geophysics, v. 18, p. 738 (Abstract).

Cassano, E., and Rocca, E., 1973, Multichannel linear filters for optimal rejections of multiple reflections: Geophysics, v. 38, p. 1053.

Cherry, J. T., and Waters, K. H., 1968, Shear-wave recording using continuous signal methods, Part I. Early development: Geophysics, v. 33, p. 229.

Churlin, V. V., and Sergeyev, L. A., 1963, Application of seismic surveying to recognition of productive part of gas-oil strata: Geolog. Nefti i Gaza, v. 7, p. 636.

Claerbout, J. F., 1970, Coarse grid calculations of waves in inhomogeneous media with application to delineation of complicated seismic structure: Geophysics, v. 35, p. 407.

——— 1971a, Numerical holography in acoustic holography, v. 3: New York, Plenum Press.

——— 1971b, Toward a unified theory of reflector mapping: Geophysics, v. 36, p. 467.

——— 1976, Fundamentals of geophysical data processing: New York, McGraw-Hill Book Co., Inc.

Claerbout, J. F., and Doherty, S. M., 1972, Downward continuation of moveout corrected seismograms: Geophysics, v. 37, p. 741.

Claerbout, J. F., and Johnson, A. G., 1971, Extrapolation of time dependent waveforms along their path of propagation: Geophys. J. Roy. Astr. Soc., v. 26, p. 285.

Cole, R. H., 1948, Underwater explosions: Princeton, Princeton Univ. Press.

Cook, E. E., and Taner, M. T., 1969, Velocity spectra and their use in stratigraphic and lithological differentiation: Geophys. Prosp., v. 17, p. 433.

Cosgrove, J. J., Edwards, C. A. M., and Grigsby, J. K., 1969, Velocity determinations and the digital computer: Publication of Olympic Geophysical Company, Houston.

Craft, C., 1973, Detecting hydrocarbons—for years: Oil & Gas J., February 19, p. 122.

Crawford, J. M., Doty, W. E. N., and Lee, M. R., 1960, Continuous signal seismograph: Geophysics, v. 25, p. 95.

Davis, T. L., 1972, Velocity variations around Leduc reefs, Alberta: Geophysics, v. 37, p. 584.

Diekmann, E., and Wierczeyko, E., 1970, A possible method for determining the extent of spread of the gas in an aquifer storage by seismic techniques: 11th Int. Gas Conf., Moscow.

Dirac, P. A. M., 1958, The principles of quantum mechanics, 4th ed.: Oxford, Clarendon Press.

Dix, C. H., 1955, Seismic velocities from surface measurements: Geophysics, v. 20, p. 68.

——— 1981, Seismic prospecting for oil: New York, Harper & Row, 1952; 2nd ed., Boston, IHRDC Publishers.

Dobrin, M. B., 1976, Geophysical prospecting: New York, McGraw-Hill Book Co., Inc.

Dobrin, M. B., Ingalls, A. L., and Long, J. A., 1965, Velocity and frequency filtering using laser light: Geophysics, v. 30, p. 1144.

Dobrin, M. B., Lawrence, P. L., and Sengbush, R. L., 1954, Surface and near-surface waves in the Delaware Basin: Geophysics, v. 19, p. 695.

Dobrin, M. B., Simon, R. F., and Lawrence, P. L., 1951, Rayleigh waves from small explosions: Trans. AGU, v. 32, p. 822.

Domenico, S. N., 1974, Effect of water saturation on seismic reflectivity of sand reservoirs encased in shale: Geophysics, v. 39, p. 759.

Embree, P., Burg, J. B., and Backus, M. M., 1963, Wide band filtering—the pie-slice process: Geophysics, v. 28, p. 948.

Erickson, E. L., Miller, D. E., and Waters, K. H., 1968, Shear wave recording using continuous signal methods, Part II. Later experimentation: Geophysics, v. 33, p. 240.

Everett, J. E., 1974, Obtain interval velocity from stack velocity in presence of dip: Geophys. Prosp., v. 22, p. 122.

Ewing, M. W., Jardetsky, W. S., and Press, F., 1957, Elastic waves in layered media: New York, McGraw-Hill Book Co., Inc.

Fail, J. P., and Grau, G., 1963, Les filters en eventail: Geophys. Prosp., v. 11, p. 131.

Farr, J. B., 1976, How high is high resolution: 46th Ann. Meeting SEG, Houston.

Farriol, R., Michon, D., Muntz, R., and Staron, P., 1970, Study and comparison of marine seismic source signatures, application to a new seismic source, the VAPORCHOC: 40th Ann. Int. SEG Meeting, New Orleans, November.

Foster, M. R., 1975, Transmission effects in continuous one-dimensional seismic model: Geophys. J. Roy. Astr. Soc., v. 42, p. 1.

Foster, M. R., Hicks, W. G., and Nipper, J. T., 1962, Optimum inverse filters which shorten the spacing of velocity logs: Geophysics, v. 27, p. 317.

Foster, M. R., Kerns, C. W., and Sengbush, R. L., 1972, Processing of geophysical data (Deconvolution): U.S. Patent 3,689,874, filed November 12, 1964, issued September 5, 1972.

Foster, M. R., and Sengbush, R. L., 1971, Optimum stack: U.S. Patent 3,622,967, filed November 7, 1968, issued November 23, 1971.

Foster, M. R., Sengbush, R. L., and Watson, R. J., 1964, Design of sub-optimum filter systems for multi-trace seismic data processing: Geophys. Prosp., v. 12, p. 173.

Foster, M. R., Sengbush, R. L., and Watson, R. J., 1968, Use of Monte Carlo techniques in optimum design of Wiener inverse filter processors that suppress distortion on seismic traces: Geophysics, v. 33, p. 945.

Fourmann, J. M., 1974, Deconvolution of a recorded signature (Vapco and Wapco process): CGG Publication, June 21.

French, W. S., 1975, Computer migration of oblique seismic

reflection profiles: Geophysics, v. 40, p. 961.
Gardner, L. W., 1947, Vertical velocities from reflection shooting: Geophysics, v. 12, p. 221.
Garotta, R., and Michon, D., 1967, Continuous analysis of velocity function and the moveout corrections: Geophys. Prosp., v. 15, p. 584.
Gassmann, F., 1951, Elastic waves through a packing of spheres: Geophysics, v. 16, p. 673.
Geertsma, J., 1961, Velocity-log interpretation: The effect of rock bulk compressibility: Soc. Petr. Eng. J., v. 22, p. 235.
Geertsma, J., and Smit, D. C., 1961, Some aspects of elastic wave propagation in fluid-saturated porous solids: Geophysics, v. 26, p. 169.
Geyer, R. L., 1970, Vibroseis parameter optimization: Oil & Gas J., Part I, v. 68, n. 15, p. 116; Part II, v. 68, n. 17, p. 114.
Godfrey, L. M., Stewart, J. D., and Schweiger, F., 1968, Application of Dinoseis in Canada: Geophysics, v. 33, p. 65.
Green, C. H., 1938, Velocity determinations by means of reflection profiles: Geophysics, v. 3, p. 295.
Guinzy, N. J., and Ruehle, W. H., 1971, Interval velocity determination: U.S. Patent 3,611,278, filed July 17, 1969, patented October 5, 1971.
Gutenberg, B., 1944, Energy of reflected and refracted seismic waves: Bull. Seismol. Soc. Am., v. 34, p. 85–102.
Hagedoorn, J. C., 1954, A process of seismic reflection interpretation: Geophys. Prosp., v. 2, p. 85.
Hansen, R. F., 1947, A new system of seismic reflection profile: Bol. Inform Petr., Buenos Aires, v. 24, p. 237.
Hicks, W. G., and Berry, J. E., 1956, Application of continuous velocity logs to determination of fluid saturation of reservoir rocks: Geophysics, v. 21, p. 739.
Hilterman, F. J., 1970, Three-dimensional seismic modeling: Geophysics, v. 35, p. 1020.
——— 1975, Amplitudes of seismic waves—a quick look: Geophysics, v. 40, p. 745.
Holzman, M., 1963, Chebyshey optimized geophone arrays: Geophysics, v. 28, p. 145.
Johnson, C. H., 1948, Remarks regarding multiple reflections: Geophysics, v. 13, p. 19.
Klauder, J. R., Price, A. S., Darlington, S., and Albersheim, W. J., 1960, The theory and design of chirp radars: Bell Sys. Tech. J., v. 39, p. 745.
Klein, M. V., 1970, Optics: New York, John Wiley & Sons, Inc.
Knopoff, L., 1956, Diffraction of elastic waves: J. Acoust. Soc. Am., v. 28, p. 217–229.
Kolmogorov, A. N., 1956, Foundations of the theory of probability (2nd Engl. ed., translation edited by N. Morrison, with an added bibliography by A. T. Bharucha-Reid): New York, Chelsea Publishing Co.
Kologinczak, J., 1974, STAGARAY system improves primary pulse/bubble ratio in marine exploration: paper OTC 2020, 6th Ann. Offshore Tech. Conf., Houston, May, p. 801.
Kramer, F. S., Peterson, R. A., and Walter, W. C., 1968, Seismic energy sources in 1968 handbook: 38th Ann. Int. Meeting of SEG, Denver, October.
Krey, T., 1952, Significance of diffraction in the investigation of faults: Geophysics, v. 17, p. 843–858.
Lamb, H., 1904, On the propagation of tremors over the surface of an elastic solid: Trans. Phil. Roy. Soc. Lond., v. 203, p. 1.
Larner, K., and Hatton, L., 1975, Wave equation migration: Two approaches: Western Geophysical Company publication.
Lee, Y. W., 1960, Statistical theory of communication: New York, John Wiley & Sons, Inc.
Letton, W., III, and Bush, A. M., 1969, Time varying velocity filters: Abst. 39th Ann. Int. Meeting of SEG, Calgary, September 14–18.
Levin, F. K., 1962, The seismic properties of Lake Maracaibo: Geophysics, v. 27, p. 35.
——— 1971, Apparent velocity from dipping interface reflections: Geophysics, v. 36, p. 510.
Levinson, N., 1947, The Wiener RMS (root mean square) error criterion in filter design and prediction: J. of Math. and Phys., v. 25, p. 261.
Lindsey, J. P., 1960, Elimination of seismic ghost reflections by means of a linear filter: Geophysics, v. 25, p. 130.
Loewenthal, D., Lu, L., Roberson, R., and Sherwood, J., 1974, The wave equation applied to migration and water bottom multiples: 44th Ann. Int. Meeting of SEG, Dallas.
Lofthouse, J. H., and Bennett, G. T., 1978, Extended arrays for marine seismic acquisition: Geophysics, v. 43, p. 3.
Macelwane, J. B., and Sohon, F. W., 1936, Introduction to theoretical seismology: New York, John Wiley & Sons, Inc.
Malinovskaya, L. N., 1957, Dynamic characteristics of longitudinal reflected waves beyond the critical angle: Bull. Acad. Sci. USSR, Geophys., no. 5, p. 22.
Marr, J. D., 1971a, Seismic stratigraphic exploration, Part 1: Geophysics, v. 36, p. 311.
——— 1971b, Seismic stratigraphic exploration, Part 2: Geophysics, v. 36, p. 533.
——— 1971c, Seismic stratigraphic exploration, Part 3: Geophysics, v. 36, p. 676.
Martner, S. T., and Silverman, D., 1962, Broomstick distributed charge: Geophysics, v. 27, p. 1007.
Mayne, W. H., 1962, Common reflection point horizontal stacking technique: Geophysics, v. 27, p. 927.
McClintock, P. L., 1975, Seismic data processing techniques for Northern Michigan reefs: East. Sec. AAPG Meeting, East Lansing, Mich., October.
McClure, C. D., Nelson, H. F., and Huckabay, W. B., 1958, Marine sonoprobe system, new tool for geological mapping: Am. Assoc. Petr. Geol. Bull., v. 42, p. 701.
McDonal, F. J., Angona, F. A., Mills, R. L., Sengbush, R. L., Van Nostrand, R. G., and White, J. E., 1958, Attenuation of shear and compressional waves in Pierre Shale: Geophysics, v. 23, p. 421.
McDonal, F. J., and Sengbush, R. L., 1966, Geological data pro-

cessing to resolve fine subsurface detail: World Petr. Conf. Mexico City.

Musgrave, A. W., 1962, Applications of expanding reflection spread: Geophysics, v. 27, p. 981.

——— (Ed.), 1967, Seismic refraction prospecting: Tulsa, SEG.

Musgrave, A. W., Ehlert, G. W., and Nash, D. M., Jr., 1958, Directivity effect of elongated charges: Geophysics, v. 23, p. 81.

Nafe, J. E., 1957, Reflection and transmission coefficients: Bull. Seismol. Soc. Am., v. 47, p. 205.

Neidell, N. S., and Taner, M. T., 1971, Semblance and other coherency measures for multichannel data: Geophysics, v. 36, p. 482.

Newman, P., and Mahoney, J. T., 1973, Patterns—with a pinch of salt: Geophys. Prosp., v. 21, p. 197.

Oppenheim, A. V., and Schafer, R. W., 1975, Digital signal processing: Englewood Cliffs, NJ, Prentice-Hall, Inc.

Ormsby, J. F. A., 1961, Design of numerical filters with applications to missile data processing: J. Assoc. Comput. Mach., v. 8, p. 440.

Pan, P.-H., and De Bremaecker, J. C., 1970, Direct location of oil and gas by the seismic reflection method: Geophys. Prosp., v. 18, p. 712.

Parr, J. O., Jr., and Mayne, W. H., 1955, A new method of pattern shooting: Geophysics, v. 20, p. 539.

Peacock, K. L., and Treitel, S., 1969, Predictive deconvolution: Theory and practice: Geophysics, v. 34, p. 155.

Peacock, R. B., and Nash, D. M., 1962, Thumping techniques using full spread of geophones: Geophysics, v. 27, p. 952.

Pennebaker, E. E., 1968, An engineering interpretation of seismic data: Soc. of Petr. Eng. of AIME Meeting, Houston, September 29–October 2.

Peterson, R. A., Fillipone, W. R., and Coker, F. B., 1955, The synthesis of seismograms from well log data: Geophysics, v. 20, p. 516.

Pfleuger, J. E., 1954, Delta-T formula for obtaining average seismic velocity to a dipping reflection bed: Geophysics, v. 19, p. 339.

Prakla Report 3-74, 1974, Vibroseis processing, p. 5.

Prakla-Seismos Brochure, Data acquisition—Offshore Geophysics, November, p. 16.

Prakla-Seismos Reports 1-72, 2-72, 3-72, 4-72, 1-74, 1-76.

Press, F., and Dobrin, M. D., 1956, Seismic wave studies over a high-speed surface layer, Geophysics, v. 21, p. 285.

Ramo, A. O., and Bradley, J. W., 1977, Bright spots, milligals, and gammas: Geophysics, v. 42, p. 1534.

Rayleigh, Lord (John WIlliam Strutt), 1885, On waves propagated along the plane surface of an elastic solid: Proc. Lond. Math. Soc., v. 17, p. 4.

Reilly, M. D., and Greene, P. L., 1976, Wave equation migration: Seiscom-Delta report.

Reynolds, E. B., 1974, Seismic interpretation for drilling: Oil & Gas J., v. 72, n. 10, p. 112.

Ricker, N., 1953, The form and laws of propagation of seismic wavelets: Geophysics, v. 18, p. 10.

Ricker, N., and Lynn, R., 1950, Composite reflections: Geophysics, v. 15, p. 30.

Rieber, F., 1936, Visual presentation of elastic wave patterns under various structural conditions: Geophysics, v. 1, p. 196–218.

Riley, D. C., 1975, Wave equation synthesis and inversion of diffracted multiple seismic reflections: Doctoral diss., Geophysics Department, Stanford Univ.

Robinson, E. A., 1957, Predictive decomposition of seismic traces: Geophysics, v. 22, p. 467.

——— 1966, Multichannel z-transforms and minimum delay: Geophysics, v. 31, p. 482.

——— 1967, Statistical communication and detection, with special reference to digital data processing of radar and seismic signals: New York, Hafner Publishing Co.

——— 1980, University course in digital seismic methods used in petroleum exploration: Houston, Pexcon International, Inc.

Rockwell, D. W., 1967, The digital computers role in the enhancement and interpretation of North Sea seismic data: Geophysics, v. 32, p. 259.

——— 1971, Migration stack: Oil & Gas J., v. 69, n. 16, p. 202.

Saugy, L., and Engels, J. P., 1975, Constant velocity analysis on growth anticline, Niger delta: Reprint, SEG convention, Denver.

Savit, C. H., Brustad, J. T., and Snyder, J., 1958, The moveout filter: Geophysics, v. 23, p. 1.

Schneider, W. A., 1978, Integral formulation of migration in two and three dimensions: Geophysics, v. 43, p. 49.

Schneider, W. A., and Backus, M. M., 1968, Dynamic correlation analysis: Geophysics, v. 33, p. 105.

Sengbush, R. L., 1960, Impulse response calculations using sinc functions: M. Sc. in Eng. report, Southern Methodist Univ., Dallas.

——— 1962, Stratigraphic trap study in Cottonwood Creek field, Big Horn Basin, Wyoming: Geophysics, v. 27, p. 427.

——— 1967, Optimum detector type and depth in marine seismic exploration: U.S. Patent 3,350,683, filed March 8, 1966, issued October 31, 1967.

Sengbush, R. L., and Foster, M. R., 1968, Optimum multichannel velocity filters: Geophysics, v. 33, p. 11.

——— 1972, Design and application of optimal velocity filters in seismic exploration: IEEE Trans. Computers, special issue, July, v. C-21, p. 648.

Sengbush, R. L., Lawrence, P. L., and McDonal, F. J., 1961, Interpretation of synthetic seismograms: Geophysics, v. 2, p. 138.

Shah, P. M., 1973, Use of wavefront curvature to relate seismic data with surface parameters: Geophysics, v. 38, p. 812.

Shanks, J. L., 1967, Recursive filters for digital processing: Geophysics, v. 32, p. 33.

Shannon, C. E., 1948, A mathematical theory of communications: Bell Sys. Tech. J., v. 27, p. 739.

Sherwood, J. W. C., and Poe, P. H., 1971, Seismic wavelet processing: 41st Ann. Int. Meeting of SEG, Houston, November 8–11.

Shugart, T. R., 1973, Deconvolution, an illustrated review: Geophysics, v. 38, p. 1221.

Slotnik, M. M., 1959, Lessons in seismic computing: Tulsa, SEG.

Smith, M. K., 1956, Noise analysis and multiple seismometer theory: Geophysics, v. 21, p. 337.

Sorge, W. A., 1965, Rayleigh-wave motion in an elastic half-space: Geophysics, v. 30, p. 97.

Steele, W. E., Jr., 1941, Comparison of well survey and reflection "time-delta-time" velocities: Geophysics, v. 6, p. 370.

Stolt, R. H., 1978, Migration by Fourier transform: Geophysics, v. 43, p. 23.

Stonley, R., 1949, The seismological implications of aeolotropy in continental structure: Monthly Notices Roy. Astr. Soc., Geophys. Suppl., v. 5, p. 343.

Summers, G. C., and Broding, R. A., 1952, Continuous velocity logging: Geophysics, v. 17, p. 598.

Swartz, C. A., and Sokoloff, V. M., 1954, Filtering associated with selective sampling of geophysical data: Geophysics, v. 19, p. 402.

Taner, M. T., Cook, E. E., and Neidell, N. S., 1970, Limitations of the reflection seismic method; Lessons from computer simulations: Geophysics, v. 35, p. 551.

Taner, M. T., and Koehler, F., 1969, Velocity spectra-digital computer derivation and application of velocity functions: Geophysics, v. 34, p. 859.

Tegland, E. R., 1972, Computer-assisted interpretation of seismic structure and velocity data: 42nd Annual SEG Meeting, Anaheim, Calif. November.

Trorey, A. W., 1970, A simple theory for seismic diffractions: Geophysics, v. 35, p. 762.

——— 1977, Diffractions for arbitrary source-receiver locations: Geophysics, v. 42, p. 1177.

Ulrych, T. J., 1971, Application of homomorphic deconvolution to seismology: Geophysics, v. 36, p. 650.

Ursin, B., 1978, Attenuation of coherent noise in marine seismic exploration using very long arrays: Geophys. Prosp., v. 26, p. 722.

Van Melle, F. A., and Weatherburn, K. R., 1953, Ghost reflections caused by energy initially reflected above the level of the shot: Geophysics, v. 18, p. 793.

Van Valkenburg, M. E., 1960, Introduction to modern network synthesis: New York, John Wiley & Sons, Inc.

Vasil'ev, Y. I., 1957, Study of alternating waves in seismic prospecting: Bull. Acad. Sci. USSR, Geophys., no. 3, p. 30.

Wadsworth, G. P., Robinson, E. A., Bryan, J. B., and Hurley, P. M., 1953, Detection of reflections on seismic records by linear operators: Geophysics, v. 18, p. 539.

Watson, R. J., 1965, Decomposition and suppression of multiple reflections: Geophysics, v. 30, p. 54.

Western Geophysical Company, 1971, The MAXIPULSE ® system: Company brochure.

White, J. E., 1978, Generation and propagation of seismic waves: Houston, Pexcon Press.

White, J. E., and Angona, F. A., 1955, Elastic wave velocities in laminated media: J. Acoust. Soc. Am., v. 27, p. 310.

White, J. E., and Sengbush, R. L., 1953, Velocity measurements in near-surface formations: Geophysics, v. 18, p. 54.

——— 1963, Shear waves from explosive sources: Geophysics, v. 28, p. 1001.

Widess, M. B., 1957, How thin is a thin bed: Proc. Geophys. Soc. Tulsa, 1957–58; reprinted in Geophysics, v. 38, p. 1176, 1973.

Wiener, N., 1930, Generalized harmonic analysis: Acta Math., v. 55, p. 117.

——— 1949, Extrapolation, interpolation, and smoothing of stationary time series, with engineering applications: Cambridge, The Technology Press of M.I.T.

Wuenschel, P. C., 1960, Seismogram synthesis including multiples and transmission coefficients: Geophysics, v. 25, p. 106.

——— 1965, Dispersive body waves—an experimental study: Geophysics, v. 30, p. 539.

Wyllie, M. R. J., Gregory, A. R., and Gardner, G. H. F., 1958, An experimental investigation of factors affecting elastic wave velocities in porous media: Geophysics, v. 23, p. 459.

Wyllie, M. R. J., Gregory, A. R., and Gardner, L. W., 1956, Elastic wave velocities in heterogeneous and porous media: Geophysics, v. 21, p. 41.

Index

Other Books in Seismic Exploration from IHRDC

Migration of Geophysical Data by E.A. Robinson, 1983, 208 pp.

Seismic Studies in Physical Modeling edited by J.A. McDonald, G.H.F. Gardner, and F.J. Hilterman, 1983, 258 pp.

Foundations of Wave Theory by T.R. Morgan, 1983, 140 pp.

The Mini-Sosie Method by M.G. Barbier, 1983, 90 pp.

Pulse Coding in Seismology by M.G. Barbier, 1983, 89 pp.

Simple Seismics by N.A. Anstey, 1982, 168 pp.

Seismic Exploration for Sandstone Reservoirs by N.A. Anstey, 1980, 138 pp.

Seismic Interpretation: The Physical Aspects by N.A. Anstey, 1977, 625 pp.

Seismic Stratigraphy by R.E. Sheriff, 1980, 227 pp.

A First Course in Geophysical Exploration and Interpretation by R.E. Sheriff, 1979, 313 pp.

Seismic Prospecting for Oil by C.H. Dix, 1981 (1952), 422 pp.